# TECHNIQUES OF CHEMISTRY

ARNOLD WEISSBERGER, *Editor*

VOLUME I

# PHYSICAL METHODS OF CHEMISTRY

PART V

*Determination of Thermodynamic and Surface
Properties*

## TECHNIQUES OF CHEMISTRY

ARNOLD WEISSBERGER, *Editor*

VOLUME I

PHYSICAL METHODS OF CHEMISTRY, in Five Parts
(Incorporating Fourth Completely Revised and Augmented Edition of
Physical Methods of Organic Chemistry)
*Edited by Arnold Weissberger and Bryant W. Rossiter*

VOLUME II

ORGANIC SOLVENTS, Third Edition
*John Riddick and William S. Bunger*

VOLUME III

PHOTOCHROMISM
*Edited by Glenn H. Brown*

# TECHNIQUES OF CHEMISTRY

VOLUME I

# PHYSICAL METHODS OF CHEMISTRY

INCORPORATING FOURTH COMPLETELY REVISED AND AUGMENTED
EDITION OF TECHNIQUE OF ORGANIC CHEMISTRY,
VOLUME I, PHYSICAL METHODS OF ORGANIC CHEMISTRY

*Edited by*

## ARNOLD WEISSBERGER
AND
## BRYANT W. ROSSITER

Research Laboratories
Eastman Kodak Company
Rochester, New York

PART V

*Determination of Thermodynamic and Surface Properties*

## WILEY-INTERSCIENCE

A DIVISION OF JOHN WILEY & SONS, INC.
New York · London · Sydney · Toronto

PLAN FOR

# PHYSICAL METHODS OF CHEMISTRY

**PART I**

Components of Scientific Instruments, Automatic Recording and Control, Computers in Chemical Research

**PART II**

Electrochemical Methods

**PART III**

Optical, Spectroscopic, and Radioactivity Methods

**PART IV**

Determination of Mass, Transport, and Electrical-Magnetic Properties

**PART V**

Determination of Thermodynamic and Surface Properties

## AUTHORS OF PART V

A. E. ALEXANDER
Department of Physical Chemistry, University of Sydney, Sydney, Australia

JOHN R. ANDERSON
Director of Research, Permutit Research Laboratories, Princeton, New Jersey

JETT C. ARTHUR, JR.
Southern Utilization Research and Development Division, Agricultural Research Division, U.S. Department of Agriculture, New Orleans, Louisiana

D. R. DOUSLIN
Thermodynamics Research Group, U.S. Department of the Interior, Bartlesville, Oklahoma

LEE T. GRADY
Senior Supervisor, Drug Standards Laboratory, American Pharmaceutical Association Foundation, Washington, D. C.

JOHN B. HAYTER
Department of Physical Chemistry, University of Sydney, Sydney, Australia

GEORGE E. HIBBERD
    *Bread Research Institute of Australia, North Ryde, Australia*

WILLIAM J. MADER
    *Director, Drug Standards Laboratory, American Pharmaceutical Association Foundation, Washington, D. C.*

J. R. OVERTON
    *Research Laboratories, Tennessee Eastman Company, Kingsport, Tennessee*

EVALD L. SKAU
    *Southern Utilization Research and Development Division Agricultural Research Division, U.S. Department of Agriculture, New Orleans, Louisiana*

JULIAN M. STURTEVANT
    *Department of Chemistry, Sterling Chemistry Laboratory, Yale University, New Haven, Connecticut*

G. W. THOMSON
    *Thomson Associates, Detroit, Michigan*

BERNHARD WUNDERLICH
    *Department of Chemistry, Rensselaer Polytechnic Insitute, Troy, New York*

NEW BOOKS AND NEW EDITIONS OF BOOKS OF THE TECHNIQUE OF ORGANIC
CHEMISTRY SERIES WILL NOW APPEAR IN TECHNIQUES OF CHEMISTRY. A LIST
OF PRESENTLY PUBLISHED VOLUMES IS GIVEN BELOW.

## TECHNIQUE OF ORGANIC CHEMISTRY
ARNOLD WEISSBERGER, *Editor*

# INTRODUCTION TO THE SERIES

Techniques of Chemistry is the successor to the Technique of Organic Chemistry Series and its companion—Technique of Inorganic Chemistry. Because many of the methods are employed in all branches of chemical science, the division into techniques for organic and inorganic chemistry has become increasingly artificial. Accordingly, the new series reflects the wider application of techniques, and the component volumes for the most part provide complete treatments of the methods covered. Volumes in which limited areas of application are discussed can be easily recognized by their titles.

Like its predecessors, the series is devoted to a comprehensive presentation of the respective techniques. The authors give the theoretical background for an understanding of the various methods and operations and describe the techniques and tools, their modifications, their merits and limitations, and their handling. It is hoped that the series will contribute to a better understanding and a more rational and effective application of the respective techniques.

Authors and editors hope that readers will find the volumes in this series useful and will communicate to them any criticisms and suggestions for improvements.

*Research Laboratories*              ARNOLD  WEISSBERGER
*Eastman Kodak Company*
*Rochester, New York*

# PREFACE

*Physical Methods of Chemistry* succeeds, and incorporates the material of, three editions of *Physical Methods of Organic Chemistry* (1945, 1949, and 1959). It has been broadened in scope to include physical methods important in the study of all varieties of chemical compounds. Accordingly, it is published as Volume I of the new Techniques of Chemistry Series.

Some of the methods described in *Physical Methods of Chemistry* are relatively simple laboratory procedures, such as weighing and the measurement of temperature, refractive index, and determination of melting and boiling points. Other techniques require very sophisticated apparatus and specialists to make the measurements and to interpret the data; x-ray diffraction, mass spectrometry, and nuclear magnetic resonance are examples of this class. Authors of chapters describing the first class of methods aim to provide all information that is necessary for the successful handling of the respective techniques. Alternatively, the aim of authors treating the more sophisticated methods is to provide the reader with a clear understanding of the basic theory and apparatus involved, together with an appreciation for the value, potential, and limitations of the respective techniques. Representative applications are included to illustrate these points, and liberal references to monographs and other scientific literature providing greater detail are given for readers who want to apply the techniques. Still other methods that are successfully used to solve chemical problems range between these examples in complexity and sophistication and are treated accordingly. All chapters are written by specialists. In many cases authors have acquired a profound knowledge of the respective methods by their own pioneering work in the use of these techniques.

In the earlier editions of *Physical Methods* an attempt was made to arrange the chapters in a logical sequence. In order to make the organization of the treatise lucid and helpful to the reader, a further step has been taken in the new edition—the treatise has been subdivided into technical families:

Part I    Components of Scientific Instruments, Automatic Recording and Control, Computers in Chemical Research
Part II   Electrochemical Methods
Part III  Optical, Spectroscopic, and Radioactivity Methods

Part IV    Determination of Mass, Transport, and Electrical-Magnetic Properties

Part V    Determination of Thermodynamic and Surface Properties

This organization into technical families provides more consistent volumes and should make it easier for the reader to obtain from a library or purchase at minimum cost those parts of the treatise in which he is most interested.

The more systematic organization has caused additional labors for the editors and the publishers. We hope that it is worth the effort. We thank the many authors who made it possible by adhering closely to the agreed dates of delivery of their manuscripts and who promptly returned their proofs. To those authors who were meticulous in meeting deadlines we offer our apologies for delays caused by late arrival of other manuscripts, in some cases necessitating rewriting and additions.

The changes in subject matter from the Third Edition are too numerous to list in detail. We thank previous authors for their continuing cooperation and welcome the new authors to the series. New authors of Part V are D. R. Douslin, Lee T. Grady, J. R. Overton, Bernhard Wunderlich, John B. Hayter, and G. E. Hibberd.

The editors sincerely regret the death of Professor A. E. Alexander just prior to the completion of his manuscripts. We are indebted to Professor Alexander's co-authors, Mr. John B. Hayter and Dr. G. E. Hibberd for contributing to and completing the work.

We are grateful to the many colleagues who advised us in the selection of authors and helped in the evaluation of manuscripts. They are for Part V: Dr. George F. Beyer, Dr. James J. Christensen, Dr. J. Robert Dann, Dr. Reed M. Izatt, Dr. Louis D. Moore, Jr., Dr. Michael W. Orem, Dr. S. Elaine B. Petrie, Mrs. Donna S. Roets, Dr. Willard R. Ruby, and Dr. Don W. Vanas.

The senior editor expresses his gratitude to Bryant W. Rossiter for joining him in the work and taking on the very heavy burden with exceptional devotion and ability.

ARNOLD WEISSBERGER
BRYANT W. ROSSITER

*January* 1970
*Research Laboratories*
*Eastman Kodak Company*
*Rochester, New York*

# CONTENTS

## DETERMINATION OF THERMODYNAMIC AND SURFACE PROPERTIES

Chapter **1**

# TEMPERATURE MEASUREMENT

Julian M. Sturtevant

## 1  INTRODUCTION

Temperature measurement and, to a lesser extent, temperature control are fundamental to nearly all experimental work in the natural sciences. Very few of the important attributes of material substances are even approximately independent of temperature. Scientists in general, and chemists in particular, must therefore be familiar with modern developments in temperature measurement and control.

In this chapter we discuss the principles and methods of temperature measurement, with attention limited for the most part to the temperature range from 90 to 1000°K within which the great bulk of chemical work is carried out. Chapter VIII, Part IB, includes a discussion of temperature control.

## 2  DEFINITION OF A TEMPERATURE SCALE [1, 2]

### International Practical Temperature Scale

Temperatures in scientific work are expressed in terms of the International Temperature Scale, adopted in 1927 by the Seventh General Conference on

Weights and Measures and revised in 1948 by the Ninth General Conference. At the Eleventh General Conference in 1960, the title of the scale was changed to the International Practical Temperature Scale, and the text of the 1948 definition was clarified by an extensive revision. Further major revisions were incorporated in 1968. A comprehensive discussion of the current International Practical Temperature Scale of 1968 (IPTS-68) is given in Chapter IV, p. 233.

## Fixed Defining Points

The fixed defining points of the IPTS-68 are the temperatures in degrees Celsius, °C, of eleven equilibria. For the temperature range $-180$ to $1000°C$ the pertinent equilibria, all except (b) at a pressure of one standard atmosphere are (a) liquid and gaseous oxygen, $-182.962°C$; (b) the triple point of water, $0.01°C$; (c) liquid water and stream, $100°C$; (d) solid and liquid zinc, $419.58°C$; (e) solid and liquid silver, $961.93°C$; and (f) solid and liquid gold, $1064.43°C$.

The temperature of equilibrium between ice and water saturated with air at one atmosphere, the "ice point," is very closely $0.00°C$ on the International Scale. The thermodynamic scale of temperature agrees within present experimental uncertainties with the Kelvin scale defined by the relation $TK = t°C + 273.15$ over the range of validity of IPTS-68. (According to IPTS-68, temperatures may be expressed in degrees Celsius, as above, or in kelvins, abbreviated K, *not* °K.) The term degree Centigrade commonly used in this country carries, for all practical purposes, the same meaning as the degree Celsius.

## Interpolation

From 0 to $630°C$ the temperature, $t$, is deduced from the resistance, $R_t$, of a platinum resistance thermometer by means of a three-constant equation (p. 237), the constants being determined by calibration at the triple point of water and at the steam and zinc points. From $-183$ to $0°C$ a four-constant interpolation equation (p. 237) is employed, with the constants determined by measurements at the boiling point of oxygen and at the three fixed points used for the range 0 to $630°C$. The purity and physical condition of the thermometer should be such that $R_{100}/R_0 > 1.3925$.

From $630°$ to the gold point, interpolation is accomplished by means of a standard platinum versus platinum-rhodium thermocouple, one junction of which is at $0°$. Above the gold point, temperatures are defined in terms of Planck's radiation formula.

The chief value of the International Practical Temperature Scale lies in the fact that its practically universal acceptance has removed the ambiguities formerly present in the specification of temperatures.

## Secondary Temperature Standards

For many applications it is convenient to have reference temperatures more closely spaced than those provided by the International Practical Temperature Scale. Temperatures defined by the equilibria between the liquid and solid forms of pure substances are usually preferable to those defined by the equilibria between the liquid and gaseous forms because of the difficulty of avoiding errors due to superheating and pressure effects in boiling-point determinations. (See, however, Chapter IV.)

One of the best-defined and most reproducible secondary standard temperatures is the freezing point of benzoic acid. Schwab and Wichers [3] have shown that pure benzoic acid (supplied by the National Bureau of Standards) sealed in a partially evacuated tube containing a thermometer well gives a thermometric standard reproducible to 0.002° and is considerably easier to apply than the steampoint. The freezing point is 122.37°.

Pure samples of tin (freezing point 231.90°) and lead (freezing point 327.4°) may also be obtained from the National Bureau of Standards, and furnish convenient thermometric standards.

Lower reference temperatures are supplied by the equilibrium between anhydrous sodium sulfate and its decahydrate (32.37°) [4], and by the equilibrium between solid and gaseous carbon dioxide ($-78.48°$) [5]. In the latter case, pressure corrections are important, and care must be taken to establish equilibrium between the solid and its pure vapor rather than between the solid and its vapor diluted with air. Commercial dry ice appears to be sufficiently pure for this purpose. Numerous additional secondary standards are listed by Coxon [6].

## 3  LIQUID-IN-GLASS THERMOMETERS [7]

### Mercury-in-Glass Thermometers

The vast majority of temperature measurements in chemical work has been done in the past, and undoubtedly will be done in the future, with mercury-in-glass thermometers. It is therefore important to understand the limitations of this type of thermometer and the errors likely to arise in its use.

Mercury thermometers allow the measurement of temperatures in the range from $-25$ to 360°, or up to 600° in the case of thermometers constructed of Supremax glass and containing mercury under nitrogen pressure. Gallium in fused silica has been used for thermometry up to 1100°. For measurements of the greatest precision mercury thermometers have been largely replaced by resistance thermometers or thermocouples. However, short-interval mercury thermometers still find some application in calorimetric [8,9] and other types of work requiring considerable precision. The familiar Beckmann thermometer can be read with a lens to 0.001–0.002°, though care has to be exercised to make the accuracy of the readings this good. Coops and van Ness [8] have used a Beckmann thermometer for readings of temperature differences with a precision of 0.0001°. So-called calorimetric thermometers in which the amount of mercury in the system cannot be changed, as it can with the Beckmann type, cover a short temperature interval of 10–20°. Barry [9] has described a thermometer for the range 15–21° which can be read to 0.0001°, and which he considers to have about the limit of precision obtainable with mercury thermometers. Thermometers of this type should have a small auxiliary scale including the ice point; errors due to irreversible or slowly reversible changes in the volume of the thermometer bulb can be minimized by the application of a correction equal to the observed deviation at the ice point.

### Sources of Error in the Use of Mercury Thermometers

#### CHANGES IN BULB VOLUME

The bulb of a (Celsius) mercury thermometer contains mercury corresponding to about 6000° of scale length. It is thus evident that small changes in bulb volume have a relatively large effect on the calibration of the thermometer. For this reason, a thermometer used to read temperatures to a few tenths of a degree or less should have a reference mark on its scale corresponding to some readily obtained fixed thermometric point such as the ice point. Since the volume of the capillary is small, it may usually be safely assumed that any error observed at the reference point may be applied as a correction to the remainder of the calibrated range of the instrument.

An improperly annealed thermometer, or one used at high temperatures, will undergo gradual and irreversible changes in bulb volume. As mentioned above, such changes can be corrected for on the basis of a reading made at a suitable reference point.

Temporary changes in bulb volume result when a thermometer is heated and then cooled; the bulb does not return to its original volume at the lower temperature for a period which may be as long as several days. With good grades of thermometers this may cause errors of as much as 0.01° for each 10° the bulb is heated above the temperature at which measurements are to be made. In contrast, on being raised to a higher temperature the bulb will undergo the appropriate change in volume within a few minutes. This effect makes it clear that considerable care must be exercised in the treatment given a thermometer employed in precise measurements.

#### PRESSURE EFFECTS

Atmospheric and hydrostatic pressures have an appreciable effect on the reading of sensitive thermometers. According to Busse [7], the pressure coefficient for thermometers having bulb diameters of 5–7 mm is of the order of 0.1°/atm. Smith and Menzies [10] found errors of as much as 0.2° at 20 mm compared with the reading at atmospheric pressure, and stated that the magnitude of the effect could not be predicted from the dimensions of the thermometer bulb, although the effect was found to be proportional to the change in pressure. Pressure changes giving significant effects can result from changing the position of a sensitive thermometer from the vertical to the horizontal.

The errors discussed here are much reduced in a mercury-in-quartz thermometer [11]. Unfortunately, such instruments do not appear to be commercially available.

EXPOSED STEM CORRECTION

Most thermometers are calibrated to read correctly when the bulb and the liquid in the stem are exposed to the temperature to be measured. Such thermometers are called "total immersion thermometers." In many laboratory operations it is inconvenient, or impossible, to use such a thermometer under conditions of total immersion. In these cases one should *always* determine the appropriate exposed stem correction, and apply the correction if it is found to be significant. The stem correction is computed by the equation

$$\Delta = kn\,(t_{\mathrm{obs}} - t_s)/(1 - kn), \tag{1.1}$$

where $\Delta = t_{\mathrm{true}} - t_{\mathrm{obs}}$ is the correction in degrees, $k$ is the differential coefficient of expansion of mercury in glass, $n$ is the number of degrees of exposed stem, $t_{\mathrm{obs}}$ is the apparent temperature of the body under measurement, and $t_s$ is the effective temperature of the exposed mercury. For most purposes $k$ (for Celsius thermometers) may be taken equal to 0.00016; in more precise work definite information should be obtained concerning the coefficient of expansion of the glass of which the thermometer stem is constructed. The effective temperature of the exposed mercury is usually estimated by placing a second thermometer with its bulb in contact with the stem in the middle of the exposed mercury. For greater accuracy a more reliable estimate of the average exposed stem temperature is obtained by using a special type of thermometer with a long thin bulb. The estimate of the exposed stem temperature will in general be more reliable for thermometers of the etched stem type than for those of the enclosed-scale and capillary type. As a matter of fact, the latter type of thermometer may have a significant, but indeterminate, stem correction unless the *entire instrument* is immersed, because of convection currents in the air inside the stem case.

The quantity $kn$ in (1.1) is usually small compared to unity, so that in many cases the simpler expression

$$\Delta \approx kn(t_{\mathrm{obs}} - t_s) \tag{1.2}$$

may be employed. For example, if $n = 200°$, the error in $\Delta$ is only 3%. It is evident from a consideration of the uncertainties inherent in the determination of the stem correction that wherever possible a thermometer should be used totally immersed.

Some thermometers, known as "partial immersion thermometers," are calibrated with the instrument immersed to a specified depth. The purpose of such calibration is to eliminate the necessity for making exposed stem corrections; however, this purpose is realized only if the conditions of stem temperature during use are sufficiently close to those during calibration.

RADIATION ERRORS [12]

If a thermometer is used in a transparent medium, its indication may be in error as a result of radiation from a nearby hot body, such as an incandescent lamp used in making observations. An effect of this nature may also invalidate a computed stem correction.

LAG ERRORS

When a thermometer at a temperature $t_0$ is immersed in a medium at a temperature $t_m$, the quantity $t - t_m$, where $t$ is the thermometer temperature, varies with time approximately according to the equation

$$t - t_m = (t_0 - t_m)e^{-K\tau}, \tag{1.3}$$

where $\tau$ is the time and $K$ is a constant depending on the type of thermometer (chiefly the bulb diameter), the thermal characteristics of the medium, and the speed of motion of the medium relative to the thermometer. Thus, after $1/K$ sec, the reading of the thermometer differs from its proper final value by about $(1/e) (t_0 - t_m) \approx 0.4 (t_0 - t_m)$ degrees; after $5/K$ sec, $t - t_m \approx 0.007 (t_0 - t_m)$. The value of $1/K$ for an ordinary laboratory thermometer in well-stirred water is about 2 sec, and, for a Beckmann thermometer, about 9 sec. For an ordinary laboratory thermometer, $1/K$ becomes as large as 10 sec in still water and 200 sec in still air. From these figures it can be seen that in most situations lag errors will not be important if readings are made a minute or two after immersing the thermometer. On the other hand, in some work attention must be paid to the thermometric lag—for example, in the observation of changing temperatures.

## Thermometer Calibration

Thermometers used for purposes requiring a moderate or high degree of precision should in all cases have their calibrations checked. Thermometers used for the routine determination of melting and boiling points can be conveniently calibrated by means of observations made on standard substances by the methods given in Chapters III and IV. If the conditions of use, including depth of immersion, are reasonably constant, the calibration can include the exposed stem correction. An obvious method of calibration is to immerse both the instrument under test and a standard thermometer, of any suitable type, in a well-stirred bath the temperature of which may be adjusted to a number of values within the range of interest. In this method changes of the bath temperature during periods when readings are made must be slow enough so that differences in lag of the two thermometers produce no significant errors.

Mercury thermometers may be calibrated by the National Bureau of Standards if thoroughly reliable values are desired.

Any calibrated thermometer should be checked at a convenient and reproducible fixed point with sufficient frequency to ensure that significant changes in bulb volume are detected and corrected for.

### Beckmann Thermometers

The foregoing discussion of mercury-in-glass thermometers also applies to Beckmann thermometers. There is an additional source of error with this type which results from the fact that there are different amounts of mercury in the measuring system at different settings of the thermometer. Beckmann thermometers are usually calibrated to indicate correctly temperature differences at 20°, provided of course, that the corrections and precautions described above are observed. Temperature differences observed at other temperatures must be multiplied by the *setting factors* listed in Table 1.1 for thermometers constructed of Jena 16$^{\text{III}}$ glass.

**Table 1.1**   Setting Factors for Beckmann Thermometers Made of Jena 16$^{\text{III}}$ Glass

| Setting (°C) | Factor | Setting (°C) | Factor |
|---|---|---|---|
| 0  | 0.9931 | 60  | 1.0105 |
| 10 | 0.9968 | 70  | 1.0125 |
| 20 | 1.0000 | 80  | 1.0143 |
| 30 | 1.0029 | 90  | 1.0161 |
| 40 | 1.0056 | 100 | 1.0177 |
| 50 | 1.0082 |     |        |

The reader is referred to the work of Coops and van Ness [8] for an interesting description of the use of a Beckmann thermometer in highly precise measurements.

### Other Liquid-in Glass Thermometers

Various liquids other than mercury find some application in thermometers. Organic liquids, such as alcohols, toluene, and pentane, are used in thermometers reading down to −100°. Such liquids have coefficients of expansion some 5 to 10 times that of mercury, so that equation (1.1) for the exposed stem correction obviously does not apply. For most thermometric liquids [13] there happens to be an approximately inverse relation between density and heat capacity, so that with bulbs of moderate diameter where the thermal conductivity of the liquid is not a controlling factor, the thermometer lag is roughly independent of the liquid used.

Mercury is generally preferred as a thermometric liquid because of its high degree of stability. Organic liquids generally require the addition of a coloring compound which may hasten polymerization and other reactions leading, usually, to a gradual contraction of the liquid.

## 4   RESISTANCE THERMOMETRY

### Platinum Resistance Thermometers

In the neighborhood of room temperature, the electrical resistivity of platinum increases approximately 0.4%/deg. Since platinum is obtainable in highly reproducible and pure condition, and is relatively unaffected by aging or by gases and fumes over a wide temperature range and since electrical resistance can be measured with a high degree of precision, this change of resistivity with temperature furnishes one of the best means of temperature measurement [14]. As stated above, interpolation in the International Practical Temperature Scale in the range of $-183$ to $630°$ is defined in terms of the platinum resistance thermometer. Many other metals and alloys have temperature coefficients of resistivity somewhat larger than that of platinum. While some of these find limited application in temperature-indicating devices, they are inferior to platinum for precise measurements because of their greater susceptibility to external physical and chemical effects.

Platinum resistance thermometers for precision measurements are usually wound of annealed wire on mica supports in such a way that the metal is subjected to as slight a strain as possible when the thermometer is heated or cooled. For most applications, the coil is enclosed in a sealed glass or silica tube; when it is desired to minimize the lag of the thermometer, as in calorimetric applications, the coil is enclosed in a flattened metal case. Thermometers are usually manufactured with a resistance of 25.5 or 2.55 $\Omega$ at $0°$, so that the resistance will change by about 0.1 or 0.01 $\Omega$/deg, respectively. Modern thermometers are provided with four leads, two from each end of the coil, the use of which is discussed below. It is evident that thermometers can be made in other forms to meet special needs. For example, a winding of platinum or other metallic ribbon around the outside of a vessel gives an integrating, low-lag measure of the temperature of the outside wall.

The temperature corresponding to the measured resistance, $R_t$, of a platinum thermometer may be calculated by empirical equations (see p. 237).

Schwab and Smith [15] have published values of certain terms which facilitate calculations based on these equations.

## Other Types of Resistance Thermometers

### Thermometers Using Metals Other than Platinum

Metals and alloys other than platinum may be used in resistance thermometers in cases where the chemical and physical stability of platinum is not required. A high grade of nickel has an advantage over platinum in that its temperature coefficient of resistivity is about 0.6%/deg at ordinary temperatures as compared with 0.4% for platinum.

### Thermometers Using Nonmetallic Materials

THERMISTORS

Methods of manufacture of semiconducting mixtures of metal oxides have been perfected to the point that such mixtures, known as thermally-sensitive resistors, or thermistors, have sufficient reproducibility and stability to be useful for a wide variety of purposes including resistance thermometry. Thermistors have *negative* temperature coefficients of resistivity which at ordinary temperatures are of the order of 10 times that of platinum. The specific resistance, $\rho$, in ohm-centimeters, of these semiconductors is given approximately by the expression

$$\rho = \rho_\infty e^{B/T}, \tag{1.4}$$

where $T$ is the absolute temperature, and $B$ and $\rho_\infty$ are empirical constants. The resistance of some thermistors is better fitted by the equation

$$\rho = \rho_\infty e^{B/(T+\theta)}, \tag{1.5}$$

where $\theta$ is another empirical constant, than by equation (1.6). A typical value of $B$ is 3500°. With this value of $B$, the temperature coefficient of resistance according to equation (1.6) is $-3.9$%/deg at 300°K. Because of this large temperature coefficient, thermistors can usually be employed with more simple auxiliary measuring equipment than is required to achieve equivalent sensitivity with metallic resistance thermometers. On the other hand, the electrical noise of thermistors is larger than that of metallic conductors, so that the ultimate noise-limited sensitivity of a thermistor thermometer is about the same as that of a metallic resistance thermometer.

In addition to their large temperature coefficients of resistivity, an outstanding value of thermistors is the wide variety of forms in which they can be manufactured, ranging from relatively large disks to very small beads. The latter are available in diameters as small as 0.2 mm, and may be obtained with glass coatings and thin noble metal leads so that they may be used in corrosive media. It is evident that such devices have extremely small heat capacities and thermal lags.

## Bolometers

Resistance thermometers of extremely low heat capacity can be constructed of very thin wires or evaporated films of metal. Bolometers (see Chapter VI, Part IB) of this type have found important application in measuring radiant energy, as in infrared spectrographs. It has been found that the stability of such a spectrograph is greatly improved if the light source is shuttered and the ac output from the bolometer is amplified by a narrow-band amplifier. It is evident that in this arrangement the bolometer should have a very short response time. Bolometers suitable for use with shuttering frequencies of 10 Hz or higher have been described.

## Measurement of Resistance

### Direct-Current Wheatstone Bridge

The most commonly used method for measuring the resistance of a resistance thermometer employs a dc Wheatstone bridge. The bridge (Fig. 1.1) generally consists of two fixed-ratio arms, $A$ and $B$, of a variable arm, $D$, and the thermometer, $X$. With a four-lead thermometer, it is possible to eliminate the resistances of the leads. If $r_A = r_B$ ($r$ = resistance), and the four leads are designated as $C$, $c$, $T$, and $t$, it is evident that, with the thermometer connected as in Fig. 1.1$a$, at bridge balance

$$(r_D)_a + r_C = r_X + r_T.$$

Correspondingly, with connections as in Fig. 1.1$b$,

$$(r_D)_b + r_T = r_X + r_C.$$

Therefore

$$r_X = \frac{(r_D)_a + (r_D)_b}{2}.$$

Thermometer bridges are frequently equipped with a mercury commutator for making this change in connections. The resistance of the leads can frequently be completely neglected in the case of thermometers such as thermistors, which have high resistances.

With a 25.5-$\Omega$ thermometer, 0.0001 $\Omega$ corresponds to 0.001°. The bridge must therefore be constructed and calibrated with great care if precise temperature measurements are to be made. The variable arm usually consists of six decades, the lowest of which is in steps of 0.0001 $\Omega$. The variable resistance of the decade contactors, which is an important source of error in such precise resistance measurements, is ingeniously eliminated in the Mueller-type bridge. The contact resistances in the lower decades are in series with relatively large fixed resistances so that they have only a negligible effect. The decade resistance values result from the shunting of a series of low re-

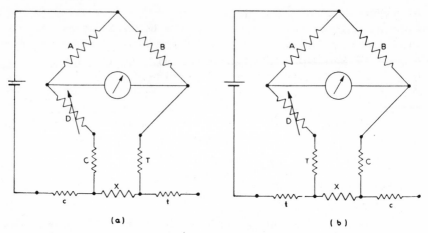

**Fig. 1.1**   Simple thermometer bridge, illustrating use of four-lead thermometer.

sistances by the large fixed resistances. Improvements on the Mueller bridge have been described by Evans [16].

Too large a current passing through the thermometer will cause fallacious readings as a result of heating of the thermometer coil. With 25.5-Ω thermometers having flattened metal protecting tubes, a current of 5 mA is usually employed, while with those enclosed in glass tubes the current should not exceed 2 mA. These currents cause a heating of a few thousandths of a degree in the thermometer. If the measuring current is kept constant, it is usually safe to assume that the heating is the same at all temperatures and in all media. This problem needs to be given particular attention in the case of a high-resistance thermometer such as a thermistor. In such cases, because of the large changes in resistance involved, it will not in general be permissible to assume that the heating effect is constant if the measuring current is held constant, so that it is important to reduce the measuring current to the lowest possible value.

Differential measurements can be conveniently carried out by using two similar thermometers in adjacent arms of a Wheatstone bridge; or, if suitable small trimming resistors are placed in series with the thermometers, greater sensitivity is obtained by using two thermometers at each of the two temperatures the difference between which is being measured. The advantages of differential measurement are fully realized only if the two temperatures being compared are rather close together.

### Potentiometric Method

The resistance of a thermometer may be determined by potentiometric observation of the voltage drops across the thermometer and a standard resistor in series with it. This method can give as good accuracy as the bridge method if a good potentiometer is employed. Its chief value will be in cases in which no thermometer bridge is available but a satisfactory potentiometer is.

## Automatic Indication and Recording of Temperature

A wide variety of commercially-available, precise, self-balancing indicating and recording potentiometers can be utilized in the continuous monitoring of temperatures measured either by resistance thermometers or thermocouples.

A Wheatstone bridge made up of a thermometer and three fixed arms gives an output which is proportional to the temperature of the thermometer over a restricted range on both sides of the balance point of the bridge. Outside this range allowance has to be made for deviations from linearity; this is easily done where necessary by direct calibration of the thermometer plus the indicating system. When the bridge output is to be continuously recorded it is impossible to eliminate the effect of thermometer lead resistances by means of a four-lead thermometer. In this case a three-lead thermometer can be employed. As seen in Fig. 1.2, two of the leads, *a* and *b*, which are as nearly identical as possible, are in adjacent arms of the bridge so that changes in their resistances cancel out.

Recording potentiometers have insufficient sensitivity to measure directly the bridge output in many applications, so that amplification is necessary. For this purpose excellent dc amplifiers are available, having sensitivities down to 1 $\mu$V or less full scale. Such amplifiers, because of their readily controllable time constants, freedom from disturbance by vibrations, and general convenience, have largely replaced galvonometers as detectors of bridge off-balance even when continuous recording is not needed.

It is possible to realize a temperature sensitivity with an ac resistance thermometer bridge and ac amplifier which approaches the fundamental limit imposed by Johnson noise in the thermometer. For example, the sensitivity

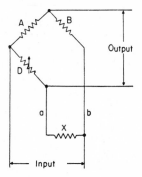

**Fig. 1.2**   Three-lead thermometer, *X*, arranged to minimize the effects of lead resistance.

obtained [17] with a 50-$\Omega$ nickel thermometer (with 2 mA measuring current) using a narrow-band amplifier was 5 $\mu$deg, whereas the noise output of the thermometer would lead one to expect a limiting sensitivity of 2 $\mu$deg. Since, with this thermometer, 5 $\mu$deg. corresponded to a resistance change of only 1.5 $\mu\Omega$ great care had to be exercised to avoid all sources of extraneous resistance changes in the bridge circuit. Such high sensitivities can in general be usefully applied only in differential thermometry.

## 5    THERMOELECTRIC THERMOMETRY [18]

If a circuit composed of two different metals (see Fig. 1.3) contains a current-indicating device, a current is found to flow in the circuit when the two junctions between the metals are maintained at different temperatures ($t_1 \neq t_2$). The electromotive force (emf) causing the current flow is a single-valued function of the temperature difference, provided that either $t_1$ or $t_2$ is held constant, and the physical or chemical character of the metals is not altered. Such a system is called a *thermocouple*. Several thermocouples may be connected in series, as in Fig. 1.4, to give an emf which is the sum of the emf values of the single couples. Such an arrangement is called a *multijunction thermocouple*, or a *thermel*, or a *thermopile*. The thermoelectric power of a thermocouple is of the order of a few to 40 or 50 $\mu$V/deg, depending on the metals and the temperature. In general, the thermoelectric power decreases with the temperature. Table 1.2 lists some important data for the four most common types of thermocouple.

Because of the fact that the emf of a couple is primarily a function of the temperature difference between its junctions, thermocouples are particularly useful in the measurement of small temperature differences. The junctions of a multijunction thermocouple may be distributed over a surface or through-

**Table 1.2**   Characteristic Data for Various Types of Thermocouples [a]

| Type | Useful Temperature Range (°C) | Emf per Junction with Reference Junction at 0°C (mV) | | | |
|------|------|------|------|------|------|
| | | −200° | +100° | +300° | +1000° |
| Platinum to platinum-rhodium | 0–1450 | — | 0.643 | 2.315 | 9.57 |
| Chromel P to alumel | −200–1200 | −5.75 | 4.10 | 12.21 | 41.31 |
| Iron to constantan | −200– 750 | −8.27 | 5.40 | 16.56 | 58.22 |
| Copper to constantan | −200– 350 | −5.54 | 4.28 | 14.86 | — |

[a] W. F. Roeser, in *Temperature*, Vol. 1, 1941, p. 180.

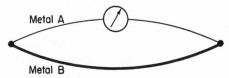

**Fig. 1.3** Simple thermocouple circuit.

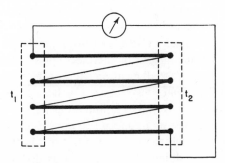

**Fig. 1.4** Thermocouples connected in series.

out a volume to average out temperature irregularities in the body under observation.

### Design and Construction of Thermocouples

Thermocouples may be constructed in a wide variety of physical forms for various applications. Most frequently wires of the two metals are soldered or welded together, and insulated or otherwise protected as required by the particular application. The metals in the neighborhood of one set of junctions may be in the form of thin ribbons in close thermal contact with a surface the temperature of which is to be measured. The heat capacity of thermocouples may be made very low by using fine wires or ribbons. The construction of thermocouples of extremely low heat capacity has received considerable attention [19] in connection with recording infrared spectrographs using shuttered radiation sources to give ac thermocouple outputs. Such a thermopile made of thin layers of metal formed by evaporation may give an output amplitude with shutter frequencies as high as several cycles per second which is not much smaller than the value without shuttering.

For temperatures below 300°, copper-constantan (or copper-Advance) thermels are usually employed. If only moderate sensitivity is sought, the number of junctions will be small and will usually be in part determined by the potentiometric and indication equipment available. In some applications physical factors limit the number of junctions to a very small number, frequently to one. In such cases the desired sensitivity of measurement will dictate the choice of the potentiometer and detector. When very high sensitivity, of the order of microdegrees, is desired, one can use either a relatively small number of junctions with a galvanometer or an amplifier having high sensitivity, or a higher-resistance thermel having as many as 10,000 junctions with less sensitive detection equipment. The size of the wires used in constructing the thermel should be carefully considered. A decision here is based on a balance [20] between low electrical resistance and low thermal conduction with due consideration being given such factors as possible restrictions on the allowable heat capacity of the measuring junctions.

One of the greatest advantages of thermocouples is the ease with which they can be constructed. We shall mention a few of the more important points to be considered in the actual construction. Further details are to be found in papers on this subject [18], as well as in numerous papers on calorimetry and related subjects. It is obviously important to approach as closely as possible the ideal situation in which the only potential in the thermocouple is that due to the temperature difference to be measured. Since thermal emf is developed when an inhomogeneous metal is exposed to a temperature gradient, the wire used in constructing a thermel should be tested for homogeneity [21]. Furthermore, the thermel should be exposed to the smallest temperature gradients possible, and these gradients should be as nearly as possible the same during calibration and subsequent use. Thermels should be constructed so as to avoid subjecting the wires to any strain, particularly after calibration. In the construction of a multijunction thermel, one should follow, if possible, a systematic procedure in which the insulation resistance of each wire against the rest of the thermel is tested before that wire is incorporated in the thermel.

Thermopiles having hundreds or thousands of junctions in series can be fabricated by an ingenious electroplating technique [22] if the physical distance between "hot" and "cold" junctions is small. Constantan (or Advance) wire is wound on an insulating form, and half of the coil is then electroplated with copper. The copper shorts the constantan and creates one hot and one cold junction per turn.

## Solid State Thermoelectric Modules

Solid state plates [23] developed primarily for Peltier cooling applications have been found to be convenient short-path indicators of heat transfer, for example to replace the electroplated thermopiles mentioned in the preceding paragraph.

## Measurement of Potential

### Direct Measurement

The output of a thermopile is in many cases measured by means of a potentiometer and galvanometer or other detector of potentiometer off-balance. The use of galvanometers has greatly decreased with the commercial development of excellent microvolt, or nanovolt, amplifiers. The potentiometer principle is discussed in Chapter III, Part IA [24]. In applications of thermopiles to the measurement of small temperature differences, it may be possible to dispense with the potentiometer; in such applications that high input impedance available in amplifiers is particularly advantageous, since the current flowing in the thermopile is much smaller than if a properly damped galvanometer is used. Amplifiers have the added advantage over galvanometers that they usually include provision of an output suitable for a strip-chart recorder.

Since the potentials developed by thermocouples are small, potentiometers of relatively high precision are in general required. When the precision must exceed 5 $\mu$V (0.125° per junction for copper-constantan couples) a slide wire is generally not used for interpolating between the steps of the lowest decade. The interpolation is accomplished by observing galvonometer deflections or amplifier outputs, or by opposing the small residual emf by a small fraction of the potential drop in a resistor carrying current read on a sufficiently sensitive ammeter (Lindeck method). This latter operation can be automated by suitable feedback circuitry.

In the measurement of potentials of the order of 1 $\mu$V, great care must be exercised to avoid interference from an emf arising either from leakage from higher potential circuits or from temperature gradients in nonhomogeneous parts of the circuit (stray thermal emf). White [25] has described methods of shielding for eliminating interference from the first cause, as well as various devices which lessen the troubles due to thermal emf. Provision is usually made for a partial elimination of thermal and leakage emf by observing the detector zero with the circuit as nearly as possible in its operating condition except for the removal of the unknown emf.

An interesting scheme for comparing two potentials, the (low impedance) sources of which may be electrically connected at any desired point or need not be connected at all, has been described by Dauphinee [26]. This method, which consists in rapidly switching a high-quality capacitor back and forth from one potential source to the other and measuring any off-balance current which flows in one of the circuits, appears to have received less application in conjunction with resistance thermometers and thermopiles than it merits.

## Calibration of Thermocouples

In many applications, thermels serve as indicators of temperature differences which may be expressed in arbitrary units, and they therefore need not be calibrated. In cases in which calibration is needed, it will frequently be found possible to perform the calibration in the apparatus for which the thermel was designed by measurements on a substance with known thermal properties (such as heat capacity or melting point). If an absolute calibration is necessary, it may be accomplished by determining the emf corresponding to a series of thermometric fixed points (with the reference junctions held at a constant temperature, usually 0°C), or by comparison with a calibrated resistance thermometer or gas thermometer.

*Reference Temperatures.*    The most important reference temperature used in thermoelectric thermometry is the melting temperature of ice. White's [27] ice-point apparatus has a Dewar flask contained within a second Dewar, which gives, with good commercial ice, a temperature constant to 0.0001° for at least a day. A good reference temperature at −78.48° is also furnished by solid carbon dioxide [28].

Whenever possible, a thermocouple should be employed as a differential temperature indicator, since small potentials can be measured with greater precision than large ones, and errors due to inhomogeneity in the thermel wires are lessened if the thermel is not exposed to any large temperature gradients. This important advantage can sometimes be secured by immersing the reference junctions in a body of high heat-capacity and thermal conductivity, such as a block of copper, in a Dewar flask which is contained in another Dewar, the whole assembly being immersed in a thermostat at a temperature close to that of the working junctions [29].

## 6    MISCELLANEOUS THERMOMETRIC METHODS

### Quartz Crystal Thermometer

The frequency of an oscillator controlled by a quartz crystal, utilizing the piezoelectric effect, is a function of temperature. This effect, coupled with the great precision available in electronic frequency measuring techniques, has been put to good use in a recently developed thermometer [30] which is finding rapidly expanding applications in scientific laboratories. This thermometer may be used in the range −80 to 250°. An absolute accuracy of ±0.02° is claimed, and a sensitivity of ±0.0001° in differential temperature measurements. A useful feature of the instrument is a digital output which is compatible with current forms of digital data processing equipment. Drawbacks for some applications are the relatively large size and heat capacity of the temperature probes, and the sampling times required (0.1 sec at a sensitivity of 0.01°, 10 sec at 0.0001°).

## Expansion Thermometers

### Gas Thermometers

Gas thermometers are of great importance in that they serve as the means for establishing the thermodynamic temperature scale, and are frequently employed in research at very low temperatures. Detailed discussions of many aspects of gas thermometry are available in the General References listed on page 22.

### Bimetallic Thermometers

Bimetallic thermometers, composed of two metals having different coefficients of expansion closely bonded together, find some application, particucularly in industrial problems, in the measurement of temperatures with moderate precision.

## Radiation Pyrometry [31]

Above about 1000° the International Practical Temperature Scale is defined in terms of Planck's expression for the amount $J_{\lambda T}$, of energy radiated at the wave length, $\lambda$, by a black body at a temperature, $T$ per unit time per unit area of the black body throughout the solid angle $2\pi$:

$$J_{\lambda T} = \frac{8\pi ch}{\lambda^5(e^{hc/\lambda kT} - 1)},\qquad(1.6)$$

where $c$ is the velocity of light, $h$ is Planck's constant, and $k$ is Boltzmann's constant. The total radiant intensity is

$$\int_0^\infty J_{\lambda T}d\lambda.\qquad(1.7)$$

### Total Radiation Pyrometry

It is evident that the relations given above can be utilized in temperature measurements based on observation either of the total radiation intensity from a body, or of radiation in a restricted wavelength region, provided the body under examination behaves like a black body or has a known emissivity. Total radiation pyrometers, as the name indicates, are instruments for estimating the total radiant intensity from a source. Actually, such instruments include optical systems which introduce some wavelength selectivity; for this and other reasons, they cannot be used for absolute measurements, but must be empirically calibrated. These instruments usually consist of a nonreflecting receiving surface, having a very small heat-capacity, the temperature of which is measured by resistance or thermoelectric thermometry, together with a suitable optical system.

## Optical Pyrometry

Optical pyrometers indicate the intensity of radiation within a more or less restricted wavelength range in the visible part of the spectrum. For best results, a filter is used to restrict and define the wavelength range. The brightness of the object under examination is usually compared with that of a filament through which a variable current flows, the current being adjusted so that the brightness of the object is the same as that of a filament. The filament current is related by empirical calibration to the temperature.

## 7   CHOICE OF METHOD OF TEMPERATURE MEASUREMENT

Some general statements can be made which may be of help in selecting the type of thermometric equipment to be used for various purposes. However, these statements cannot be expected to be fully applicable to many situations which arise in specialized problems. In such situations there may be unusual factors influencing the choice of thermometer which will completely override the considerations given below.

The majority of temperature measurements in chemical practice can be most conveniently and inexpensively carried out with mercury-in-glass thermometers. This is true of ordinary melting-point (Chapter III) and boiling-point (Chapter IV) determinations, observations of the temperature of reaction mixtures, and so on. All temperatures of significance in characterizing substances or in defining their properties should, of course, be measured with thermometers calibrated with an accuracy at least equal to the apparent accuracy of the reported data.

In chemical kinetic measurements, since the rate of many reactions increases by about 10%/deg, temperatures should be specified with an accuracy of 0.05° or better. In many thermodynamic measurements accuracies of 0.01° or better are required. In such work, temperatures are frequently measured by standard platinum resistance thermometers which give results directly expressed on the International Practical Temperature Scale. The equipment involved is, however, rather expensive, so that there is some advantage in using Beckmann or other short-interval, mercury-in-glass thermometers, provided that these are carefully calibrated against a standard resistance thermometer by the experimenter or by a standardizing laboratory.

The measurement with high precision of small temperature differences is conveniently carried out by a differential resistance thermometer bridge (two thermometers in adjacent arms of the bridge, one at each temperature), or by thermocouples. The construction of thermocouples is considerably easier than the construction of reliable resistance thermometers, so that in cases where thermometers of special form are required, thermocouples will usually be found to be more convenient. There is not much difference in the cost and complexity of the subsidiary equipment required for these two types of thermometers.

Temperature measurements outside the range of mercury-in-glass thermometers are for the most part carried out with resistance thermometers or thermocouples, except at extremely high temperatures, where radiation methods are used.

Requirements of very low thermometer heat capacity or high speed of response cannot be met by mercury-in-glass thermometers. Resistance thermometers and thermocouples can be made to satisfy rather extreme requirements in either of these directions.

## References

1. H. F. Stimson, in *Temperature: Its Measurement and Control in Science and Industry*, Vol. 3, Part 1, Reinhold, New York, 1962, p. 59. In future references to this useful collection of volumes, the title will be shortened to *Temperature*, and the publisher's name and address will be omitted. See General References, p. 22.
2. H. F. Stimson, D. R. Lovejoy, and J. R. Clement, in *Experimental Thermodynamics*, Vol. 1, J. P. McCullough and D. W. Scott, Eds., Butterworths, London, 1968, Chap. 2; C. R. Barber, *Nature*, **222**, 929 (1969).
3. F. W. Schwab and E. Wichers, *J. Res. Natl. Bur. Std.*, **34**, 333 (1945).
4. J. Hole, *Chem. Met. Eng.* **52**(6) 115 (1945).
5. R. B. Scott, in *Temperature*, Vol. 1, 1941, p. 212.
6. W. F. Coxon, *Temperature Measurement and Control*, Macmillan, New York, 1960.
7. For detailed discussions see J. Busse, in *Temperature*, Vol. 1, 1941, p. 228; H. F. Stimson, D. R. Lovejoy and J. R. Clement, Ref. 2; articles by R. D. Thompson and by J. A. Hall and V. M. Leaver, in *Temperature*, Vol. 3, Part 1, 1962.
8. J. Coops and K. van Ness, *Rec. trav. chim.* **66**, 142 (1947).
9. F. Barry, *J. Amer. Chem. Soc.* **42**, 1911 (1920).
10. A. Smith and A. Menzies, *J. Amer. Chem. Soc.* **32**, 905 (1910).
11. H. Moreau, J. A. Hall, and V. M. Leaver, *J. Sci. Instr.* **34**, 147 (1957).
12. C. E. Arregger, *Can. Chem. Process. Ind.* **29**, 269 (1945).
13. R. L. Weber, *Temperature Measurement and Control*, Blakiston, Philadelphia, 1941, p. 13.
14. E. F. Mueller, in *Temperature*, Vol. 1, 1941, p. 162; H. J. Hoge and F. G. Brickwedde, *J. Res. Natl. Bur. Std.* **28**, 217 (1942); H. F. Stimson, D. R. Lovejoy, and J. R. Clement, Ref. 2.
15. F. W. Schwab and E. R. Smith, *J. Res. Natl. Bur. Std.* **34**, 360 (1945).
16. J. P. Evans, *Temperature*, Vol. 3, Part 1, 1962, p. 285.

17. A. Buzzell and J. M. Sturtevant, *J. Amer. Chem. Soc.* **73,** 2454 (1951).
18. For thorough discussions of this subject, see the articles by W. F. Roeser, R. B. Scott, J. G. Aston, W. P. White, and H. T. Wensel, in *Temperature*, Vol. 1, 1941; P. H. Dike, *Thermoelectric Thermometry*, Leeds and Northrup, Philadelphia, 1954; H. F. Stimson, D. R. Lovejoy, and J. R. Clement, Ref. 2.
19. L. C. Roess and E. N. Dacus, *Rev. Sci. Instr.* **16,** 164 (1945); D. F. Hornig and B. J. O'Keefe, *ibid.* **18,** 474 (1947).
20. B. Whipp, *Phil. Mag.* **18,** 745 (1934).
21. W. F. Roesser and H. T. Wensel, in *Temperature*, Vol. 1, 1941, p. 284.
22. H. Wilson and T. D. Epps, *Proc. Phys. Soc. (London)* **32,** 326 (1919); T. H. Benzinger and C. Kitzinger, in *Temperature*, Vol. 3, Part 3, 1961, p. 43; W. J. Evans and W. B. Carney, *Anal. Biochem.* **11,** 449 (1965).
23. Cambion, Cambridge, Massachusetts; MCP Electronics Ltd., Wembley, England.
24. T. M. Dauphinee, in *Temperature*, Vol. 3, Part 1, 1962, p. 269.
25. W. P. White, in *Temperature*, Vol. 1, 1941, pp. 265, 279.
26. T. M. Dauphinee, *Can. J. Phys.* **31,** 577 (1953); *Temperature*, Vol. 3, Part 1, 1962, p. 269; E. E. Stansbury, E. B. Nauman, and C. R. Brooks, *Rev. Sci. Instr.* **36,** 480 (1965).
27. W. P. White, *J. Am. Chem. Soc.* **56,** 20 (1934). See also J. L. Thomas, in *Temperature*, Vol. 1, 1941, p. 159.
28. R. B. Scott, in *Temperature*, Vol. 1, 1941, p. 212.
29. W. P. White, *The Modern Calorimeter*, Chem. Catalog Co., New York, 1928, p. 131.
30. Hewlett-Packard Co., Palo Alto, California.
31. See, for example, H. T. Wensel, in *Temperature*, Vol. 1, 1941, p. 3; W. E. Forsythe, *ibid.*, p. 1115; various authors in *Temperature*, Vol. 3, Part 1, 1962, section IX (pp. 421 ff.); H. F. Stimson, D. R. Lovejoy, and J. R. Clement, Ref. 2.

### General

American Institute of Physics, *Temperature: Its Measurement and Control in Science and Industry*, Reinhold, New York; Vol. 1, 1941; Vol. 2, H. C. Wolfe, Ed., 1955; Vol. 3, C. M. Herzfeld, Ed.-in-chief, Part 1, F. G. Brickwedde, Ed.; part 3, J. D. Hardy, Ed., 1962.

Aronson, M. H., Ed., *Temperature Measurement and Control Handbook*, Instruments Publishing Co., Pittsburgh, Pennsylvania, 1964.

Coxon, W. F., *Temperature Measurement and Control*, Macmillan, New York, 1960.

National Bureau of Standards, U.S., *Bibliography of Temperature Measurement, January 1953 to June 1960*, N.B.S. Monograph 27; *Bibliography of Temperature Measurement, July 1960 to December 1965*, N.B.S. Monograph 27 Supplement.

Reilly, J., and W. N. Rae, *Physico-Chemical Methods*, 3rd ed., Van Nostrand, Princeton, New Jersey, 1939.

Weber, R. L., *Temperature Measurement and Control*, Blakiston, Philadelphia, Pennsylvania, 1941.

Chapter **II**

# DETERMINATION OF PRESSURE AND VOLUME

G. W. Thomson and D. R. Douslin

# 1  MEASUREMENT OF PRESSURE

## Introduction

Measurement of pressure and volume often presents special problems to the organic chemist because the procedures are by nature quantitative and require methods and equipment that must be selected carefully for the degree of accuracy required, the state of aggregation and volatility of the compound, and the compatibility of the compound with the materials of construction of the apparatus. These factors, which greatly influence the choice of method or apparatus, are discussed in general throughout the text rather than under separate subject headings. Whenever possible the accuracy and range of a classical method are given; however, the possibility of extending them by improving the classical method should be investigated.

Equilibrium temperature measurement (Chapter III) presents no difficulties except perhaps in comparative ebulliometry or simple boiling-point determinations where the likelihood of superheating of some liquids must be recognized and the apparatus designed to eliminate this possibility.

The vapor pressure methods discussed in this chapter cover most of the range from a fraction of a micron to the critical pressure.

## Terminology, Units, and Conversion Factors

Pressure is defined as force per unit area. Total pressure at any point is the *absolute* pressure, while the difference in pressure between any two points is simply *differential* pressure. The term *gage* applies to a pressure measured with respect to atmospheric pressure as base. Therefore, atmospheric pressure must be added to gage pressure to obtain absolute pressure.

By definition [1], the normal atmosphere unit is exactly 1,013,259 dyne/cm$^2$ or in International System of Units (SI) [2] 101.3250 kN/m$^2$. Table 2.1, based on the above definition, is a convenient source of conversion factors for pressure in different units. To convert a value of pressure given in units of the left column to units of the top row simply multiply the given pressure by the common factor.

## Barometers and Manometers

For most vapor pressure measurements an open-end mercury manometer, a diaphragm gage, or some other variety of pressure transducer is sufficient. If absolute pressure is desired, the atmospheric or barometric pressure must be determined at the same time. Marvin [3] discussed the theory and construction of barometers and the proper methods of handling, cleaning, and reading them. Also, in a recent monograph [4] Brombacher, Johnson, and Cross have described, in considerable detail, various designs of mercury barometers and manometers and have presented extensive correction tables for the effects of temperature, gravity, capillarity, and other errors applicable to portable instruments. Many commercial, laboratory barometers are manufactured according to these recommendations. The manufacturer usually supplies a brochure with each instrument containing instructions for its installation, use, and maintenance. Another readily available source of information for reduction of barometric pressure readings is the *Handbook of Chemistry and Physics* [5].

## Fundamental Considerations

The basic principle involved in barometric pressure measurement is the balancing of the weight of a mercury column against the weight of a column of air extending from the barometer to the upper limit of the atmosphere. The height of the mercury column so obtained is a direct measure of the atmospheric pressure at that point. Meteorologists frequently correct this

**Table 2.1** Conversion Factors for Pressure Units

| Units | dyne/cm² | normal atm | bar | kN/m² | mm Hg or torr | in. Hg | lb/in.² |
|---|---|---|---|---|---|---|---|
| dyne/cm² | 1 | $0.9869233 \times 10^{-6}$ | $10^{-6}$ | $10^{-4}$ | $7.500617 \times 10^{-4}$ | $2.952993 \times 10^{-5}$ | $1.4503830 \times 10^{-5}$ |
| normal atm | 1,013,250 | 1 | 1.013250 | 0.01013250 | 760 | 29.92120 | 14.696006 |
| bar | $10^6$ | 0.9869233 | 1 | 0.01 | 750.0617 | 29.52993 | 14.503830 |
| kN/m² | $10^4$ | 101.3250 | 100.0000 | 1 | 7.50062 | 0.2952993 | 0.14503830 |
| mm Hg or torr | 1,333.224 | $1.3157895 \times 10^{-3}$ | $1.333224 \times 10^{-3}$ | 0.1333224 | 1 | 0.03937 | 0.0193685 |
| in. Hg | 33,863.95 | 0.03342112 | 0.03386395 | $3.386395 \times 10^{-4}$ | 25.40005 | 1 | 0.4911570 |
| lb/in.² | 68,947.31 | 0.06804570 | 0.06894731 | $6.894731 \times 10^{-4}$ | 51.71473 | 2.036009 | 1 |

value to sea level, but in most laboratory applications only local barometric pressure is needed.

If the space in the tube above the mercury column were a perfect vacuum and there were no nongravitational forces, the atmospheric pressure, $P_a$, in dyne/cm$^2$ (see Table 2.1 for conversion to mm Hg or other units), would be related to the height of the mercury column, $h$, in centimeters, by

$$P_a = h\rho g, \tag{2.1}$$

where $\rho$ is the density of mercury in g/cm$^3$, and $g$ is the acceleration of gravity in cm/sec.$^2$

However, actual barometers do not come up to these ideal conditions because capillary forces depress the level of the mercury surface and residual gas is always in the space above the mercury column. To allow for them, commercial barometers are usually adjusted against a certified standard at the factory. Thus for compensated barometers no further corrections for capillarity or residual gas are needed unless improper usage has allowed air to enter the evacuated leg of the barometer. Many high-grade barometers have an index mark engraved on a small plate fastened permanently to the case, the mark being exactly 30 in. above the point of the ivory pin in the cistern. Comparison of this mark with the scale will show the size of the manufacturer's correction.

Sometimes allowance must be made for difference in elevation between the reservoir of the barometer and a point on the experimental apparatus. A rule useful for short distances is that each foot of elevation corresponds to a differential pressure of 0.027 mm Hg. If air is flowing between the barometer and the experimental apparatus, as is usually the case with central air conditioning or effective hoods, either correct the measured barometric pressure or move the barometer closer to the apparatus.

The effect of altitude on the gravitational constant is usually small: barometric pressure readings are seldom corrected. For example, at 1000 m above sea level and 700 mm Hg observed pressure, the correction amounts to negative 0.22 mm Hg.

In a check of several commercial, laboratory barometers, up to 0.5 mm Hg of residual gas was found in the vacuum space; as much as 4 mm Hg of residual gas has been reported [6]. The usual magnitude of this error in a new commercial barometer is closer to 0.25 mm Hg. The vapor pressure of mercury is negligible at room temperature (0.0012 mm Hg at 20°C). If necessary a correction, $P_g$, for noncondensable residual gas can be made to the barometric height, $h_t$, by application of the gas laws. If the bore of the tube above the mercury is uniform and the scale reading at the top of the closed end, $h_E$, is known, then

$$P_g = \frac{C(t + 273)}{h_E - h_t}, \tag{2.2}$$

where $C$ is an empirical factor obtained by calibration against a standard barometer. The residual gas correction should be added to the barometric pressure reading.

*Capillary Depression of Mercury.* The capillary depression of mercury is not treated adequately in some handbooks, and the majority of readily available tables are based on meager experimental evidence. The error is one of the most serious in accurate barometric and manometric measurements because it is difficult to measure or compute accurately. Its magnitude is independent of the height of the mercury column, the pressure, and, to a large extent, the temperature, but is affected greatly by the condition of the surface of the tube and the mercury.

*Contact Angle and Capillary Constant.* Capillary depression is a function of the angle at which the mercury contacts the walls of the tubing. Unfortunately, the contact angle is not constant: it can be varied at will from about 30 to 90° and is affected greatly by the nature of the walls, moisture in the gas space, and traces of impurities in the mercury. These factors frequently cause unsymmetrical menisci, to which standard capillary depression tables do not apply.

The basic mathematical analysis of the shape of the meniscus was made by Laplace [7] in 1812. Unfortunately, application of Laplace's analysis to practical problems requires accurate values of the surface tension of mercury or, more conveniently, of the "capillary constant," which is usually denoted by $a$ (cm) and defined by

$$a = \left(\frac{2\gamma}{(\rho_2 - \rho_1)g}\right)^{1/2} \tag{2.3}$$

where $\gamma$ is the surface tension in dyne/cm, and $\rho_2$ and $\rho_1$ are the density of the fluids separated by the meniscus. The value of $a$ is very sensitive to all the factors affecting capillary depression and especially to trace amounts of impurities in the mercury. For fresh, clean mercury, $a$ should be 0.266 cm at room temperature; slight contamination will drop it to 0.248 or lower [8]. Blaisdell [9] presented tables for the capillary depression, $h_0$, in terms of tube radius $x$ and meniscus height $y$, all in terms of dimensionless ratios involving $a$. In general the tables are useful for 11- to 22-mm-diameter tubes.

*Published Tables.* The tables of capillary depression corrections we have found in laboratory handbooks have a curious history. Mendeleef and Gutkowski [10] published the basic values in 1876 but gave no experimental details. The paper, presented orally, was said to be from Mendeleef's book, *On Barometric Leveling and Reconciliation of Column Readings.* Further information about these and other measurements was promised in a forthcoming second volume of his book, *On the Elasticity of Gases.* A careful search did not disclose either of these books, and a similar experience has been noted in the literature [11]. The original table listed results for tubes of 4.042-, 5.462-, 8.606-, and 12.717-mm diam. These values, as interpolated by Kohlrausch [12], provide the present standard set of values for capillary corrections [13–15]. Thus, even the present capillary correction tables rest on the word of Mendeleef that the work was very carefully done and is highly reliable.

The values in the International Critical Tables [6] are based on Süring's [16] revision and interpolation of the Delcros tables (which were in turn calculated from theoretical formulas derived by Schleiermacher), for 434.0 dyne/cm for the surface tension of mercury at 20°C or a capillary constant of 0.2556 cm. The results are tabulated against tube radius from 1 to 7 mm.

Probably the best experimental work was done by Cawood and Patterson [17], who paid particular attention to purity of the mercury, errors of refraction, and exclusion of moisture from the system. They found that small amounts of water vapor *reduced* the capillary depression greatly. Values for capillary depression are given as functions of tube diameter, 10 to 19 mm, and meniscus height, 0 to 1.8 mm. They are a little higher than those in the International Critical Tables, and are very different from handbook values based on Mendeleef's work. A few comparisons are shown in Table 2.2.

*New Table.* The experimental results of Cawood and Patterson can be reconciled with Blaisdell's tables [9] if a high value, 0.266 cm [8, 18], is used for *a*. However, Cawood and Patterson observed a small, but definite, irregularity, making their values for tube diameters between 13 and 17 mm appear somewhat too high. While this irregularity has little effect on the general utility of their results, it precludes accurate extrapolation to

**Table 2.2**  Comparison of Capillary Depression by Cawood and Patterson and Mendeleef

| Tube diameter (mm) | Capillary Depression at Meniscus Height Shown (mm) | | | | | |
|---|---|---|---|---|---|---|
| | 0.8 mm | | 1.2 mm | | 1.6 mm | |
| | C&P[a] | M[b] | C&P | M | C&P | M |
| 10 | 0.298 | 0.15 | 0.416 | 0.25 | 0.521 | 0.33 |
| 11 | 0.220 | 0.10 | 0.319 | 0.18 | 0.411 | 0.24 |
| 12 | 0.153 | 0.07 | 0.230 | 0.13 | 0.306 | 0.18 |
| 13 | 0.112 | 0.04 | 0.168 | 0.10 | 0.224 | 0.13 |

[a] See Ref. 17.      [b] See Ref. 10.

smaller tubes. For this reason new values (Table 2.3) were computed for tube diameters of 11 to 20 mm based on Blaisdell's theoretical tables with $a = 0.266$ cm. No irregularity was noted, and we believe that the new table is better than Cawood and Patterson's. Table 2.3 was extended down to 2-mm diam. by scaling up the values in the International Critical Tables by an empirical factor based on the tabular values in the overlapping range (11- to 14-mm diam.). The factor requires an increase of 1.2%/mm diam. for all meniscus heights. For example, the new values for tubes of 8-mm bore were obtained by multiplying the International Critical Tables values by 1.096. The error in this method of correction is probably less than $\pm$ 0.02 mm or 5%, whichever is larger.

*Effect of Tube Size.* Table 2.3 illustrates that small tubing will cause high capillary depressions strongly dependent on the height of the meniscus. Therefore, tubes smaller in diameter than 7 mm are not recommended for manometers or barometers. Even for a 7-mm tube an error of 0.1 mm in the height

**Table 2.3** Recommended Capillary Depression for Mercury in Glass Tubes

| Tube diameter (mm) | Meniscus Height (mm) | | | | | | | | |
|---|---|---|---|---|---|---|---|---|---|
| | 0.2 | 0.4 | 0.6 | 0.8 | 1.0 | 1.2 | 1.4 | 1.6 | 1.8 |
| 2 | 2.52 | 4.51 | | | | | | | |
| 3 | 1.14 | 2.13 | 2.93 | | | | | | |
| 4 | 0.63 | 1.22 | 1.73 | 2.12 | 2.47 | | | | |
| 5 | 0.39 | 0.76 | 1.10 | 1.41 | 1.64 | 1.84 | | | |
| 6 | 0.26 | 0.51 | 0.75 | 0.96 | 1.15 | 1.30 | 1.42 | | |
| 7 | 0.18 | 0.37 | 0.53 | 0.69 | 0.82 | 0.94 | 1.04 | 1.13 | |
| 8 | 0.13 | 0.26 | 0.38 | 0.50 | 0.61 | 0.70 | 0.78 | 0.84 | 0.90 |
| 9 | 0.10 | 0.20 | 0.29 | 0.38 | 0.45 | 0.52 | 0.59 | 0.64 | 0.69 |
| 10 | 0.08 | 0.15 | 0.21 | 0.28 | 0.34 | 0.39 | 0.45 | 0.49 | 0.53 |
| 11 | 0.06 | 0.11 | 0.16 | 0.21 | 0.26 | 0.31 | 0.35 | 0.38 | 0.41 |
| 12 | 0.04 | 0.08 | 0.13 | 0.16 | 0.20 | 0.24 | 0.27 | 0.29 | 0.32 |
| 13 | 0.03 | 0.07 | 0.10 | 0.13 | 0.16 | 0.18 | 0.21 | 0.23 | 0.25 |
| 14 | 0.03 | 0.05 | 0.07 | 0.10 | 0.12 | 0.14 | 0.16 | 0.18 | 0.19 |
| 15 | 0.02 | 0.04 | 0.06 | 0.08 | 0.09 | 0.11 | 0.12 | 0.14 | 0.15 |
| 16 | 0.02 | 0.03 | 0.05 | 0.06 | 0.07 | 0.09 | 0.10 | 0.11 | 0.12 |
| 17 | 0.01 | 0.02 | 0.04 | 0.05 | 0.06 | 0.07 | 0.08 | 0.08 | 0.09 |
| 18 | 0.01 | 0.02 | 0.03 | 0.04 | 0.04 | 0.05 | 0.06 | 0.07 | 0.07 |
| 19 | 0.01 | 0.01 | 0.02 | 0.03 | 0.03 | 0.04 | 0.05 | 0.05 | 0.06 |
| 20 | 0.01 | 0.01 | 0.02 | 0.02 | 0.03 | 0.03 | 0.04 | 0.04 | 0.04 |

NOTE: This correction in millimeters must be added to the observed height of the top of the meniscus.

of the meniscus will cause an error of 0.1 mm in the pressure. Unfortunately, many commercial, laboratory barometers are constructed of much smaller tubing; one particular brand is only 3.7 mm.

*Density of Mercury and Value for the Acceleration of Gravity.* There is little chance that the density of mercury will be significantly different from 13.5951 $g/cm^3$ at 0°C since the presence of even traces of impurities would so befoul the mercury surface that it could not possibly be used for precise barometric measurements.

Recently Cook and Stone [19] and Stone [20] determined the density of mercury at 20°C and 1 atm as 13.545884 $g/cm^3$. The corresponding value at 1 atm and 0°C calculated by the expansion formula of Beattie et al. [8], is 13.595080 $g/cm^3$. Beattie et al. [8], also give the thermal dilation and density of mercury over a wide temperature range.

The local value of the acceleration of gravity, $g$, may be computed from the formula [21]

$$g = 978.0490(1 + 0.0052884 \sin^2 \varphi - 0.0000059 \sin 2\varphi) - 0.0003086\, H$$

$$(2.4)$$

where $\varphi$ is the latitude and $H$ is the height above sea level in meters. However, the value of $g$ calculated from (2.4) should be reduced by 0.014 cm/sec² to account for the latest absolute determination of the gravity at Potsdam [22], 981.260 cm/sec² instead of 981.274 cm/sec². A relation recommended by Brombacher, Johnson, and Cross [4] for the variation of gravity with latitude and elevation differs slightly, but not significantly, from (2.4) above.

The length of the mercury column for local gravity may be corrected to the value for standard gravity by the ratio $g/980.665$.

*Recommended Procedure for Mercury Manometers.* This section provides directions for measuring the correct height of a manometer column at a known temperature and for converting the measured height to standard units of pressure.

1. Before taking any readings, see that the manometer (or barometer) is exactly vertical because a deviation of 1° will cause an error of 0.015% or about +0.11 in 760 mm Hg pressure.

2. Tap the manometer until the menisci are symmetrical and their bases horizontal. If the tube sizes are equal, both menisci should have the same shape and height; often the menisci can be improved by moving them up and down. Read the positions of the crowns of both menisci; the difference in readings is the uncorrected pressure difference. If the tube sizes are equal and the menisci have the same height, no additional observations are necessary; however, when the tube sizes are unequal, or the heights of the menisci are different, the position of the bases of the menisci should also be observed.

Parallax and errors caused by improper illumination should be avoided during observation of the menisci. For accurate work a cathetometer is necessary, and the usual precautions for the use of this instrument must be observed. The cathetometer is best used as a null instrument for comparing column levels against a thermostated scale placed in the bath, thus errors that may result from sighting through the liquid are minimized. The cathetometer should be checked against a suitable standard scale and proper leveling techniques used. Proper lighting of the meniscus will avoid highlights that could be mistaken for the crown of the meniscus. Brombacher et al. [4], recommended miniature electric lamps with a green translucent shield between lamp and mercury column. For illuminating the mercury meniscus, Beattie et al. [8], and Collins and Blaisdell [23] describe an optical train which can be directed from behind the manometer tube into the telescope of the cathetometer.

3. Apply capillary corrections to both menisci, unless the tube sizes are equal and the menisci are the same height. The total correction is the difference between the corrections for the upper and lower menisci. Procedures for making capillary corrections are described in the section on Barometers and Manometers.

4. Reduce the above readings to standard conditions (0°C and standard acceleration of gravity 980.665). To correct for the variation of the density of mercury multiply the measured height of the mercury column by the ratio, $\rho/\rho_s$, where $\rho$ is the density at the average temperature of the manometer and $\rho_s$ is the density at 0°C. Values of the density of mercury can be found in most handbooks or in a publication by Beattie et al. [8]. To correct for standard gravity multiply the measured height of the mercury column by $g/g_s$ where $g$ is the value of the acceleration of gravity at the location of the experiment and $g_s$ is 980.665 cm/sec². If $g$ has not been determined for the geographical area of the experiment, it can be calculated with sufficient accuracy for most purposes by (2.4) which takes into account the altitude as well as the latitude. In most commercial instruments the scale can be assumed to read correctly at 0°C (or other calibration temperature so stated in the specifications of the instrument), but a correction for thermal expansion to the experimental temperature is necessary. To make the scale correction simply multiply the measured or nominal mercury height by the ratio $1/[1 + \alpha(t - t_s)]$ where $\alpha$ is the mean linear thermal coefficient of expansion of the scale material, $t$ is the temperature of the manometer, and $t_s$ is the temperature at which the scale was calibrated. The following values of of $\alpha/°C$ are widely accepted [4]: aluminum, $24.5 \times 10^{-6}$; low-carbon steel, $11.5 \times 10^{-6}$; and Invar, 0 to $5 \times 10^{-6}$.

Often, corrections for linear expansion of the scale and change in density of the mercury are combined; they are tabulated in handbooks and monographs [4].

5. Determine if a significant gas or liquid head exists between the surface of the mercury and the point where a true reading of pressure is desired such as on the surface of a liquid sample. If the mercury meniscus is below the surface of a liquid sample, the pressure at the surface of the sample will be less than the manometer reading, and vice versa. To calculate the corrrection in equivalent units of mm Hg multiply the difference in height (cm) by the density of the separating gas or liquid ($g/cm^3$) and divide the product by the density of mercury at 0°C (13.5951). If this correction is large, it should be adjusted for gravity by the ratio $g/g_s$.

6. When one column of the manometer is open to the atmosphere, determine if a pressure difference exists between the barometer and manometer due to a difference in elevation or from air flow in the laboratory. The possible magnitudes of these pressure effects were discussed (see p. 27).

7. When an open-end manometer is maintained much above room temperature, the vapor pressure of mercury becomes important. In this case, the vapor pressure of mercury should be subtracted from the observed height. If the manometer is the closed-end type and both columns are at system temperature, no correction is necessary if the mercury vapor is insoluble in the sample material and the van der Waals interactions between mercury vapor and sample vapor are negligible. Van der Waals interaction can be ignored except at high pressure. Selected values of mercury vapor pressure are given in [4]; however, at high pressures these values must be adjusted for the Poynting effect of total pressure, $P_g$, on the liquid mercury which increases the vapor pressure according to the relation

$$2.303 \; RT \log \frac{p_{Hg}}{p_{Hg0}} = V(P_g - p_{Hg0}) \tag{2.5}$$

in which $V$ is the molal volume of liquid mercury at $T$, K, and $p_{Hg0}$ is the normal vapor pressure of mercury under its own equilibrium pressure taken from [5].

### Mercury Manometers

The simplest mercury manometer is the U-tube, either open at both ends or closed at one end. Most of the precautions and directions stated in the preceding section also apply to this type of manometer. The diameter of the columns should be wide enough to minimize capillarity errors. At least 15-mm diameter is suggested for accurate work; however, if the manometer is quite long, smaller tubing can be used without causing large percentage error.

*Conventional Mercury Manometers.* The open-end, U-tube mercury manometer consists of a wide-bore tube (approximately 15 mm), which is not sealed at either end. Sometimes a constriction is placed near the bend to dampen large pressure fluctuations and prevent loss of mercury during an unusual surge of pressure. The open-end manometer is frequently used for measuring gage pressures near atmospheric; one side is left open to the atmosphere, and the other is connected to the system. The pressure in the system is then the algebraic sum of the atmospheric pressure, as read on a barometer nearby, and the manometer reading. In accurate work a correction for the elevation of the barometer must not be ignored (see p. 27). Open-end manometers are also of value when low pressures must be measured, but in this case the leg which is not connected to the system is connected to a high-vacuum line. The manometer then indicates the absolute pressure in the system. But, for most routine measurements at moderately low pressures, the closed-end manometer is more convenient than the open-end type.

The closed-end, U-tube manometer is standard equipment obtainable from laboratory supply houses. It is the most commonly used instrument for measuring pressures from about 5 to 300 mm Hg. It is almost impossible to seal off one leg of the manometer to enclose a perfect vacuum. Usually the tube is boiled out with mercury until all air is expelled. A convenient and simple test for a good vacuum is the sharpness of the "click" heard when the mercury runs back and strikes the closed end of the tube. Even a manometer which produces an audible click when filled may contain a minute bubble of air in the vacuum leg after a few days, but for reasonably high pressures this can usually be ignored. In any event, the manometer should be checked by carefully comparing the levels of the mercury when the system side is connected to a high vacuum.

To avoid the difficulty of maintaining a perfect vacuum in the closed leg, a stopcock can be sealed to the top of the normally closed end. Unfortunately, there is no simple way to detect a small leak. A practical solution is to place a small U-tube loop containing mercury between the closed end and the stopcock.

The closed-end manometer exists in various designs and adaptations, many of which facilitate readings below 5 mm Hg. In Rayleigh's manometer [24, 25], Fig. 2.1, two pointers, P, are placed above the mercury levels in the legs. The legs are made of wide glass bulbs to obviate capillarity errors. A zero reading is obtained by connecting both sides to a good vacuum and rotating the manometer until the pointers just touch, simultaneously, the surfaces of the mercury. When the gage is used to measure pressure, one side is maintained at a high vacuum and the other is connected to the system. The gage is then rotated until the pointers again just touch the mercury levels. The motion is magnified by an optical lever with mirror, M, attached to the pointers. A serious disadvantage of all optical-lever manometers is their high sensitivity to vibration [26].

Mündel's manometer [27] also has wide bulbs in the manometric legs, but by being fixed vertically avoids the inconvenience of rotating as in the Rayleigh gage. Instead, the pointers are connected to micrometer screws, and the

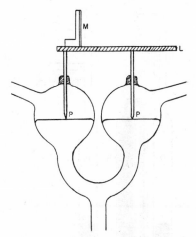

**Fig. 2.1** Rayleigh's manometer, showing the pointers, *P*, and mirror, *M*, of the optical lever. The tilting mechanism is not shown.

contact between the mercury and the pointers is observed through microscopes. The difference in mercury levels is determined from the difference in the readings on the micrometer sleeves.

Willingham, Taylor, Pignocco, and Rossini [28] used a closed-end manometer for pressure measurements at fixed points by placing tungsten electrodes along one of the columns. The pressures were manostatically controlled at the fixed pressure points determined by the position of the electrodes. The mercury heights at the electrode contacts were determined from the boiling points of water (see p. 60, for the adaptation of this apparatus to accurate vapor pressure measurements).

*Zimmerli Gage.* The Zimmerli gage [29] obviates many of the usual difficulties with closed-end manometers. The principal feature is the use of three wide-bore arms (Fig. 2.2). *A* and *B* are the columns of the U-tube, each at least 16-mm diam. Tube *A*, the indicating column, is connected to the vacuum line. Tube *B*, the reference column, instead of being sealed at the top is connected to a capillary, *C*, which is joined to a wide tube, *D*, at the bottom. The connection of both indicating and reference columns to the same vacuum line is the fundamental difference between the Zimmerli gage and the usual closed-end manometer. As soon as the pressure is reduced to a value corresponding to the difference in heights of the mercury columns in *A* and *B*, the mercury will separate at the top of the bend, between *B* and *C*, and as the pressure diminishes each part will recede in *B* and *C* until the levels become constant. The difference in height of the mercury levels in *A* and *B* indicates the pressure.

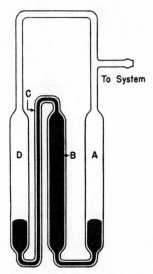

To System

**Fig. 2.2**  Zimmerli gage: *A* and *B* are the columns of the U-tube, *C* is a capillary tube, and *D* is a mercury reservoir.

*Dubrovin Gage.*    The Dubrovin gage is one of several whereby the reading of a manometer is multiplied to extend the pressure range and increase the accuracy. The gage, illustrated schematically in Fig. 2.3, magnifies the effect of the pressure by as much as 25 times [30, 31]. The principle is similar to that of the floating-tube barometer [32]. An inverted cylindrical tube, closed at the upper end, floats freely on the mercury. The lateral motion is constrained by three point-guides lightly touching the inside walls of the container. The gage is placed on its side and a high vacuum is applied until all air and absorbed gases are removed. It is then set on end, and atmospheric pressure is slowly restored. On progressively reducing the pressure the following events occur: (*a*) the tube floats while the inside remains entirely filled with mercury; (*b*) the tube remains stationary until the external pressure is equal to the height of the mercury column inside the tube above the outside level; and (*c*) the mercury column drops while the tube itself rises [31]. This condition is shown in Fig. 2.3.

The theory was presented in detail by Germann and Gagos [31], who derive the following equation for the magnification factor:

$$-dH/dN = D_1^2(D_2^2 - D_1^2) \qquad (2.6)$$

where $D_1$ and $D_2$ are the internal and external diameters of the tube. The absolute value of $N$ depends on the pressure, $P$, in the outer tube, and on

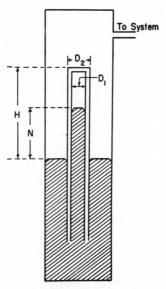

**Fig. 2.3** Dubrovin multiplying manometer: $D_1$ and $D_2$ are the internal and external diameters of the float, and $H$ and $N$ are the heights of the tube and mercury column, respectively.

capillary forces. The absolute value of $H$ depends on the densities of the tube and the liquid, the pressure, $P$, and surface forces. It is evident that the ratio of the absolute value of $H$ to $N$ is meaningless. On the other hand, a decrease in the pressure, $P$, of 1 mm Hg is accompanied by a decrease in $N$ of 1 mm Hg if the confining liquid is mercury. If simultaneously the tube rises 10 mm, then $H$ has increased by 10 mm and the change of $H$ with respect to $N$ is 10, which is then the true magnification of the gage. The apparent density of the tube can be changed either by making the lower submerged part of the tube a double-walled, sealed cylinder or by attaching a separate glass float at its base. The lower range of the instrument with mercury as the liquid is about 0.01 mm Hg. This limit can be extended to 0.0006 mm Hg by the use of a nonvolatile oil such as Apiezon-B [31] or a silicone such as Dow-Corning 704 fluid [33]. The silicone is about 7% denser than water, but has a very high boiling point and about the same viscosity as a light machine oil. In addition, it is extremely stable to oxidation and can be degassed by boiling at close to atmospheric pressure.

The Dubrovin gage is available commercially. One model has a magnification factor of 9 and covers the range from 0 to 20 mm Hg. This model may be calibrated against a closed-end mercury manometer at higher pressures and against a McLeod gage at lower pressures.

*Other Multiplying Manometers.*   A sloping mercury manometer is superior to the Dubrovin gage when low pressures of the order of 1 mm Hg are subject to some variation. Burton [34] has shown a convenient design, which has been used with a 25-to-1 magnification. This particular model has an ingenious trap (Fig. 2.4), which obviates difficulties caused by pulling over bubbles into the closed leg when gases are suddenly introduced into the sloping manometer. Any bubbles pulled over collect just below the inner seal and have no effect on the pressure reading.

Roberts [35] has described a very sensitive U-tube compensated micromanometer that consists of two wide-bore legs of slightly different diameter, *A* and *B*, connected by a horizontal capillary that contains a bubble of air which serves as an index. A slight pressure difference between *A* and *B* causes an appreciable motion of the bubble. If the length of the bubble is approximately equal to the distance between the axes of *A* and *B*, the gage is almost independent of small changes of position or level.

Differential gages that use two liquids of different density usually have limited utility for vapor pressure measurements because of the solubility of the substance being measured in the liquids. An important exception is the Young and Taylor [26] vacuum micromanometer, which has a multiplying factor of 1000 and an accuracy of about 1%. The particular instrument illustrated by the inventors covered the range from 1 to 100 $\mu$. The principle is essentially that of the Roberts capillary bubble method previously described. The high viscosity and surface tension of mercury make the instrument somewhat sluggish. In another modification a second liquid, such as *n*-pentane, used in the capillary is sealed from the system by two barometric legs of mercury. The designers have shown that the sensitivity, or multiplying factor, of the gage can be closely approximated by

$$C = \frac{D_2{}^2}{2D_1{}^2} \left( 2 - \frac{\rho_P}{\rho_M} \right), \tag{2.7}$$

where $\rho_P$ is the density of *n*-pentane or similar liquid, $\rho_M$ the density of mercury at the same temperature, $D_1$ the diameter of the capillary containing the reference bubble, and $D_2$ the diameter of the mercury reservoir connected to the barometric leg on the system side. The value of *C* in the instrument de-

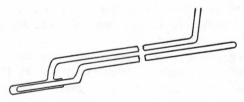

**Fig. 2.4**   Burton's sloping manometer.

scribed ($D_2$ = 81 mm, $D_1$ = 1.26 mm) was 1056. A check value of 1015 was obtained from 40 readings on a McLeod gage. This reproducibility was considered satisfactory in view of the uncertainties in measuring the diameters of the capillary and reservoir and in determining the constant of the McLeod gage.

Details of the design, operation, and construction are given at length in the original article. To measure molecular weights of volatile liquids [26] only 3 to 4 min was required per determination.

*McLeod Gage.*   The McLeod gage [36] is most useful from 1 $\times$ 10$^{-5}$ to 10 mm Hg. Application is based on Boyle's law. A measured volume of the gas, $V$, at the pressure of the system is compressed to a smaller volume, $v$, at a pressure which can be measured directly. The ratio of the volumes, $V/v$, determines the sensitivity of the gage.

A typical McLeod gage is shown in Fig. 2.5, but the part below $C$ takes a form based on the method used for handling mercury. The gage is connected to the system at $S$. In operation, the mercury from reservoir $R$ is forced upward into the gage by opening the three-way stopcock $T$ to the atmosphere. As the mercury rises, it traps at $C$ a gas volume, $V$, which fills bulb $D$ and capillary $E$. The usual operating practice is to permit the mercury to rise in capillary $F$ until it reaches a position $A$ opposite the top of $E$. The mercury in $E$ has then reached a point such as $B$ while the gas is compressed to small volume $v$. The pressure of the compressed gas is then equal to the original unknown system pressure plus the pressure of the mercury column between $A$ and $B$. If $l$ is the length of the mercury column (the reading) and the system pressure, $P$, is negligible compared to $l$, then, from Boyle's law,

$$P = (v/V) \, l. \tag{2.8}$$

For a capillary of constant cross section, $v = Al$, where $A$ is the volume per unit length. Hence

$$P = (A/V) \, l^2 \tag{2.9}$$

The ratio, $A/V$, must be determined from careful measurements of the capillary volume and the total volume. After the reading has been taken, $T$ is opened to the vacuum line, $V$, and the mercury is permitted to drop below $C$.

A nomograph by Lott [37] is available for estimating the capillary bore and the bulb volume for a given pressure range.

Commercial McLeod gages are usually calibrated at the factory. They are available in a variety of modifications that differ in minor details from the type described. Several tilting models of McLeod gage are also available commercially. They are very compact and are suitable for all applications not requiring the highest accuracy.

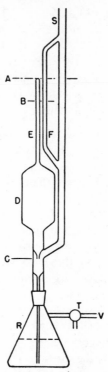

**Fig. 2.5** McLeod gage: *A* and *B* are mercury levels during reading in capillary tubes *F* and *E*, respectively; *D* is gas-volume bulb; *C* is the bulb throat; *R* is a mercury reservoir; *S* is a connection to the system; *T* is a three-way valve or stopcock to atmosphere or compressed-air supply; and *V* is a vent.

For a description of some of the improvements which have been suggested see papers by Booth [38], Gaede [39], Flosdorf [40], Hickman [41], Pfund [42], and Rosenberg [43].

The McLeod gage gives erratic results not only when condensable gases are present but also when ammonia, carbon dioxide, traces of water vapor, and similar materials which do not follow Boyle's law are in the system. For this reason, the gage is frequently isolated by a liquid air trap.

### Oil Manometers

The advantage gained at low pressures by using a manometer fluid one-thirteenth as dense as mercury does not, unfortunately, mean a 13-fold gain in accuracy. Some of the errors inherent in closed-end "oil" manometers have been discussed by Hickman [44]. These include faulty vacuum in the closed limb caused by dissolved air (−0.01 mm), products of cracking during the boiling-out process (−0.01 mm) or volatiles carried around from the working limb (−0.05 mm), and differences in specific gravity of the liquid in the two limbs caused by a difference in either the temperature (±0.02 mm) or the composition (±0.01 mm) of the liquid in the limbs. Thus, the total errors ordinarily encountered by Hickman equal −0.04 to −0.10 mm Hg.

Using butyl phthalate, Hickman devised a somewhat complicated setup [45] in which a good reference vacuum was maintained by a small condensation pump. The results were accurate but the apparatus was so troublesome to operate that numerous improved models were tried. The best model out of 15 was described in 1934 [43] and has proved useful for measurements down to a few microns. A similar apparatus by Malmberg and Nicholas [46] (Fig. 2.6) can be boiled out under vacuum by simply inverting it. Having been boiled out, it is stood right side up and is ready for use. The ebullition tube, $E$, serves to prevent bumping and promotes smooth boiling. Degassing should be done at a temperature low enough to avoid decomposition of the liquid. No difficulties from this source have been noted with butyl phthalate.

Saylor and co-workers [47] used a Hickman oil manometer in their studies on the vapor pressures of many compounds between 0 and 80°C.

The butyl phthalate manometer used by Bradley and Swanwick [48] includes reservoirs which prevent contamination of the manometric liquid with silicone stopcock grease. In the same paper, Bradley and Swanwick have included valuable notes on the use of oil manometers in high-precision vapor-pressure work.

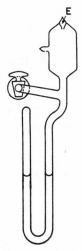

**Fig. 2.6**   Malmberg and Nicholas oil manometer showing ebullition tube, $E$.

## Pressure Transducers

The Bourdon tube, quartz spiral, and elastic diaphragm, although not recent inventions, are the most common pressure-sensitive elements used in modern pressure transducers. Recent improvements in pressure-transducer instrumentation involve better devices for sensing and magnifying the deflections into reliable readout systems. In the area of low pressures, instruments with very small leak-through rates are now available. Even a cursory review of the trade literature reveals a good variety of instruments for any sensitivity and pressure from 1 $\mu$ to 100 kbar and for almost any conceivable laboratory application in which the transducer is operated as a null, differential, or absolute-pressure measuring device.

The Bourdon tube gage (Fig. 2.7) is the most commonly used pressure measuring device. The pressure sensitive element of the gage consists of a semicircular section of tube flattened to a nearly ellipical cross-section with one end connected to a pressure system and the other end free to deflect when pressure is applied. Gages are available commercially in pressure ranges up to 100 kbar with an accuracy of 0.1% of the calibration. Although some gages are fine precision instruments, their accuracy must be established by calibration against a standard deadweight gage (see p. 43), and even though the Bourdon-tube gage is largely trouble-free, a few precautions are necessary in its operation.

The gage should be mounted in an upright position, particularly if it is connected to a liquid system. Also, the liquid in the gage and connecting lines should be freed of air or gas bubbles. Pressure surges which cause the pointer to whip could well destroy the validity of the calibration or damage the mechanism. Use of the gage at extreme temperatures requires an applied correction if the gage does not have a thermal compensator. As a final precaution, do not use a gage for oxygen unless it is recommended for this service and check to determine if the tube is free of oil and other organic matter.

A system that employs a quartz helix as the primary pressure sensor (Fig. 2.8), with an optical secondary transducer and servo amplifier has the advantage, in the organic laboratory, of being compatible with a wide variety of corrosive chemical vapors. The manufacturer of one line of quartz helix gages describes how they work.

"The optical transducer is mounted on a gear that travels concentrically around the Bourdon tube. In operation the deflection of the pressurized Bourdon tube is found by rotating the gear until the microammeter reading is zero, i.e., light reflected from the tube mirror falls equally on a pair of balanced photocells. The digital counter reading associated with the null meter reading is then multiplied by a scale factor to determine the pressure." A calibration chart is supplied with each capsule [49].

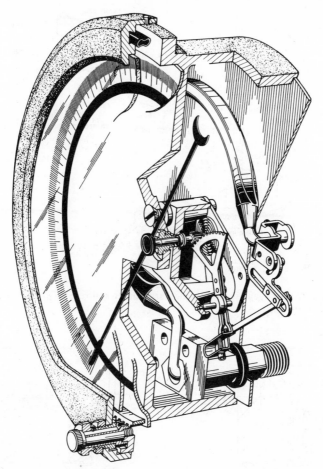

**Fig. 2.7** Bourdon tube gage. (Courtesy of the Heise Bourdon Tube Company, Inc.)

Many of the pressure transducers available commercially are based on the principle of the simple elastic diaphragm or bellows (Fig. 2.9). When pressure is applied to one side of the diaphragm, the physical movement is detected by a secondary sensing element that can be designed to respond to changes in either optics, capacitance, magnetic reluctance, inductance, resistance, or electromechanical resonance.

## The Piston Manometer

The piston manometer (Fig. 2.10), commonly called a deadweight gage, is an accurate pressure-measuring instrument that consists essentially of a piston in a close-fitting cylinder mounted so that the hydrostatic head on the

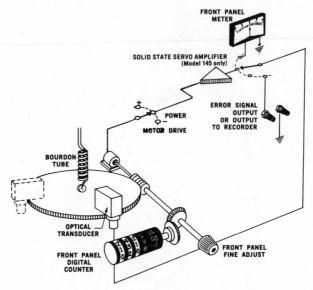

**Fig. 2.8**  Quartz spiral gage showing optical transducer and reading system. (Courtesy of Texas Instruments, Inc.)

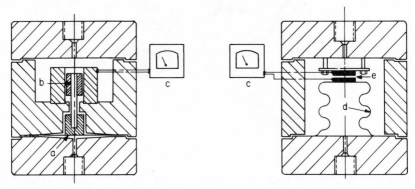

**Fig. 2.9**  Diaphragm and bellows transducers. *a*, diaphragm; *b*, differential transformer as secondary sensor; *c*, readout and recording; *d*, bellows; *e*, capacitor as secondary sensor.

underside of the piston is balanced by weights brought to bear on the top or exposed end of the piston. It is a primary instrument that can be used to measure pressure in terms of the weight and the measured area of the piston. In practice, however, it is often calibrated against a pressure standard, such as the vapor pressure of carbon dioxide at 0°C [50],* which is equivalent to

---

* One of the first accurate determinations was 26,144.7 mm Hg [50]. A more recent value by Dadson, 26,137.6 mm Hg, was determined with high accuracy [50]; however, an intermediate value (26,140.8 mm Hg) has been reported [95].

determining the effective area, $A_e$, of the piston in the equation

$$P_p = \frac{F_e}{A_e},\qquad(2.10)$$

where $P_p$ is the pressure at the piston and $F_e$ is the force due to the total load. The total load includes the weight of the piston proper, equivalent barometric pressure, hydrostatic heads, and weights.

Determining the pressure at an internal point, such as on the mercury surface of the cell (Fig. 2.11), usually involves pressure-transmitting fluid heads that must be added to $P_p$, (2.10), which is conveniently referred to the bottom of a regular piston. When the deadweight gage is calibrated on location against carbon dioxide as standard, the effective area, $A_e$, or the effective pressure per unit weight so evaluated, cancels variation from standard gravity, air buoyancy on the weights, the effects of the lubricating film, and deviation of the piston from vertical. Corrections for the distortion of the piston-cylinder by the applied pressure and the expansion and contraction of the piston-cylinder from temperature changes must be applied if accuracy better than a few one-hundredths of one per cent is sought. In the pressure range 10–500 atm the linear increase in the effective area of a 0.05 in.² piston cylinder is given by Beattie [51] and by Dadson [52] as very nearly 0.02% of the pressure. Variation of the gage constant with temperature due to thermal expansion is given by the relation

$$C_{t_1} = C_{t_2}\,[1 + \alpha(t_2 - t_1)],\qquad(2.11)$$

where $t_2 - t_1$ is the difference between the applied temperature and the calibration temperature and $\alpha$ is twice the mean thermal coefficient of linear expansion for the metal of the piston. The value of $\alpha$ is about $2.2 \times 10^{-5}/°C$

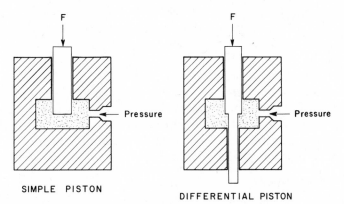

F        F

Pressure       Pressure

SIMPLE PISTON

DIFFERENTIAL PISTON

**Fig. 2.10** Schematic drawing of a simple piston and a differential piston in which the pressure acting on the piston area balances the applied force, $F$.

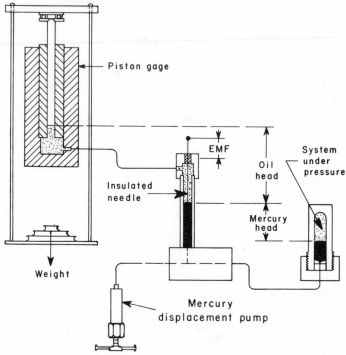

**Fig. 2.11**  Pressure measuring system illustrating use of the deadweight or piston gage.

for steel and can be assumed to apply without significant error to metals such as Carpenter 440 C unless the temperature difference is unduly large. If the gage constant was determined by direct measurement of the piston and cylinder diameters or if the calibration against a pressure standard was made at some location far removed from the place where the gage is being used, an appropriate correction for change of gravitational constant is necessary.

Examples of piston gage calculations and simplified working equations for the reduction of the effects of gravity, temperature, pressure, and several other variables have been published by Cross [53].

A variation of the piston manometer principle applied to low-pressure measurements, 0.1–40 mm Hg, is used in the inclined-piston gage in which declination of the axis of the piston from the horizontal provides a sensitive means of changing the force exerted by the piston. A detailed description of the inclined-piston gage is given in the section on vapor pressure (p. 50).

## Low Pressure Gages

### Pirani, Thermocouple, and Thermistor Gages

The *Pirani gage* [54, 55] consists of a heated filament of metal that has a high coefficient of electrical resistance. The temperature and therefore the resistance of the filament depend on the thermal conductivity of the residual gas, which at low pressures is linearly dependent on the pressure. The range of the gage is 1–1000 $\mu$, but it must be calibrated for each residual gas and so can only be used indirectly in vapor pressure measurements. In practice, two bulbs containing identical filaments are used; one is evacuated to $1 \times 10^{-6}$ mm Hg and sealed, and the other is connected to the system. The two filaments are connected to the arms of a Wheatstone bridge, and any change in the temperature of the gage bulb is shown by an imbalance of the bridge. Many variations of the Pirani gage have been designed and several excellent models are on the market.

*Thermocouple gages*, in which the thermocouple is welded to the center of a heated wire, and the *thermistor gage* [56], which depends on the resistance of a bead of semiconducting material, operate on the same principle as the Pirani gage.

### Knudsen-Effusion and Knudsen-Torque Gages

Application of the Knudsen gage to measure vapor pressures in the range from a few hundred microns to a few tenths of a micron is discussed later (p. 74).

### Inclined-Piston Manometer

Since the inclined-piston manometer has been used mostly to measure vapor pressures over the range 0.1–40 mm Hg, it is described below (p. 50).

## Vapor Pressure

Because the vapor pressure of a substance is a unique function of its temperature, control and measurement of temperature, as well as pressure, must be considered when deciding upon a program of vapor pressure determinations. Although modern techniques in thermometry are treated thoroughly in Chapter I, the choice of a proper temperature-measuring device and the manner in which it is used is an explicit consideration for each vapor pressure apparatus. Mercury-in-glass thermometers are adequate for ordinary measurements, but definitive vapor pressure–temperature relations can only be obtained with accurately calibrated platinum resistance thermometers or thermocouples. Also, proper placement of the thermometer or thermocouple is very important because every apparatus will contain temperature gradients. For example, in a static system the fluid should surround the thermometer if possible, and in an ebulliometric system care must be taken to place the thermometer at a point where pressure–temperature equilibrium exists.

### Static Methods

In the classical method two barometric tubes stand together in the same cistern. The substance is introduced into the Torricellian vacuum of one of them. The depression of the mercury caused by the vapor pressure of the substance is read by comparison with the other tube. In practice the tubes are jacketed; the mercury is thoroughly boiled out and the substance de-gassed. At low pressures, the method is very sensitive to the presence of dis-solved and adsorbed gases and to impurities which can give trouble even after careful boiling of the mercury and the sample [57]. Therefore, the classical static method is not recommended for vapor pressures below 1 atm; however, it is useful at high temperatures and pressures or when combined with pressure–volume–temperature measurements (see p. 50).

In a different modification of the simple classical method the substance is contained in a small bulb sealed to a gage and connected to a vacuum line by a mercury cutoff. The gage, bulb, and trap are immersed in a controlled tem-perature bath, after the substance has been degassed by freezing under a high vacuum. The gage then gives the vapor pressure of the material at the tem-perature of the bath. This method was used to determine the vapor pressure of tetraethyllead [58], nickel carbonyl [59], and mustard gas [60].

Static vapor pressure measurements are often made in connection with calorimetric studies. The apparatus used by Aston [61] and Giauque [62] are typical.

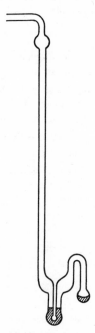

**Fig. 2.12**  Early model of the isoteniscope.

THE ISOTENISCOPE

The Smith-Menzies isoteniscope [57] obviates many of the difficulties from adsorbed or dissolved gas. The original design is shown in Fig. 2.12 and a later modification [63] in Fig. 2.13.

In use, the isoteniscope is connected to a manometer, surge tank, and pressure-control system. The sample is first boiled under a slight vacuum until air ceases to escape through the U-tube liquid. Air is slowly admitted into the system until the liquid levels in the U-tube are alike, and the temperature and pressure on the manometer are read. The sample is then boiled out again and the readings are repeated. After a few such repetitions the dissolved air and adsorbed gases are completely removed and successive readings are virtually identical. The sample is thus purified in the equipment until the measured vapor pressure is constant. The procedure does not remove higher-boiling impurities, decomposition products, or compounds that boil close to or form azeotropes with the material under test. Less than 1 g of material is needed.

The isoteniscope method can be adapted to solids if a noninteracting, non-solvent, nonvolatile confining liquid can be found. Suggested liquids are

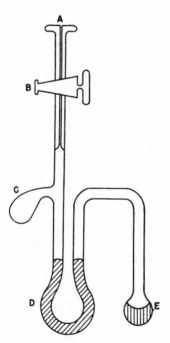

**Fig. 2.13** Modified isoteniscope showing flat joint connector, A; stopcock, B; reservoir, C; mercury manometer, D; and sample, E.

mercury and melted paraffin wax at lower temperatures and fusible alloys and molten salt mixtures at higher temperatures.

A recent modification [63] (Fig. 2.13), is particularly convenient when Hg is used as the containing liquid. Bulb $C$ is large enough to hold all the mercury. The isoteniscope can be easily attached to the rest of the apparatus by the flat joint, $A$.

First the isoteniscope is evacuated, flamed, and rinsed with dry air or nitrogen. Clean mercury is introduced, and the isoteniscope is reconnected to the vacuum line and is warmed until no more gas bubbles appear. It is then detached and tilted until all the mercury is in $C$. The sample is then distilled in by immersing bulb $E$ in liquid air or nitrogen. Next, part of the sample is boiled off to remove dissolved air. Then the stopcock, $B$, is closed, and the isoteniscope is detached and tilted until all the mercury is in manometer $D$.

The apparatus is now ready for measurements and can be connected to the auxiliary system and operated as described above.

With this modification the vapor pressures of numerous compounds [64–68] have been determined. The auxiliary equipment used has been described by Booth, Elsey, and Burchfield [69]. The average pressure deviation from a smooth curve was about 2 mm Hg, with maximum deviations as high as 6 mm Hg. The stopcock can be troublesome because many materials are soluble in stopcock grease, they are difficult to maintain vacuum-tight, and consequently a leak may develop when the isoteniscope is placed in a thermostatic bath. Stopcock lubricants should be tested by immersing the evacuated isoteniscope in the bath for about 1 hr. The test should be repeated with a small amount of the substance in the isoteniscope. The stopcock should not leak when the liquid is slowly distilled into a cold trap. Soap, graphite, or a silicone lubricant, such as Dow Corning stopcock grease, may be helpful.

PRESSURE–VOLUME–TEMPERATURE CELL

Vapor pressure measurements in the range from several atmospheres to the critical pressure are usually made during pressure–volume–temperature measurements. The operation of a variable volume P–V–T apparatus, shown schematically in Fig. 2.11, for accurate determination of vapor pressure has been described in considerable detail [51, 70–74]. The organic substance is placed in a temperature-controlled, variable-volume pressure chamber, and the volume is adjusted by adding or withdrawing mercury until liquid and vapor phases are present in equilibrium. A determination made over the complete two-phase range from dew point to bubble point (Fig. 2.14), should produce a horizontal isothermal line on the P–V plot if the sample is pure and the determined pressures have been corrected to the surface of the liquid phase.

INCLINED-PISTON

The inclined-piston gage [75–77] is especially adapted for measuring pressure in the range 0.1–40 mm Hg, which is below the lower limit of the

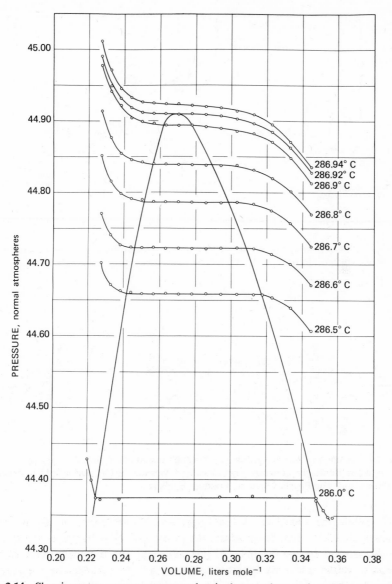

**Fig. 2.14**  Showing pressure measurements taken in the two-phase region of fluorobenzene. (Reproduced from *J. Amer. Chem. Soc.*)

ebulliometer or boiling-point apparatus and above the upper limit of the Knudsen-effusion or Knudsen-torque method.

The method of operation can be followed conveniently from the schematic drawing, Fig. 2.15. Detailed features of the construction are in Figs. 2.16 and

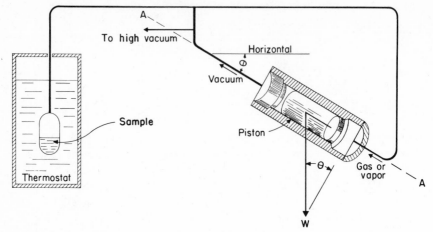

**Fig. 2.15** Vapor-pressure measurement with an inclined-piston gage showing axis, $A$--$A$, of piston; angle of declination, $\theta$; and vertical component of piston weight, $W$.

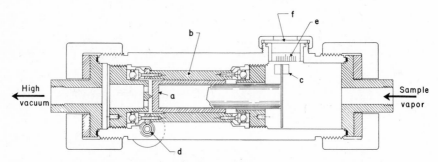

**Fig. 2.16** Section of inclined-piston gage through piston and cylinder: $a$, piston; $b$, cylinder; $c$, eccentric weight; $d$, worm gear; $e$, scale; $f$, viewing port. (Courtesy of The Institute of Physics and The Physical Society, London, England.)

2.17. While the chamber at the upper end of the piston is continuously evacuated to $10^{-5}$ mm Hg or better, the piston and cylinder are declined from the horizontal by some angle, $\theta$, just sufficient for the component of the piston's weight along its longitudinal axis, $AA$, to balance the pressure of the vapor in the lower chamber. The smallest increment in the declination of the piston is found, by trial and error, that will reverse the direction of axial travel of the piston. The equilibrium value of $\theta$ is taken at the midpoint of the angular increment. Motion of the piston is observed through the window $f$, Fig. 2.16. Provided the pressure due to vapor head is negligible, the pressure at the liquid surface of the sample in the thermostat can be calculated from the equation

$$p = \frac{g}{g_{std}} \frac{W \sin \theta}{A} \tag{2.12}$$

which relates the angle of declination $\theta$, weight $W$, and area $A$ of the end of the piston, and the local and standard acceleration of gravity $g$ and $g_{std}$, respectively. Sensitive response of the piston to pressure is obtained by the use of an eccentric weight which produces an oscillation of the piston relative to the cylinder without disturbing pressure equilibration. An inert oil lubricant on the piston provides a positive gas seal.

If the sample vapor is inert to the materials of construction in the lower gage chamber and the lubricant on the piston, the vapor may be transmitted directly to the gage, provided that the sample temperature is kept below ambient temperature to prevent condensation of the sample in the gage and lines. When vapor pressure measurements above ambient temperatures are required or when samples are corrosive or highly reactive to the lubricant on the piston, the vapor must be isolated in the thermostat by a sensitive, inert

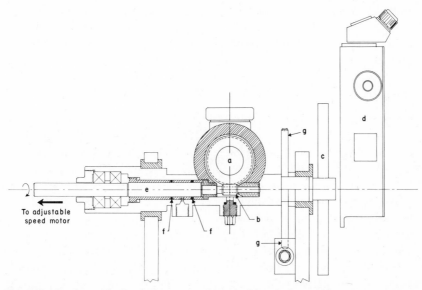

To adjustable speed motor

**Fig. 2.17** Section of inclined-piston gage through drive shaft: $a$, piston; $b$, worm gear; $c$, inscribed goniometer disk; $d$, goniometer optical readout; $e$, shaft; $f$, rubber "O" rings; $g$, elevating gear. (Courtesy of The Institute of Physics and The Physical Society, London, England.)

diaphragm with nitrogen or some other suitable gas used to transmit pressure from the diaphragm to the piston. In this mode of operation the diaphragm serves as a null-point indicator.

The precision of the gage is easy to determine because the piston area and weight are constant, leaving the angle $\theta$ as the only variable. If the gage is furnished with a goniometer that can be read to 1 sec, then the factor controlling precision will be the smallest practical increment in the declination of the piston, $\pm 5$ sec, that will reverse the direction of its axial travel. In terms of pressure, this increment is conservatively set at 0.001 mm Hg. An estimate of absolute, overall accuracy of vapor pressures measured on a correctly constructed piston gage was estimated to range from 0.001 to 0.012 mm Hg between total pressures of 0.01–40 mm Hg. Table 2.4 shows representative measurements on ice, water, and solid and liquid phases of hexafluorobenzene. Additional information about the construction and operation of the inclined-piston gage is given in [75] and [76].

### Boiling-Point Methods

In contrast to static-sample methods, discussed above, boiling-point methods involve a dynamic system in which the sample is boiled and condensed as a steady-state equilibrium process, all of which takes place under a controlled pressure from an inert gas blanket. If a thermometer is placed in the boiling liquid, the temperature registered is far from being a reliable indication of the temperature at which the liquid and vapor are in equilibrium because liquids often superheat. If, however, the thermometer is placed in the condensing vapor, the measured *temperature of condensation* is close to the true boiling point at the prevailing pressure.

#### HOOVER, JOHN, AND MELLON EBULLIOMETER

The *semimicro ebulliometer* devised by Hoover, John, and Mellon [78, 79], Fig. 2.18, provides a remarkably simple and rapid method of obtaining vapor pressures with an accuracy adequate for most purposes. The sample size is only 1–2 ml. The useful range of pressures is 5–760 mm Hg. A standard taper joint can be used to support the thermometer. The authors believe that the only critical dimension is that which places the thermometer bulb just below the lip of the boiling pot. The lower portion of the side arm acts as a condenser. A stream of water is run onto a piece of cloth or asbestos cord which is placed over the upward portion of the side arm and is drained through a funnel placed below. A length of tungsten wire is sealed through the bottom of the pot to project through a sheet of asbestos paper which is supported just below the apparatus. When heated by a microburner, this wire serves as a point source of heat that minimizes bumping. A glass chimney around the flame is helpful, although further shielding may be required

**Table 2.4**   Vapor Pressures Measured With an Inclined-Piston Gage

| Temperature (°C) | Pressure (mm Hg) | | Temperature (°C) | Pressure (mm Hg) | |
| | Present | Wash-burn[a] | | Obs | Obs-Calc[b] |
| --- | --- | --- | --- | --- | --- |
| *Ice* | | | *Hexafluorobenzene, solid* | | |
| −31.426 | 0.249 | 0.247 | −57.544 | 0.070 | 0.000 |
| −24.868 | 0.484 | 0.483 | −52.699 | 0.124 | +0.002 |
| −20.072 | 0.770 | 0.771 | −47.858 | 0.211 | +0.002 |
| −15.195 | 1.221 | 1.220 | −43.003 | 0.359 | +0.004 |
| −10.307 | 1.897 | 1.898 | −38.139 | 0.607 | +0.014 |
| −5.012 | 3.010 | 3.010 | −33.277 | 0.988 | +0.010 |
| −3.039 | 3.559 | 3.557 | −28.403 | 1.592 | +0.002 |
| −2.062 | 3.873 | 3.860 | −23.518 | 2.538 | −0.010 |
| | | | −18.624 | 4.023 | −0.001 |
| *Water* | | | −13.725 | 6.257 | 0.000 |
| −2.500 | 3.816 | | −8.818 | 9.570 | −0.006 |
| +0.000 | 4.586 | | −3.868 | 14.463 | −0.002 |
| 5.000 | 4.548 | | −1.898 | 16.965 | +0.001 |
| 10.000 | 9.219 | | +4.945 | 28.877 | −0.007 |
| 15.000 | 12.801 | | 5.045 | 29.103 | +0.001 |
| 20.000 | 17.548 | | | | |
| | | | *Hexafluorobenzene, triple-point* | | |
| | | | (5.17) | (29.377) | 0.000 |
| | | | *Hexafluorobenzene, liquid* | | |
| | | | 5.200 | 29.423 | −0.005 |
| | | | 5.329 | 29.608 | −0.041 |

[a] E. Washburn, *Mon. Weath. Rev.*, **52**, 488 (1924).
[b] Cox equations for hexafluorobenzene:

*Solid*

$\log (p/29.377) = A_s(1 - 278.32/T)$
with $\log A_s = 0.402338 + 4.50698 \times 10^{-3}T - 9.01548 \times 10^{-6}T^2$

*Liquid*

$\log (p/29.377) = A_l(1 - 278.32/T)$
with $\log A_l = 1.004282 - 7.85236 \times 10^{-4}T + 7.68853 \times 10^{-7}T^2$

around the apparatus for work with high-boiling materials or in drafty or cold rooms.

The sample is added to the pot to about 10-mm depth but should not touch the thermometer bulb even after expansion on heating. The apparatus is

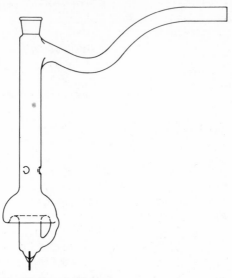

**Fig. 2.18**   Hoover-John-Mellon ebulliometer.

attached to an accurate manometer and a pressure manostat capable of ±0.1 mm Hg. After the pressure has been controlled to a suitable level, the flame is adjusted to form a ring of condensing vapor just above the bulb of the thermometer. About 10 min should be allowed for thermal equilibration.

A few successful measurements have been made on compounds which are slightly unstable. The materials tested have degassed quickly and the time needed for the measurement was so short that a sample was introduced, measured, and removed before appreciable decomposition had occurred. A few moments of boiling removed volatile impurities which could be distilled through the side arm. The accuracy is estimated at 0.5°C.

RAMSAY AND YOUNG APPARATUS

In Ramsay and Young's apparatus [80, 81], shown schematically in Fig. 2.19, the liquid is allowed to trickle onto the thermometer bulb, which is covered with a thin layer of absorbent cotton, glass wool, asbestos, or similar material. The bath is maintained 10–15° above the value read on the thermometer. The apparatus is connected to a cold trap, a manometer, and a surge tank, which, in turn, is connected by stopcocks to a vacuum pump and a supply of air or other inert gas.

The apparatus is evacuated thoroughly, then the bath is brought to temperature, a little air is admitted, and enough liquid is allowed to enter to thoroughly wet the cotton. The temperature and pressure are read as soon

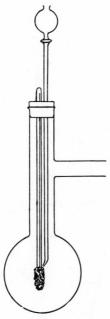

**Fig. 2.19** Ramsay and Young's apparatus. The side arm is connected to a cold trap, manometer, surge tank, and manostat; the flask is immersed in a heating bath.

as they become constant, a little more air is admitted, and the process is repeated. When the supply of liquid on the cotton is exhausted, more is admitted from the reservoir. If desired, the results can be checked by beginning at a high pressure and gradually lowering the pressure with the vacuum pump. Liquid must be kept out of the flask below the bulb. A series of measurements [82] on α-pinene and β-pinene were made on a modified Ramsay and Young apparatus using a constant temperature vapor bath [83]. These measurements extended from 1.9 to 760 mm Hg, with an average precision of about 1%.

CONVENTIONAL BOILING-POINT APPARATUS

A convenient apparatus for measuring boiling points at reduced pressure, which has been used in Ethyl Corporation laboratories for some time, is especially applicable to compounds that boil from 10 to 760 mm Hg and from 30 to 180°C. The liquid is boiled in a modified Claisen flask, Fig. 2.20, which is connected to a condenser, receiver, manometer, cold trap, surge tank, pressure manostat, and vacuum pump. Bumping can be decreased by use of boiling stones or glass wool or by preheating the bottom of the flask. Glass wool [84] packed into the flask so that some extends above the surface

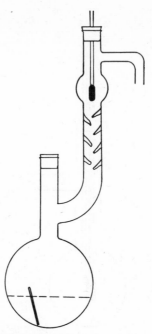

**Fig. 2.20** Boiling-point apparatus. The exit tube is connected to a condenser, receiver, cold trap, manometer, surge tank, and manostat; heating unit and insulation are now shown.

of the liquid is an effective but laborious way to prevent bumping. A simple alternative is the Hnizda float, Figs. 2.20 and 2.21. It was adapted by V. Hnizda from the microchemical boiling-point apparatus of Siwoloboff [85]. It consists of a glass tube sealed off as shown with the end *A* rough ground. The hollow end floats upright in the liquid while the roughened end promotes smooth boiling. The Siedeglocke [86] boiling bell is similarly constructed, but the lower end is flared into a bell instead of being rough ground. The float designed by Boegel [87] does not float upright and is less efficient.

The procedure for the measurements is as follows. Evacuate and test the system for leaks. Fill the Claisen flask one-half full with the fluid to be tested; if necessary it can be distilled in under vacuum through a connection to the back of the flask. Dissolved air, if present, can be removed without much loss of sample by gently boiling the liquid at low temperatures at the highest vacuum feasible. But this extra precaution is seldom required. To begin measurements, reduce the presssure in the system and set the manostat control at the lower limit of the intended range of measurements. Heat the flask until the material distills slowly into the cold trap at a constant rate. For most of the compounds tested the preferred rate is 1–3 ml/min, but the

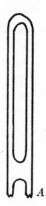

**Fig. 2.21** Hnizda float showing rough ground end at *A*.

proper minimal value should be determined by test. As soon as a steady state has been reached, read the temperature of the vapor and the pressure, simultaneously. An emergent stem correction (see Chapter I) should be made if needed. Several sets of data should be taken for the next few minutes until successive readings are constant. If further measurements are desired at higher pressures, the heating is discontinued, the pump is shut off, and a little air or inert gas is bled in until the pressure is slightly above the desired value. The manostat is readjusted, the pump and heater are turned on again, and the entire process is repeated. To ensure a successful set of measurements, superheating of the vapor and excessive condensation must be prevented by appropriately insulating the system. The manometer connection should be close to the thermometer, and all connections should be wide bore.

The accuracy of the boiling-point method described above becomes very poor at pressures below 10 mm Hg. In an excellent analysis of the method Hickman and Weyerts [88] have blamed the pressure reading and have shown several methods for improvement. They recommended a specially-designed still head that contains a built-in automanometer which uses the liquid itself as a manometric fluid. The reflux condenser acts as a simple condensation pump for maintaining a reference vacuum on one side of the automanometer. Construction of the equipment requires a certain amount of glass-working skill, but they have also shown a simpler model. The reader is referred to Hickman and Weyerts' paper for a detailed description of the equipment.

The acute problem of superheating of the liquid at very low pressures has been discussed; however, many boiling points have been measured in the difficult range 0.5–10 mm Hg by a special type of equilibrium still based on a design by Fenske [89–91].

PRECISION BOILING-POINT

Routine, but highly-precise vapor-pressure measurements are facilitated greatly by the boiling-point method of Willingham and co-workers [92]. A specially-designed boiler is connected to the mercury manometer (Fig. 2.22), which is provided with electrical contacts for maintaining the pressure at 20 fixed points. In operation, the pressures corresponding to the contacts are determined by calibration with water. A platinum resistance thermometer is located in the vapor space, which is connected to a manometer and to a special valve for precise control of the reflux ratio.

Several minor improvements were made by Stull [93], but the basic principles were unchanged.

HYPSOMETER

The hypsometer was used by Stimson [94] to study the boiling points of water and sulfur as fixed points on the International Practical Temperature Scale. With a similar hypsometric apparatus (Fig. 2.23), Edwards and John-

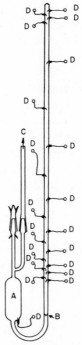

**Fig. 2.22** Contact manometer: *A*, mercury reservoir; *B*, constriction; *C*, to boiler; *D*, electrical lead wires.

son [95] determined the vapor pressure of $CO_2$ at 0°C as a fixed point on the pressure scale (see p. 44).

Carbon dioxide vapor is generated in the heater section of the hypsometer and is condensed on a brass "cold finger" which is cooled with a mixture of dry ice and freon 11. As the carbon dioxide condensate flows down the thermometer well it comes into equilibrium with rising vapor at the pressure indicated by a piston gage. Since the vapor–liquid surface on the thermometer well is large, the attainment of equilibrium is relatively easy. Furthermore, the condensate that flows down the thermometer well should be free of permanent gas impurities, which are concentrated at the top of the hypsometer, and of nonvolatile impurities, which remain in the liquid at the bottom of the hypsometer.

### Gas-Saturation Method

In the gas-saturation method, first used by Regnault [96] in 1845, a current of inert gas is passed through or over the material slowly enough to ensure

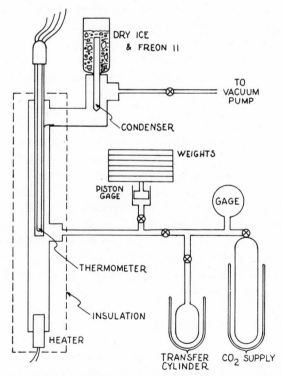

**Fig. 2.23**   National Bureau of Standards hypsometer. (Reproduced from *J. Res. Nat. Bur. Std.*)

saturation. The vapor pressure of the material is then computed on the assumption that the total pressure of a mixture of gases is equal to the sum of the pressures of the separate or component gases when each is at the temperature and each occupies the total volume of the mixture (Dalton's law). The partial pressure of vapor must be computed by the real gas laws from the total gas volume and the weight of material vaporized, which is usually obtained either by condensing and weighing or by the loss of weight of sample in the saturator. The method is thus limited by the incorrect assumption that there is no interaction between the vaporized sample and the inert gas. Fortunately, this assumption usually causes no large error. Gerry and Gillespie [97] have presented an accurate method for calculating vapor pressures by the gas-saturation method based on the Beattie-Bridgeman equation of state. Errors as large as 1.6% were noted for iodine when no corrections were made.

*Apparatus.* The particular arrangement of apparatus in the gas-saturation method depends largely on the problem at hand, but all arrangements have some features in common. Laboratory low-pressure air is usually a good source for the inert gas current if the compound is stable in the presence of oxygen. If not, either nitrogen or helium may be substituted. However, varied demands on the air supply in a large laboratory usually cause wide fluctuations in pressure. Consequently, to overcome variations in flow rate, an automatic pressure regulator and a surge tank reservoir of several liters capacity are essential. The gas is dried before it passes to the thermostated saturator. The evaporated material is removed in a trap, if necessary, and the inert gas is measured in a collecting bottle or, usually, in a standard gas meter. When the amount of gas is very large it may be measured by determining the loss in weight of a thermostated saturator that contains water. The total pressure at the saturator can be measured quite satisfactorily with an open-end manometer that contains a suitable light manometric liquid. The barometric pressure is corrected to the level of the saturator manometer (see p. 27).

Before final assembly, preliminary runs on the equipment are needed to establish the greatest flow rate that will completely saturate the gas stream with the sample. If complete saturation is not achieved at low flow rates, the saturator should be redesigned. No doubt the sample will be saturated if saturation was easily obtained with a test liquid of equal vapor pressure. A suitable saturator for liquids is shown in Fig. 2.24. The fritted-glass disk does not give an abnormally high pressure drop. A constant gas flow is not necessary if the rate is kept below the maximum flow rate for complete saturation because only the total amount of the saturated gas needs to be known.

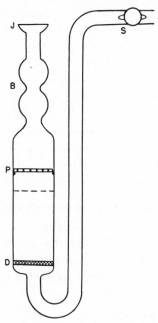

**Fig. 2.24** Saturator. $D$ is a sintered-glass disk; $P$ is a perforated porcelain disk; $B$ are bulbs filled with glass wool; $S$ is a stopcock; and $J$ is a spherical ground-glass joint. The saturator is immersed until the level of the bath is between $J$ and $B$; the broken line shows the initial liquid level.

Usually the amount of vaporized sample can be measured accurately by absorbing the material in a suitable liquid and analyzing the solution. For example, iodine solutions have proved useful for absorbing tetraethyllead from a gas stream. When determining the vapor pressure of $\beta,\beta'$-dichloro-diethyl sulfide by the gas-saturation method, Balson, Denbigh, and Adam [98] found that the best method was to absorb the gas in 25% acetic acid, completely liberating the chlorine. The solution was titrated potentiometrically with silver nitrate.

The equipment used by these authors merits special attention. Unusual features included a separate preheating coil identical to the spiral saturator, placed vertically instead of horizontally, and an auxiliary, preheated air stream which mixed with the saturated vapor and prevented premature condensation of the liquid. The air to the saturator was metered by displacing it from a 6-liter bottle with water under a constant head. The air was thoroughly dried before it entered the thermostated zone. The precision was about 3% from 0.05 to 3.5 mm Hg.

Recently, Overberger, Steele, and Aston [99] used the gas-saturation method for determining the vapor pressure of solid hexamethylbenzene. Since attempts to condense hexamethylbenzene from an inert carrier gas resulted in an unfilterable aerosol, pure oxygen was substituted for the inert carrier gas, and the hexamethylbenzene in it was determined by continuous combustion in a quartz chamber. Carbon dioxide and water from the combustion of hexamethylbenzene were collected and weighed in absorption tubes packed with anhydrous magnesium perchlorate and soda-asbestos.

*Calculations.* Since the calculations are frequently misunderstood they are treated here in detail. In the example chosen the evaporated sample is removed in a cold trap and the inert gas is saturated with water in a bubbler before it enters a wet-test gas meter. This procedure prevents contamination of the meter by the sample and ensures saturation of the gas stream with water in the wet-test meter. Regardless of the sequence of operations, the important fact for the calculations is that the amount of dry gas remains the same at every step. Avogadro's law and the ideal gas laws are assumed to be valid.* A more rigorous treatment based on Dalton's law and the Beattie-Bridgeman equation of state has been discussed by Gerry and Gillespie [97].

*Notation*

$T_S$ = temperature of saturator, K

$T_M$ = temperature of wet-test meter, K

$P_S$ = total pressure at saturator, mm Hg, obtained from barometric pressure corrected to manometer level, open-end manometer at saturator

---

* A common misconception is that $P$ (total) $= \Sigma_i p_i$ is an expression of Dalton's law when the partial pressures, $p_i$'s, are related to the number of molecules, $n_i$'s, through the ideal gas law, $n_i = p_i V/kT$, where $k$ is Boltzman's constant. Dalton's law applies to real gases with the reservation that unlike gas molecules do not interact. More correctly, $P$ (total) $= \Sigma_i p_i$ is a corollary of Avogadro's law: "Those quantities of different perfect gases which have equal values of $PV/T$ contain an exactly equal number of molecules." Thus, when a volume, $V$, of ideal gas contains two or more species, $n = \Sigma n_i$, and $P$ (total) $= nkT/V = (kT/V)\Sigma n_i$; Avogadro's law states that if

$$\frac{p_j V}{T} = \frac{p_k V}{T} = \cdots \frac{p_i V}{T} \cdots = \frac{PV}{T},$$

then $n_j = n_k = \cdots n_i \cdots = n$; and also the ratios

$$\frac{p_j V}{Tn_j} = \frac{p_k V}{Tn_k} \cdots \frac{p_i V}{Tn_i} \cdots = \frac{PV}{Tn} = \frac{PV}{T\Sigma n_i} = \text{constant, } k,$$

from which

$$n_i = \frac{p_i V}{kT} \quad \text{and} \quad \Sigma n_i = \frac{\Sigma p_i V}{kT} = \frac{PV}{kT}.$$

Thus, by cancellation of $V/kT$, $\Sigma p_i = P$ is seen to be a consequence of Avogadro's law.

$P_M$    = total pressure at wet-test meter, mm Hg
$p_S$     = vapor pressure of sample, mm Hg, at $T_S$
$p_W$    = vapor pressure of water, mm Hg, at $T_M$
$g$      = weight of sample evaporated, grams
$(MW)$ = molecular weight of sample
$V_M$    = volume from wet-test meter, liters
$V_A$    = volume of dry gas measured at $T_S$, $P_S$, liters
$V_S$    = total volume at saturator, liters

$V_A$ is easily established from the measured volume, $V_M$, by correcting to the saturator temperature and pressure using the ideal gas law

$$V_A = \frac{V_M (T_S/T_M)(P_M/P_S)(P_M - p_W)}{P_M}. \tag{2.13}$$

Since $P_S V_A = (P_S - p_S)V_S$, substituting the expression from equation (2.13) for $V_A$ and $gRT_S/(MW)p_S$ for $V_S$ gives

$$p_S = \frac{gRT_M(P_S - p_S)}{(MW)V_M(P_M - p_W)} \tag{2.14}$$

Since the volume of the vaporized substance at $T_S$, $p_S$ is $gRT_S/(MW)p_S$ according to the ideal gas law; then

$$\frac{p_S}{gRT_S/(MW)p_S} = \frac{P_S - p_S}{V_A} \tag{2.15}$$

Substitution of the value of $V_A$ from (2.13) gives

$$p_S = \frac{gRT_M}{(MW)V_M} \frac{(P_S - p_S)}{(P_M - p_W)} \tag{2.16}$$

$$= \frac{(P_S - p_S)}{K}, \tag{2.17}$$

where

$$K = \frac{(MW)V_M(P_M - p_W)}{gRT_M} \tag{2.18}$$

whence

$$p_S = \frac{P_S}{K + 1} \tag{2.19}$$

If the dry gas is metered either before it enters the saturator, or after the cold trap,

$$V_A = V_M\left(\frac{T_S}{T_M}\right)\left(\frac{P_M}{P_S}\right), \tag{2.20}$$

and

$$p_S = \frac{gRT_M(P_S - p_S)}{(MW)V_M P_M} \qquad (2.21)$$

$$= \frac{(P_S - p_S)}{K} \qquad (2.22)$$

where

$$K = \frac{(MW)V_M P_M}{gRT_M}. \qquad (2.23)$$

Most wet-test meters should be calibrated at regular intervals.

*Accuracy.* For most organic compounds with vapor pressures below 10 mm Hg, an accuracy of 3% is easily obtained. The use of a vigorously stirred thermostated bath is necessary for maintaining adequate temperature control. The accuracy can often be improved by a careful analysis of the possible experimental errors in various parts of the procedure before the experiment is performed. For this purpose, the use of an approximate value of the vapor pressure will help to decide whether the amount evaporated is most accurately determined from the loss in weight of the saturator vessel, by weighing a trap containing the condensate, or by analysis of an absorbent. The use of semimicro analytical methods can frequently shorten the time of a vapor pressure determination.

The gas-saturation method has been extensively employed by Baxter and his colleagues [100–103]; their papers contain much that is of interest to the organic chemist although their principal studies were on hydrated salts. In a series of measurements [102] on chloropicrin, cyanogen bromide, chloroarsines, and arsenic trichloride, the method was reliable to within a few per cent. The pressures measured ranged from 224 mm Hg to a few microns. The accuracy was poorest in the extremely low pressure range.

## Comparative Ebulliometry

One of the most satisfactory methods of determining vapor pressure relations above 15 mm Hg is by comparative ebulliometry [104–106]. In this method the boiling temperatures of sample and reference compound are measured at the same time and at the same pressure by connecting the ebulliometers to a common manifold kept at nearly constant pressure by a blanket of inert gas, preferably helium. Since pressure in the manifold is calculated from the boiling temperature and the accurately known vapor pressure relations of the reference compound, it need not be measured. Water is often selected as the reference compound because it is easily purified and the vapor pressure–temperature relations are accurately known [107–110]. However, the tendency for water to superheat and boil erratically at low pressure limits the ebulliometric method to not less than 150 mm Hg unless a boiler of a special design is used. The upper pressure limit is usually determined by the strength of the glass ebulliometer; however, if metal is used throughout, the upper pressure is only limited by the critical point or by the thermal stability of the sample. Also, an upper temperature limit may be imposed by the design of the heating element in the boiler. Although a twin ebulliometric apparatus is most commonly used (Fig. 2.25) a third ebulliometer for a secondary standard sample may be added. The auxiliary equipment needed in an ebulliometric apparatus includes a high-vacuum system, means for purifying the blanketing gas, ballast tanks to eliminate pressure surges, and a mercury manometer for nominal indication of pressure. The differential ebulliometer (Fig. 2.26) is provided with reentrant thermometer wells for measuring

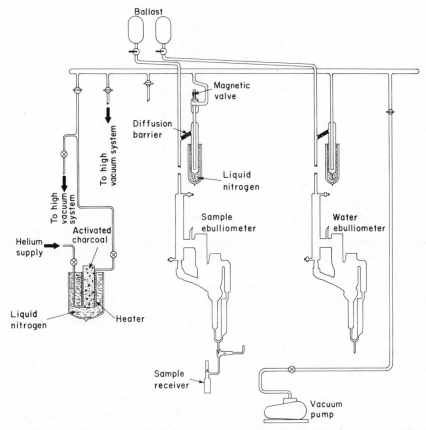

**Fig. 2.25**  Comparative ebulliometric apparatus showing twin ebulliometers and common pressure manifold. (Reproduced from *J. Chem. Eng. Data, Amer. Chem. Soc.*)

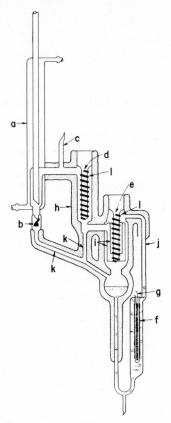

**Fig. 2.26** Differential ebulliometer: *a*, condenser; *b*, drop counter; *c*, sample seal-off; *d*, condensation temperature well; *e*, boiling temperature well; *f*, glass thread; *g*, electrical heater; *h*, asbestos insulation; *i*, glass shield (baffle); *j*, percolator tube; *k*, condensate return tubes; *l*, glass spiral. (Reproduced from *J. Chem. Eng. Data, Amer. Chem. Soc.*)

ebullition and condensation temperature; thus a continual check of sample purity can be made if thermal decomposition is suspected. Temperatures are readily measured with a precision of 0.001°C if platinum resistance thermometers are employed.

Herington and Martin [104] describe a comparative ebulliometric apparatus in which simple ebulliometers are used with excellent results. The latest major improvement to this apparatus is a bubble-cap boiler described by Ambrose [105] (Fig. 2.27) which produces smooth boiling of water down to 15 mm Hg and thus considerably extends the lower range of the ebulliometric method. Osborn and Douslin [106] report vapor pressure measurements on 36 sulfur compounds by a comparative ebulliometric method that they describe in detail.

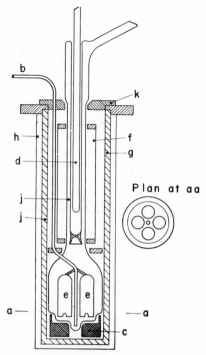

**Fig. 2.27** Ambrose boiler: *aa*, liquid level; *b*, filling tube; *c*, heater; *d*, thermometer; *e*, bubble caps; *f*, radiation shield; *g*, heated jacket; *h*, outer canister; *j*, differential thermocouples; *k*, Sindanyo plate and lid. (Courtesy of The Institute of Physics and The Physical Society, London, England.)

In contrast to the relatively complex ebulliometric apparatus just described one can always make comparative measurements by connecting the material and the reference to the opposite arms of some type of manometer. These methods have been most widely used in the study of the vapor pressure of solutions and binary mixtures, but problems that involve mixtures are outside the scope of this chapter.

A simple comparative apparatus of this nature is known as a Bremer-Frowein tensimeter [111]. This is shown schematically in Fig. 2.28. Bulbs *A* and *B* are connected to the limbs of a manometer containing a suitable manometric liquid. *A* contains the substance under test and *B* the reference compound. It is good practice to distill these compounds into the bulbs by the connections shown, freeze if convenient, and remove the adsorbed and dissolved gases with a vacuum pump. Moisture should be rigorously excluded from the material by passing the vapor through an adequate drying agent. Water is frequently used as a reference compound since its vapor pressure is well known and it can be easily prepared in high purity.

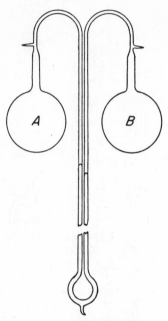

**Fig. 2.28**   Bremer-Frowein tensimeter: $A$, substance; $B$, reference compound.

The work of Calingaert and associates [112] on hexamethylethane is cited as an example of this technique. The vapor pressure of a purified sample of the hydrocarbon over the temperature range 0–110°C was determined by measuring the difference in pressure exerted by the hydrocarbon and water at the same temperature. In general, this method is not recommended except for special problems because the vapor pressure can be determined as accurately by other means with much less trouble. However, the procedure for filling the tensimeter is of wide interest because similar techniques are required whenever liquids must be put into a closed system. The water and hydrocarbon were charged into the apparatus as follows.

An auxiliary bulb was sealed to the bottom of the U-tube to hold the mercury while bulbs $A$ and $B$ were being filled. A second auxiliary bulb was sealed to $A$, and a third to $B$ through a stopcock, the auxiliary bulb being sealed to the bottom outlet of the barrel of the stopcock. A second stopcock was also sealed to the third bulb for use in evacuating the system. After enough mercury for the manometer was placed in the lower bulb, about 20 ml of distilled water was placed in the bulb attached to $B$ through the barrel of the stopcock. The system was then evacuated, and about 15 ml of the water was distilled into the auxiliary bulb attached to $A$ by cooling the bulb in ice water. The water in this bulb was then frozen, and the system was filled with nitrogen. The residual water in the bulb attached to $B$ was removed, and the bulb was dried by heating and blowing nitrogen into it. About 15 g of hexamethylethane was then charged into the bulb through the barrel of the stopcock, and the system was evacuated while the bulbs containing the water and hydrocarbon were cooled in dry ice baths. The mercury was run into the manometer tube by tilting the apparatus, and the lower bulb was sealed off. The ice in the axuiliary bulb attached to $A$ was melted and distilled into $A$. When most of it had distilled, the distillation was reversed by cooling the auxiliary bulb in ice water. When about one half of it had distilled, the auxiliary bulb was sealed off. A similar procedure was used for introducing hexamethylethane into $B$.

## Methods for Very Small Vapor Pressures

Measuring very small vapor pressures presents especially difficult problems. If the compound can be properly degassed, a static method with a multiplying manometer, a Bourdon gage, a diaphragm gage, or a Knudsen gage may be entirely adequate. The gas-saturation method is also suitable but may require too much time for each measurement unless microanalytical methods are applicable. Some of the methods for determining low vapor pressures are described below.

### Hickman's Method

Hickman and associates [88, 113] described a direct boiling-point method which gave consistent results down to 0.03 mm Hg. In an earlier paper Hickman and Weyerts [88] discussed various kinds of errors that are encountered in distilling high-boiling, "phlegmatic" liquids. Their discussion is recommended highly for its critical comments and helpful suggestions on the determination of boiling points below 15 mm Hg.

To measure very low vapor pressures by a boiling-point method is impossible because the velocity of the stream of molecules from the boiling liquid to the vapor thermometer bulb is comparable to the random velocity of the molecules, which determines its equilibrium pressure. The thermometer therefore does not register accurately the condensation temperature of the saturated vapor. The apparatus devised by Hickman [114] can be used either as a *tensimeter*, for measuring vapor pressure, or as a *hypsometer* (secondary manometer), by measuring the boiling point of a liquid with known vapor pressure. The model shown in Fig. 2.29 has a cup on the thermometer which increases its sensitivity when the pressure is lowered. A similar apparatus, without the cup around the thermometer, was entirely satisfactory except that the response was slow to a decrease in pressure.

The tensimeter is essentially a well-insulated, short, wide boiler connected to a water-jacketed condenser. The vapors pass the thermometer at low

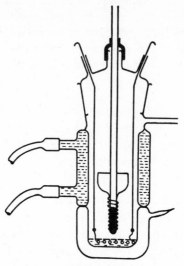

**Fig. 2.29**  Hickman's tensimeter-hypsometer.

velocity with very little obstruction. Hickman [113] used either an oil manometer [114] (see p. 40) or Pirani gage (see p. 46) to measure the low pressures involved.

In operation the liquid is boiled with various pressures of residual air or inert gas in the apparatus. For reliable results, it is necessary to establish at each pressure a temperature plateau independent of the amount of heat used. The useful range appears to be 0.2 to 4 mm Hg, although Hickman claims that the lower end of the useful range is 0.03 mm Hg. At the lowest pressures optimum operating conditions are difficult to determine, vapor and residual gas interdiffuse, and the random velocity of the molecules is small in comparison with their forward velocity. Accurate readings above the useful range are prevented by bumping and splashing.

In the low range Hickman [112, 115] has recommended a different apparatus in which the pressure of the vapor issuing from a wide orifice is balanced against a long, light-weight pendulum. The results obtained were consistent down to 0.7 $\mu$. Complete details are presented in Hickman's article, including results on three solids that melt above 148°C.

A slightly improved model [116] is shown in Fig. 2.30. The substance is placed in the boiler, $B$, which is submerged in a heating bath, and the side tube, $V$, is connected to a vacuum system capable of maintaining a pressure of 0.1 $\mu$ or less. The whole assembly can be rotated on a horizontal axis which is not shown. The pendulum, $D$, is a flat duralumin disk suspended by a duralumin wire from a shaft riding in jeweled bearings. At the start the

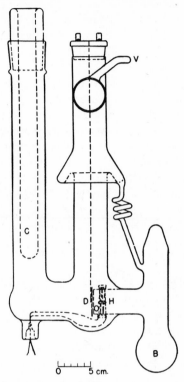

**Fig. 2.30**  Pendulum tensimeter showing boiler, $B$; vacuum connection, $V$; platinum resistance heater, $H$; pendulum, $D$; and cold finger, $C$.

apparatus is rotated until the orifice of the boiler is just closed. As soon as the substance reaches thermal equilibrium, the issuing vapor blows the pendulum away from the orifice. The apparatus is then rotated by an angle $\theta$ until the orifice is again closed. Hickman recommended [113] that the vapor should stream past the open pendulum for some time before taking readings in order to thoroughly degas the sample. The cold finger, $C$, filled with liquid nitrogen or carbon dioxide snow, is intended [116] to eliminate possible back pressure on $D$. The small platinum resistance heater, $H$, was used [116] occasionally to vaporize any material that condensed near the orifice. The vapor pressure, $P$, is calculated from the relation

$$P\,(\mu) = 736(4W/\pi D^2) \sin \theta \qquad (2.24)$$

where $W$ is the weight of the pendulum in grams, and $D$ the diameter of the orifice in centimeters. For Hickman's model [113] $W = 2.405$ g, $D = 3.30$ cm, $P/\sin \theta = 207$. Both papers contain detailed discussion of the precision and accuracy, but it appears that the authors' estimates are optimistic. The wide differences between the results of Hickman, Hecker, and Embree [113], Verhoek and Marshall [116], and Perry and Weber [115] on dibutylphthalate are of interest: at 50° C, 0.54, 0.55, 0.65, and at 90°C, 21.3, 30.6, 21.2 $\mu$, respectively.

Alignment of the disk and orifice is critical. To get the disk absolutely flat and placed so that it will hang in the exact plane of the orifice is difficult. Also, the establishment of the closure points requires a delicate touch, because at lower pressure the tendency is to overclose and at higher pressures the vapor stream causes the pendulum to bump. In general, it is helpful to approach balance always from the same side, a point between absolute closure and balancing or floating the pendulum on the vapor stream, and a very slow approach to the balance point. In the Ethyl Corporation laboratory a cooling collar was used above the condensate collector and a platinum resistance thermometer was placed in a well in the boiler in order to avoid the uncertainty of stem corrections.

### Effusion

*Knudsen-Effusion Apparatus.*   In the pressure range from a few hundred microns to a few tenths of a micron, the equilibrium pressure of a contained substance can be measured by its rate of effusion through a small orifice. When the orifice configuration is short, round tubular, as is usually the case, and the capsule containing the condensed sample is in an evacuated space, the pressure inside the capsule can be calculated by the relation

$$P\ (\mu) = 17{,}144(m/aKs)\ \sqrt{T/M}. \qquad (2.25)$$

In this relation, $s$ is the time in seconds during which $m$ grams of sample material escapes through a small circular orifice, area $a$, from a capsule kept at temperature $T$, K. The effective molecular weight, $M$, in grams refers to the vapor species, which often is a monomer but in some instances occurs as a dimer or other multiple of the chemical molecular weight of the substance. For a vanishingly thin orifice with diameter less than one-tenth of the mean free path of the molecules, the dimensionless "Clausing factor" [117], $K$, is exactly 1; however, the practical orifice is in reality a short tube for which theoretically derived values of $K$, Table 2.5, are shown as a function of the length to radius ratio, $l/r$, of the tubular part of the orifice. A schematic drawing of the Knudsen-effusion apparatus (Fig. 2.31), shows the principal parts of a device that might be operated as a loss-weight experiment in which the capsule, $a$, is weighed before and after a time interval, or alternately as a weight-collecting experiment in which a known fraction of the effused material is collected on a cold target, $d$, which can be removed and weighed. Although descriptions of both kinds of experiments with the Knudsen-effusion

**Table 2.5**  Clausing's Correction Factors for Knudsen-Effusion Emperiments

| $l/r$ [a] | $K$ | $l/r$ [a] | $K$ |
|---|---|---|---|
| 0 | 1. | 1.2 | 0.6320 |
| 0.2 | 0.9092 | 1.4 | 0.5970 |
| 0.4 | 0.8341 | 1.6 | 0.5659 |
| 0.6 | 0.7711 | 1.8 | 0.5384 |
| 0.8 | 0.7177 | 2.0 | 0.5136 |
| 1.0 | 0.6720 | | |

[a] Ratio of length of orifice tube to its radius.

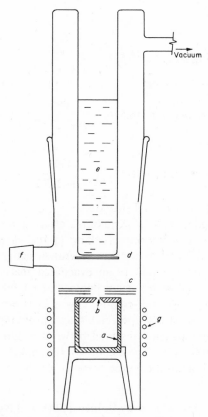

**Fig. 2.31**  Knudsen effusion apparatus: $a$, capsule; $b$, orifice; $c$, thermal shields; $d$, target; $e$, cold finger; $f$, port; $g$, heater.

apparatus are rather plentiful in the chemical literature, one can avoid extensive reading on the subject by consulting, for example, the classical work of Robson and Gillis [118] on the target collecting system used for measuring the vapor pressure of boron carbide in the region $4$–$80 \times 10^{-6}$ atm and $2300$–$2500°$K, or the work of Smith, Gorin, Good, and McCullough [119] on the loss-weight method used to obtain vapor-pressure results for four dihalobiphenyls at pressures varying from 2 to 40 $\mu$ and temperatures from 294 to 360°K.

There is little doubt that a correction must be made for orifice geometry, but the way to carry this out remains controversial. See, for example, the papers by Whitman [120], Rossman and Yarwood [121], and Whitman [122]. In view of these differences of opinion, some workers "calibrate" their effusion equipment by the use of a reference material. The data of Neumann and Völker [123] in benzophenone have been used for this purpose. These are represented by

liquid:   16–55°C

$$\log p\,(\mu) = 13.521 - \frac{4021}{t + 273.2} \tag{2.26}$$

solid, stable form:   16–42°C

$$\log p\,(\mu) = 16.516 - \frac{4983}{t + 273.2} \tag{2.27}$$

solid, unstable form:   11–26°C

$$\log p\,(\mu) = 16.111 - \frac{4796}{t + 273.2} \tag{2.28}$$

Calculated values are shown in Table 2.6.

The principal disadvantage of Knudsen's effusion method, which is described above, is the troublesome breaking of the vacuum at intervals in order to determine the loss in weight. Several investigators have designed balances which can be operated in an evacuated space. Volmer [124] and Dietz [125] used a conventional beam balance and Ubbelohde [126] used a quartz spring balance. The self-cooling effects in these "hanging pots" or "hanging buckets" are usually appreciable and probably account for some of the systematic discrepancies observed between results on the same compound by different observers. The use of several buckets with different sizes of effusion holes will demonstrate the presence or absence of self-cooling.

## Torsion

Another approach introduced by Volmer [124], and Neumann and Völker [123], and lately called the "Torker" technique by Freeman [127] as a mnemonic for "torsion-Knudsen effusion-recoil," utilizes a measurement of

**Table 2.6**   Vapor Pressure of Benzophenone

| Vapor Pressure ($\mu$) | Temperature (°C) | | |
|---|---|---|---|
| | Liquid | Metastable[a] | Stable[a] |
| 0.1 | (3.7) | (7.1) | (8.7) |
| 0.2 | (9.6) | (12.1) | 16.3 |
| 0.3 | (13.1) | (15.1) | 19.2 |
| 0.5 | 17.7 | 19.0 | 23.1 |
| 0.7 | 20.8 | 21.6 | 25.7 |
| 1.0 | 24.2 | 24.5 | 28.5 |
| 1.2 | 25.9 | 25.9 | 30.0 |
| 2.0 | 31.0 | — | 34.1 |
| 3.0 | 35.1 | — | 37.5 |
| 5.0 | 40.4 | — | 41.8 |
| 7.0 | 44.0 | — | 44.8 |
| 10.0 | 47.9 | — | 47.9 |
| 20.0 | 55.9 | — | — |
| 30.0 | (60.7) | — | — |
| 50.0 | (66.9) | — | — |
| 70.0 | (71.2) | — | — |
| 100.0 | (75.8) | — | — |

[a] The temperatures in parentheses are extrapolated. Melting points of the two forms are 26.1 and 48.0°C.

the recoil force from a stream of effusing molecules to determine vapor pressure in the range $10^{-3}$ to $10^3$ dyne/cm². A typical Torker system (Fig. 2.32), as described in detail by Freeman, consists of a double effusion cell, having antiparallel horizontal orifices, suspended by a vertical torsion wire. The pressure, $P_T$, at cell temperature $T$

$$P_T = \frac{2K\varphi}{A_1 f_1 d_1 + A_2 f_2 d_2} \tag{2.29}$$

is a function of $\varphi$, the angular displacement of the cell, the torsion constant $K$, the orifice areas $A_1$ and $A_2$, the perpendicular distances, $d_1$ and $d_2$, of the orifice axes from the axis of suspension, and factors, $f_1$ and $f_2$, which correct for the nonidealities of the orifices. Values of $f$ are given by Freeman and Edwards [128] as functions of the length $l$ and radius $r$ of the orifice, and the temperature. The factors should not be confused with the "Clausing" factors, Table 2.5, that apply to the simple Knudsen-effusion experiments described earlier.

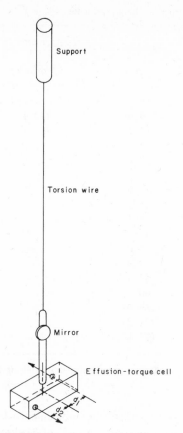

**Fig. 2.32** Torker effusion apparatus.

A variation of the Torker method by Rodionov [129] utilizes a magnetic suspension which provides increased sensitivity for low pressure measurements.

### Absolute Manometers

Sometimes a highly sensitive manometer, which contains no manometric liquid, is convenient. Quartz spirals and glass Bourdon gages are not satisfactory below about 10 $\mu$. A different principle is involved in manometers in which the force required to lift a lid off a seating (with a high vacuum below and the vapor pressure above) is measured. Originally these "absolute" manometers [130,131] were used to determine the vapor pressures of metals. Sometime later, Dietz [132] used a similar method on crystalline potassium chloride and cesium iodide with results that were accurate to 3% in the 1–20-$\mu$ range. A modification by Balson [133] enabled him to measure the vapor pressure of bromobenzyl cyanide between 1.5 and 270 $\mu$; the maximum sensitivity was 1$\mu$. The all-glass equipment requires careful construction. The reader is referred to the article for detail.

An absolute piston manometer has been used by Ernsberger and Pitman [134] and by Kaufman and Whittaker [135] in the range 0.5–20 $\mu$. The piston is made of concentric mica disks suspended freely on a quartz spring in an accurately ground cylinder. The pressure to be measured is on one side and a reference vacuum on the other. The precision is better than 1%; the accuracy, although difficult to assess, is probably better than 5%.

An inclined-piston method useful in the range 0.1–40 mm Hg was discussed (p. 50).

### Micro and Semimicro Methods

Special techniques are employed for measurements on very small samples. These include the semimicro ebulliometer [78, 79] (p. 54), submerged bulblet method, García's modification of Emich's boiling-point method, and Rosenblum's adaptation of a boiling-point method by Shriner and Fuson, described below. Frequently, however, the methods used on larger samples can be adapted easily to semimicro quantities. Only a few milliliters of sample were needed in the Hoover-John-Mellon ebulliometer and in the isoteniscope. Also, the methods described above for very low pressures require very small samples.

Most of the microchemical techniques for determining boiling point at atmospheric pressure are adaptable to subatmospheric pressure. The methods, adequately described in standard texts on microchemistry, include: Emich's capillary bubble method [136–139], which does not require elaborate equipment; Schleiermacher's method [140, 141], which employs a special massive aluminum block for heating a sample enclosed over mercury in a particular type of boiling-point tube; Siwoloboff's method [142, 143], which uses a capillary tube containing another inverted capillary immersed in the liquid; and Niederl and Routh's scheme [144] for heating the sample adsorbed on a piece of tile under mercury.

### Submerged Bulblet

The submerged bulblet method of Smith and Menzies [145], used up to 400°C, is a simple adaption of their apparatus for determining the boiling points of small amounts of liquids or nonfusing solids at atmospheric pressure. The vapor pressure apparatus is shown in Fig. 2.33. The L-shaped tube at the top leads to a surge tank, manometer, manostat, and connections for air and vacuum lines.

A small glass bulb is prepared with a capillary tube 3- to 4-cm long and at least 1-mm bore. The capillary is bent close to the bulb, and a small amount

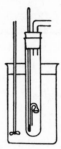

**Fig. 2.33**  Submerged-bulblet vapor-pressure apparatus.

of liquid is introduced. Solids are more conveniently added before the tube is bent. The bulb and its contents, with the capillary pointing down, as shown, are then attached by string or asbestos cord to a thermometer.

To obtain the boiling point at atmospheric pressure the thermometer is immersed in a tall beaker containing a transparent liquid and fitted with a glass stirrer (as in a conventional melting-point apparatus) or a small mechanical stirrer. The bath is heated and, as the boiling point is approached, stirred vigorously. At first a few bubbles of air are expelled from the bulb, but when the boiling point is reached a steady stream of bubbles emerges. The substance must be nearly insoluble in the bath liquid. Upon removing the flame, the stream of bubbles stops sharply when the temperature drops a fraction of a degree. The flow of bubbles can be resumed by restoring the temperature level.

In making a boiling-point determination the bubbles are allowed to escape until the dissolved air and adsorbed gases are driven off. The temperature is then lowered very slowly while the bath is stirred vigorously. The boiling point is taken when the stream of bubbles ceases. The barometric pressure and the depth below the surface of the liquid to the end of the capillary are also recorded. The entire process is repeated until the readings are constant. The thermometer and bulblet are then lifted out of the bath in order to avoid contaminating the sample with the bath liquid.

This procedure is applicable only if the sample is insoluble in the bath liquid. If the substance is soluble in the bath liquid, the temperature at which the bubbles stop is not sharply defined. In this case the temperature reading is taken when the bath liquid has ascended to a predetermined point, 5–10 mm Hg from the opening. The boiling point is not affected by the vapor pressure of the bath liquid.

The pressure at the end of the capillary is the barometric pressure plus the head of the bath liquid. The head is obtained by multiplying the density of the bath liquid by its depth to the end of the capillary, and dividing by the density of mercury at 0°C (13.595).

Smith and Menzies give tables for the density of several suitable bath liquids: 92.75 wt % sulfuric acid, paraffin wax (mp 53°C), the eutectic mixture of potassium and sodium nitrates (45.5 wt % $NaNO_3$), and a mixture of these nitrates with lithium nitrate in the following proportions: 18.18 wt % $NaNO_3$, 54.55 $KNO_3$, and 27.27 $LiNO_3$. This mixture melts at 120°C but damages thermometer glass above 250°C.

The procedure for making vapor pressure measurements below the normal boiling point is similar to the operation described above. The thermometer and bulblet are immersed in a test tube that contains the bath liquid, which, in turn, is immersed in a second bath (Fig. 2.33), that contains the same liquid. The second bath is well stirred and maintained at the desired temperature. This time the liquid is slowly boiled out of the bulblet by reducing the pressure. Careful manipulation is needed because a slight increase in pressure will stop the flow of bubbles. Smith and Menzies have recommended closing the rubber tube at two places with the fingers; thus very little air is admitted at a time. When the stream of bubbles stops (or if the substance is somewhat soluble in the bath liquid, when the bath liquid ascends the required distance from the end of the capillary) the pressure and temperature are read. After several repetitions, the calculations are made as before, allowing for the submerged depth of the bulblet.

### Rosenblum's Method

In Rosenblum's method [146] a drop or two of liquid is placed in a side-arm test tube (Fig. 2.34). A small, inverted, closed-end capillary tube is attached to the thermometer bulb and lowered until the open end of the capillary is below the surface of the liquid. The side-arm test tube is sealed to a condenser and is placed in a standard Thiele melting-point tube bath. The apparatus is connected to a manometer, surge tank, and manostat, as with the submerged bulblet method.

The pressure is set on the manostat and the bath liquid is heated until a steady stream of bubbles emerges from the capillary. When the liquid rising in the cooling capillary is level with the outside liquid, the boiling point at the pressure of the system has been reached, and the pressure and temperature should be read.

Rosenblum stated that he modified the method from a normal boiling-point scheme described by Shriner, Fuson, and Curtin [147], but it appears to have closer kinship with the micro boiling-point method of Siwoloboff [85].

A method similar to Rosenblum's has been described by Hays, Hart, and Gustavson [148].

### García's Micromethod

The standard capillary bubble method of Emich for determining boiling points on a microscale is not readily adaptable for measurements under reduced pressures. García [149] has improved Emich's original procedure

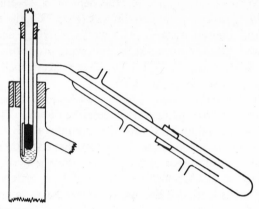

**Fig. 2.34**  Rosenblum's micromethod.

for boiling points at atmospheric pressure and has provided a simple adaptation for determining vapor pressures on small samples (0.002 ml).

A capillary boiling-point tube is prepared by drawing a piece of 5- to 7-mm-
bore tubing to a capillary about 7 cm long and 1-mm bore. One end of the
capillary is sealed, and a drop of liquid is introduced by means of a capillary
tube. The liquid is then forced to the bottom by centrifuging. The tube is
then placed next to a thermometer in a standard Thiele melting-point bath,
as in Fig. 2.35.

In operation, the capillary is connected to the usual manometer, surge tank,
and manostat, and the pressure is adjusted to the desired value. The bath is

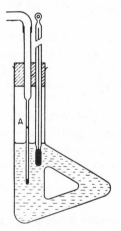

**Fig. 2.35**  Garcia's micromethod showing droplet in capillary tube at *A*.

then heated slowly (near the boiling point) until a little of the material condenses in the cooler part of the capillary tube above the liquid level of the bath forming a droplet, $A$. The bath is allowed to cool slowly. The temperature and pressure are read when the droplet descends below the level of the liquid in the bath. If the bath is heated too rapidly, two or more droplets might form, and the capillary tube must be recentrifuged. García says, "The striking difference between the above procedure and that of Emich is that no air bubble is left at the bottom of the capillary below the sample; the space which exists below the condensed droplet is completely filled with the vapors of the pure organic liquid. Theoretically this gives a more accurate relation between changes in pressure of the pure vapors and corresponding changes in temperature such as occur when the droplet descends as the temperature of the bath is lowered. Since the boiling point is related to these effects and Emich's method does not have pure vapor below the droplet, it might be advisable to follow this procedure even in the determination of the boiling point at ordinary conditions." For these reasons we have not described Emich's capillary bubble method.

## Vapor Pressure-Temperature Relations

Equations relating vapor pressure and temperature are principally employed in smoothing, interpolating, and extrapolating data, and in computing the differential coefficient, $dP/dT$, which occurs in the expression for the heat of vaporization [150]. Numerous equations are available for representing vapor pressure–temperature relations; however, we have found that the Antoine and Cox equations, which are discussed in detail below, are the most useful. The Rankine or Kirchoff equation, also discussed below, is sometimes preferred because of its theoretical significance. A comprehensive survey of vapor pressure equations is given in Partington's treatise [151]. The large number of functional forms, which are at least partly successful, is noteworthy. Evidently the true relation must be an exceedingly complex one which, even if it were known exactly, would have limited utility for practical purposes.

### Preliminary Examination of Vapor Pressure Data

Before attempting to fit vapor pressure data with a correlating equation a preliminary, critical examination should determine if any points are greatly in error and if trend of the data with temperature follows the well-established pattern. Also, when pressures are being measured experimentally, time can often be saved by plotting each point as it is determined. Pressure measurements should proceed along a regular series of increasing temperatures, beginning with the lowest; thus thermal decomposition can be detected at the earliest time and the sample can often be removed from the controlled temperature environment before it is ruined.

As an aid to the critical examination of vapor pressure data and as a guide in the selection of a correlating equation, the results may be compared with the established characteristics of the schematic log $P$ versus $1/T$ plot (Fig. 2.36). Vapor pressure data are shown (Fig. 2.37), for the solid and liquid phases of hexafluorobenzene, 1,4-dimethylbenzene, and 1,2-diaminoethane. The break in each curve occurs at the melting point; the change in slope from solid to liquid phase is related thermodynamically to the heat of fusion. The most important general feature of the solid-phase portion of the curves is the concave shape of the lines, a characteristic that can be observed with accurate data. Upon first crystallizing a new compound the solid phase obtained might well be in a metastable state, as shown for 1,2-diaminoethane, with vapor pressures significantly higher than the stable crystalline form. However, as the melting temperature is approached the metastable form will usually transform to the equilibrium crystal, and an apparent offset in the vapor pressure curve will be obtained. When this happens, cool the sample to the lowest temperature again, before allowing the sample to melt, and establish a new vapor pressure curve for the equilibrium crystalline phase. In the liquid region, the shape of the curve will most probably be

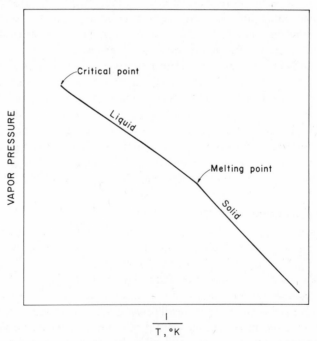

**Fig. 2.36**  Schematic plot of vapor pressure–temperature relations for solid and liquid phases.

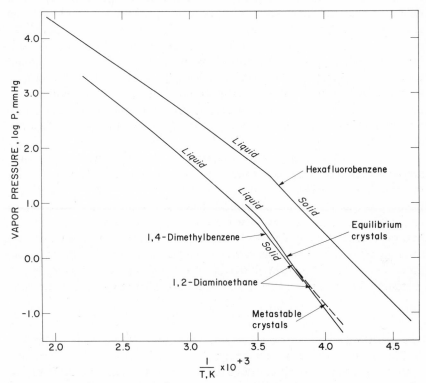

**Fig. 2.37**  Experimental vapor pressures including results for a metastable crystalline phase of 1,2-diaminoethane.

convex over a large temperature range from the melting point to just below the critical temperature. For many compounds, as for hexafluorobenzene, the curvature changes from convex to concave several degrees below the critical temperature. Even though a particular set of vapor pressure data might not show a characteristic curvature, either because the data are in a short temperature range or are perhaps imprecise, a knowledge of the usual vapor pressure behavior of organic compounds, gained by studying Figs. 2.36 and 2.37, will be valuable when correlating vapor pressure results with an analytic function.

### Antoine Equation

One of the most successful, simple equations for representing the vapor pressure–temperature relation was proposed by Antoine [150, 152],

$$\log P = A - B/(t + C), \qquad (2.30)$$

in which $A$, $B$, and $C$ are empirical constants. A significant feature of the Antoine equation is the shape of its trace on a $\log P$ versus $1/T$ plot. For $C = 273.15$ the plot is a straight line; for $C < 273.15$ the curve is convex; and for $C > 273.15$ the curve is concave. Thus, the Antoine equation is useful for solids, which generally show concave behavior, as well as for liquids, which show convex behavior from the melting point up to a reduced temperature, $T/T_c$, of 0.75. However, the Antoine equation is not applicable in the region of reverse curvature which occurs somewhat below the critical temperature, nor is it recommended for long extrapolations into regions of low pressure.

As a preliminary exercise, Antoine constants might be derived by the simultaneous solution of three equations that represent selected points taken near the middle and extremes of the temperature range of the data. Such preliminary constants are generally unsatisfactory for representing precise data, and often give a bad statistical distribution of the deviations from data points. The only satisfactory method for deriving the constants is a least-squares solution with the use of appropriate weighting factors. The least-squares procedure as described by Willingham et al. [92] involves rearranging and redefining the constants of the Antoine equation (2.30) to give the linear form,

$$at + b + c \log P - t \log P = 0 \tag{2.31}$$

used in setting up the normal equations. (When the Willingham et al. method is followed, note that terms to the right of the equal signs in the three simultaneous equations (10) of [92] should carry a negative sign.) The least-squares calculations can be done readily on a desk calculator, but when many compounds are to be studied, the calculations should be performed with a computer. Examples of the use of the Antoine equation can be found for a wide variety of compounds, but perhaps the largest application is in the field of hydrocarbons [153, 154]. Also, the Antoine equation has been successfully applied to many nitrogen and sulfur compounds [155–157]. The "Cox chart" used by Stull [159] for interpolating a wide variety of organic and inorganic compounds is based on the Antoine equation with $C = 230$ [150].

A study of the results mentioned in the above references should provide an excellent background for planning vapor pressure correlations.

## Cox Equation

Although the Antoine equation is entirely satisfactory for many applications, it is sometimes unsuitable for vapor pressure data when they extend over a wide temperature range and are highly accurate or precise. Also, the inability of a single Antoine equation to reproduce a change in curvature in a $\log P$ versus $1/T$ plot was mentioned previously. In this case, the data can be broken up into two or more overlapping ranges and separate Antoine equations derived for each range,* or a more complicated vapor pressure–temperature relationship such as the Cox equation [158]

$$\log (P/P_{\text{ref}}) = A(1 - \Phi/T) \tag{2.32}$$

where $\log A = a + bT + cT^2 + \dots$, can be used. Not only can the Cox equation follow a change in curvature of the $\log P$ versus $1/T$ plot, but it extrapolates to low pressures with good accuracy. In most applications of the Cox equation the reference pressure, $P_{\text{ref}}$, and temperature, $\Phi$, are taken at the normal boiling point, but any other convenient point can be used. For example, in the case of hexafluorobenzene [76], the triple-point pressure and temperature, 29.377 mm Hg and 278.32°K, respectively, provide a common expression

$$A(\text{liquid or solid}) = \frac{\log (P/29.377)}{1 - 278.32/T} \tag{2.33}$$

for both solid and liquid phases. If the Cox equation is written in linear form,

$$\log \left[ \frac{\log (P/P_{\text{ref}})}{1 - \Phi/T} \right] = a + bT + cT^2 \tag{2.34}$$

the derivable constants $a$, $b$, and $c$ can be determined by a least-squares procedure similar to that outlined for the Antoine equation by Willingham et al. [92]. Each point is given a weighting factor, $w_i = P_i \ln P_i/\sigma_i$, where $\sigma_i$ is the estimated uncertainty in the pressure. In comparative ebulliometric or boiling-point measurements the experimental uncertainty can be conveniently assumed proportional to pressure, $\sigma_i = s + tP_i$, where $s$ and $t$ are estimated from a critical analysis of the experimental method. The Cox equation has been used extensively to correlate vapor pressure values for sulfur compounds [157], nitrogen compounds [155], and unsaturated hydrocarbons [154].

### Semitheoretical Equations

Many of the simpler relations are based on the exact Clapeyron equation [160]:

$$\frac{dP}{dT} = \frac{\Delta H}{T\Delta V}, \tag{2.35}$$

where $T$ is the absolute temperature, $P$ is the vapor pressure, $\Delta H$ is the heat of vaporization per mole, and $\Delta V = V_g - V_l$ is the change in volume per mole

---

* This procedure was applied to water by G. W. Thomson, Ethyl Corp. *Report LTD 50-32*, April 12, 1950, and used later by B. J. Zwolinski in *American Petroleum Institute Research Project 44* tables, "Selected Values of Properties of Hydrocarbons and Related Compounds," and the Manufacturing Chemists Association tables, "Selected Values of Properties of Chemical Compounds."

accompanying the vaporization, where $V_g$ and $V_l$ are the molar volumes of the gas and liquid phases. This exact relation can also be written in terms of $\Delta z = z_g - z_l = P\Delta V/RT$ [150]:

$$\frac{d \ln P}{d(1/T)} = -\frac{\Delta H}{R\Delta z}. \tag{2.36}$$

If the molar volume of the liquid is negligible and the vapor is considered to be a perfect gas (i.e., $PV_g/RT = 1$), (2.36) becomes

$$\frac{d \ln P}{d(1/T)} = -\frac{\Delta H}{R}, \tag{2.37}$$

which may be integrated (assuming $\Delta H$ is constant over the given temperature interval) to give

$$\log P = A - \frac{B}{T}$$

$$= A - \frac{B}{t + 273.15}, \tag{2.38}$$

where $t$ is in degrees centigrade, $B = \Delta H \,(\text{cal}/\text{mole})/2.30258\,R$.

The three assumptions involved in deriving (2.38) may be replaced by the assumption that $\Delta H/\Delta z$ is constant [150], because $\Delta H$ and $\Delta z$ have similar temperature dependence.

Although equation (2.38) gives an approximate picture of the data, systematic deviations are usually encountered, except for very low-boiling compounds and for pressure ranges somewhat below 1 mm Hg. A possible solution is to substitute a polynomial in the temperature, $T$, for the quantity $\Delta H/\Delta z$ on the right-hand side of (2.36); the result is the Rankine or Kirchoff equation

$$\log P = A - B/T + D \log T. \tag{2.39}$$

If a linear temperature term is added to (2.39) and $D$ is taken equal to 1.75, the Nernst equation is obtained,

$$\log P = A - B/T + 1.75 \log T + ET \tag{2.40}$$

The constants of equations of this type should have reasonable values. An excellent guide to the interpretation of the magnitude of these constants is found in Van Laar's editorial comments in the vapor pressure sections of *Tables Annuelles* [161]. Some useful notes have also been presented elsewhere [150].

Relating vapor pressure measurements to other properties by means of the Clapeyron equation is extraordinarily difficult. The complete closure of a "thermodynamic vaporization network" requires highly precise vapor pressure data over a wide temperature range, liquid and gaseous heat capacities, $P-V-T$ information (so the effect of pressure on the thermal properties can be determined), and the heat of vaporization as a function of temperature over a broad range of temperatures. Information on accurate $P-V-T$ correlations below the critical point is meager, and the exact temperature dependence of the heat of vaporization is not known at low pressures. Since this topic presents too many difficulties for an elementary presentation, it is not considered further.

## 2  MEASUREMENT OF VOLUME

### Introduction

The volumetric methods described in this chapter mostly concern measurements on solids, liquids, or gases at ordinary temperatures and pressures. For accurate measurements at high pressures and extreme temperatures, the effect of these variables on the instruments themselves must be taken into account. Distortion of a thin-walled metal or glass vessel by hydrostatic head of the calibrating or contained fluid can be determined by a method described below. Expansion or contraction of the volume, $V$, of a container due to temperature change can be determined by a simple relation, $V = V_0(1 + \alpha t)^3$, in which $\alpha$ is the linear coefficient of expansion of the vessel material determined for the same temperature base as $V_0$.

### Routine Volumetric Determinations

#### Standard Markings and Volume Calibration

Although the reader is probably familiar generally with standard volumetric glassware, cylinders, pipettes, graduated flasks, and burettes, the tolerances in the specifications [162] for the manufacture of such equipment may also be of interest. Most precision volumetric glassware is calibrated at a standard temperature, 20 or 25°C, which is inscribed. Sometimes the marking also indicates whether the volume is contained or delivered. Infrequently the vessel will be calibrated both ways. The line width is held at 0.3 mm for subdivided apparatus and 0.4 mm for single-line apparatus. When water or other wetting liquids are used in an apparatus where volume is limited by a meniscus, the reading is made at the bottom of the meniscus. When reading a mercury meniscus, the highest point of the meniscus is taken. The following procedure for filling flasks and cylinders has been recommended by J. C. Hughes [162]:

"In filling flasks, the entire interior of the flask below the stopper will be wetted; in the case of cylinders, the liquid is allowed to flow down one side only. After filling to a point slightly below the graduation line, the instrument is allowed to drain for about 2 min. It is then placed below a burette having a long delivery tube and a bent tip, and the filling is completed by discharging water from the burette against the wall of the flask or cylinder about 1 cm above the graduation line, and rotating the receiving vessel to rewet the wall uniformly.

"Flasks and cylinders which are to be used to deliver are filled approximately to the test point, then emptied by gradually inclining them, avoiding . . . agitation of the contents and re-wetting of the walls. Allow half a minute for emptying. When the continuous outflow has ceased, the vessel should be nearly vertical and should be held in this position for another half-minute. The adhering drop is removed by contact with the wetted wall of the receiving vessel."

When determining the true capacity of a vessel by weight of water or mercury, various corrections must be applied for the effect of temperature on the density of the calibrating fluid, for distortion of the vessel under the weight of the calibrating fluid, for distortion of the vessel due to internal and external pressure, and for compression of the vessel material and the calibrating fluid. Often the distortion of the vessel under pressure or vacuum can be evaluated easily and directly by applying a small suction on the filled vessel through a capillary that is filled part way with the calibrating fluid, Fig. 2.38. Movement of the fluid level from $a$ to $b$ in the known capillary, as the pressure changes from $p_a$ to $p_b$, can be converted directly to a volume correction factor for distortion from pressure and fluid head. Obviously, values obtained for volume distortion by this method are reliable only at or near the conditions of the calibration.

### Volumetric Micrometer

Since a capability for transferring very precisely known volumes of fluid is often fundamental to determining the volume of solids, gases, immiscible liquids, and voids, the volumetric micrometer is a common instrument in laboratories where state properties of a substance are investigated. It works on the displacement principle. Usually a piston is driven axially from one end by means of a calibrated micrometer screw while the other end moves through a gland or seal into a vessel, usually cylindrical, which contains the fluid, without void, forcing the fluid through an external orifice as the piston advances. Volumetric micrometers are available commercially for a wide variety of sizes, operating conditions, and accuracies. The trade literature provides an abundance of specifications on materials of construction, pressure range, displacement per screw revolution, backlash, total displacement, method of readout, and so on.

A continuous sampling device, Fig. 2.39, based on the piston-cylinder principal of the volumetric micrometer was described by Thompson, Mueller, Coleman, and Rall [163] to effect the near quantitative separations of ma-

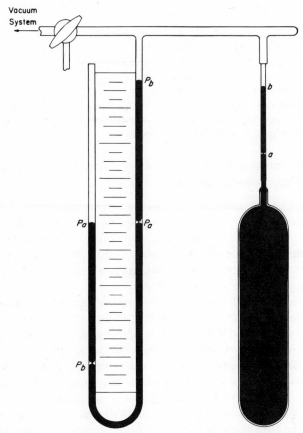

**Fig. 2.38** Method for determining the effect of pressure head on calibrated volume showing rise of mercury from $a$ to $b$ as a result of pressure decrease from $P_a$ to $P_b$.

terial at the extreme ends of a thermal diffusion column. The operation of the device is described as follows:

"Piston 7, attached to piston rod 8, moves within the precision-bore glass cylinder. The rod is actuated by sleeve 5, which travels over the $\frac{3}{8}$-inch-diameter drive screw having 40 threads per inch. The sleeve is prevented from turning by a groove in its underside which engages a pin in the sleeve mount. Cylinders (precision-bore tubing) of different diameters, with their accompanying pistons, are interchangeable in the mount, 9. The cylinder entrance is connected through a three-way stopcock to the proper part of the thermal diffusion column by means of small-bore, stainless steel hypodermic needle stock. The cylinder can be emptied by rotating the cock to the exhaust port

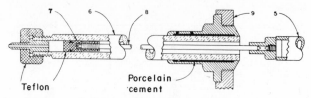

**Fig. 2.39** A continuous volumetric sampling device showing cylinder, piston, and piston rod, 6, 7, and 8, respectively; drive sleeve, 5; and support, 9. (Courtesy of Industrial and Engineering Chemistry, American Chemical Society.)

and manually returning the piston to 'empty' position, collecting the expelled sample in an appropriate container.''

Application of the volumetric micrometer for continuously recording the volume of adsorbed or evolved gas in reaction kinetics studies has been described by Mahoney et al. [164]. This apparatus, an improved version of one described by Edgecombe and Jardine [165, 166], is equipped with a capacitive micromanometer which is matched to a servo-system for controlling and recording the rate of evolution of gas.

Another modification of the volumetric micrometer is the quantitative displacement mercury pump, Fig. 2.40, which can be used in a variety of applications (see p. 46) up to 10,000 psi, where liquid or gas volumes are controlled and determined by the volume of mercury which operates as an immiscible confining fluid. The volume of mercury delivered under system pressure is to a first approximation equal to the volume of the piston, $b$, which is forced into the cylinder, $d$, through the packing gland, $i$, according to the number of turns taken on the micrometer screw, $a$. With this design the micrometer screw does not turn but is forced forward or backward by the capstan and nut, $f$, which are confined to rotate between a pair of thrust bearings, $e$. The capstan can be motor driven or hand operated. Calibration of the micrometer screw by weighing successive portions of mercury delivered can be done readily with an absolute accuracy of 0.004 ml. However, to retain this degree of accuracy in experiments at high pressure, correction terms must be applied for the compression of the mercury and piston as well as for the expansion of the cylinder. The corrected volume is usually computed in terms of the density of mercury at one atmosphere and at the temperature of the pump.

## Bulk Volume

When bulk volume of a solid specimen is wanted, a simple method that involves weighing displaced mercury from a conventional pycnometer is the most accurate procedure to follow. The wide-mouth pycnometer [167], (Fig.

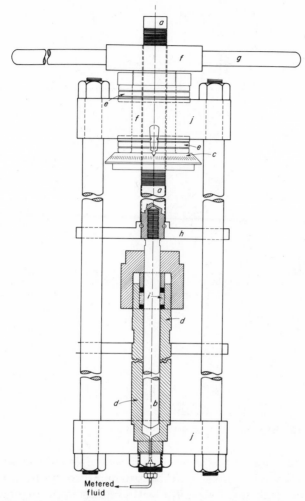

**Fig. 2.40** Volumetric micrometer for use at high pressures: *a*, calibrated micrometer screw; *b*, piston; *c*, scale and counter; *d*, cylinder; *e*, thrust roller bearings; *f*, nut; *g*, handles; *h*, guide; *i*, packing; *j*, headers.

2.41), is filled with mercury and the top is inserted causing excess mercury to overflow through the capillary opening. After excess mercury is brushed off the pycnometer is placed in a clean container, the top of the pycnometer is removed, and the specimen is immersed in the mercury by pushing it down with the top. To avoid trapped air between the top and the specimen, a piece of small-diameter wire can be run into the capillary until all the air is expelled. The product of the weight of the displaced mercury and its specific

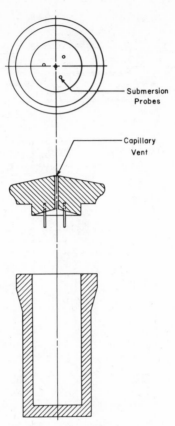

Submersion
Probes

Capillary
Vent

**Fig. 2.41**  Pycnometer for bulk volume determination.

volume at the temperature of the measurement gives the volume of the test specimen. If several samples are to be determined routinely, a bulk-volume cell apparatus [168], Fig. 2.42, will allow the measurements to proceed rapidly and with an accuracy that often approaches that of the pycnometer method. The use of the bulk-volume cell, as described by Hamontree and Rall, involves filling the mercury reservoir and the sample cylinder with mercury until the micrometer screw makes an electrical contact, thus establishing a zero setting. A solid sample is then held beneath the mercury in the sample cylinder and a new reading of the contact point of the micrometer is obtained. Calibration of the device with the aid of cylinders of accurately known volumes will allow the volume of the sample to be determined from the established calibration curve. When operated with care the bulk-volume cell apparatus gives results reproducible to 0.1 % on samples of approximately 8 ml.

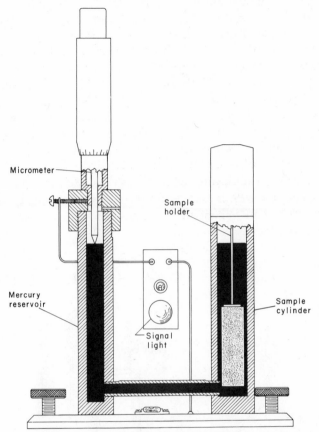

**Fig. 2.42**   Bulk volume cell and auxiliary apparatus. (Courtesy of the Oil and Gas Journal.)

For ordinary bulk-volume determinations the method described above is quite satisfactory; however, when the surface of the sample is rough, air will become trapped in the indentations, thus preventing the entry of mercury and causing a positive error in the bulk-volume determination. If accuracy in bulk volume is extremely critical, the source of error just described can be eliminated by carrying out the measurements in an evacuated enclosure.

### Porosity

The measurement of porosity is essentially a determination of volume and is distinguished from permeability, which is a measure of the flow characteristic of a fluid through a porous medium. An aggregate of crystals or solid particles that are cemented through pressure, recrystallization, or interstitial material exhibits a pore volume that is usually taken to mean all interconnected pores that ultimately communicate with the surface of the sample. Obviously, isolated voids exist but the volume of these cannot be estimated without crushing the sample .Thus, a determination of porosity by measuring the volume of gas that is required to fill the pore spaces of a solid sample includes only interconnected pores. The classical method described by Washburn and Bunting [169] involves the simple apparatus, Fig. 2.43, which may be immersed in a constant temperature bath if the highest accuracy is desired. The solid sample having a bulk volume, $V_{Bl}$, is placed in cell $A$ with a mass of glass, $m_g$, of known density, $D_g$, which fills most of the remaining volume, $V_A$, of cell $A$. Cell $A$ is next evacuated to some small initial pressure, $P_{A0}$, (about 20 mm Hg). Cell $B$, which contains a known volume, $V_B$, of dry inert gas at pressure $P_{B0}$, is opened to cell $A$ through the interconnecting stopcock and the final pressure, $P_f$, is read. From these data, the porosity, defined as the fraction of the bulk volume that is interconnected pore volume, $V_{pore}$, is given by the following relation

$$\text{Porosity} = \frac{V_{pore}}{V_B} = \left[\frac{(P_{B0} - P_f)V_B}{(P_f - P_{A0})V_{Bl}} - \frac{V_A - m_g/D_g - V_{Bl}}{V_{Bl}}\right]. \quad (2.41)$$

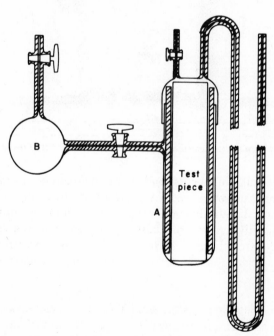

**Fig. 2.43** Apparatus for determining porosity showing sample holder, $A$, and mercury reservoir, $B$.

## Volume Changes of Liquids under Pressure-Bulk Modulus

In many applications where organic fluids are used in hydraulic systems, a knowledge of the change in volume of the fluid under an applied pressure is needed in order to design for optimum size and for eliminating harmonic vibrations that sometimes result from an unfortuitous combination of pump characteristics, volume of the system, and the elasticity characteristic of the fluid. The elasticity characteristic of a static, isothermal fluid is best expressed in terms of the linear secant modulus equation [170].

$$\overline{K} = \frac{V_0 P}{V_0 - V} = K_0 + mP \tag{2.42}$$

in which $V_0$ is an initial volume (at 1 atm, essentially), $P$ and $V$ are final pressure and volume, respectively, $K_0$ is the modulus at zero pressure, and $m$ is the slope of the bulk-modulus–pressure curve. Hayward has pointed out [170] that the linear secant modulus (2.42) is the inverted form of Tait's equation [171] and that the relation commonly known as Tait's equation

$$\frac{V_0 - V}{V_0} = C \ln \left( \frac{B + P}{B} \right), \tag{2.43}$$

in which $B$ and $C$ are determined constants, is in reality a misquotation and inferior to the linear secant modulus equation for representing the compressive properties of liquids.

Under applied conditions of rapid pressure variation where the entropy, $S$, is nearly constant throughout a cycle of pressure change, the compressive characteristics of a fluid are represented thermodynamically by the isentropic tangent bulk modulus relation

$$K_s = - V \left( \frac{\partial P}{\partial V} \right)_s, \tag{2.44}$$

which is related through classical fluid mechanics to the velocity of sound. Wostl, Buehler, and Dresser [172] have used the relationship, $K_s = dC^2/g$, where $d$ is the density of the liquid, $C$ is the velocity of sound, and $g$ is the gravitational constant, to determine $K_s$ from ultrasonic measurements of the velocity of sound. A schematic drawing of their apparatus is shown in Fig. 2.44. The ultrasonic method is under consideration by ASTM Committee D-2 on Petroleum Products and Lubricants [173] for certification as a standard method.

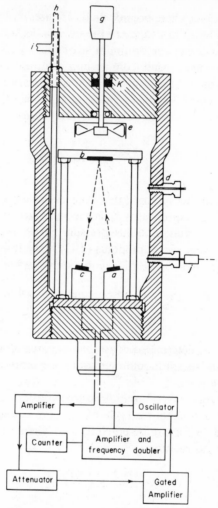

**Fig. 2.44** Ultrasonic interferometer for determining bulk modulus of liquids: *a*, oscillating crystal; *b*, reflector; *c*, receiving crystal; *d*, thermocouple; *e*, fan; *f*, filling and emptying tube; *g*, stirring motor; *h*, *i*, tubing connections; *j*, amplifier and temperature readout.

## References

1. F. D. Rossini, *Pure Appl. Chem.* **9**, 453 (1964).
2. M. L. McGlashan, *Physico-Chemical Quantities and Units*, Royal Institute of Chemistry, Monograph for Teachers No. 15, London, England, 1968.

3. C. F. Marvin, *Barometers and the Measurement of Atmospheric Pressure*, 7th ed., Circular F, Instrument Division, U.S. Weather Bureau, 1941.

4. W. G. Brombacher, D. P. Johnson, and J. L. Cross, *Nat. Bur. Std. Monogr. 8* on Mercury Barometers and Manometers, 1960. Available from the Superintendent of Documents, Government Printing Office, Washington, D. C., 40 cents.

5. R. C. Weast, Ed., *Handbook of Chemistry and Physics*, 50th ed., The Chemical Rubber Co., Cleveland, Ohio, 1967–1968.

6. H. H. Kimball, *International Critical Tables*, Vol. I, McGraw-Hill, New York, 1926, pp. 65–70.

7. P. S. Laplace, *Mécanique Céleste* (Bowditch translation), Vol. 4, 1812, p. 737.

8. J. A. Beattie, D. D. Jacobus, J. M. Gaines, Jr., M. Benedict, and B. E. Blaisdell, *Proc. Amer. Acad. Arts Sci.* **74**, 327 (1941).

9. B. E. Blaisdell, *J. Math. Phys.* **19**, 217 (1940).

10. D. I. Mendeleef and E. K. Gutkowski, *J. Russ. Phys. Chem. Soc.* **8**, 212 (1876).

11. E. T. Levanto, *Ann. Acad. Sci. Fennicae, Ser. AI*, No. 8 (1941).

12. F. W. Kohlrausch, *Lehrbuch der praktischen Physik*, Teubner, Leipzig, 1901, p. 579.

13. F. E. Fowle, Ed., *Smithsonian Physical Tables*, 8th ed., Smithsonian Institution, Washington, 1934, p. 187.

14. N. A. Lange, Ed., *Handbook of Chemistry*, 10th ed. revised, McGraw-Hill, New York, 1967, p. 1694.

15. C. D. Hodgman, Ed., *Handbook of Chemistry and Physics*, 42d ed., The Chemical Rubber Co., Cleveland, Ohio, 1960–1961, p. 1677.

16. R. Süring, *Ber. Taetigkeit Kgl. Preuss. Meteor. Inst.* **24** (1916); Landolt-Bornstein Tabellen, Main vol., p. 72.

17. W. Cawood and H. S. Patterson, *Trans. Faraday Soc.* **29**, 514 (1933).

18. Note also the value of 0.2701 at 25°C obtained by C. Kimball, *Trans. Faraday Soc.* **42**, 526 (1946).

19. A. H. Cook and N. W. B. Stone, *Phil. Trans. Roy. Soc. London* **250A**, 279 (1957).

20. N. W. B. Stone, *Phil. Trans. Roy. Soc. London* **254A**, 1038 (1961).

21. E. U. Condon and H. Odinshaw, Eds., *Handbook of Physics*, McGraw-Hill, New York, 1958, p. 59, eq. 7.42.

22. G. W. C. Kaye and T. H. Laby, *Tables of Physical and Chemical Constants*, Wiley, New York, 1966, pp. 15–16.

23. S. C. Collins and B. E. Blaisdell, *Rev. Sci. Instr.* **7**, 213 (1936).

24. Lord Rayleigh, *Trans. Roy. Soc. (London)* **196A**, 205 (1901).

25. J. Reilly and W. N. Rae, *Physico-chemical Methods*, Vol. 1, 5th ed., Van Nostrand, Princeton, N. J., 1953, pp. 247–248.

26. W. S. Young and R. C. Taylor, *Anal. Chem.* **19**, 133 (1947).

27. C. F. Mündel, *Z. Physik. Chem. (Leipzig)* **85**, 435 (1913).

28. C. B. Willingham, W. J. Taylor, J. M. Pignocco, and F. D. Rossini, *J. Res. Nat. Bur. Std.* **35**, 219 (1945).

29. A. Zimmerli, *Ind. Eng. Chem., Anal. Ed.* **10**, 283 (1938).

30. J. Dubrovin, *Instr.* **6**, 194 (1933).

31. F. E. E. Germann and K. A. Gagos, *Ind. Eng. Chem., Anal. Ed.* **15**, 285 (1943).

32. Caswell, *Phil. Trans. Roy. Soc. London* **24**, No. 290, 1597 (1704).
33. Dow Corning Corporation, Midland, Michigan.
34. M. Burton, *Ind. Eng. Chem., Anal. Ed.* **9**, 335 (1937).
35. B. J. P. Roberts, *Proc. Roy. Soc.* (*London*) **78A**, 410 (1906).
36. See, for example, J. Reilly and W. N. Rae, *Physico-chemical Methods*, Vol. 1, 5th ed., Van Nostrand, Princeton, New Jersey, 1953, pp. 249–251; F. Daniels, J. H. Mathews, and J. W. Williams, *Experimental Physical Chemistry*, 3rd ed., McGraw-Hill, New York, 1941; Ostwald-Luther, *Hand- und Hilfsbuch zur Ausführung physiko-chemischer Messungen*, p. 215; S. Dushmen, *The Scientific Foundations of High Vacuum Technique*, Wiley, New York, 1962, pp. 225–234; J. Strong and collaborators, *Procedures in Experimental Physics*, Prentice-Hall, Englewood Cliffs, New Jersey, 1938, pp. 137–143; S. Jnanananda, *High Vacua*, Van Nostrand, New Jersey, 1947, pp. 157–171, 242.
37. P. Lott, *Anal. Chem.* **28**, 276 (1956).
38. H. S. Booth, *Ind. Eng. Chem., Anal. Ed.* **4**, 380 (1932).
39. W. Gaede, *Ann. Physik.* **41**, 289 (1913).
40. E. W. Flosdorf, *Ind. Eng. Chem., Anal. Ed.* **10**, 534 (1938).
41. K. C. D. Hickman, *J. Opt. Soc. Amer.* **18**, 305 (1929).
42. A. H. Pfund, *Phys. Rev.* **18**, 78 (1921).
43. P. Rosenberg, *Rev. Sci. Instr.* **10**, 131 (1939), Dr. B. B. Dayton has pointed out in *Vide* **2**, 349 (1947) that the grinding of the capillary suggested by Rosenberg may cause a large systematic error in the McLeod gage reading.
44. K. C. D. Hickman, *Rev. Sci. Instr.* **5**, 161 (1934).
45. K. C. D. Hickman, *J. Phys. Chem.* **34**, 627 (1930).
46. C. G. Malmberg and W. W. Nicholas, *Rev. Sci. Instr.* **3**, 440 (1932).
47. J. M. Stuckey and J. H. Saylor, *J. Amer. Chem. Soc.* **62**, 2922 (1940); F. H. Field and J. H. Saylor, *ibid.* **68**, 2649 (1946).
48. D. C. Bradley and J. D. Swanwick, *J. Chem. Soc.* **1958**, 3207.
49. *Precision Pressure Instruments*, Bulletin 021, Texas Instruments Incorporated, Stafford, Texas.
50. O. C. Bridgeman, *J. Am. Chem. Soc.* **49**, 1174 (1927); R. G. P. Greig and R. S. Dadson, *Brit. J. Appl. Phys.* **17**, 1633 (1966).
51. J. A. Beattie, *Proc. Amer. Acad. Arts Sci.* **69**, 369 (1934).
52. R. S. Dadson, "The Accurate Measurement of High Pressures and the Precise Calibration of Pressure Balances," *The Inst. Mech. Eng., Intern. Union Pure Appl. Chem., Joint Conf. on Thermodynamics and Transport Properties Fluids*, July 10–12, 1957, Inst. Mech. Eng., London.
53. J. L. Cross, "Reduction of Data for Piston Gage Pressure Measurements," *Natl. Bur. Std. Monog.* **65**, August 1964.
54. M. von Pirani, *Verhandl. Deut. Physik. Ges.* **4**, 686 (1906).
55. C. F. Hale, *Trans. Amer. Electrochem. Soc.* **20**, 243 (1911).
56. J. A. Becker, C. B. Green, and G. L. Pearson, *Trans. Amer. Inst. Elec. Eng.* **65**, 711 (1946).
57. A. Smith and A. W. C. Menzies, *J. Amer. Chem. Soc.* **32**, 1412 (1910).
58. E. J. Buckler and R. G. W. Norrish, *J. Chem. Soc.* **1936**, 1967.
59. J. S. Anderson, *J. Chem. Soc.* **1930**, 1653.

60. E. W. Balson, K. G. Denbigh, and N. K. Adam, *Trans. Faraday Soc.* **43,** 42 (1947).

61. J. G. Aston and co-workers, *J. Amer. Chem. Soc.* **58,** 2354 (1936); **62,** 886 (1940); **68,** 52 (1946).

62. W. M. Jones and W. F. Giauque, *J. Amer. Chem. Soc.* **69,** 983 (1947).

63. H. S. Booth and H. S. Halbedel, *J. Amer. Chem. Soc.* **68,** 2652 (1946).

64. H. S. Booth and P. H. Carnell, *J. Amer. Chem. Soc.* **68,** 2650 (1946).

65. H. S. Booth and W. F. Martin, *J. Amer. Chem. Soc.* **68,** 2655 (1946).

66. H. S. Booth and J. F. Suttle, *J. Amer. Chem. Soc.* **68,** 2658 (1946).

67. H. S. Booth and D. R. Spessard, *J. Amer. Chem. Soc.* **68,** 2660 (1946).

68. H. S. Booth and A. A. Schwartz, *J. Amer. Chem. Soc.* **68,** 2662 (1946).

69. H. S. Booth, H. M. Elsey, and P. E. Burchfield, *J. Amer. Chem. Soc.* **57,** 2064 (1935).

70. D. R. Douslin, R. H. Harrison, and R. T. Moore, *J. Chem. Thermodyn.* **1,** 305 (1969).

71. D. R. Douslin, R. T. Moore, J. P. Dawson, and G. Waddington, *J. Amer. Chem. Soc.* **80,** 2031 (1958).

72. S. Young, *Sci. Proc. Roy. Dublin Soc.* **12,** 374 (1909–1910).

73. C. de la Tour, *Ann. Chim. Phys.* **21**(2), 127 (1822).

74. C. de la Tour, *Ann. Chim. Phys.* **22**(2), 410 (1823).

75. D. R. Douslin and J. P. McCullough, *Report of Investigations* 6149, U. S. Bureau of Mines, 1963, 11 pp.

76. D. R. Douslin and A. Osborn, *J. Sci. Instr.* **42,** 369 (1965).

77. V. O. Hutton, *J. Res. Natl. Bur. Std.* **63C,** 47 (1959).

78. S. R. Hoover, H. John, and E. F. Mellon, *Anal. Chem.* **25,** 1940 (1953).

79. E. F. Mellon, D. J. Viola, and S. R. Hoover, *J. Phys. Chem.* **57,** 607 (1953).

80. W. Ramsay and S. Young, *J. Chem. Soc.* **47,** 42 (1885).

81. F. H. Field and J. H. Saylor, *J. Amer. Chem. Soc.* **68,** 2649 (1946); J. M. Stuckey and J. H. Saylor, *ibid.* **62,** 2922 (1940). The second paper includes valuable operational data.

82. J. E. Hawkins and G. T. Armstrong, *J. Amer. Chem. Soc.* **76,** 3756 (1954).

83. R. W. Ryan and E. A. Lantz, *Ind. Eng. Chem.* **20,** 40 (1928).

84. A. Angeli, *Gazz. Chim. Ital.* **23**(II), 104 (1893).

85. A. Siwoloboff, *Ber.* **19,** 795 (1886).

86. C. Weygand, *Organisch-Chemische Experimentierkunst*, Barth, Leipzig, 1938, 1st part, p. 41, Fig. 23.

87. J. W. Boegel, *Ind. Eng. Chem., Anal. Ed.* **8,** 476 (1936).

88. K. C. D. Hickman and W. Weyerts, *J. Amer. Chem. Soc.* **52,** 4714 (1930).

89. M. R. Fenske, "Laboratory and Small Scale Distillation," *Science of Petroleum*, Oxford, London, 1938.

90. R. W. Schiessler and F. C. Whitmore, *Ind. Eng. Chem.* **47,** 1660 (1955).

91. H. S. Myers and M. R. Fenske, *Ind. Eng. Chem.* **47,** 1652 (1955).

92. C. B. Willingham, W. J. Taylor, J. M. Pignocco, and F. D. Rossini, *J. Res. Natl. Bur. Std.* **35,** 219 (1945).

93. D. R. Stull, *Ind. Eng. Chem., Anal. Ed.* **18,** 234 (1946).

94. *Natl. Bur. Std. Handbook* **77,** 40 (1964).

95.  J. L. Edwards and D. P. Johnson, *J. Res. Natl. Bur. Std.* **72**, 27 (1968).
96.  H. V. Regnault, *Ann. Chim.* **15**, 129 (1845).
97.  H. T. Gerry and L. J. Gillespie, *Phys. Rev.* **40**, 269 (1932).
98.  E. W. Balson, K. G. Denbigh, and N. K. Adam, *Trans. Faraday Soc.* **43**, 42 (1947).
99.  J. E. Overberger, W. A. Steele, and J. G. Aston, *J. Chem. Thermodyn.* **1**, 535 (1969).
100.  G. P. Baxter, C. H. Hickey, and W. C. Holmes, *J. Amer. Chem. Soc.* **29**, 127 (1907).
101.  G. P. Baxter and J. E. Lansing, *J. Amer. Chem. Soc.* **42**, 419 (1920).
102.  G. P. Baxter, F. K. Bezzenberger, and C. H. Wilson, *J. Amer. Chem. Soc.* **42**, 1386 (1920).
103.  G. P. Baxter and W. C. Cooper, Jr., *J. Amer. Chem. Soc.* **46**, 923 (1924).
104.  E. F. G. Herington and J. F. Martin, *Trans. Faraday Soc.* **49**, 154 (1953).
105.  D. Ambrose, *J. Sci. Instr.* (*J. Phys. E.*) **1**, 41 (1968).
106.  A. G. Osborn and D. R. Douslin, *J. Chem. Eng. Data* **11**, 502 (1966).
107.  International Steam Tables, *Mech. Eng.* **57**, 710 (1935).
108.  N. S. Osborne, H. F. Stimson, and D. C. Ginnings, *J. Res. Natl. Bur. Std.* **23**, 261 (1939).
109.  H. F. Stimson, *J. Res. Natl. Bur. Std.* **73A**, 493 (1969).
110.  M. R. Gibson and E. A. Bruges, *J. Mech. Eng. Sci.* **9**, 24 (1967).
111.  P. C. F. Frowein, *Z. Physik. Chem.* **1**, 5 (1887); **17**, 52 (1895).
112.  G. Calingaert, H. Soroos, V. Hnizda, and H. Shapiro, *J. Amer. Chem. Soc.* **66**, 1389 (1944).
113.  K. C. D. Hickman, J. C. Hecker, and N. D. Embree, *Ind. Eng. Chem., Anal. Ed.* **9**, 264 (1937).
114.  K. C. D. Hickman, *Rev. Sci. Instr.* **5**, 161 (1934).
115.  E. S. Perry, W. H. Weber, and B. F. Daubert, *J. Amer. Chem. Soc.* **71**, 3720 (1949); E. S. Perry and W. H. Weber, *ibid.* **71**, 3726 (1949).
116.  F. H. Verhoek and A. L. Marshall, *J. Amer. Chem. Soc.* **61**, 2737 (1939).
117.  P. Clausing, *Z. Physik.* **66**, 471 (1930); *Ann. Physik.* **12**, 961 (1932). See also the extensive treatment in S. Dushman and J. M. Lafferty, *The Scientific Foundations of Vacuum Technique*, Wiley, New York, 1962.
118.  H. E. Robson and P. W. Gillis, *J. Phys. Chem.* **68**, 983 (1964).
119.  N. K. Smith, G. Gorin, W. D. Good, and J. P. McCullough, *J. Phys. Chem.* **68**, 940 (1964).
120.  C. I. Whitman, *J. Chem. Phys.* **20**, 161 (1952).
121.  M. G. Rossman and J. Yarwood, *J. Chem. Phys.* **21**, 1406 (1953).
122.  C. I. Whitman, *J. Chem. Phys.* **21**, 1407 (1953).
123.  K. Neumann and E. Völker, *Z. Physik. Chem.* **161A**, 33 (1932). The equations above were obtained from a new examination of their experimental data. The equations for the solid forms intersect the liquid equation at the melting points used by these authors: 48.0° and 26.1°C.
124.  M. Volmer, *Z. Physik. Chem., Bodenstein-Festband* **1931**, 863.
125.  V. Dietz, *J. Amer. Chem. Soc.* **55**, 472 (1933).
126.  A. R. Ubbelohde, *Trans. Faraday Soc.* **34**, 282 (1938).

127. R. D. Freeman in *The Characterization of High-Temperature Vapors*, J. L. Margrave, Ed., Wiley, New York, 1967, Chap. 7, "Momentum Sensors."
128. R. D. Freeman and J. G. Edwards in *The Characterization of High Temperature Vapors*, J. L. Margrave, Ed., John Wiley, New York, 1967, Appendix C, "Transmission Probabilities and Recoil Force Correction Factors for Conical Orifices."
129. A. V. Rodionov, *Vestn. Mosk. Univ. Ser. II* **22**(6), 48 (1967).
130. W. H. Rodebush and C. C. Coons, *J. Amer. Chem. Soc.* **49**, 1953 (1927).
131. W. H. Rodebush and W. F. Henry, *J. Amer. Chem. Soc.* **52**, 3159 (1930).
132. V. Dietz, *J. Chem. Phys.* **4**, 575 (1936).
133. E. W. Balson, *Trans. Faraday Soc.* **43**, 48 (1947).
134. F. M. Ernsberger and H. W. Pitman, *Rev. Sci. Instr.* **26**, 584 (1955).
135. M. H. Kaufman and A. G. Whittaker, *J. Chem. Phys.* **24**, 1104 (1956).
136. F. Emich and F. Schneider, *Microchemical Laboratory Manual*, Wiley, New York, 1932.
137. F. Emich, *Monatsh.* **38**, 219 (1917).
138. A. D. Gettler, J. B. Niederl, and A. A. Benedetti-Pichler, *Mikrochemie* **11**, 174 (1932).
139. A. A. Benedetti-Pichler and W. F. Spikes, *Introduction to the Microtechnique of Inorganic Qualitative Analysis*, Microchemical Service, Douglaston, New York, 1935.
140. A. Schleiermacher, *Ber.* **24**, 944 (1891).
141. F. Pregl and H. Roth, *Quantitative Organic Microanalysis*, 3rd English ed., Blakiston, Philadelphia, 1937, p. 225.
142. A. Siwoloboff, *Ber.* **19**, 795 (1886).
143. A. A. Morton, *Laboratory Technique in Organic Chemistry*, McGraw-Hill, New York, 1938.
144. J. R. Niederl and I. R. Routh, *Mikrochemie* **11**, 251 (1932).
145. A. Smith and A. W. C. Menzies, *J. Amer. Chem. Soc.* **32**, 897 (1910).
146. C. Rosenblum, *Ind. Eng. Chem., Anal. Ed.* **10**, 449 (1938).
147. R. L. Shriner, R. C. Fuson, and D. Y. Curtin, *The Systematic Identification of Organic Compounds*, 4th ed., Wiley, New York, 1956.
148. E. E. Hays, F. W. Hart, and R. C. Gustavson, *Ind. Eng. Chem., Anal. Ed.* **8**, 286 (1936).
149. C. R. García, *Ind. Eng. Chem., Anal. Ed.* **15**, 648 (1943).
150. G. W. Thomson, *Chem. Rev.* **38**, 1 (1946).
151. J. R. Partington, *An Advanced Treatise on Physical Chemistry, Vol. II, The Properties of Liquids*, Longmans, Green, London and New York, 1951, pp. 265–274.
152. C. Antoine, *Compt. Rend.* **107**, 681, 836, 1143 (1888).
153. B. J. Zwolinski, "Selected Values of Properties of Hydrocarbons and Related Compounds," *American Petroleum Institute Research Project 44*, Thermodynamics Research Center, Texas A&M University, College Station, Texas. (Looseleaf data sheets, extant, 1968).
154. A. G. Osborn and D. R. Douslin, *J. Chem. Eng. Data* **14**, 208 (1969).
155. A. G. Osborn and D. R. Douslin, *J. Chem. Eng. Data* **13**, 534 (1968).

156. P. T. White, D. G. Barnard-Smith, and F. A. Fidler, *Ind. Eng. Chem.* **44**, 1430 (1952).
157. A. G. Osborn and D. R. Douslin, *J. Chem. Eng. Data* **11**, 502 (1966).
158. E. R. Cox, *Ind. Eng. Chem.* **28**, 613 (1936).
159. D. R. Stull, *Ind. Eng. Chem.* **39**, 517 (1947).
160. B. P. E. Clapeyron, *J. École Polytech. (Paris)* **14**, No. 23, 153 (1834).
161. J. J. Van Laar, *Tables Annuelles Internationales des Constantes et Données Numériques*, Gauthier-Villars, Paris, 1913–1930, Vol. 4, pp. 288 *et seq.*; Vol. 5, pp. 228 *et seq.*; Vol. 6, pp. 149 *et seq.*; Vol. 7, pp. 196 *et seq.*; Vol. 8, pp. 233 *et seq.*; Vol. 9, pp. 158 *et seq.*; Vol. 10, pp. 157 *et seq.*
162. J. C. Hughes, "Testing of Glass Volumetric Apparatus," in *Apparatus in Precision Measurement and Calibration, Handbook 77*, Vol. 3, National Bureau of Standards, 1961, pp. 726–739.
163. C. J. Thompson, F. G. Mueller, H. J. Coleman, and H. T. Rall, *Ind. Eng. Chem.* **52**, 53A (1960).
164. L. R. Mahoney, R. W. Bayma, A. Warnick, and C. H. Ruof, *Anal. Chem.* **36**, 2516 (1964).
165. F. H. C. Edgecombe and D. A. Jardine, *Can. J. Chem.* **39**, 1728 (1961).
166. F. H. C. Edgecombe and D. A. Jardine, *Rev. Sci. Instr.* **33**, 240 (1962).
167. D. B. Taliaferro, Jr., T. W. Johnson, and E. J. Dewees, *U.S. Bureau of Mines Report of Investigation 3352*, 1937.
168. H. C. Hamontre and Cleo G. Rall, *Oil Gas J.*, 51, No. 37, 94, Jan. 19, 1953.
169. E. W. Washburn and E. N. Bunting, *J. Amer. Chem. Soc.* **5**, 112 (1922).
170. A. T. J. Hayward, "Compressibility Equations for Liquids," *Report 295*, National Engineering Laboratory, June 1967.
171. P. G. Tait, "Report on Some of the Physical Properties of Fresh Water and Sea Water," *The Voyage of HMS Challenger*, Vol. 2, Part 4, Neill, Edinburgh, 1889.
172. W. J. Wostl, R. J. Buehler, and T. Dresser, *Rev. Sci. Instr.* **37**, 1665 (1966).
173. *1968 Book of ASTM Standards*, Part 17, American Society for Testing and Materials, 1916 Race St., Philadelphia, Pa. 19103, p. 1074.

Chapter III

# DETERMINATION OF
# MELTING AND FREEZING TEMPERATURES

Evald L. Skau and Jett C. Arthur, Jr.

---

## 1  INTRODUCTION

The freezing point or melting point of a pure compound is the temperature at which solid crystals of the substance are in equilibrium with the liquid phase at atmospheric pressure. If this equilibrium condition is approached by cooling the liquid, the temperature is commonly called the freezing point, and if approached by heating the solid, it is designated as the melting point. The triple point of a compound is its melting temperature measured under its own vapor pressure. The difference between the melting and triple points depends upon the pressure difference and the effect of dissolved air. It is negligible for most practical purposes (0.0099°C in the case of water) [1, 2], and too small to be measured by the ordinary types of melting-point apparatus.

According to thermodynamic definition, freezing point and melting point, terms frequently used interchangeably, designate the same temperature. For solid substances which melt without decomposition, the melting point has become the most important physical property in practical organic chemistry. It is usually the first property determined on a newly synthesized compound, being a basis for characterization, identification, and estimation of purity.

Numerous reasons can be cited why the organic chemist has chosen the melting point rather than another property for this purpose. As generally determined, only small amounts of material are needed for the melting-point determination, and the procedure and equipment required are very simple. The melting point is practically independent of pressure and can, therefore, be taken without making extra physical measurements or corrections, as is necessary, for example, with the boiling-point determination. The change from solid to liquid usually occurs without such troublesome effects as superheating and its accompanying sources of error. Relatively few substances, furthermore, decompose when heated to the melting point. The "mixed melting-point" method of identification is especially effective since the melting point is not, as a rule, an additive property.

The use of melting point and melting and freezing behavior by the inorganic chemist is confined chiefly to the estimation of purity, determination of solubilities, and the study of high-temperature solid–liquid phase equilibria, usually of multicomponent mixtures. Information obtained is of both theoretical and practical importance in the fields of metallurgy, geophysics, ceramics, refractories, and space travel, particularly in exit and reentry problems.

One of the most important applications of the melting point is its use as a criterion of purity. If a substance is pure, the melting point will usually be sharp. With mixtures, melting usually occurs over a range of temperatures; and, conversely, melting over a range practically always indicates that the material is a mixture.

In the case of some isomeric substances, a mere comparison of the melting points can be used in the identification of the isomers. For example, it is possible to distinguish the para from the ortho or the meta isomers by the fact that the para has the highest melting point of the three.

In spite of the advantages enumerated above, the determination of melting point presents to the observer many occasions for misinterpretation of data. The most useful applications can be made only if the organic chemist has a thorough knowledge of the possibilities involved. This chapter is concerned with the uses to which the melting-point method may be applied with a description of methods of measurement and with a discussion of sources of error which may arise in the interpretation of melting points observed by various experimental methods. For additional details on the subjects discussed in this chapter consult the General References at the end of the chapter.

The precision with which the melting point should be determined varies with the information desired. For ordinary purposes of characterization or identification, the capillary-tube method gives results of adequate accuracy—in fact, if the compound is known to be one of several possibilities, an approximate melting point may be sufficient to establish its identity. If the compound is a new one, greater precision is required because the melting point reported should be sufficiently accurate to distinguish the material from others which melt at approximately the same temperature. For very high precision, the cooling-curve method may be resorted to for obtaining true equilibrium between solid and liquid.

Discretion must also be used in the choice of the temperature-measuring device so that it may be appropriate for the accuracy desired. The careful investigator uses well-calibrated instruments. He is always cognizant of the sources of error and, if they are significant, either eliminates them or, when possible, applies a correction. Some of these factors will be discussed in the chapter on temperature measurement.

When the melting point is used as an indication of purity, some knowledge of the behavior of binary mixtures is valuable for proper interpretation of the data obtained. Some of the factors affecting the melting point of such mixtures will be considered.

## 2   THEORETICAL CONSIDERATIONS

### Energy and Volume Changes at Melting Point

When a pure crystalline solid melts, the molecules, arranged in a high degree of order in a crystal lattice, are separated by thermal forces to form the liquid state, in which they are in almost complete disorder. For an ideally pure substance this transformation is accompanied by an abrupt increase in entropy, in heat content, and usually in volume. This is shown schematically by the full curves in Fig. 3.1, in which the ordinate, $y$, may represent either entropy, heat content, or volume/mole. The difference, $y_2 - y_1$, represents the molar entropy change, the molar heat of fusion, or the change in molar volume, as the case may be, at the melting point, $T_m$. The heat of fusion is also called the heat of crystallization.

The broken lines in Fig. 3.1 show the curve for the same substance containing an impurity. Except for a small abrupt increase at the eutectic point, $T_e$, the change now occurs gradually over a temperature range instead of at a constant temperature. It starts at $T_e$, and reaches completion at a temperature slightly lower than the melting point of the pure substance.

In a few exceptional cases, as for water, bismuth, gallium, and germanium, $y_1$, the volume of the solid at the melting temperature, is greater than $y_2$, the volume of the liquid. Thus, though Fig. 3.1 would still represent the change in entropy and heat content with temperature, the volume change for the pure substance would be represented by a gradual increase in the volume of the solid up to $T_m$, an abrupt decrease in volume at constant temperature at $T_m$, and then another gradual increase as the temperature is further increased.

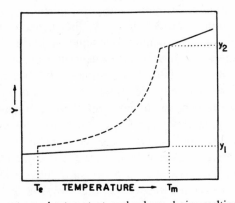

**Fig. 3.1**   Change in entropy, heat content, and volume during melting.

## Effect of Pressure

The effect of pressure, $p$, on the melting temperature in degrees absolute, $T$, of a pure substance can be expressed quantitatively by the Clausius-Clapeyron equation:

$$\frac{dT}{dp} = \frac{T(V_l - V_s)}{\Delta H},$$

where $V_l$ and $V_s$ are the molar volumes of the liquid and solid, respectively, and $\Delta H$ is the molar heat of fusion. Since, with few exceptions, the density of a solid is greater than that of the liquid and, therefore, $V_l - V_s$ is positive, an increase in the pressure usually raises the melting temperature.

Considerable experimental data have been reported showing the effects of high pressures on melting temperatures, heats of fusion, and molar volumes of many common organic compounds [3]. In general, the effects on different compounds are qualitatively similar; for example, with an increase in pressure there is an increase in melting temperature and heat of fusion and, at constant temperature, a decrease in volume. This is shown schematically in Fig. 3.2 for ethyl alcohol, calculated from the data of Bridgman [3, 4], in which the ordinate represents melting temperature, heat of fusion, or specific volume. The compression curve was run at 50° C with the ethyl alcohol freezing at 24,400 kg/cm².

The following are a few of the rather startling melting temperatures at 35,000 kg/cm² (33,800 atm) pressure reported by Bridgman [3, 4]: for water 166.6°C instead of the usual 0°C; for ethyl alcohol 109°C instead of −117.3°C; for $n$-butyl alcohol 155°C instead of −89.8°C; for $n$-propyl bromide 169°C instead of −110°C; and for carbon disulfide 209°C instead of −111.6°C.

For water at low pressures, $V_l - V_s$ is negative, so that its melting temperature is lowered by an increase in pressure at a rate of 0.0075°C/atm. At higher pressures, however, more dense polymorphic forms of ice appear so that $V_l - V_s$ becomes positive and a further increase in pressure raises the melting temperature. Germanium is unique in that its melting temperature continues to decrease linearly with increasing pressure from 936°C, at 1 atm, to 347 ± 18°C, at 180,000 atmospheres [5]. This indicates that the liquid remains denser than the solid even at this extreme pressure.

The melting and freezing of organic compounds and their polymorphic transitions can be initiated by changing the pressure isothermally. Thus it is apparent from the specific volume–pressure curve in Fig. 3.2 that liquid ethanol at 50°C would be expected to freeze at a pressure of 24,400 kg/cm². Actually, this does not usually take place until this pressure is exceeded, a phenomenon corresponding to supercooling. However, in contrast, crystalline solid, with few exceptions, begins melting immediately on reaching the

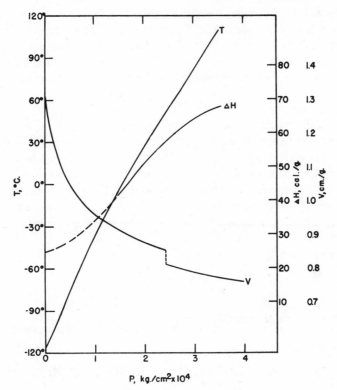

**Fig. 3.2** Effect of pressure on melting temperature, $T$; heat of fusion, $\Delta H$; and specific volume, $V$, of ethanol.

domain of thermodynamic stability of the liquid. The pressures required to initiate freezing of the following compounds at 25°C are as follows: benzene (standard mp 5.5°C) 680 kg/cm²; chloroform ($-63.5$°C) 5,550 kg/cm²; chlorobenzene ($-45.2$°C) 7,500 kg/cm²; and $n$-heptane ($-90.5$°C) 11,450 kg/cm².

### Ideal Systems

In the case of a mixture containing a large amount of component $A$ and a small amount of a second component, $B$, as for example a nearly pure compound, the melting point of the mixture will usually be lower than that of pure $A$. The extent of the lowering depends upon a number of variables and may be estimated by calculation on the basis of certain assumptions. If the two components form an "ideal solution," the additive properties of the solution are linear functions of its composition. Thus the partial vapor pressure, or the activity, of each component is, as defined by Raoult's law, proportional to its mole fraction. Using this relation with the familiar Clausius-Clapeyron equation as applied to a liquid mixture in equilibrium with crystals of pure $A$, it is possible to derive the freezing-point law.

Starting from the Clausius-Clapeyron equation or from the second law of thermodynamics it can be shown [6] on the basis of certain assumptions that

$$d \ln p/p_0 = (\Delta H/RT^2) \, dT,$$

where $p$ is the partial pressure of component $A$, $p_0$ is the vapor pressure of pure $A$ at the same temperature, $\Delta H$ is its molar heat of fusion, $R$ is the constant of the ideal gas law, and $T$ is the equilibrium temperature in degrees absolute. The lowering of the melting point or freezing point of component $A$ can be determined by integrating this equation between the limits $T_0$ to $T$, $T_0$ being the freezing point of pure $A$. Thus, if $\Delta H$ is constant, we obtain

$$\ln (p/p_0) = (\Delta H/RT_0) - (\Delta H/RT) = - \Delta H(T_0 - T)/RTT_0.$$

Raoult's law may be written

$$p = N_A p_0,$$

where

$$N_A = \text{mole fraction of } A = \frac{\text{moles of } A}{\text{total number of moles in mixture}}.$$

Then we may write

$$\ln N_A = 2.303 \log N_A = - \Delta H(T_0 - T)/RTT_0$$

or

$$\log N_A = (\Delta H/2.303RT_0) - (\Delta H/2.303RT). \tag{3.1}$$

It follows from (3.1) that, if $\Delta H$ is constant over the temperature range considered, the plot of $\log N_A$ against $1/T$ should give a straight line whose slope is $(-\Delta H/2.303R)$ and whose upper limit is $\log N_A = 0$ at the freezing point of pure $A$. Thus the complete solubility curve in any ideal solvent may be found for a given solute either by calculation or graphically, if its melting point is known together with either its heat of fusion or a single solubility value.

If $\Delta T = T_0 - T$, we obtain an expression for the freezing-point lowering, $\Delta T$,

$$\Delta T = - (RTT_0/\Delta H) \ln N_A = - (2.303RTT_0/\Delta H) \log N_A. \tag{3.2}$$

The derivation of these simplified equations describing the extent of freezing-point lowering caused by impurities in $A$ involves the following assumptions: (a) that the components in question are mutually ideal; (b) that the vapor over the liquid and solid at equilibrium behaves as an ideal gas within the temperature and pressure ranges involved; (c) that the volumes of the liquid and solid phases are negligible in comparison with the corresponding volume of vapor. From (3.2) it is possible to calculate $\Delta T$ with fair accuracy even in concentrated solutions for systems of two components which behave in an ideal manner. In some instances, where the necessary data are available, it is possible to obtain more accurate values by taking into account the slight differences due to the assumptions mentioned [7].

From (3.2) it is apparent that the freezing-point (melting-point) lowering observed for substance $A$ is a direct function of the mole fraction of $A$ in the mixture, that it is inversely proportional to the molar heat of fusion of $A$, and that it is directly proportional to the product of the freezing points of the mixture and of pure $A$ in degrees absolute. The depression is thus independent of the kind of impurity present provided ideal solutions are formed. Thus 1 mole % of any substance or mixtures of substances forming an ideal solution will lower the freezing point of benzene by 0.65°C. Naphthalene, biphenyl, and benzene form mutually ideal solutions of this nature, and the complete binary freezing-point-lowering curves show, as expected, that this relation holds even in concentrated solutions. In extremely dilute solutions all substances are said to exhibit ideal behavior. However, this generalization does not hold without qualification. For instance, in the classical case of electrolytes in ionizing solvents, the deviation from the behavior predicted by the ideal freezing-point equation increases with the dilution.

The magnitude of the molar heat of fusion of component $A$ is an important factor when the melting point is used as a criterion of purity. Substances such as camphor or cyclohexanol, which have very low molar heats of fusion, will exhibit a large lowering of the melting point with only very small mole fractions of impurity. On the other hand, a solvent with a relatively high molar heat of fusion such as stearic acid will show very little change in melting point with even a high concentration of solute. Table 3.1 shows a comparison of the molar heats of fusion for a number of substances and the freezing-point lowering ($\Delta T$) for one-tenth mole fraction of impurity calculated from (3.2) by the method of successive approximations.

The extent of melting-point lowering of a given substance is a function also of its melting point in degrees absolute. A compound with a high melting point will show a greater $\Delta T$ for a given mole fraction of impurity than a solvent with a low melting point, other things being equal. Thus anthracene ($\Delta H = 6890$ cal) and capric acid ($\Delta H = 6680$ cal) (see Table 3.1) have nearly identical molar heats of fusion, but, since their melting points are 490° and 304°K, respectively, anthracene will in the ideal case show a lowering about 2.5 times as great as that of capric acid for 1 mole % of impurity.

In establishing purity or identity, it is therefore evident that the mere comparison of melting points will afford more valid conclusions if the heat of fusion of the substance is also available for the interpretation. A freezing

**Table 3.1** Calorimetric Heats of Fusion

| Substance | Mp (°C) | Mol wt | cal/g | $\Delta H$ (cal/mole) | $\Delta T$ for 0.1 Mole Fraction of Impurity (calc) |
|---|---|---|---|---|---|
| Acetic acid | 16.55 [a] | 60.05 | 46.7 [c] | 2800 | 6.14 |
| Aniline | −6.15 [a] | 93.12 | 27.1 [e] | 2520 | 5.79 |
| Anthracene | 216.5 [e] | 178.22 | 38.7 [f] | 6890 | 7.18 |
| Benzene | 5.45 [a] | 78.11 | 30.40 [b] | 2370 | 6.69 |
| Benzoic acid | 122.45 [a] | 122.12 | 33.9 [g] | 4140 | 7.76 |
| Benzophenone | 47.85 [a] | 182.21 | 23.5 [g] | 4280 | 4.96 |
| Biphenyl | 68.6 [f] | 154.20 | 28.8 [f] | 4440 | 5.42 |
| Butyl alcohol (tert) | 25.4 [b] | 74.12 | 21.43 [b] | 1590 | 11.29 |
| Camphor | 178.4 [b] | 152.23 | 10.74 [b] | 1650 | 24.27 |
| Capric acid (n) | 31.3 [a] | 172.26 | 38.8 [g] | 6680 | 2.88 |
| Cetyl alcohol | 49.10 [a] | 242.44 | 33.8 [g] | 8190 | 2.63 |
| Cyclohexanol | 25.46 [a] | 100.16 | 4.27 [d] | 428 | 38.03 |
| Dibenzyl | 51.2 [a] | 182.25 | 30.7 [f] | 5600 | 3.88 |
| Dichlorobenzene (p) | 53.2 [a] | 147.10 | 29.7 [g] | 4370 | 5.02 |
| Ethylene dibromide | 9.97 [a] | 187.88 | 13.5 [g] | 2540 | 6.45 |
| Methyl cinnamate | 34.5 [g] | 162.18 | 26.5 [g] | 4300 | 4.54 |
| Naphthalene | 80.22 [a] | 128.16 | 35.8 [e] | 4590 | 5.60 |
| Nitrobenzene | 5.65 [a] | 123.11 | 23.53 [e] | 2900 | 5.50 |
| Stearic acid | 68.8 [g] | 284.47 | 47.5 [g] | 13510 | 1.80 |
| Trichloroacetic acid | 59.1 [g] | 163.4 | 8.6 [g] | 1405 | 15.67 |
| Triphenylmethane | 92.1 [f] | 244.32 | 21.1 [f] | 5160 | 5.33 |
| Water | 0.0 | 18.00 | 79.67 [f] | 1434 | 10.45 |

[a] Landolt-Börnstein, *Physikalisch-Chemische Tabellen*, Erg. IIIa, Table 84.
[b] *Ibid.*, Erg. IIIc, Table 305.
[c] *Ibid.*, Erg. IIb, Table 305.
[d] *Ibid.*, Erg. I, Table 305.
[e] *Ibid.*, Erg. IIIc, Table 313.
[f] G. S. Parks and H. M. Huffman, *Ind. Eng. Chem., Ind. Ed.* **23,** 1138 (1931).
[g] *International Critical Tables*, Vol. 5, McGraw-Hill, New York, 1929, pp. 132–134.

point which is "within a few degrees of the true value" does not in all cases indicate the same degree of purity. Consider, for instance, samples of camphor and trichloroacetic acid, each melting 2.0°C below its true melting point. Calculations from (3.2) and Table 3.1 indicate that this lowering corresponds to only 0.8 mole of impurity in the camphor but to 13 mole % in the trichloroacetic acid.

In this connection, it is interesting to note that the use of melting points has been suggested as a method of analyzing binary fatty acid mxtures. The accuracy required by this application is shown by the fact that 1.0 g of palmitic acid in the presence of 9.0 g of stearic acid would lower the melting point of the stearic acid from 69.4 to 67.1°C, a difference of only 2.3° [8]. The depression is small because stearic acid has a relatively low melting point and a high molar heat of fusion. These factors conspire to cause a large amount of impurity to have relatively little effect on the melting point of the stearic acid.

It should be pointed out that even dissolved air, like any other impurity, lowers the freezing point. For practical purposes this effect is negligible since the solubility of air is slight. Richards and co-workers [9], for example, showed that dissolved air in benzene depressed its freezing point by 0.031°C. In actual practice this effect was offset, however, by the effect of atmospheric pressure so that the observed difference between the freezing point in air and and the triple point was only 0.003°C.

### Systems that Deviate from Ideality

In contrast to the binary systems considered above, most of the mixtures encountered by the organic chemist are not ideal in that they do not follow Raoult's law. The lowering of the freezing point is a colligative property calculable from the freezing-point equation for ideal cases; it is possible in some instances, therefore, to explain the observed deviations from ideality in terms of known characteristics of the substances involved. A few of the more common causes for such deviations are mentioned in the paragraphs below. The term "solvent" is used here with reference to the substance which first freezes out of solution.

1. If the solute is known to associate to form dimers or trimers, the number of effective molecules in the system will be reduced accordingly, and the observed lowering of the freezing point will be less than that predicted for the ideal case. Acetic acid is such a substance, and its lowering of the melting point of naphthalene, indicated graphically in Fig. 3.3, is less than that of benzene or biphenyl, both of which show essentially ideal behavior in naphthalene.

2. Another very common form of deviation occurs when the solvent and solute, $A$ and $B$, combine partially in the liquid phase to form a third substance, $AB$. In such cases, the deviation from ideality is often said to be due to solvation, or molecular compound formation. The formation of $AB$ does not increase the number of solute molecules in the solvent but actually decreases the number of molecules of free solvent remaining. The mole fraction of solvent is, therefore, smaller, and the mole fraction of solute

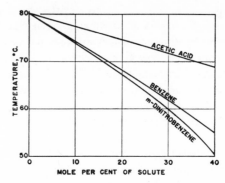

**Fig. 3.3**  Lowering of the freezing point of naphthalene (data from Ward [10] and Skau [11]).

greater than expected, so that the depression of the freezing point is more than that predicted by the freezing-point equation. *m*-Dinitrobenzene forms such a molecular compound with naphthalene [11]; the resulting melting-point curve of mixtures of these two substances is shown in Fig. 3.3. Picric acid in naphthalene would be expected to show a similar deviation, since the compound naphthalene picrate is readily formed, having a melting point of 149°C [12].

3. In some cases, the solute molecules dissociate to form two or more molecules, thus increasing the effective mole fraction of the solute and giving rise to freezing-point depressions greater than those calculated for the ideal case. Hexaphenylethane dissolved in naphthalene dissociates partially to form the triphenylmethyl radical, and the observed melting points for the mixture are lower than those for benzene in naphthalene [13]. This behavior for the analogous case of tetraphenyl-di-$\beta$-naphthylethane in benzene will be discussed in connection with Fig. 3.20. Similarly, naphthalene picrate has been shown to dissociate into naphthalene and picric acid when dissolved in alcohol, acetic acid, or benzene [14]. This phenomenon corresponds to the well-known abnormal depression of the freezing point of water by the addition of electrolytes.

The above deviations represent typical cases which may be explained in a simple manner. Often, however, an explanation is not possible because the system is too complicated or because the behavior of the system is not sufficiently known. In determining the molecular weight of a substance by melting-point or freezing-point depression, the experimenter must always keep in mind possible causes of deviation and their effect on the results obtained. A solute which dissociates does not always cause an abnormally high freezing-point depression. In naphthalene, for example, naphthalene picrate dissociates to some extent to form naphthalene and picric acid [15].

The solvent, naphthalene, will therefore be increased in mole fraction by an amount depending upon the extent of dissociation, so that the observed depression of the melting point of naphthalene will not be greater, but less, than that predicted from ideality.

Other possible causes for deviations from the freezing-point solubility law have been the subject of investigations by Hildebrand [7] and others [10].

## Cooling and Heating Curves

Extending the original definition of the freezing or melting point of a pure compound to include impure compounds or mixtures, one may define the freezing point, also called the primary freezing point, as the temperature at which a liquid of given composition is in equilibrium with the type of crystals which form on initial crystallization. When heat is added to such an equilibrium mixture, more crystals will melt (or dissolve in the liquid phase); and when heat is removed, more crystals will form. The equilibrium temperature, that is, the freezing point, will depend only upon the composition of the liquid phase at any moment, and not on the relative proportions of liquid and solid. In general, whenever the crystals have the same composition as the liquid phase, no change in the liquid composition takes place during freezing, and, consequently, the equilibrium temperature remains constant; the loss of heat to the surroundings will be exactly counterbalanced by the heat of fusion of the crystals formed. If, on the other hand, the composition of the crystals differs from that of the liquid, the freezing temperature will change as more and more of the solid forms. This change is the fundamental principle behind cooling-curve and heating-curve behavior, a knowledge of which is essential to a full understanding of melting- and freezing-point phenomena, abnormalities in behavior, and their significance.

In the phase diagram for mixtures of phthalic anhydride and naphthalene shown in Fig. 3.4, the temperature of the freezing point is plotted against the mole % of naphthalene in the mix. The curve *AC* therefore indicates the freezing point observed for various concentrations of naphthalene. Point *A* represents the melting point of pure phthalic anhydride, namely 130.8°C. When pure liquid anhydride is cooled to this temperature, solid anhydride forms and, according to the phase rule, the temperature must remain constant as long as the two phases exist. Adding or subtracting heat changes only the amounts of solid and liquid present but does not change the equilibrium temperature.

Consider the behavior of a mixture of phthalic anhydride and naphthalene having the composition represented by point *B*. If liquid of this composition is cooled, the initial, or primary, freezing point will be 124°C, at which temperature crystals of pure phthalic anhydride will begin to form. This removal of solvent from the solution will result in an increase in the con-

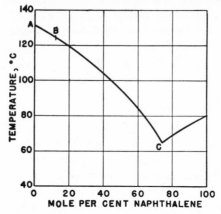

**Fig. 3.4**  Freezing points of mixtures of naphthalene and phthalic anhydride [16].

centration of the naphthalene and a further lowering of the freezing point as predicted by the freezing-point equation. Thus, as the concentration of naphthalene increases, the equilibrium temperature gets lower and lower (that is, $\Delta T$ of (3.2) becomes larger and larger) following curve *BC*. According to the phase rule, the presence of the additional component (naphthalene) permits an additional degree of freedom; hence the temperature changes.

Cooling (or heating) curves are obtained by permitting the system under investigation to lose heat to (or gain heat from) the surroundings at a uniform rate. The temperature of the sample is then measured at convenient time intervals during the phase change of the system. The temperature is plotted against the time to obtain either the heating or cooling curve depending upon whether heat was added to or withdrawn from the sample.

Schematic cooling and heating curves for the system benzoic acid-cinnamic acid are shown in Fig. 3.5 along with the binary diagram for that system. (The actual experimental curves are not rectilinear, but show gradual rather than sharp changes in slope.) The cooling curve for pure benzoic acid (curve *A*) would exhibit a flat region or "halt," *ab*, at 121.5°C, the freezing point, after which the temperature would drop quite rapidly. Heating the pure solid benzoic acid (curve *D*) would cause the temperature to rise at an even rate until 121.5°C, the melting point, is reached. At this point, the halt *ab* would again appear, while the temperature would remain constant, until all of the benzoic acid has been changed into the liquid state. Actually, the degree of flatness of region *ab* in the heating or cooling curve would be limited by the ability to obtain good equilibrium between the solid and liquid states, a condition which is particularly difficult to attain when only a small fraction of the material is liquid. The heating and cooling curves for pure

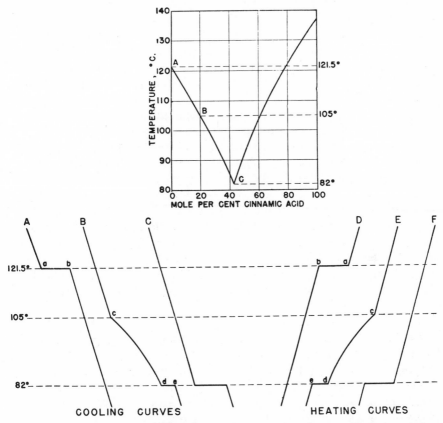

**Fig. 3.5** Freezing points and heating and cooling curves for mixtures of benzoic and cinnamic acids [16].

cinnamic acid would be similar to those obtained with pure benzoic acid with the exception that the halt would occur at 136.8°, the melting point of the pure cinnamic acid.

Cooling a liquid solution of benzoic acid containing 20 mole % of cinnamic acid, one would obtain schematically a curve similar to curve *B* in Fig. 3.5. In this case, the temperature would drop at a uniform rate to 105°C. At this point, *c*, the slope of the cooling curve would change, as shown, since here pure benzoic acid would start to crystallize from the solution. As mentioned in the case of naphthalene in phthalic anhydride, the temperature would continue to change because the concentration of solute in the solution would increase with the separation of solid benzoic acid. Thus the cooling curve would indicate that the temperature is still dropping, but at a different rate, as shown by the slope of the section *cd*. Finally, at 82° both the cinnamic acid and benzoic acid would crystallize from the solution, each in its own crystalline form, and in a ratio equal to that of the solution composition. The composition of the liquid would, therefore, not change during this part of the freezing process, and the temperature would remain constant. The cooling curve would exhibit a brief halt, *de*, at this temperature until all the liquid is converted into solid. This temperature is known as the eutectic temperature of the mixture. The eutectic composition for the system shown in Fig. 3.5 is 57 mole % benzoic acid and 43 mole % cinnamic acid.

On heating a solid with the over-all composition of 20 mole % cinnamic acid in benzoic acid, the curve obtained would be the reverse of the cooling curve (curve *E*, Fig. 3.5). The first liquid formed at the eutectic temperature of 82°C would have the eutectic composition indicated above. The heating curve would exhibit a brief halt, *ed*, during which interval the eutectic liquid would continue to form until all the cinnamic acid is in the liquid solution and only pure benzoic acid remains in solid form. The temperature would then gradually rise, following curve *dc* as the benzoic acid dissolves in the liquid mixture until the sample is all liquid. The temperature would then have reached 105°, at which point the heating curve would exhibit an increase in slope as shown.

If the over-all composition of the benzoic acid-cinnamic acid mixture were the same as the eutectic composition, the cooling and heating curves would be like those shown in curves *C* and *F* of Fig. 3.5. The halt obtained would be similar in shape to those observed for the pure substances, but would occur at the eutectic temperature of 82°C. The composition of the liquid would not change during the crystallization process and the temperature, therefore, would remain constant until no more liquid remains.

The behavior described above is typical for any system, ideal or not, which exhibits a simple eutectic point.

### Compound Formation with Congruent Melting Point

Many binary mixtures are more complicated than those discussed above, in that components *A* and *B* are in equilibrium with one or more loosely combined compounds (e.g., $AB$, $AB_2$, $A_2B$) which usually dissociate to a greater or lesser extent in the liquid state. The compound formed may be one of two types according to whether it has a congruent or an incongruent melting point. A compound with a congruent melting point behaves as a pure substance since, at the melting point, both the liquid and the solid phase have exactly the same composition as the compound. If the compound has an incongruent melting point, it decomposes upon melting to form another solid and a liquid, each having a composition different from that of the original compound.

Phenol and α-naphthylamine combine to form a molecular compound which has a congruent melting point, that is, one in which the liquid form has the same composition as the solid (Fig. 3.6). This diagram can be divided by a vertical line at 50% composition to give two ordinary eutectic systems of the type discussed above. In one case, we would have α-naphthylamine and the α-naphthylamine-phenol compound, and, in the second system, phenol and the compound. The system *syn*-trinitrobenzene and β-naphthylamine is one in which the 1–1 compound formed has a congruent melting point 40°C higher than either of the pure components (Fig. 3.7).

### Compound Formation with Incongruent Melting Point

In the case of incongruently melting compounds, the melting point of the compound represents a condition of unstable equilibrium. The solid compound, upon melting, decomposes reversibly into a new solid phase and a liquid. This liquid is saturated both with respect to the compound and the new solid.

Such a system is illustrated by acetic acid and dimethylpyrone, which form a 1–1 compound with an incongruent melting point (Fig. 3.8). Upon cooling a system of composition $B$, pure dimethylpyrone would be deposited between 94 and 25°C; and, at 25°, a liquid of composition $D$ would dissolve the dimethylpyrone, producing crystals of the compound represented by a 1–1 mixture, the composition of the liquid phase remaining constant meanwhile until all of the liquid is exhausted. From the over-all composition of the system, it is evident that excess dimethylpyrone would remain. The same three phases would be in equilibrium at 25°C in a system having the original composition $C$; but, since more acetic acid would be present than is required to form the compound, some of the liquid phase would remain after all the

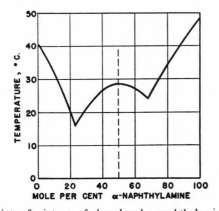

**Fig. 3.6**  Freezing points of mixtures of phenol and α-naphthylamine [16].

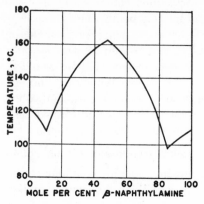

**Fig. 3.7**   Freezing points of mixtures of trinitrobenzene and $\beta$-naphthylamine [16].

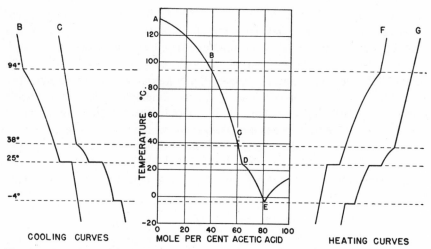

**Fig. 3.8**   Freezing points and heating and cooling curves of mixtures of acetic acid and dimethylpyrone [16].

solid dimethylpyrone has dissolved. Crystals of the 1–1 compound would then separate until the eutectic composition $E$ is reached, when both the compound and acetic acid would crystallize out together as in a simple eutectic system. The heating curves are the reverse of the cooling curves just described, as can be seen from a study of curves $F$ and $G$ in Fig. 3.8.

### Solid Solutions

The above systems have been mixtures from which a crystalline substance in definite proportions separates on freezing. Frequently this is not the case, and the organic chemist may often prepare a homogeneous crystal phase containing both components but in indefinite proportions. These solid solutions, or mixed crystals as they are sometimes called, may form either in all proportions or with limited solubility. Examples of these two types are illustrated by Figs. 3.9 and 3.10, in which the upper curves represent the liquid composition and the lower, broken curves the solid composition in equilibrium with the liquid at the same temperature. Thus points *a* and *b* correspond to liquid and solid compositions occurring at equilibrium conditions.

There are three types of systems exhibiting solid solutions in all proportions. Mixtures of naphthalene and β-naphthol (Fig. 3.9, I) illustrate the type in which the freezing point of all mixtures are between the freezing points of the pure components. The second type (Fig. 3.9, II) is one in which the freezing-point curve passes through a maximum. Very few systems of this type are known, the system *d*-carvoxime-*l*-carvoxime being probably

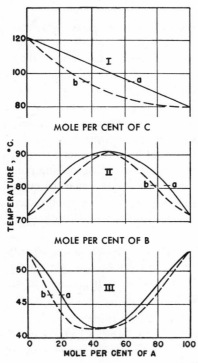

**Fig. 3.9**  Freezing points of solid solutions [17]. Curve I: component *A*, naphthalene; *B*, β-naphthol. Curve II: *A*, *d*-carvoxime; *B*, *l*-carvoxime. Curve III: *A*, *p*-chloroiodobenzene; *B*, *p*-dichlorobenzene.

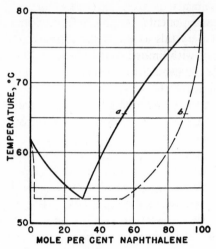

**Fig. 3.10**    Freezing points of mixtures of naphthalene and monochloroacetic acid [18].

the best example. In the third type (Fig. 3.9, III) the freezing point passes through a minimum, as illustrated by the system p-chloroiodobenzene-p-dichlorobenzene. It should be noted that in the first two types the freezing and melting points exhibit a rise due to the presence of the second component, contrary to expectations based upon the freezing-point equation.

Monochloroacetic acid and naphthalene form a system (Fig. 3.10) in which a discontinuous series of solid solutions are formed upon cooling. The first liquid formed upon melting any mixture containing between 2 and 53 mole % of naphthalene would have the eutectic composition of 30 mole % naphthalene. Outside this range, the initial melting point as determined by a refined method of measurement would not be the eutectic temperature, but some higher temperature as indicated by the broken curve, and the first liquid would have a composition indicated by the corresponding point on the liquidus curve.

Mixtures also exist which form a series of solid solutions with the same melting point. In such a case, since the cooling curves for all compositions would be identical, it is impossible to distinguish between the two pure constituents and their mixtures by means of the melting point. An example of this type of system is that of d-camphor-l-camphor, mixtures of which have melting points as shown in Table 3.2. Systems of this type are also found with the d- and l-forms of camphoroxime, borneol, camphoric anhydride, and camphene [19].

It is evident from this discussion of cooling and heating curves that, with some exceptions, the observation of sharp melting points is limited to pure

Table 3.2 Melting Points of Mixtures of *d*- and *l*-Forms of Camphor [19]

| *d-Camphor* (%) | *Mp* (°C) |
|---|---|
| 100 | 178.6 |
| 86.2 | 178.8 |
| 81.0 | 178.6 |
| 70.8 | 179.1 |
| 57.9 | 178.7 |
| 48.7 | 178.6 |
| 30.1 | 178.3 |
| 19.1 | 177.8 |
| 11.3 | 178.5 |
| 0.0 | 177.7 |

substances or eutectic mixtures. In actual practice, however, the melting points for substances which are nearly pure appear sharp because, by the usual methods, the observer is unable to detect the initial formation of the small amount of eutectic liquid.

Consider, for example, a mixture of 1% naphthalene and 99% phthalic anhydride (see Fig. 3.4). By means of the ordinary melting-point determination in a capillary tube, the mixture would be found to melt very close to the melting point of pure phthalic anhydride (130.8°C). An accurate heating curve would show a slight halt during the melting of the eutectic mixture at the eutectic temperature (64.9°C), at which all the naphthalene would liquefy. The remaining solid phthalic anhydride would then dissolve in the mixture (if true equilibrium were preserved) until the entire system has liquefied at a temperature just below the melting point of pure phthalic anhydride. Thus the sample would actually melt over the range of 64.9 to almost 130.8°, even though the melting would not be perceptible by ordinary techniques until the higher temperature is almost attained.

Figure 3.11 shows the observed initial and final melting temperatures observed in capillaries for mixtures of naphthalene and *p*-nitrophenol with different compositions. The initial melting temperatures were obtained by a special technique for observing the "softening point" [21]. The melting ranges are indicated by the dotted lines. It will be noted that, for mixtures near the middle of the diagram, the initial melting temperature is the same as that of the eutectic mixture; but, whenever nearly pure naphthalene or nearly pure *p*-nitrophenol is melted, the apparent melting range as observed in a capillary tube is small, although the range actually starts at the eutectic temperature. The very nature of the phenomenon is such that the appearance of the liquid is gradual and the choice of the lower point of the range is highly subjective, especially for wide melting ranges.

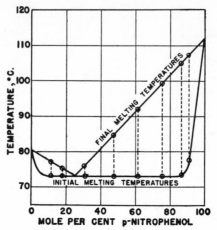

**Fig. 3.11**    Observed melting ranges for mixtures of naphthalene and *p*-nitrophenol [20].

Strictly speaking, no really pure substance has ever been prepared; if the melting point could be determined with sufficient accuracy, any substance would exhibit a certain amount of premelting [22]. In connection with the determination of heats of fusion, the purest water prepared by Dickinson and Osborne of the National Bureau of Standards exhibited premelting to the extent of 0.0005°C, probably because of the presence of slight impurities [23]. It is thus evident that sharp melting points are reported for substances of ordinary purity only because the observer with the usual apparatus is unable to see the small amount of liquid formed during the premelting stage.

### Steric Relationships from Freezing-Point Curves

Optical antipodes often crystallize together to form a racemic compound; that is, they give a binary freezing-point diagram which shows 1–1 compound formation similar to those illustrated in Figs. 3.6 and 3.7 except in that they are, of course, perfectly symmetrical. Thus combinations of the compounds $A+$ and $A-$, $B+$ and $B-$, and $C+$ and $C-$ form the racemic compounds $A + A -$, $B + B -$, and $C + C -$, respectively. It was found by Centnerzwer [24] that a "quasiracemic" [25] compound of the form $A + B -$ was formed between $(+)$ chlorosuccinic acid and $(-)$ bromosuccinic acid. This fact shows, presumably, that $B-$ has the same configuration as $A-$ since it is to be expected that only similar compounds of opposite configuration form the quasiracemic compound. Timmermans [26] and Fredga [27] have developed the use of this principle to identify relative steric configurations. Thus, it is desired to establish the relative configurations of the optical antipodes of three substances, $A$, $B$, and $C$, this may be accomplished by obtaining binary freezing-point diagrams of various combinations of the antipodes and classifying the systems on the basis of whether or not they form a quasiracemic compound. Suppose, for example, the following results were obtained:

| Binary Freezing-Point Diagrams Showing Simple Eutectics | Binary Freezing-Point Diagrams Showing Compound Formation |
|---|---|
| A+ and B+ | A+B− |
| A− and B— | A−B+ |
| A+ and C− | A+C+ |
| A− and C+ | A−C− |
| B+ and C− | B+C+ |
| B− and C+ | B−C− |

Then $A+$, $B+$, and $C-$ can be said to have the same steric configuration on the one hand, and $A-$, $B-$, and $C+$ on the other. This is true, however, only when the compounds $A$, $B$, and $C$ are very similar. Often it is necessary to use derivatives of the original antipodes to obtain conclusive results.

## Melting Temperatures and Molecular Constitution

### Homologous Series

The phenomenon of melting a crystalline solid involves the transition of the molecules from an ordered, crystalline lattice in which each atom or molecule occupies a definite mean position to a liquid state of more or less complete disorder. This transition occurs when the thermal vibrations of the atoms overcome the intermolecular forces holding the molecules in the solid lattice. The intensity of thermal vibration, or temperature, required to melt the solid depends, therefore, on those factors involved in crystal structure and interatomic and intermolecular attraction in the solid. Thus, although certain general relations may be observed between the melting point and molecular constitution of organic compounds, the many factors involved make the relationships necessarily complex and obscure.

In a homologous series, the melting points are related to the molecular weight. It has been suggested [28, 29] that the melting points in a homologous series tend to approach a common limit, or "convergence temperature," of of about 117°C as the molecular weight increases. Generally the first members of a series melt below 117°C, and, when the chain length is increased, by addition of —CH$_2$— groups, the melting points of the compounds increase, rapidly at first and then gradually to the common limit; for example, the paraffins, fatty acids, nitriles, ketones, and the odd-numbered members of the oxalic acid series. Sometimes the melting points of the first members of a series are higher than 117°C, and, on increasing the chain length, the melting points decrease to the common limit; for example, the even-number members of the oxalic acid series. If the melting points of the first members of a series are near the common limit, then there is practically no change in the melting point with chain length; for example, the amides of fatty acids and the monoalkyl derivatives of urea and thiourea. There are numerous exceptions to these generalities: substances containing an acidic and a basic radical, substances containing two amide groups or analogous groups, and other isolated series [29].

For the higher members of a homologous series, Austin [30] has suggested the empirical formula $\log M = A + BT_m$, where $M$ is the molecular weight, $T_m$ is the melting point in degrees absolute, and $A$ and $B$ are constants characteristic of the true homologous series, normal and secondary paraffins having different constants. A modified relationship, $\log M = A_0 + B_{t_m}$, in which $t_m$ is the melting point in degrees centigrade, and $A_0$ is the value of $\log M$ at 0°C, may also be written. Figure 3.12 illustrates the conformity of a number of homologs to the modified formula; and Table 3.3 gives the $A_0$ and $B$ constants for a number of series which follow the formula.

For lower members of a series, the correlation between observed melting temperature and that predicted by this equation is far from satisfactory. In

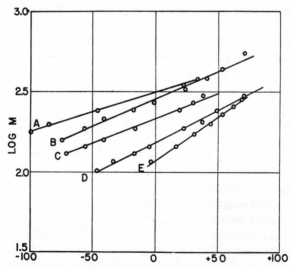

**Fig. 3.12**  Relation between melting points and log $M$ for homologous series: $A$, $n$-alkyl iodides; $B$, acid anhydrides; $C$, methyl esters of $n$-fatty acids; $D$, $n$-aliphatic alcohols; $E$, $n$-fatty acids (even). The abscissa is temperature ($t_m$) in degrees centigrade.

**Table 3.3** $A_0$ and $B$ Constants in the Equation Relating Melting Point and Molecular Weight for Homologous Series

| Series Description | $A_0$ | $B$ | Minimum Number of C in the Chain for Which Formula Applies |
|---|---|---|---|
| Normal alkanes | 2.337 | 0.0038 | 15 |
| Normal alkenes-1 | 2.333 | 0.0040 | 10 |
| Normal alkynes-1 | 2.290 | 0.0036 | 7 |
| Normal alkadiynes-1 $(n - 1)$ | 2.235 | 0.0036 | 11 |
| 1-phenyl-$n$-alkanes | 2.410 | 0.0031 | 3 |
| 1-iodo-$n$-alkanes | 2.495 | 0.0024 | 4 |
| Normal alkanols-1 | 2.185 | 0.0040 | 6 |
| Normal alkanals | 2.245 | 0.0039 | 4 |
| Normal alkanones-2 | 2.185 | 0.0039 | 7 |
| Normal alkanoic acids | | | |
| Odd | 2.143 | 0.0047 | 5 |
| Even | 2.070 | 0.0053 | 6 |
| Methyl esters of normal alkanoic acids | 2.332 | 0.0031 | 8 |
| Ethyl esters of normal alkanoic acids | 2.382 | 0.0031 | 5 |
| Normal alkanoic acid anhydrides | 2.456 | 0.0035 | 4 |
| Normal alkanoic acid chlorides | 2.400 | 0.0034 | 6 |
| Normal alkane nitriles | 2.258 | 0.0035 | 5 |

fact, in the case of certain series, such as the fatty acids, normal paraffins, glycols, and alkyl malonic acids, the phenomenon of alternation occurs—with each additional $CH_2$ alternately larger and smaller increases of the melting point are observed, depending on whether the number of carbons in the chain is even or odd. Nevertheless, the Austin equation may frequently be used with considerable accuracy to predict the melting points of new or unknown homologs of high molecular weight.

Compounds with normal chains, in general, have higher melting points than their isomers. The branched molecules of high symmetry, on the other hand, usually have exceptionally high melting temperatures; for example, $(CH_3)_4C$, mp $-16°C$, and $(CH_3)_3CC(CH_3)_3$, mp $101°C$. Timmermans [31] has suggested that these molecules can absorb a considerable amount of rotational energy before their thermal vibrations cause a disruption of the crystal lattice.

## Effect of Substituents

Certain generalities have also been noted in the relations between the melting points of kindred compounds. For compounds, which are isomeric or of similar structure with the exception of a single grouping, the following general melting-point relations hold:

| Compound with Higher Melting Point | Corresponding Compound with Lower Melting Point | Compound with Higher Melting Point | Corresponding Compound with Lower Melting Point |
|---|---|---|---|
| Trans | Cis | Ketone | Hydrocarbon |
| Para | Ortho and meta | Alcohol | Hydrocarbon |
| Iodide | Bromide | Alcohol | Ether |
| Bromide | Chloride | Alcohol | Mercaptan |
| Chloride | Fluoride | Amide | Amine |
| Alkyl chloride | Acid chloride | Diamide | Diamine |
| Fatty acid | Acid chloride | Amine | Hydrocarbon |
| Fatty acid | Nitriles | Dialkylamine | Trialkylamine |
| Fatty acid | Alcohol | Azo | Azoxy |

It has long been known that a para-disubstituted derivative of benzene has a higher melting point than the corresponding ortho or meta isomers. Recently a more comprehensive relationship has been suggested [32] for the melting points of such position isomers as summarized in the following rules which hold in a large number of cases but to which numerous exceptions are cited.

*Rule 1.*  When the disubstituted benzene derivative contains *one* meta orienting group and *one* ortho-para orienting group, the order of the melting points for the isomers is: ortho < meta < para.

*Rule 2.*  When the disubstituted benzene derivative contains *only* ortho-para orienting or *only* meta orienting groups, the order of the melting points for the isomers is: meta < ortho < para.

## Globular Compounds

Timmermans [33] was the first to note that compounds that have a "globular" molecular structure and also an unusually low entropy of fusion (below about 5 cal/(deg)(mole)) form "plastic crystals." This is a phase of matter quite distinct from ordinary crystals. These compounds have abnormally high melting points. They are called globular, because the molecules either are symmetrical around their centers ($CH_4$, $CCl_4$, pentaerithritol, etc) or would give a sphere by rotation around an axis (cyclohexane, camphor, etc).

Methylcyclohexane has a relatively high entropy of fusion, 11.0 cal/(deg) (mole), and is nonglobular, presumably because the $CH_3$ prevents rotation. Its melting point is $-129.7°C$, as compared with 6.5C for that of cyclohexane. Timmermans has suggested that the molecules in plastic crystals can absorb a considerable amount of rotational energy before their thermal vibrations cause a disruption of the crystal.

Plastic crystals have a number of unusual properties [34–36]. They are tacky and easily deformed. They will flow through a small hole under relatively low pressure. Their triple-point pressure is high, so that some, for example, $C_2Cl_6$, sublime below the melting point. They have one or more transition points at lower temperatures. Their specific heats and usually their dielectric constants rise on lowering the temperature, instead of falling as usual. They have high molal freezing-point depression constants. This property is the basis for Rast's capillary method for determining molecular weight using camphor as a solvent.

Timmermans [34] lists over 100 compounds which form plastic crystals, representing aliphatic derivatives, heterocyclics, cyclics, condensed cyclics, bridged cyclics, and camphors. The list also includes inorganic compounds, for example, HCl, $H_2S$, $H_3P$, $H_4Si$, and argon. HF, $H_2O$, $H_3N$, and $Cl_4Si$, on the other hand, do not form plastic crystals.

## Correlation and Prediction of Solubility Data

The solubility curve for a solute in a given solvent represents the solid-liquid equilibrium temperatures over a wide range of concentrations. Solubility curves can therefore be constructed from primary freezing-point data for the binary system.

A number of graphical methods have been reported for correlating solubility data for homologous and analogous compounds. Each is based on an equation or relationship developed from thermodynamic considerations. These methods afford a basis for predicting new solubility data from existing data. They fall into two categories depending upon whether they involve (a) correlation with the number of carbon atoms, $n$, in the members of a homologous series or (b) correlation with the solubility data of a reference compound. In all of these procedures the concentration of the solute is expressed as mole fraction, $N$, and temperature is expressed in degrees Centigrade, $t$, or in degrees Kelvin, $T$.

### Correlation of Homologous Series

#### ISOTHERM METHOD

The isotherm method [37] is based on a correlation of the solubilities of a homologous series of compounds in a given solvent *at a given temperature* with the number of carbon atoms in the molecule. The values of log $N$ for each temperature are plotted against $n$. A comprehensive survey [38] of the extensive literature data for long-chain homologous series indicates that the isotherms are usually straight lines or smooth curves which can be used for prediction. For example, the published solubility data for the homologous symmetrical aliphatic secondary amines included only the members containing 16, 24, 26, 28, 30, and 36 carbon atoms. From the smooth log $N$ versus $n$ isotherm plots for each solvent (for example, see Fig. 3.13) it was possible by graphical interpolation to read off the solubilities of the missing members of the series, containing 18, 20, 22, 32, and 34 carbon atoms, at each temperature. For homologous series which exhibit alternation of melting points, separate families of isotherms are obtained for the even- and odd-carbon members of the series.

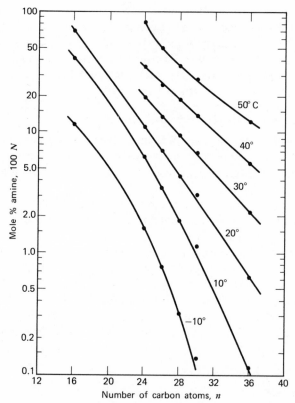

**Fig. 3.13**   Solubility isotherms for the symmetrical normal secondary amines in chloroform [37].

ISOPLETH METHOD

The isopleth method [39] is based on the linear or almost linear relation between the reciprocal of the number of carbon atoms in the members of a homologous series and the reciprocal of the temperatures in degrees Kelvin at which their saturated solutions in a given solvent have the *same concentration*. Isopleth plots based on the solubilities of a few members of a series can be used to predict the solubilities of the intermediate members. The isopleth plots for the symmetrical normal secondary amines in *n*-hexane, based on the solubilities of the members containing 16, 24, 28, and 36 carbon atoms, are shown in Fig. 3.14. From these isopleths it is possible to predict the values of $1/T$ corresponding to the various solubilities of the missing homologs containing 18, 20, 22, 26, 30, 32, 34, and 38 carbon atoms. Here again the odd- and even-carbon homologs must be considered separately, unless the series exhibits no alternation of melting point.

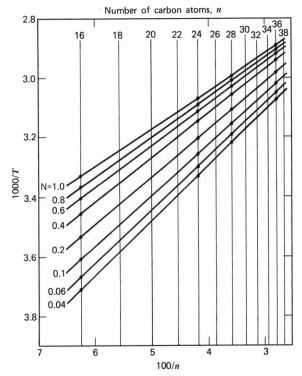

**Fig. 3.14** Solubility isopleths for the symmetrical normal secondary amines in *n*-hexane [39].

The isopleth plots are much less likely to show marked curvature than the corresponding isotherms. In general, the isopleth method has a wider range of application. In fact, it can often be used when the isotherm method is inapplicable.

## Correlation with a Reference Compound

### ANALOGOUS COMPOUNDS IN THE SAME SOLVENT

A number of graphical methods of prediction have been reported based on a linear correlation with the solubility data of a reference compound for which solubility data in the same solvent are available. Complete solubility data for a compound can be read from the correlation plot constructed from the melting point and one or two solubility determinations.

Harris [40] was apparently the first to use a reference compound for correlation. He deduced from Dühring's relation that there should be a linear relationship between the aqueous solubilities of salts of the "same chemical nature." He showed that straight lines are obtained by plotting the temperatures, $t$, at which $AgNO_3$, $RbNO_3$, $NH_4Cl$, $CsNO_3$, $Sr(NO_3)_2$, and $Pb(NO_3)_2$ have certain solubilities, $N$, against the temperatures, $t_{ref}$, at which the reference substance, $KNO_3$, has the same solubilities, see Fig. 3.15.

Othmer and Thakar [41] obtained linear correlation by plotting solubility "against the vapor pressure of the solvent" using log-log paper. The vapor pressure is plotted on the $x$-axis and calibrated with the corresponding temperatures. The values for the solubilities are then plotted on the temperature lines against the corresponding position on the $y$-axis. Straight lines were obtained for the solubilities of urea, sucrose, and a number of inorganic salts in water from 0 to 100°C (Fig. 3.16). The method has apparently not been tested over wider ranges of concentration or in other solvents.

Johnson et al. [42] derived an improved linear correlation from the equation of Othmer and Thakar [41]. Straight lines are obtained by plotting solubilities of a compound at various temperatures on log-log paper against the corresponding values for an analogous reference compound at the same temperatures. The data for the odd- and even-carbon fatty acids from capric to stearic acids in benzene, cyclohexane, and butanol were shown to give linear plots when palmitic acid was used as the reference compound. The method is valid for the solubilities in water of: (a) sucrose, lactose, and glucose with maltose as the reference compound; and (b) the sulfates of ammonium, cobalt, copper, and potassium with aluminum sulfate as the reference compound (Fig. 3.17).

The *isopleth reference method* [43] is based on a correlation derived from the freezing-point-depression equation. A linear or almost linear isopleth reference plot is obtained by plotting $1/T_a$ versus $1/T_{ref}$, that is, the reciprocals of the temperature in °K at which a compound, $a$, has the same mole-

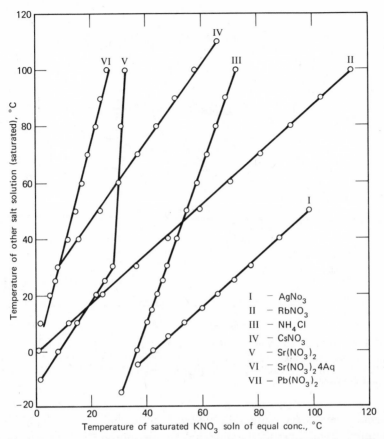

**Fig. 3.15** Saturation temperatures in water of various salts vs. KNO₃. I, AgNO₃; II, RbNO₃; III, NH₄Cl; IV, CsNO₃; V, Sr(NO₃)₂; VI, Sr(NO₃)₂.4 H₂O; VII, Pb(NO₃)₂[40].

percent solubility in a given solvent as an analogous reference compound in the same solvent.

The validity of this correlation was established by applying it to much of the extensive literature data for long-chain compounds. Most of the reference isopleths for the systems investigated were straight or slightly curved lines as illustrated in Fig. 3.18. Their linearity or degree of curvature can be established by two solubility determinations and the melting point of the compound in question. The complete solubility curve can then be obtained by interpolation. The following fatty acids were found to behave as analogous compounds: behenic, stearic, heptadecanoic, palmitic, oleic, elaidic, petroselinic, petroselaidic, linoleic, erucic, and brassidic acids. Palmitamide and stearamide also belong to this group. A satisfactory correlation was

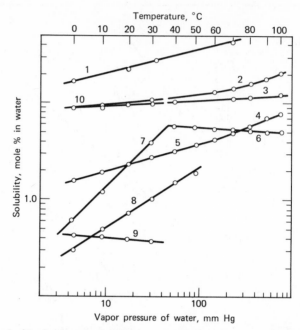

**Fig. 3.16**   Logarithmic plot of solubility in water versus vapor pressure of water. 1, urea; 2, sucrose; 3, ammonium sulfate; 4, copper sulfate ($\alpha$); 5, copper sulfate ($\beta$); 6, sodium sulfate (anhydrous); 7, sodium sulfate (decahydrate); 8, potassium permanganate; 9, calcium hydroxide [41].

obtained for dodecyl chloride, dodecyl iodide, and tetradecyl bromide in chloroform, with hexadecyl iodide in the same solvent as a reference system. The method is also applicable to aromatic isomers. $p$-Nitroaniline can be used as a reference compound for its ortho and meta isomers dissolved in acetone or chloroform.

ANALOGOUS COMPOUNDS IN RELATED SOLVENTS

More recently [44] it was found that the isopleth reference method of plotting can be used to correlate and predict the solubilities of analogous compounds in "related" solvents. Four families of related solvents were found: (*a*) benzene, hexane, cyclohexane, chlorobenzene, *o*-xylene, and toluene; (*b*) acetone, butanone, ethyl acetate, and butyl accetate; (*c*) isopropanol, 95% ethanol, butanol, *p*-dioxane, and diethyl ether; and (*d*) carbon tetrachloride and 1,2-dichloroethane. A typical plot, illustrating the correlation obtained between the solubilities of a number of analogous compounds in a number of related solvents, is shown in Fig. 3.19.

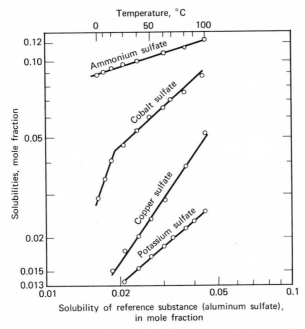

**Fig. 3.17** Solubilities in water of inorganic sulfates versus aluminum sulfate as reference substance [42].

# 3 TECHNIQUES AND APPARATUS

## Time–Temperature Curves

### Cooling and Heating Curves

The essential parts of an apparatus for the determination of a cooling curve are a jacket to permit cooling at a slow, steady rate; an inner sample tube which can be heated to melt the sample; a stirrer to maintain equilibrium throughout the sample; and a thermometer or thermocouple of low heat capacity to measure the temperature of the sample. A simple modified Beckmann apparatus which will meet these conditions and give rough cooling curves may be used to good advantage when 2- to 5-g samples are available. This apparatus, similar in design to that shown in Fig. 3.20 consists of two concentric test tubes, the inner one being 15–20-mm outside diameter and 8–10 cm in length, the outer one about 30-mm outside diameter and 12 cm in length. The sample is placed in the inner tube, which is fitted with a cork holding a calibrated thermometer and a wire stirrer. The sample is melted by the application of heat to the inner tube, and allowed to cool again with the outer tube in place to prevent excessive heat loss. Time and

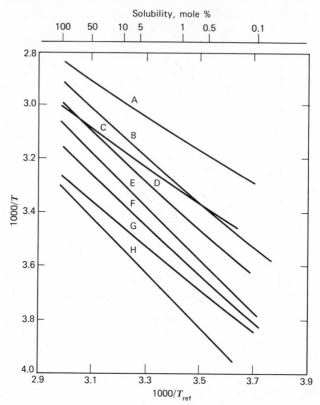

**Fig. 3.18** Isopleth reference plots for a number of fatty acids in acetone with palmitic acid in acetone as the reference system. *A*, behenic; *B*, stearic; *C*, brassidic; *D*, heptadecanoic; *E*, petroselaidic; *F*, elaidic; *G*, erucic; *H*, petroselinic [43].

temperature data are taken as the sample is allowed to solidify with stirring. A constant temperature bath around the outer tube helps to control the rate of heat loss and the rate of crystallization, and increases the time during which the solid–liquid equilibrium temperature may be determined.

Lynn [45] has described a more precise apparatus of this sort that is simple to construct and easy to operate, requiring only 1 g of sample for the determination of suitable cooling curves (see Fig. 3.20). The jacket consists of a small unsilvered Dewar flask (15-mm inside diameter × 150-mm deep) inside of which is a Pyrex heater tube (13-mm outside diameter × 90-mm long) wound externally with fine Nichrome wire. About 1.5 m of no. 36 wire wrapped on in turns about 2 mm apart are suitable for this purpose. The ends of the wire are fastened at small holes in the tube and attached to copper-wire leads. A small amount of glass wool is placed at the bottom of

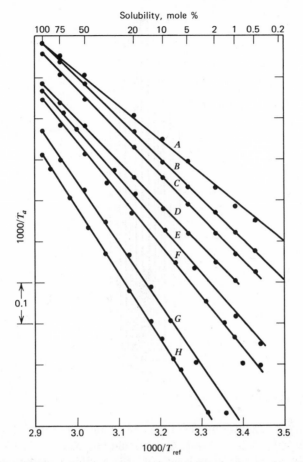

**Fig. 3.19** Isopleth reference plots: *A*, stearic acid in hexane; *B*, stearic acid in *o*-xylene; *C*, heptadecanoic acid in cyclohexane; *D*, stearic acid in chlorobenzene; *E*, elaidic acid in toluene; *F*, oleic acid in hexane; *G*, linoleic acid in hexane; *H*, oleic acid in *o*-xylene. Reference system (abscissa), stearic acid in benzene [44].

the heater tube to support the thin-walled test tube (9-mm inside diameter × 90-mm long) holding the sample. Heating and cooling rates may be controlled by a rheostat or variac in series with the heater. The thermometer or thermoelement placed in the sample or a separate ring stirrer may be used for stirring.

In making the observations for the cooling curve, the sample is first melted by means of the heater and then allowed to cool slowly by readjusting the current through the heater. Some supercooling of the liquid should take place so that, when crystallization starts, crystals are formed quickly through-

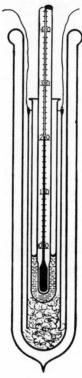

**Fig. 3.20**   Cooling-curve apparatus [45].

out the sample; equilibrium between the liquid and solid phases is then retained by slow, steady stirring. Care should be taken, however, that super-cooling not be too great, especially if the sample is small; otherwise the heat evolved by the formation of crystals (heat of fusion) may be too small to raise the temperature of the sample and the thermometer back to the freezing point before the entire mass solidifies. Time–temperature observations should be taken throughout the cooling process and plotted to show the shape of the curve. Often crystallization of the supercooled liquid may have to be induced by seeding or by tapping the thermometer or tube, or by stirring vigorously momentarily.

The time–temperature diagrams obtained by this method afford a good basis for determination of the freezing point of a substance. If the substance is pure, the observed temperature will remain constant at the freezing point for about one-half the time required for complete freezing. In any case, the maximum temperature attained after crystallization starts is taken as the freezing (or melting) point of the sample.

If considerable impurity is present, the flat portion of the experimental curve will be much shorter, and may never be horizontal at all, in which case the freezing point obtained will be considerably lower than the true value (see p. 165). This apparatus is therefore not applicable to the determination of binary freezing-point diagrams.

In determining cooling curves as outlined above, it is essential that the temperature gradient between the sample and the surroundings be not too high. If heat is being removed from the sample by the bath at too great a rate or if the rate of crystallization is too low, the observed freezing point may be lower than the true temperature of solidification, because the heat loss from the sample to the surroundings is greater than the heat of fusion supplied by the formation of crystals. This difficulty can sometimes be overcome by increasing the initial degree of supercooling, since it has been found that the rate of crystallization increases, within certain limits, with the extent of supercooling. The fact that the temperature is constant during the halt does not necessarily mean that errors due to the above sources have been eliminated. Hence it is always desirable to repeat the determination using different degrees of supercooling; the highest value observed will be the most nearly correct.

Many substances crystallize too slowly to give a satisfactory cooling curve. In fact, many tend to form viscous, supercooled liquids and, on further cooling, amorphous solids or "glasses." If they can be made to crystallize at all, however, their melting points can be obtained from heating curves. Tammann showed that the ease of crystallization depends upon the rate of formation of crystal nuclei, the rate of crystal growth, and the viscosity, each of which is affected independently by temperature. The interplay of these factors results in a different optimum temperature for each specific compound at which the tendency to form crystals will be at a maximum. To induce crystallization the sample should be intensively cooled and then allowed to warm very slowly so as to pass slowly through the optimum temperature. The warming may also be carried out in such a manner that there is a temperature gradient within the sample. Once a few crystals have formed the sample should be alternately cooled and warmed just below its melting point until crystallization is complete. For a more detailed account of the factors and the techniques involved in inducing crystallization the reader is referred to the chapter on "Crystallization and Recrystallization" in Volume III, Part I, of this series.

A more refined cooling- and heating-curve apparatus has been described by Skau [46]. A 0.5-g sample is hermetically sealed in a thin glass tube supplied with a thermocouple well and then suspended in an air cryostat the temperature of which can be accurately measured and controlled from 200°C down to liquid-air temperatures. Time–temperature data are plotted for both the sample and the surroundings. The sample can be handled entirely *in vacuo* if desired, and contamination is thus impossible. This is of great importance not only in the case of compounds which are hygroscopic (like ethyl alcohol), or which absorb carbon dioxide from the atmosphere (like certain amines), or which oxidize in air (like polyphenols), but also in the case of liquids freezing below 0°C, since at low temperatures the condensation of moisture and of carbon dioxide as well as the solution of air in the sample is a very grave source of contamination. Complete heating and cooling curves can be obtained without stirring and are thus readily reproducible. Freezing points of pure substances can be determined with an accuracy and a precision of about 0.03°C [47*a*]. As suggested by Keyes [47*b*], sealed tubes of this design, i.e., with a built-in thermocouple or thermometer well, are ideal as permanent sample tubes for secondary melting-point or freezing-point standards.

This apparatus can also be used for the accurate determination of transition points [47*a*, 48] and of the temperature at which solids melt to form liquid crystals and at which liquid crystals transform to the clear liquid state [49]. Conversely, it is applicable to the calibration of single-junction thermocouples [47*a*]. For a more detailed description of this apparatus and technique and of its use in determining heating and cooling curves as a criterion of purity, the reader is referred to the chapter on calorimetry.

It has been pointed out [46] that controlled heating curves, such as are obtained with this apparatus, are much more sensitive in the detection of impurities than cooling curves, especially since small samples are used. The slightest impurity causes premelting and has a major effect on the first part of the heating curve. This was confirmed by Smit [50]. Mathieu [51] using a slightly modified apparatus showed that a quantitative estimate of the percentage of impurity and the freezing point at zero impurity can be obtained from these heating curves by applying the Taylor and Rossini method of analysis [52]. A similar apparatus for determining the melting curves of 0.3-ml samples and its use for calculating purity, heat of fusion, and heat capacity were described by Gunn [53]. Measurements were made on samples of ammonia, water, and benzene with known amounts of contaminants. A sensitivity of the order of $10^{-5}$ mole fraction impurity and an accuracy of 10–20% was reported. Smit [54] used heating curves obtained by a "thin-film" apparatus as a criterion of purity. The diameter of the sample tube was such that the immersion thermometer bulb was surrounded by a thin layer of the liquid sample. A modification of this apparatus was tested by Carlton [55] on samples containing known amounts of impurity.

An apparatus for determining freezing points on 50-ml samples with a precision of 0.001–0.003°C, developed by Rossini and co-workers [56], is shown schematically in Fig. 3.21. The evacuated double-walled vessel *A*

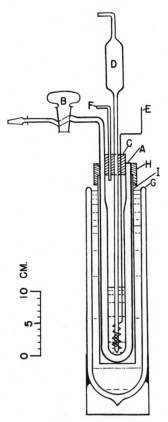

**Fig. 3.21** Freezing-point apparatus of Mair, Glasgow, and Rossini [56].

containing the substance under investigation and supplied with a stopcock, *B*, is centered in the brass cylinder, *I*, by means of the asbestos collar, *H*. The cork stopper, *C*, is supplied with holes to accommodate the resistance thermometer, *D*; a stirrer, *E*; an inlet for dry air, *F*; and with a small hole, not shown, for the "seed wire." The liquid in the Dewar flask, *G*, acts as a cooling or warming bath.

Temperature as measured by means of the platinum resistance thermometer is plotted against time as illustrated for benzene in Fig. 3.22. Curve *GHI* represents the equilibrium part of the curve. From such a curve it is possible to calculate with high precision not only the freezing point of the impure sample but also the percentage of impurity and the freezing point of the substance with zero impurity, provided that the curve represents true equilibrium between solid and liquid phases while a large fraction crystallizes, and provided that the fraction of impurity in the sample tested

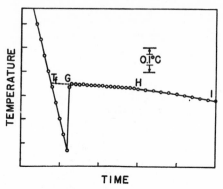

**Fig. 3.22**    Cooling curve for determining the freezing point of benzene [56].

is small [57]. These calculations can also be made from the corresponding heating-curve data. This apparatus and method have been adopted as test methods for high-purity hydrocarbons by the American Society for Testing and Materials [58, 59].

The apparatus shown in Fig. 3.21, developed for determining freezing points in air at atmospheric pressure, was subsequently remodeled to permit determination of triple points; that is, freezing points at saturation pressure, the condition in which the substance is under its own vapor pressure in the absence of air or other gases. This modified apparatus is particularly suited for determining more accurately the freezing points of compounds having a small heat of fusion and a normal or large solubility for air at the freezing point [60]. Special apparatus and techniques were required for highly reactive substances, such as $TiCl_4$ [61]. Additional modifications and refinements have been desired [62].

The Skau-Mathieu heating-curve method for determining purity has been found to compare favorably with the Rossini method [63a] and has now been fully automated to give excellent results as a routine method [63b].

Stull [64] has developed an automatic temperature recorder activated by a small platinum-resistance thermometer and has shown that it is applicable to the accurate recording of cooling curves. This has been further developed by Witschonke [65] as the basis for an automatic freezing-point apparatus in which a small constant temperature differential is maintained between the sample and the surroundings, so that constant controlled rates of heat transfer are obtained. The time–temperature curve of the sample is plotted automatically with an accuracy of ±0.01°C over the range −40°C to +200°C. Herington and Handley [66] described a temperature recorder which makes use of a thermistor as the temperature-sensitive element. In applying this to the automatic recording of freezing-point curves [67], they pointed out that for any desired temperature sensitivity a more rugged recorder can be used with a thermistor than is required for a platinum resistance thermometer and that the heat capacity of the thermistor is small.

### Differential Thermal Methods

Differential thermal analysis (DTA) and differential scanning calorimeters (DSC) can be used to measure the energy changes accompanying phase changes. See chapters on these methods.

*Differential thermal analysis* gives an automatic graphical record of the difference in the temperature of a compound and that of an inert reference substance against time or temperature as the two specimens are heated in an environment which is heated at a constant rate. Commercially-available DTA instruments operate over a range from $-150$ to $+1500°C$. In general, melting points determined by this method lack precision; a precision of $\pm0.5°C$ is claimed in the range 0–300°C. Herington [68] and Handley [69] have, however, designed an apparatus, particularly suitable for use in industrial quality-control laboratories, for comparing the purities of two or more samples.

The *differential scanning calorimeter* gives a plot of the energy necessary to establish zero temperature difference between a substance and a reference substance against either time or temperature as the two specimens are subjected to a controlled constant rise in the environmental temperature. The Perkin-Elmer DSC Model 1B has been described as being useful for many varied thermal measurements including the measurement of purity of organic compounds [70]. The DSC scan obtained for a highly pure crystalline compound can be used to determine its heat of fusion and the fraction of the sample which has melted up to or at specific temperatures. The mole % of impurity and the melting point at zero impurity can then be calculated. The investigation and evaluation of this method by several workers [71–74] have resulted in the development of a simple routine procedure using 1- to 3-mg samples, including inorganic (metals) materials [74]. The accuracy of purity values by this technique has been shown to fall off rapidly below purities of 99 mole %. The impurities must be insoluble in the solid phase and soluble in the melt. Expressions for treating solid solutions have, however, been proposed [71]. A computer program was developed to simplify and improve the precision of the purity calculations [71]. Using the computer calculation, a standard deviation of 0.02 mole % impurity was found for 20 determinations involving eight different compounds, as compared to greater than 0.04 mole % by ordinary calculations.

The DSC method for purity is less accurate than the sophisticated methods based on heating and cooling curves or on adiabatic calorimetry but has the advantage of convenience, speed, and small sample size. It affords a general purpose means of estimating purity for comparison of sample purity in quality control.

## Calorimetric Methods

The ultimate degree of accuracy of the freezing point and the percentage of impurity obtained by cooling- and heating-curve methods depends upon the validity of certain assumptions [75], and is limited by the fact that the measurements are made on a dynamic system. A higher degree of accuracy in determining the freezing point, the percentage of purity, and the correct freezing point of the pure substance can be attained by means of accurate adiabatic calorimetric measurements of the heat capacity during the melting process. Observations can thus be made under static conditions of the equilibrium temperatures corresponding to various fractions of the substance melted [76]. A detailed discussion of such measurements will be found in chapter on calorimetry.

Cines [77] and Mathieu [78] pointed out that the solid–liquid ratio values, and therefore the calculated percentages of impurity, by the cooling-curve or thermometric method were apparently not in accord with those determined by the calorimetric method. This was confirmed by McCullough and Waddington [79]. Glasgow et al. [80], in an unusually rigorous and objective comparison of benzene samples containing known amounts of *n*-heptane, concluded that, for these samples, the results by the two methods agreed within statistical uncertainty. An international project for comparing thermal methods of determining the purity of organic compounds was organized in 1957 by the National Bureau of Standards at the recommendation of the IUPAC Commission on Physico-Chemical Data and Standards. The results of the investigation have not yet been published, but a preliminary report was presented before the Calorimetry Conference in August 1961 at Ottawa, Canada [81].

## Other Static Methods

The melting point and the proportion of solid and liquid phase at specific temperatures over the melting range can also be precisely determined by nuclear magnetic resonance, dilatometric, or dielectric measurements under static conditions. It is thus possible to calculate the purity of a highly pure compound as well as the melting point at zero impurity by the same methods as were used above for the corresponding calorimetric data. The measurements are time-consuming because long periods of equilibration are required at each temperature.

### Nuclear Magnetic Measurements

At the phase transition from solid to liquid there is a narrowing of the line width and a change in the spin echo amplitude of the nuclear magnetic resonance (NMR) spectrum of a substance. If impurity is present, premelting occurs at the impurity centers in the sample. A partly liquid and partly solid sample exists just below the apparent melting temperature. The fraction of liquid can be measured as a function of temperature by pulsed NMR for any substance which contains nuclei with spin greater than zero. The spin echo amplitude is directly proportional to the fraction of the liquid (see chapter on NMR). It is, therefore, possible from NMR measurements to calculate the percentage of impurity present and, if this is small, the melting point of the pure substance [82]. Burnett and Muller [83] observed the melting of two samples of ethane containing 0.03 and 0.80% impurity, and recorded the apparent melting temperatures, 89.81 and 89.65°K, and the true melting temperatures, 89.82 and 89.84°K, respectively.

Herington and Lawrenson [84] simplified the method by using the motional narrowing from a relatively wide line for solid phase, typically $10^4$ Hz, to a relatively narrow line for liquid phase, typically 30 Hz. The proton spectrum of the substance being investigated was initially recorded at about 20° below the melting temperature. The magnetic sweep of the instrument was set to record the narrow line for the liquid phase. The area under the narrow line was proportional to the fraction of the substance melted. For systems forming no solid solutions, the area under the line was linearly proportional to the temperature; the slope of the plot was proportional to purity; and the intercept on the temperature axis was the melting temperature of pure material. Karagounis et al. [85] have used the changes in line widths of NMR spectra with change in phase to record the lowering of melting points of substances spread in molecular layers over surfaces.

### Dilatometric Measurements

Melting point and purity can also be determined from the change in volume with temperature over the melting range (see Fig. 3.1). The ratio of solid to liquid phase at specific temperatures can be calculated from dilatometric measurements. Swietoslawski and Plebanski [86] have developed a dilatometric technique capable of high precision. The highest precision is attained with a dilatometric cryometer designed to cover a relatively small temperature range. Purity determinations with a sensitivity of 0.0001 mole % were made on a number of highly pure benzene samples. Plebanski [87] reported a statistical study of the results obtained on some benzene samples as determined by static, calorimetric, and dilatometric methods and dynamic methods.

### Dielectric Measurements

A method for purity determination, developed by Ross and Frolen [88], takes advantage of the large change in the dielectric constant of many substances on passing from the solid to the liquid state. A sample of the substances is melted to fill a capacitor homogeneously, and then is frozen by quenching to preserve the random dispersion of the impurities. The change in electrical capacitance with temperature over the melting range is correlated with the fraction of the substance melted. Purity can then be calculated by the above procedures. The melting process is more reproducible than the freezing process which could involve supercooling. The method has been applied to: (*a*) benzene, naphthalene, and 2-methyl naphthalene which have small dipole moments and little polarizability; (*b*) *p*-dichlorobenzene which has no dipole moment but is easily polarized; and (*c*) nitrobenzene which exhibits a ten-fold increase in dielectric constant on melting. Applications are restricted to samples which have high electrical resistivity, a purity of greater than 98.5 mole %, and impurities closely related to the major component. A precision of determination of melting temperature by dielectric measurements greater than that obtained by time–temperature curves is claimed.

### Sealed-Tube Method

A convenient method for determining freezing points of pure compounds, capable of a precision of ±0.1°C, is the static, sealed-tube method described on page 167. This method is particularly adapted to compounds which absorb or react with moisture, carbon dioxide, or oxygen.

### Capillary Method

The melting point most commonly reported by the organic chemist, and probably the most conveniently determined, is that obtained by packing the finely divided sample in a glass capillary, immersing the tube in a bath whose temperature is gradually raised, and noting the temperature of the bath adjacent to the capillary when the substance is seen to melt. This method has two advantages: only a small amount of material is required; and a fair accuracy may be obtained with very simple equipment. The bath should be well stirred, heated at an even rate—about 1°C/min in the vicinity of the melting point for ordinary purposes—and should permit unrestricted visibility of the melting-point tube and the thermometer. Usually the capillary tube is not placed in the bath until the temperature is about 5–10°C below the melting point.

The temperature at which the last crystal dissolves is usually taken as the melting point. As pointed out previously, if the substance appears to melt over a range, the initial melting is not observable, and the initial melting point reported varies greatly with different experimenters. Many substances when heated in a capillary tube exhibit a "sintering" effect at some temperatures a little below the melting temperature. This sintering, or settling of the substance in the capillary tube, which may be due to partial melting of the solid at or above the eutectic temperature, gives notice that the sample is approaching the melting temperature.

Certain precautions should be observed in constructing and filling the capillary tubes. They should be drawn down from thin-walled soft glass tubing to a diameter of about 1 mm. Smaller tubes are difficult to fill; and, in fact, with waxy or similar materials it may be necessary to use tubes of slightly larger diameters. Thermal equilibrium throughout the sample is more easily obtained with smaller tubes.

If large tubes are used for impure substances or binary mixtures, the substance next to the walls may liquefy first and allow the solid to float or sink. The resultant segregation between the solid and liquid may cause the observed melting point of the mixture to be too high. Resolidification and remelting would tend to increase this segregation and magnify the error involved.

The tubing from which the capillaries are made should be washed and dried before drawing. Alkali and products of devitrification on the walls of the tube are known to lower the melting point, especially in the case of substances such as those having free aldehydic or ketonic groups [89]. Likewise, Dieckmann [90] and others [91] have shown that slightly different melting points are obtained with capillaries of different glasses. After the capillaries are drawn, they should be sealed at both ends to keep them clean until they are to be used.

In filling the tube, a small amount of the material to be tested is scooped into the open end. It is then worked into the bottom of the tube. This can be accomplished by a number of methods, for example, by rasping the bottom of the tube with a file, or by dropping the capillary a foot or more inside a larger tube so that the bottom of the capillary strikes the desk or table top. In any case, it is important that the substance under examination be well packed into the bottom of the tube to ensure the maximum contact between the sample and the walls of the tube. It is advisable to seal the tube when the sample is subject to change or decomposition through continued exposure to air, or when it is desired to retain the sample for some time before using. Such samples should be introduced through a long thin "funnel" so that no organic material will be in the neck of the tube to decompose during the sealing. Upon decomposition additional components would form which would tend to lower the melting point.

The ideal type of bath for melting-point determinations is one in which the temperature of the liquid surrounding the thermometer and capillary tube can be closely controlled at all times. This can be accomplished only with efficient stirring and a readily controlled heat source, preferably one with a minimum lag. It should be possible to maintain the temperature fairly constant for some time since, in accurate determinations, it is frequently desirable to set the temperature at some point just below or above the melting point. The rate of response of different thermometers to a change in temperature of the bath will differ depending upon their heat capacities. In order to compensate for this factor, the rate of heating must be low if the melting-point determinations are to be accurate.

The capillary and the thermometer bulb should be contiguous so the temperature registered by the thermometer will be as nearly as possible the actual temperature of the sample. Common practice is to fasten the capillary to the thermometer by means of a rubber band or similar device above the surface of the bath liquid. If the capillaries are long and straight, they will adhere to the thermometer by capillary action of the liquid. The use of thermometers which can be completely immersed in the bath is preferred because they eliminate the necessity of troublesome stem corrections. Thermometers and their calibrations and stem corrections are discussed in detail in Chapter I.

### Liquids for Heating Baths

The bath liquid should be relatively nonvolatile and inert. The liquids chosen may vary for the range of melting points most likely to be encountered. Liquid paraffins, such as commercial clear mineral oil, are excellent for moderate temperatures provided that their viscosities are not too high. At higher temperatures, they tend to fume and discolor. For high temperatures (above 200°C), concentrated sulfuric acid, phosphoric acid, or a mixture of six parts of sulfuric acid and four of potassium sulfate has been recommended [92, 93]. If the concentrated sulfuric acid should become discolored at high temperatures, it can be clarified by adding a crystal of potassium nitrate and heating. A solution (10–11%) of lithium sulfate in sulfuric acid can be used as a bath liquid. The lithium sulfate decreases the volatility of the bath but does not tend to solidify at room temperature, this being an advantage over using potassium sulfate. This bath liquid will tolerate small amounts of sodium or potassium nitrate, added from time to time, to remove color [94]. Phosphoric acid may give off water between 100 and 215°C, and, after each use, water should be added to restore the acid to its original volume.

Another common satisfactory bath liquid for temperatures above 150°C is dibutyl phthalate. Its viscosity at the lower ranges, however, is somewhat high to permit good control of the bath temperatures. Diphenyl ether is also commonly used. Probably the most satisfactory high-temperature bath liquids, from the point of view of both viscosity and decomposition, are the organosilicon compounds. White [95] has shown that certain silicone fluids, organosilicon oxide polymers, are very satisfactory for use in baths up to 425–440°C.

An evaluation of 62 organic substances, selected from among widely distributed types of compounds, indicated that Aroclor 1248, a halogenated hydrocarbon containing 48% chlorine, and a few specific silicone oils were superior to other compounds tested for use as heat-transfer liquids in apparatus for determining melting points [96]. These substances are colorless, relatively heat-stable, noncorrosive, and nonhygroscopic. The Aroclor could be used at temperatures as high as 340°C, and the silicone oils, 400°C.

Safety precautions, such as the wearing of eye-protective devices, are important when using hot or corrosive bath liquids. Cognizance should also be taken of the possible toxicity of the fumes from the bath liquid used.

### Calibration of Thermometers

It is not sufficient to use a thermometer with a certified calibration. For maximum accuracy the calibration should be by means of melting-point determinations on pure reference standards in the very apparatus in which the thermometer or the thermocouple is to be used. All conditions, such as rate of heating and size of sample, employed during the calibration should be as close as possible to those used in the melting-point determinations. The melting points of the standards should be in the same range as the expected melting points of the unknown compounds.

It should be mentioned that thermometers have the disadvantage of a relatively high heat capacity; they also require a stem correction, and, when used for cooling curves, a larger sample is necessary. Thermocouples, on the other hand, have a very low heat capacity, a very rapid response, and are capable of high precision. When they are used for cooling curves, therefore, much smaller samples are required. The highest precision can be obtained with the resistance thermometer, but only with large samples.

### Thiele Apparatus

The principle of the Thiele tube is the basis of a variety of types of capillary melting-point baths. In this apparatus, the bath is in the shape of a vertical loop in which the liquid is heated on one side and the capillary and thermometer suspended on the other. The liquid in the heated side rises, causing the liquid in the capillary side to descend, thus setting up a circulation of the fluid by convection. A stream of rising air bubbles is frequently introduced at the bottom of the heated side to increase the rate of circulation of the bath liquid. The simplicity and ease of construction of the Thiele tube melting-point bath have made it one of the most common pieces of equipment in the organic chemistry laboratory.

Of the many modifications [97, 98] of the Thiele tube principle, probably the best is that of Markley [98] as improved by Hershberg [99], a diagram of which appears in Fig. 3.23. Here the liquid is kept moving around the loop at a rapid rate by means of a stirrer mounted on a ball-bearing assembly.

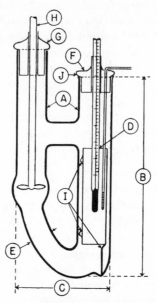

**Fig. 3.23**  The Markley-Hershberg modification of the Thiele apparatus [98, 99]. *A*, 28-mm outside diameter and 25-mm inside diameter. *B*, 17 cm; *C*, 8.5 cm; *D*, sleeve, 19-mm outside diameter, 17-mm inside diameter, 9 cm long; loops No. 26 B and S gage platinum wire; *E*, 18-mm outside diameter, wound with electrical heating element. *F*, thermometer cap; thermometer tube, 7-mm inside diameter. *G*, stirrer cap; *H*, stirrer, 5-mm outside diameter glass tubing; ball bearings with 0.61-cm (0.25-in.) hole and 2.2-cm (0.875-in.) outside diameter, unground; *I*, knobs to center sleeve; *J*, lip and wedge to prevent rotation of cap.

The sample capillary and a total-immersion thermometer are placed in an adiabatic zone inside the tube in which the deviation in temperature is said not to exceed 0.025°C. The heater can be regulated so that the rate of temperature rise can be reduced to about one-tenth of a degree per minute. The top of the thermometer side of the tube is so constructed that capillaries can be inserted and withdrawn with little difficulty. A number of refined instruments based on these principles are now available at laboratory supply houses.

Tseng [100] and Francis and coworkers [101, 102] made a rather extensive study of the capillary melting-point method and found that for different types of apparatus discrepancies of as much as 2°C were observed. With a given apparatus individual determinations usually agreed very closely, but with different observers the divergence sometimes reached 1°C. Francis and Collins developed a precise method for capillary melting-point determination of highly pure substances employing a bath which can be maintained within ±0.02°C. The approximate melting point is first determined using a rapid rate of heating. After resolidification the accurate value is found by raising the temperature rapidly at first, and finally at a rate of about 0.1°C in 4 min. An optical system is employed whereby the sample and the thermometer may be viewed through a telescopic eyepiece. The temperature at which a clear liquid first appears on the surface of the solid is taken as the melting point. It was found that if maintained at this temperature the whole contents of the tube would liquefy. The method is, of course, not applicable to impure substances, which would melt over a range. After the melting-point determination, when there is still some solid left in the capillary, the "resolidification point" can be determined with equal precision by allowing the temperature to fall *very* slowly and noting the temperature at which resolidification commences.

In an improvement of this technique for determining melting and resolidification points [102] a fraction of a milligram of the crystalline substance is deposited at a convenient height on the inside surface of a 3–4-mm glass tube closed at the lower end. It is immersed in a bath and the crystal mass, preferably in circular form, about 1.5–2.0 mm in diameter, is brought into the focus of a telescope. The melting and resolidification points are then observed in the usual manner.

Francis and his co-workers also described a modified Beckmann cooling-curve apparatus with which "setting points" can be determined on a 2-g sample. They concluded from data obtained for a number of pure substances that the setting point is closest to the true solid–liquid equilibrium temperature and is reproducible by different observers to within ±0.02°C. The melting and resolidification points showed a reproducibility of better than ±0.05°C, but the melting point was always 0.1–0.6°C above the setting point. The resolidification point was found to be close to the setting point; the extreme divergence was ±0.2°C but was usually much less.

## Copper Block

The Thiele tube method has disadvantages, especially at high temperatures: the bath liquid discolors and produces fumes; and there may be unequal heat distribution due to streamline currents and high heat losses to the surroundings. These difficulties are overcome to some extent by replacing the heating liquid with a metal block made of a good heat conductor such as copper. This scheme was originally devised by Thiele [103] and has been modified by Berl and Kullmann [104]. The apparatus consists of a copper block or cylinder in which holes are bored for two thermometers and for two capillary tubes. The holes are of such sizes that the thermometers and capillary tubes fit snugly and make good thermal contact with the block. A hole perpendicular to the capillary wells, and intersecting them, is cut through the block for observing the bottoms of the capillaries. This opening may be closed with a strip of mica to prevent access of air currents through the slit or hole and yet retain visibility. The block is then covered with asbestos and resistance wire so that it may be heated electrically. In using the apparatus the block is heated and the melting point observed with the aid of a light on the far side of the block. Various optical arrangements have been proposed to aid in observation of the capillaries when melting takes place. A good account of these and further details of the copper block method can be found in the original papers [105].

## Hot Stages

Melting temperatures are frequently determined by one of several hot-stage methods. In such a method, the sample is placed on a plate or bar which is heated, and the temperature at which melting occurs is noted by means of a thermometer or thermocouple arrangement. Hot stages present the advantages of rapid heating, elimination of a liquid bath, and an ease of operation not usually obtained by other methods of melting-point measurement, and are particularly adaptable to substances which melt with decomposition. They lack, however, the accuracy and precision attainable with a refined capillary apparatus. They should be extensively calibrated with compounds of known melting point.

An interesting application of the hot-stage method is the melting-point apparatus described by Dennis and Shelton [106]. In this apparatus, the powdered sample is sprinkled in a thin line on a copper bar heated at one end. A temperature gradient exists on the surface of the bar from the hot to the cold end, and the substance melts instantaneously where the temperature of the bar exceeds the melting point. A line of demarcation between the solid and liquid indicates the point at which melting first takes place. The temperature is usually determined by touching the bar at this point with constantan wire to form a thermocouple with a definite electromotive force corresponding to the melting temperature.

The Kofler hot bar uses two pieces of metal of different heat conductivity, heated electrically, with the bar designed so that the drop in temperature is almost linear along its length. The temperature of the hot bar can range from 10 to 260°C, with a special temperature-reading device comprising a runner with a pointer and tab designed for the specific bar. To determine a melting point, the substance is laid in a thin layer directly on the surface of the hot bar, and in a few seconds a sharp dividing line between the fluid and solid phase develops. The temperature at the dividing line is read by adjusting the pointer to rest at the line [107]. The Kofler hot bar can also be used to obtain reproducible melting points of substances that decompose [108]. Kofler hot bars, operating to 260°C, are used with an accuracy of ±0.5°C in determination of melting points.

In the Fisher-Johns melting-point apparatus, the sample is placed between cover glasses on an aluminum stage which is electrically heated. A magnifier is placed above the sample to facilitate ready observation of the crystals at the melting temperature. A thermometer with its bulb embedded in the stage immediately below the sample records the temperature of the stage at the melting point. The copper bar and various hot stages may be purchased from laboratory supply houses.

Several hot stages for microscopes have been designed for the determination of melting points with very small quantities of material. In most of these, the temperature is measured with a sensitive thermocouple, although modifications permitting the use of a mercury thermometer have also been described. For all compounds except those which are isotropic or become so on heating, the melting point can best be observed by means of a polarizing microscope, since the temperature at which the color disappears and the space lattice is ruptured is the true melting point. With an instrument of this sort Zscheile and White [109] claimed a precision of ±0.04°C. The Kofler [110, 111] microscope hot-stage melting-point apparatus has found wide application. This is a heating chamber in the form of a flat box the size of a microscope stage. The base of this heating chamber contains a metal plate upon which the sample whose melting point is to be determined is placed on a slide. The center of the metal plate contains a hole permitting the entrance of light from the illuminating mirror of the microscope. When in use, the chamber is closed by a glass plate to cut off air from the working area. The optimum rate of heating, regulated by a rheostat, depends on the properties of the sample. The rate of heating may be fairly rapid up to about 10° below the melting point of the sample. At this point the heating is discontinued until the temperature begins to drop, and then heating is resumed at about 2°/min. Near the melting point range of the sample (3°C) the rate of heating should be about 1°C/3–5 min. Temperature may be measured by a thermometer the bulb of which is placed on the base plate as near to the sample as possible. Observations are generally made with a magnification between 60✕ and 160✕. Illumination is by transmitted light. In special cases, polarized light may be used for optically anisotropic substances.

Kofler hot stages for microscopes are commonly manufactured for use in two temperature ranges: room temperature to 350°C and room temperature to 750°C, with special temperature-reading devices designed for each stage and for specific temperature ranges. For the hot stage operating to 350°C, determination of melting points with an accuracy of ±0.2°C is possible. The hot stage operating to 750°C differs only in that the tube containing the lens of the microscope, to minimize overheating, is in a raised position except during actual observation of melting. A Kofler hot stage operating to 1500°C, with provision for vertical illumination through the stage and for the tube containing the lens of the microscope to swing to one side when not in use, has also been manufactured.

Mettler [112] hot stages, operating to 300°C, are automated instruments using electronic sensing devices to determine melting temperatures of substances with a precision of ±0.1°C. A beam of light is directed through the substance on the stage to a precisely calibrated photocell. The optical properties of the substance change from opaque to transparent when it changes from a solid to a liquid on heating. Therefore, as melting occurs, the intensity of light reaching the photocell increases and sends a stop signal to the digital indicator reading out the temperature of a platinum resistance thermometer located in the stage.

Other hot and cold stages covering ranges from −100 to 1000°C are available [111]. Standard compounds of known melting point are used for calibration [111].

## Mixed Melting Points

The method of mixed melting points may be used in the identification of organic substances. Approximately equal amounts of the unknown and a substance thought to be the same compound are pulverized and intimately mixed in a mortar or on a watch glass. Capillary melting points are then taken of the known, the unknown, and the mixture. If all three melting points are essentially the same, or, if the melting point of the mixture lies between that of the two components, the identity of the unknown with that of the known is fairly certain. If the substances are not identical, the melting point of the mixture is usually 10–30° or more below that of the components and the mixture melts over a range [113]. These melting points should all be taken in the same bath simultaneously, since factors which affect the melting point, such as decomposition and rate of heating, will then be the same in all three cases.

It is not necessary to have a very pure substance in order to identify it by the method of mixed melting points. If the melting point is a few degrees too low, addition of the known will give a melting point which is between the known and the unknown when the substances are identical.

The method of identification by mixed melting points is based on the assumption that a mixture of any two substances will have a melting point which is appreciably lower than that of either alone. That this assumption does not hold in all cases is apparent by a study of typical binary system diagrams already mentioned. As illustrated by the systems in Figs. 3.7 and 3.9 b and by Table 3.2, it is quite possible for two different substances to have mixed melting points which are either higher than or equal to those of the two components. For example, if tricosanoic acid is mixed with an equimolecular proportion of tetracosanoic acid, its melting point is raised by 0.2°C and is very sharp. This and a number of similar cases of fatty-acid mixtures have been described [114]. When d-dimethyl tartrate and l-dimethyl tartrate (mp, 43.3°C) are mixed in equal proportions, a melting point of 89.4°C is obtained [115]. Gibby and Waters [116] reported that, for pure 3-bromo-5-iodo-4-aminobenzophenone (mp, 145.9°C) and pure 3,5-dibromo-4-aminobenzophenone (mp, 146°C), the mixed melting point is only about 0.5° lower, whereas in a previous paper it had been concluded that these two compounds were identical. Lock and Nottes [117] have cited many other cases where mixed melting-point behavior is abnormal. Failure to observe a lowered melting point for the mixture would thus not be absolute proof of the identity of the known and unknown. Additional means of identification, for example, crystal structure, optical rotation, and chemical analysis, should be applied in cases in which the identity of the unknown is still in question.

When mixed melting-point determinations are made on substances which decompose on melting, the information gained may not be conclusive. If the decomposition point of the mixture is not lower, the two constituents may or may not be identical; if the decomposition point is lower, however, the two constituents are shown to be different compounds.

As long as the possibility of the anomalous cases mentioned above is kept in mind, the mixed melting-point technique is useful, especially when one is employing very small quantities of the unknown. The fact that this method of identification does not require a highly purified sample eliminates the necessity for the final purification and the resultant loss of material. The method may even be used for substances which are liquid under ordinary conditions by measuring the temperature of the melting point of the frozen mixture [118]. Successful application of the method requires some knowledge of the melting-point behavior of binary mixtures, including a familiarity with the various possible types of binary freezing-point diagrams.

### Melting Points of Substances That Decompose

If a substance melts with decomposition, the melting temperature observed will be lower than the melting point of the pure substance because of the presence of decomposition products. This is especially true if decomposition occurs in the solid state before the substance actually melts; and, since premelting is invariably present to some extent, the melting point of a substance which decomposes is usually considerably lower than the true melting point. The amount of decomposition products formed depends upon the length of time the sample has been heated near the decomposition temperature before the melting point is reached. It is apparent, then, that the melting point of a substance which decomposes will depend upon the rate of heating of the melting-point bath. Thus pure tyrosine melts at 280°C when heated slowly [119] and at 314–318°C when heated rapidly [120].

Therefore, if the substance shows any indication of decomposing, the melting-point sample should be heated rapidly. This can be accomplished to the best advantage by the use of the Dennis bar or the Kofler hot bar described on page 154 or by constructing the bath so that the temperature rise is as rapid as 1° for every 2 sec. Another method of heating the sample rapidly is immersion of the capillary in the preheated bath so that the bath temperature has to be raised only a few degrees before the melting point is reached [121]. When melting points with decomposition are reported in the literature, it is essential that complete details of the method and rate of heating be given in order that the melting point can be reproduced by another experimenter.

Organic substances which decompose when heated near their melting points may exhibit any one of a number of types of decomposition. These may frequently be distinguished simply by watching the sample closely with a magnifying glass during melting. The substance may merely darken, as in the case of 1,4-anthraquinone [122]. It may undergo the loss of certain products. Thus dibromomalonic acid gives off carbon dioxide [123]; nitroguanidine evolves ammonia [124]; diiodomalonic acid decomposes to give off iodine [123]; and aniline sulfate loses water and sulfur dioxide. Some substances, such as iodoxybenzene, explode at the melting temperature [125]. In some cases, the compound may melt and form a new compound which is solid at the high temperature so that the observer may note two melting points. This is usually true of substances which form anhydrides easily [126]. In general, although melting points with decomposition are in many cases uncertain, in some cases they are fairly reliable and reproducible.

The determination of eutectic temperatures with standard compounds on a microscope hot stage affords a method for identification of organic compounds [127]. This method is especially appropriate for substances such as the amino acids which decompose at or below the melting point. Lacourt and Delande [128] reported the eutectic temperatures of 16 amino acids with 21 different partners. The well-mixed powdered samples of the two components are placed between glass covers on the hot stage under a polarizing microscope at a temperature as near as possible to the eutectic. Between crossed Nichols the crystals appear bright. Once the eutectic temperature is reached dark spots appear in the field. This is repeated to reach a very narrow interval and reproducible results.

## Anomalous Melting Points

### Polymorphism

Pure substances may exist in more than one crystalline or polymorphic form, with two or more melting points, and possibly one or more transition points. A substance is termed enantiotropic when the transition temperature is below the melting temperature of either form, as when one crystalline form is transformed to another without melting. If it is necessary to melt one form to obtain the other polymorph, the substance is monotropic.

An enantiotropic substance existing in two forms exhibits a transition point below which one form ($\alpha$) and above which the other form ($\beta$) is stable. When the $\alpha$ form is heated slowly, it transforms to the $\beta$ form at the transition temperature. This transition is reversible so that, if the $\beta$ form is cooled below the transition point, the $\alpha$ form is again formed. If the $\beta$ form is heated, it exhibits a normal melting point. When the temperature of the $\alpha$ form is raised through the transition temperature so rapidly that transition fails to occur, the melting point of the $\alpha$ form is observed. The liquid obtained by melting is metastable with respect to the solid $\beta$ form, however; and, if the temperature is not allowed to rise above the melting temperature of the latter, the liquid may solidify. The solid so formed will on further heating exhibit the melting point of the $\beta$ form.

Carbon tetrabromide is an example of a substance with exhibits enantiotropism. At ordinary temperatures, the stable form (solid II) is monoclinic [129]. When heated to 46.9°C (the transition point), a cubic form with tetrahedral symmetry (solid I) is formed which melts at 90.1°. The velocity of transformation of the metastable to the stable form above 46.9° is probably very rapid, so that the melting point of the pure solid II has probably never been observed. In the well-known case of sulfur, however, the transformation from the rhombic to the monoclinic form above the transition point (95.5°) is so slow that the melting point of the rhombic form can easily be observed to be 112.8°. At this temperature, both rhombic sulfur and the liquid are metastable. The melting point of the monoclinic form is 119.25° [130].

A monotropic substance is characterized by the fact that the transformation from the unstable to the stable crystalline modification is irreversible. Thus, if the unstable form, which always has the lower melting point, is melted, transition to the solid stable form may occur. This solid would melt at some higher temperature. Resolidifying the liquid would give either the unstable form which melts as described above or the stable form, which would melt at the higher temperature. The unstable form may transform to the stable form on standing.

Menthol [131] crystallizes in at least four different forms, $\alpha$, $\beta$, $\gamma$, and $\delta$, only one of which ($\alpha$) is stable between zero and its melting temperature, 42.5°C. The other three forms are monotropic and have lower melting points as follows: $\beta = 35.5°$, $\gamma = 33.5°$, and $\delta = 31.5°$. All these unstable forms finally revert to the stable $\alpha$ form on standing. The melted menthol may be undercooled to 32° or lower before it crystallizes in any reasonable length of time. The unstable forms are obtained by rapid cooling of the melt to different extents. If a mixture of the unstable and stable forms is heated to, say, 36°C the unstable forms liquefy. This liquid then begins to crystallize in the stable $\alpha$ form which will melt when the temperature of 42.5°C is reached.

### Dynamic Isomerism

Another interesting case of anomalous melting points is caused by "dynamic isomerism." On melting either of a pair of tautomers or geometric isomers, for example, a more or less rapid transformation sometimes takes place into a definite equilibrium mixture of the two forms. The substance now melts over a range, and since it is a simple binary mixture, it gives cooling and heating curves of types $B$ and $E$ in Fig. 3.5, respectively.

The two isomeric benzaldoximes are an interesting example of this type of "pseudobinary system" [132]. The cis modification has a melting point of 34–35°C, whereas the trans modification melts at 130°C. The system exhibits a eutectic point at about 25° at a composition corresponding to 92% of the cis form. The "natural" equilibrium mixture freezes at 27.7°, however, and will be found to melt at the same temperature immediately after solidifying. No matter what the composition of the original mixture of cis and trans forms, if it is once melted and resolidified, the melting point of 27.7° will be observed.

The cis- and trans-anisaldoximes show the same type of behavior; but in this case the rate of transformation to the equilibrium mixture is relatively slow, so that the third melting point determined would be lower than the second, the fourth lower than the third, and so on until a minimum is reached [133]. For a more detailed discussion of systems of this sort, the reader is referred to Findlay [134].

Acetoacetic ester is perhaps the most common example of dynamic isomerism. The enol and keto forms of this compound have been isolated. They are fairly stable at very low temperatures and in quartz vessels at higher temperatures, but both forms revert to an equilibrium mixture which has been shown to consist of about 93% of the keto and 7% of the enol form [135–137]. The freezing-point behavior of this pseudobinary system has never been investigated.

In order to identify or prove dynamic isomerism in such instances it is necessary to establish the fact that only two substances are present in the equilibrium mixture. The system crotonic acid–isocrotonic acid, for example, which was reported by Morrell and Hanson [138] to exhibit this phenomenon, proved on further investigation [139] to be much more complex in that secondary reactions (polymerization) also take place on melting. In true dynamic isomerism, the eutectic temperature as found by cooling and heating curves should be that of the binary system in question and should not change as the composition changes progressively toward the equilibrium mixture.

## Liquid Crystals

Many substances form liquid crystals upon heating [140] and may appear to have melting ranges instead of sharp melting points. The formation of a turbid liquid will usually be noticed at some temperature below that at which the liquid becomes clear. Upon cooling, this pehenomenon will be found to be reversible. Actually, no solid remains in the turbid liquid but the substance is an anisotropic liquid-crystalline state.

p-Methoxycinnamic acid, one of the common examples of such a substance, has been shown to transform from the solid-crystalline into the liquid-crystalline state at 172.1°C and from the liquid-crystalline into the clear liquid state at 187.3°C. A highly purified specimen of this compound gave "flats" on the cooling curves and heating curves at both of these temperatures [141], showing that each of the phase changes took place reversibly at constant temperature.

The effect of a second component on the melting point of a substance forming liquid crystals is of interest. Since the heat of "fusion" of the liquid crystals is much smaller than that of the solid crystals [141], it would be expected (see p. 113) that the freezing-point depression for a given mole % of impurity would be greater. This is confirmed by the binary freezing-point data of de Kock [142]. The liquid–crystal clarification point should therefore be a more sensitive criterion of purity than the melting point of the solid crystals.

Dupont [143] therefore suggested the use of p-azoxyanisole as a cryoscopic solvent in a modified Rast micromethod for determining molecular weight using the clarification point of the liquid crystals.

De Kock [142] also investigated the addition of p-methoxycinnamic acid to p-azoxyanisole. Each of these substances forms liquid crystals but the addition of the second component results in a lowering of both the solid crystal and liquid crystal melting points by about the same amount. De Kock suggests that this is caused by the formation of a complete series of liquid mixed-crystals showing a minimum clarification point at about 90% azoxyanisole.

## Glassy State

Certain systems in the liquid state, when cooled rapidly or in some cases even slowly, do not crystallize for one or more of the following reasons: low nucleation or crystallization rates, high liquid–crystal surface tensions, and low energies of crystallization. They become more and more viscous, and finally a glassy substance is obtained which is hard and rigid like a solid and transparent and isotropic like a liquid. On long standing, especially at an elevated temperature, they may undergo devitrification, and crystals separate. Materials which occur in the glassy state are boron oxide, albite feldspar, and silicon dioxide. By rapid cooling of molten materials, such as copper, iron, nickel, selenium, metal oxides, sulfur, glycerine, ethylene glycol, toluene, isobutyl bromide, methanol, and ethanol, products in the glassy state can be prepared. Materials in the glassy state are used in glass-ceramics, laser technology, fiber optics, and products where high compressive strength is required [144].

## Melting Points Requiring Special Techniques

The ordinary methods of melting-point determination must be modified for many substances because of their special nature. In most of these cases simple changes in technique, some of which will be discussed below, may be employed to accomplish the desired measurement. For greater detail concerning many of these methods, the reader is referred to Houben [145].

If the substance is highly colored and melting is not visible, Piccard [146] has suggested forming a plug of the material in a capillary by melting and resolidification and noting the temperature at which the plug moves along as the result of air pressure on one end of the tube.

For substances likely to explode, a very small amount of the material is laid on a liquid metal surface or on a cover glass floated thereon. Mercury is used for moderate temperatures, and soft solder in a nickel crucible for temperatures between 170 and 450°C. The sample is covered with a watch glass or funnel, and the temperature of the thermometer immersed in the bath is observed when melting occurs. Serious accidents might occur if explosive substances are handled in the usual way, so any questionable compound should be tested first by heating a small amount on a spatula in order to determine whether precautions are necessary. All observations of substances which might explode should be made from behind a transparent safety screen.

The melting temperature of compounds which sublime, or which absorb or react with moisture, carbon dioxide, or oxygen, can best be obtained by the sealed-tube thermostatic method. They can also be determined with less accuracy by the capillary method.

Hygroscopic substances, such as benzenesulfonic acid, tend to take up water, which changes the metlting point. Such materials are probably best treated by placing them in a capillary, drying, and sealing the tube to prevent reabsorption of moisture. Drying may be accomplished by heating in an Abderhalden drying apparatus or by passing warm dry air over the sample before sealing.

Some substances, such as chloranil and hexamethylethane, tend to sublime below their melting temperatures. This type of material is handled by testing in a short, sealed capillary totally immersed in the bath. Heating the entire capillary prevents sublimation and keeps the solid confined to the bottom of the tube.

If the material is sensitive to the presence of air at high temperatures, the capillary should be evacuated before sealing or the air should be replaced by an inert gas. Strychnine derivatives do not generally melt sharply when air is present. In all cases in which the capillary is sealed, as pointed out above, the sample should be introduced through a long funnel to avoid decomposition of the substance, which might otherwise stick to the neck of the tube.

Fats, waxes, and paraffin-like materials usually do not melt sharply and therefore exhibit melting ranges which may not be too reliable. One difficulty in handling these materials is the problem of filling the capillary tube. This may be done in most cases by melting the substance and sucking it up into the bottom of long, thin-walled U-tube which is then cooled well below the melting point and allowed to stand for several hours. The melting point is taken by immersing the bottom of the U in the bath alongside the thermometer bulb.

Many high polymers, such as the polyamides, are crystalline materials which exhibit more or less definite melting temperatures. Although the crystallinity of these substances usually manifests itself in opacity, the determination of an accurate melting point is often difficult owing to uncertainties in identifying the point at which phase change takes place. Biggs, Frosch, and Erickson [147] solved this problem by preparing threads of the solid polymer about $\frac{1}{16}$ in. in diameter and clamping them in a melting-point bath near the thermometer bulb. When the temperature of the bath reached the melting temperature of the polymer, the thread lost shape and was swept away in the oil stream of the rapidly stirred bath. The phenomenon of melting a crystalline organic polymer can be observed with polarized light since crystallites grow from a common nucleus to form aggregates which can be seen in the polarizing microscope. On very slow heating, the crystallinity of the polymer disappears. The temperature at which crystallinity disappears is defined as the melting temperature [148]. The phenomenon of melting of organic polymers can also be observed by dilatometric measurements of the volume as a function of temperature. The temperature is raised in small increments, about 1°C, and held constant until the volume becomes constant. It may require 24 hr for this equilibration to occur. An abnormal increase in volume, occurring usually within a range of 10°C, indicates that the polymer has melted [149]. The various physical transitions and the melting behavior of polymers depend heavily upon the history of the sample. They can best be measured and characterized by means of differential thermal analysis [150].

## Freezing Temperatures of Binary Mixtures

The determination of the freezing point of binary mixtures presents many difficulties and sources of error that are not encountered in the case of pure compounds. Some of the methods ignore one or more of these sources of error and, therefore, as will be pointed out below, they cannot be expected to give reliable results.

## Contact Method

A rather ingenious qualitative method and technique of determining which type of diagram (Fig. 3.5, 3.6, 3.7, or 3.9) a binary system will give has been suggested by Kofler [151] and Kofler [152]. This method, called the "contact method," consists essentially of bringing the two substances together between the surfaces of two cover glasses from opposite edges, heating them to the melting point, and causing them to diffuse into each other. Thus the cooled slide has supposedly all ranges of concentration from pure $A$ on the left to pure $B$ on the right. When this is observed under a polarizing microscope in the solid state and heated at a slow rate on a carefully controlled hot stage, the composition having the lowest melting point liquefies first and appears as a vertical dark line across the slide. As the temperature is raised, this line of liquid widens and, in the case of a simple eutectic system such as Fig. 3.5, it merely spreads until it reaches the width of the slide. In the case of a system exhibiting molecular compound formation (Fig. 3.6), when the temperature of the second eutectic is reached a second band of liquid forms some distance from the first. On continued heating, the liquid bands spread in both directions until they meet at the melting point of the compound and reach the outer edges at the melting points of the two pure constituents. Similarly, for a system which forms a continuous series of mixed crystals with a maximum (Fig. 3.9 $b$), the melting bands start at the two edges and move toward each other.

## Thaw-Point Method

The capillary method of determining "thaw points" and melting points developed by Rheinboldt [20, 21, 153] cannot be relied upon to give data of suitable accuracy for binary diagrams. In the first place, it does not readily permit the attainment of a true equilibrium between the solid and liquid phases, and, secondly, the technique for the preparation of specimens of known composition involves a considerable source of error. The probable inaccuracy in the determination of "thaw points" (i.e., secondary melting points) or the temperatures at which fusion begins as shown by the heating curves in Figs. 3.5–3.11 has already been discussed. It is admitted [153] that one must have considerable experience with the method on known systems before the method can be used confidently with a *reproducibility* of 0.5°C on the same sample. The limits of error for the capillary method are said to be about 1 or 2°C [154].

Grimm [154] and Lettré [155] used microscopic technique for determining thaw points and melting points with a claimed accuracy of ±2°C. Samples of various compositions are observed while melting between two cover glasses on a hot stage under a polarizing microscope.

The "contact method," the capillary method, and the microscopic method have been used mainly to distinguish the system type and have been especially useful when only small amounts of material were available. It is usually recommended that two or in some cases all three of these methods be used simultaneously on the same system for verification of the conclusions [152, 153, 155]. Thus the capillary and microscopic methods were used by Fredga [27] and by Lettré [156] in establishing the relative steric configuration of optical isomers by the procedure on page 126, in which it is necessary merely to determine whether a binary system is of the simple eutectic or of the molecular-compound type. These methods have also proved useful in showing mixed crystal formation [157] and in establishing incongruent melting points of molecular compounds [153, 158].

### Beckmann Method

A large number of binary systems have been constructed with the Beckmann apparatus described on page 174 using an ordinary thermometer. It has been shown, however, that the errors involved in this method are sometimes so large that false conclusions may be drawn even as to the type of freezing-point diagram involved [159]. The method is reliable only for dilute mixtures, and then only when each composition is made up separately to avoid cumulative errors in composition and errors due to losses or segregation of sample on the stirrer and sides.

### Cooling-Curve Method

Andrews, Kohman, and Johnston [160] have developed a method which eliminates most of these errors, but its applicability is limited to systems with favorable crystallization characteristics. In the apparatus they employed, a small, thin, glass tube containing the sample is suspended in a copper cylinder within a Dewar tube. The temperature of the cylinder is carefully controlled to fall at a uniform rate, as represented by the lower curve Fig. 3.24. The temperature of the sample is measured by means of a thermocouple made of very fine wire, so that there is very little loss of heat by conduction and very little lag. The temperature of the cylinder is started high enough so that, when the freezing point is reached, the temperature fall of sample $ABC$ is almost parallel to that of the surroundings, $LJ$. The maximum temperature, $D$, reached after crystallization sets in, is the temperature at which solid and liquid are in equilibrium. This will differ under given conditions with the degree of supercooling, in other words, with the amount of crystals formed when equilibrium has been attained. Point $D$ is therefore not the true freezing point because the composition of the liquid in equilib-

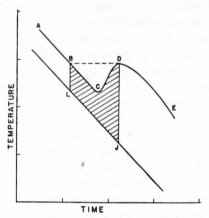

**Fig. 3.24** Typical controlled cooling curve for a binary mixture [160].

rium with crystals at $D$ differs from that of the original solution, owing to the separation of the crystals of pure solvent.

It can be shown from Newton's law of cooling that the amount of crystals formed is proportional to the area $BCDJL$. The true freezing point of the original solution can be obtained by plotting the area against corresponding values of $D$ for a number of such curves, and extrapolating to find the value of $D$ for zero area or for no crystal formation. Figure 3.25 shows three actual curves for the same sample, with different degrees of supercooling, superimposed on each other. The change of the maximum can be seen to be such that direct extrapolation of the maxima back to the curve is uncertain.

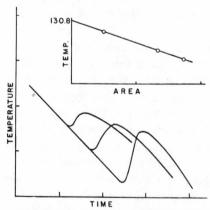

**Fig. 3.25** Cooling curves for different degrees of supercooling and corresponding area—maximum graph [160].

Plotting the area against the temperature of the maximum, however, permits a correct extrapolation to the true freezing point of the solution, in this case 130.8°C.

## Sealed-Tube Method

The simplest method for obtaining primary freezing-point data for binary or multicomponent freezing-point diagrams is the sealed-tube method. It has the decided advantage of being independent of the ease or rate of crystallization of the sample. For each composition, weighed amounts of the two components are sealed in a 15-mm glass tube using special techniques with liquid or volatile components. Two solid-glass beads are included to ensure efficient stirring. In the conventional dynamic version of the method the sample is then subjected to a very gradual increase in temperature while being turned end over end [161], shaken [162], or otherwise vigorously agitated in a liquid bath. The temperature at which the last crystal goes into solution is taken as the saturation temperature or primary freezing point. The chief objection to this procedure is that the rate of change of temperature is not slow enough to ensure attainment of equilibrium of composition and temperature.

This objection was overcome by the static, or thermostatic, version of the method [163]. Here two temperatures, a few tenths of a degree apart, are found, one at which the last crystals disappear, and the other at which a few finely divided crystals remain after prolonged agitation in a constant temperature bath. The freezing point or true solid–liquid equilibrium temperature is taken as the mean of these two temperatures corrected for both thermometer calibration and emergent stem.

An investigation by Satterfield and Haulard [164] showed that the dynamic method overshot the equilibrium temperature, as determined by the static method, by as much as 0.19°C with a heating rate of 0.02°C/min and 1.20°C with 0.10°C/min.

More recently [165] an apparatus has been described which extends the temperature range of the static method down to −60°C and the concentration range down to 0.01 mole % or lower. The liquid bath, contained in a 10-liter unsilvered Dewar, serves either as a thermostat or a cryostat. The sample tube is caused to turn end over end by a rotating horseshoe magnet mounted outside the closed system. Rotation can be interrupted for short intervals as desired. A focusable microscope light is positioned behind the apparatus so as to shine through the sample tube, to aid the visual observation of the solute crystals.

The sealed-tube method has broad applications. For concentrated mixtures it is much more accurate and reliable than cooling- or heating-curve methods. It can be used for any system in which crystals can be caused to form. It is especially adapted to systems which contain components which are volatile or absorb or react with moisture, carbon dioxide, or oxygen. An inert atmosphere or partial vacuum may be used. Multiple determinations can be made on the same sample. In solubility determinations correction can be made, when pertinent, for the slight change in composition caused by the presence of some of the solvent in the vapor phase. The method is especially effective for the detection and investigation of polymorphic forms because of the visibility of the crystalline phase, ease of temperature manipulation, and because the sample can be readily removed from the bath for the other thermal treatments required [166] to cause the formation of the different crystal modifications. The method can also be used for determining the freezing points of pure substances.

### Eutectic Temperature and Composition

The eutectic temperature can be obtained from cooling curves, heating curves, or by one of the micromethods. Both the eutectic temperature and the eutectic composition can be found accurately by extrapolation of the two branches of the freezing-point diagram. A rapid method of estimating the eutectic composition for binary solutions of solid organic compounds in liquid solvents [167] involves chilling a weighed amount of a mixture of known composition in a centrifugal filtration tube (Fig. 3.21) to well below the eutectic temperature; for example, in a dry-ice or liquid-air bath. It is then allowed to warm up to room temperature while being centrifuged, so that, as the eutectic melts, it is immediately separated from the solid and collects in the lower chamber, in which it can be weighed. The eutectic composition can then be calculated and a cooling curve for the liquid will give the eutectic temperature. This procedure is particularly suited for determining cryohydric points.

### Melting Points at High Temperatures

The melting temperatures of substances, too high to be determined by means of the fusion techniques described above [168], are determined under special conditions [169–173] which minimize oxidation or change in composition of the substances, in an apparatus which minimizes thermal gradients in the sample. Special methods of detecting melting are used, for example, ratio of brightness of heated metals [169], structural changes in metals on heating [171, 174], and special methods of thermometry. External conditions used include: reducing atmospheres, such as hydrogen; inert atmospheres, such as helium and argon; high pressure; and high vacuum.

### Dynamic Methods

The various dynamic methods are based on measurement of a discontinuous change in energy content or on some other physical property of the substance when the substance melts or freezes [171, 175, 176]. The time-temperature-curve method, frequently referred to as the *thermal analysis method*, is the method most commonly used and measures the change in heat content during the transitions from solid to liquid phase. The experimental apparatus usually consists of an electrically heated furnace, crucible, or hot stage with a centrally positioned temperature measuring device. A heating or cooling curve of the sample, obtained during controlled heating or cooling of the sample, is recorded [175, 176]. The melting temperature of barium molybdate, cooled at 7°C/min has been determined at 1480°C with an accuracy of ±5°C [175]. Supercooling, particularly for inorganic substances with high melting temperatures, such as ceramics, silicates, and oxides, can be so pronounced that melting temperatures obtained by cooling curves are much too low. The melting temperature of aluminum oxide obtained from cooling curves may be as much as 200°C lower than that observed by other techniques [177]. Heating curves for inorganic substances are generally less well-defined than cooling curves [177].

Other dynamic methods developed for determining melting temperatures of inorganic substances are based on the *visual observation* of a sample during heating or cooling. For example, a black-body sight tube is immersed into a melted substance, and the changes in temperature during the freezing of the substance are observed by means of an optical pyrometer [178].

The melting temperature of metals has been determined by drilling a hole in a sample of the metal, then heating the sample, and recording the temperature at which the hole in the sample fills with liquid metal. The melting temperatures of metals have also been determined by heating a finely divided sample until small beads of metal are formed and then recording the temperature [179].

A special technique for determining the melting temperature of samples as small as 10 $\mu$g was developed for transuranium elements [180]. The convergence of tungsten jaws, holding the sample, was microscopically observed with a precision for determination of the melting temperature of $NpF_3$ of 640 ± 1°C.

### Static Methods

Various types of adiabatic calorimeters for use at high temperatures have been described in a review by Westrum [181]. The static method, commonly known as the *quenching method*, is used mostly for binary or multicomponent systems. It takes into account the fact that phase changes in many substances, such as the silicates, are relatively slow and may be stopped or "frozen in" by rapid cooling of a heated sample. For example, a sample is heated at a constant temperature in a furnace, until equilibrium is attained. This may take days, weeks, or even months. The heated sample is then immersed in a cooled liquid or is cooled rapidly by lowering the temperature of the furnace. A special hydrothermal quenching technique, operating at pressures as high as 30,000 psi, has been developed for investigating systems containing volatile components [184]. The quenched sample is now examined for evidence of melting; it may be prepared, by the special technique required, for microscopic examination. The crystal type of crystal types present can be identified and, for glass-forming substances, the liquid present at equilibrium will appear as an isotropic material. By examining numerous specimens of the same sample material quenched from selected temperatures, extending over a suitable range, one can determine not only the temperature at which melting was complete but also the temperatures at which specific crystal phases disappeared as the equilibrium temperature was raised [173, 182].

A generalized classification of techniques and apparatus for determining melting temperatures of substances over a range from about $-200$ to $4000°C$ is summarized in Table 3.4.

**Table 3.4**  Classification of Techniques and Apparatus

| Melting Temperature Range (°C) | Heating Technique | Temperature Measurement | Environmental Conditions | Special Application |
|---|---|---|---|---|
| $-200$ to 25 | Liquid heat transfer [a] | Platinum resistance | Vacuum | Substances which are gases or liquids at room temperature |
| 25 to 300 | Liquid heat transfer [b] | Hg in glass | Air | — |
| 25 to 1500 | Hot stages [b] | Thermocouple | Air | — |
| 500 to 1500 | Resistance-type furnace [c] | Platinum resistance | He, $N_2$, $H_2$, vacuum | — |
| 500 to 2500 | Gas heating [d] | Optical pyrometer | No control over atmosphere | Good thermal contact |
| 1000 to 3500 | Induction-type furnace [e] | Optical pyrometer | Inert atmosphere or vacuum | Rapid heating, noncontaminating, concentrated in small area |

**Table 3.4** Continued

| Melting Temperature Range (°C) | Heating Technique | Temperature Measurement | Environmental Conditions | Special Application |
|---|---|---|---|---|
| 3000 to 4000 | Electron beam [f] <br> Plasma heating [g] <br> Solar furnace [h] | Optical pyrometer | Inert atmosphere | Rapid heating |
| 3500 to 4000 | Electric arc [i] <br> Radiant heating [j] | Optical pyrometer | Inert atmosphere or vacuum | Rapid heating |

[a] D. N. Glew and N. S. Rath, *J. Sci. Instr.* **42,** 665 (1965); O. Maas and D. McIntosh, *J. Am. Chem. Soc.* **34,** 1273 (1912); *ibid.* **35,** 535 (1913); O. Maas and C. H. Wright, *ibid.* **43,** 1098 (1921); H. Tchejeyan, *Chem. Ind. London* **1969,** 133.

[b] See Table 3.7.

[c] See Table 3.7; R. F. Domagala and E. Heckenbach, *Rev. Sci. Instr.* **35,** 1663 (1964); T. Fohl and R. W. Christy, *J. Sci. Instr.* **36,** 98 (1959); S. D. Gromakov, V. N. Kurinaya, Z. M. Latypov, and M. A. Chvala, *Russ. J. Inorg. Chem. Eng. Transl.* **9,** 712 (1964); E. F. G. Herington, R. Handley, and A. J. Cook, *Chem. Ind. London* **1956,** 292; W. D. Lawson and S. Nielson, *Preparation of Single Crystals*, Butterworth, Washington, D. C., 1958; R. H. McFee, *J. Chem. Phys.* **15,** 856 (1947); D. C. Stockbarger, *Rev. Sci. Instr.* **7,** 133 (1936); R. F. Domagala and E. Heckenbach, *Rev. Sci. Instr.* **35,** 1663 (1964).

[d] F. Brown and W. H. Todt, *J. Appl. Phys.* **35,** 1594 (1964); F. R. Chawat and R. M. Youmans, *Am. Ceram. Soc. Bull.* **44,** 409 (1965); E. F. Herington, *Sch. Sci. Rev.* **43,** 35 (1961).

[e] P. G. Simpson, *Induction Heating*, McGraw-Hill, New York, 1960.

[f] R. Bakish, Ed., *First International Conference on Electron and Ion Beam Science and Technology*, Wiley, New York, 1965; A. Calverley, M. Davis, and R. F. Lever, *J. Sci. Instr.* **34,** 142 (1957); A. Lawley, in *Introduction to Electron Beam Technology*, R. Bakish, Ed., Wiley, New York, 1962, Chap. 8.

[g] M. B. Gottlieb, *Internatl. Sci. Technol.* **44,** 44 (1965); L. Spitzer, Jr., *Physics of Fully Ionized Gases*, 2nd ed., Wiley, New York, 1962; T. B. Reed, "Plasmas for High Temperature Chemistry" in *Advances in High Temperature Chemistry*, L. Eyring, Ed., Academic, New York, 1967, pp. 260–316.

[h] T. Noguchi, "High Temperature Phase Studies with a Solar Furnace," *ibid.*, Vol. 2, 1969, pp. 235–262.

[i] R. D. Burch and C. T. Young, U.S. Atomic Energy Commission, N.A.A., SR-1735, 23 (1957); O. N. Carlson, F. A. Schmidt, and W. M. Paulson, *Trans. A.S.M. Quart.* **57,** 356 (1964); G. A. Geach and F. O. Jones, *Metallurgia* **58,** 209 (1958).

[j] L. R. Weisberg and G. R. Gunther-Mohr, *Rev. Sci. Instr.* **26,** 896 (1955).

## 4   FREEZING-POINT MOLECULAR WEIGHT DETERMINATIONS

**Theory**

Since the lowering of the freezing point of a solvent by a solute is proportional to the logarithm of the mole fraction of the solvent, it is possible to determine the molecular weight of the solute in a binary mixture from the freezing-point equation.

By definition, $N_A = 1 - N_B$, where $N_A$ is the mole fraction of the solvent and $N_B$ is the mole fraction of the solute present in the mixture. By substitution, it follows from (3.2) on page 112 that

$$\Delta T = - (RTT^0/\Delta H) \ln (1 - N_B).$$

For dilute solutions in which $N_B$ and $\Delta T$ are small, we can make the mathematical assumptions that $-\ln (1 - N_B) = N_B$ and $TT_0 = T_0{}^2$ because in

$$-\ln (1 - x) = x + \frac{x^2}{2} + \frac{x^3}{3} + \dots$$

$x^2/2$, $x^3/3$, . . . may be neglected if $x$ is small enough. Then

$$\Delta T = (RT_0{}^2/\Delta H) N_B.$$

This equation may also be expressed in terms of molality, $m$, that is, in terms of the number of moles of solute/1000 g of solvent. By definition,

$$N_B = \frac{m}{m + 1000/M_A}$$

where $M_A$ is the molecular weight of the solvent. Hence

$$\Delta T = \frac{RT_0{}^2}{\Delta H} \frac{m}{m + 1000/M_A}.$$

For very dilute solutions, where $m$ may be neglected in comparison with $1000/M_A$, we may write

$$\Delta T = \frac{RT_0{}^2}{\Delta H} \frac{mM_A}{1000} = \frac{M_A RT_0{}^2 m}{1000 \Delta H} = Fm,$$

where $F$ is a constant characteristic of the solvent and is defined as

$$F = \frac{M_A RT_0{}^2}{1000\Delta H}.$$

Now, if $W_B$ is the number of grams of solute/1000 g of solvent, $m = W_B/M_B$, in which $M_B$ is the molecular weight of the solute. Substituting this expression for $m$, we have $\Delta T = FW_B/M_B$, and, solving for the molecular weight,

$M_B = FW_B/\Delta T$. This is the equation used in the determination of molecular weight by means of freezing-point depression.

The use of this equation requires knowledge of the molar freezing-point depression constant, $F$, the concentration involved, and the measured depression of the freezing point. Although it is possible to calculate $F$ by substitution of the factors in the defining equation shown above, values of this constant determined experimentally using a substance of known molecular weight are much to be preferred.

Although $F$ is considered constant, it may vary considerably depending upon the substances used to determine it. Factors which contribute to lack of ideality would also tend to cause variation in experimentally determined values of the constant. These factors would result in an incorrect molecular weight, as will be discussed in connection with Fig. 3.27. The accuracy which one may anticipate from a freezing-point molecular-weight determination is also limited by the accuracy with which the constant, $F$, is known.

Table 3.5 presents a comparison of theoretical and experimental constants. The theoretical values were calculated from measured heats of fusion, whereas the experimental values were obtained using known substances in actual determinations. For constants of additional solvents the reader is referred to the *International Critical Tables* or to Landolt-Börnstein.

**Table 3.5**   Freezing-Point Molecular-Weight Constants

| Compound | Mp (°C) | F (calc) | F (obs) |
|---|---|---|---|
| Acetic acid [a] | 16.55 | 3.57 | 3.9 |
| Benzene [b] | 5.45 | 5.069 | 5.085 |
| Borneol [c] | 204.0 | — | 35.8 |
| Camphor [c] | 178.4 | 37.7 | 40.0 |
| Dioxane [c] | 11.7 | 4.71 | 4.63 |
| Ethylacetanilide [c] | 52.0 | 8.7 | 8.58 |
| Indene [c] | −1.76 | 7.35 | 7.28 |
| Naphthalene [a] | 80.1 | 6.98 | 6.899 |
| Nitrobenzene [d] | 5.82 | 6.9 | 8.1 |
| Trimethylcarbinol [e] | 25.1 | 8.15 | 8.37 |
| Water [a] | 0.0 | 1.859 | 1.853 |

[a] Landolt-Börnstein, *Physikalisch-Chemische Tabellen*, Erg. IIb, p. 1468.
[b] B. C. Barton and C. A. Kraus, *J. Am. Chem. Soc.* **73**, 4561 (1951).
[c] Landolt-Börnstein, *op. cit.*, Erg. IIIc, pp. 2667–2669.
[d] *Ibid.*, Hw. II, pp. 1427–1430.
[e] F. H. Getman, *J. Am. Chem. Soc.* **62**, 2179 (1940).

## Apparatus and Technique

The procedure for the experimental determination of molecular weights by freezing-point depression, sometimes known as the Beckmann method, is described in detail in many elementary physical chemistry manuals [185]. The apparatus consists essentially of a cooling bath, $B$, covered with a metal lid, $L$ (Fig. 3.26). Through a hole in the center of the lid passes a large glass test tube, $T$, held in place by a tightly fitting cork. Held inside the test tube is another smaller tube containing the freezing mixture, a Beckmann differential thermometer, and a small stirrer. The freezing-point tube is therefore surrounded by an air layer through which heat must be transferred, thus ensuring a low and uniform rate of cooling of the solution.

After the apparatus is assembled, a weighed amount of pure solvent is placed in the sample tube and the system allowed to come to thermal equilibrium with the sample cooling at a steady rate. Temperature measurements to the nearest 0.001°C are made as the liquid is supercooled and caused to crystallize with stirring. The final equilibrium temperature between the liquid and a small proportion of crystals is taken as the freezing point. If the

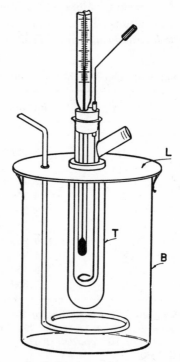

**Fig. 3.26**  Beckmann freezing-point-depression apparatus [186].

temperature were to be plotted against the elapsed time, a curve similar to the cooling curves for a pure substance as described on page 144 would be obtained. The solvent is then remelted by brief removal of the inner test tube, and the process repeated using different degrees of supercooling. A weighed amount of the unknown is then added to the solvent, and the freezing temperature is again determined as before. From the calculated concentration of solute and the observed difference in freezing point of the pure solvent and the solution, the molecular weight of the solute is computed.

In the freezing-point molecular-weight determination, the precautions to be observed are similar to those for cooling curves. The temperature of the cooling bath should not be more than about 3° below the normal freezing point of the solvent; otherwise the rate of heat loss may exceed the rate at which heat is furnished the solution by the heat of fusion of the freezing solvent, and the observed temperature will be too low. The amount of supercooling should not be greater than about one-half a degree, so that the amount of solvent freezing out will not appreciably alter the concentration of the solute. Stirring should be just rapid enough to maintain the contents of the tube at a uniform temperature, and the thermometer should be tapped throughout the determination to prevent the mercury from sticking in the fine capillary of the Beckmann thermometer. For a systematic and concise discussion of the sources of error involved in the determination of the molecular weight by freezing-point depression, the reader is referred to the comprehensive treatment by Findlay [186].

Freezing-point lowering offers one of the most convenient means of molecular weight determination. The depressions are easy to measure experimentally because the molal freezing-point depression is relatively large and the freezing point is not affected by such external factors as changes in the atmospheric pressure. The measurement can be made at low temperatures at which volatility of the solvent or the solute is not likely to cause a change in the composition of the mixture. The extreme accuracy attainable by this method employing special refinements of technique is illustrated by the data of Batson and Kraus [187], as shown in Table 3.6. Data such as these indicate that inaccuracies may in many cases be due to experimental error.

The accuracy of the freezing-point depression method may be even less than that expected from the experimental accuracy of the measurements. It should be recalled that the derivation of (3.2) involved a number of assumptions, one of which was the mutually ideal behavior of solvent and solute. It was also pointed out on page 115 that truly ideal behavior between organic substances is the exception rather than the rule. Although the behavior of most solutes approaches ideality in extremely dilute solutions, the concentrations involved in freezing-point experiments are such that, in

**Table 3.6**    Freezing Points of Triphenylmethane in Benzene [187]

| Molal Concentration | Freezing-Point Depression (°C) | Molecular Weight |
|---|---|---|
| 0.000313 | 0.00158 | 244.5 |
| 0.000634 | 0.00322 | 243.5 |
| 0.000986 | 0.00497 | 245.4 |
| 0.004096 | 0.02082 | 243.5 |
| 0.0248 | 0.1263 | 243.1 |
| 0.04375 | 0.2214 | 244.6 |
| | Theoretical: | 244.32 |

most cases, abnormalities due to nonideality are not eliminated. Any factors causing deviation from the freezing-point law would cause corresponding errors in molecular weights by freezing-point depression. Thus, using benzene as a solvent (Fig. 3.27), the molecular weight determinations would be expected to give: the correct value in the case of naphthalene or tetrachloroethane, which behave ideally with benzene; too high a value for propionic acid, phenol, or aniline, which in the concentration involved tend to associate in benzene; and too low a value for hexaphenylethane or tetraphenyldi-$\beta$-naphthylethane, which dissociate in benzene into triphenylmethyl or diphenyl-$\beta$-naphthylmethyl, respectively. The degree of association or dissociation of a solute at each concentration can be calculated from the experimental and the true values for the molecular weight.

It will be noted from Fig. 3.27 that the molecular weight obtained for certain organic compounds by freezing-point depression increases linearly with solute concentration. This phenomenon suggests extrapolation of the observed values back to zero concentration of solute to obtain the molecular weight in a very dilute and more nearly ideal solution, a procedure frequently employed in determining the molecular weight of heavy petroleum oils. Thus Rall and Smith [188] estimated molecular weights as high as 1746 for isobutene polymers in benzene solution by the freezing-point depression method.

Although the freezing-point depression method has been used rather extensively by numerous investigators for compounds of high molecular weight and for high polymers [190] the results cannot usually be considered reliable. These cryoscopic measurements yield ordinary number-average molecular weights [191]. The values obtained often vary considerably with the concentration and the solvent used, and they are often not in agreement with those obtained by other methods. Meyer and Mark [192] and Staudinger [193] consider the method inapplicable to rubber preparations, and

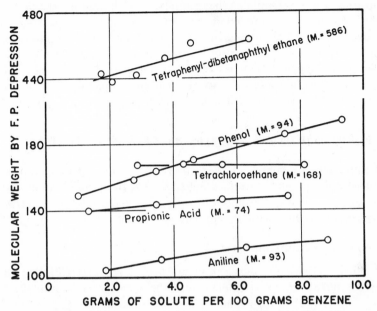

**Fig. 3.27**  Freezing-point molecular weights in benzene solutions as affected by concentration [189]; $M$, true molecular weight.

Freudenberg [194] thinks that it should be discarded in the field of polysaccharides and their derivatives. Hess [195], on the other hand, seems to prefer this method in the study of naturally occurring high polymers.

Recent improvements in semimicro- and microtechniques have made it possible to determine freezing-point molecular weights on extremely small samples of organic substances. A method with an accuracy of about 0.002° when 0.1–1 $\mu$g of solution is used has been reported [196].

The method of Rast [197] employs camphor as a solvent because of its unusually high molal freezing-point constant (see Table 3.5). The melting point of the pure camphor is compared with that of the camphor–unknown mixture, and the molecular weight of the unknown is calculated in the usual manner. The melting points are determined by means of a capillary method essentially the same as that described above. The procedure has been modified [198] so that one can determine the molecular weight with a sample weighing a fraction of a milligram. Additional cryoscopic solvents have been suggested which in some cases may have advantages over camphor [199].

Details of this micromethod may be found in any microchemical test or laboratory manual [199, 200]. In its application it is essential that the sample and the camphor be thoroughly mixed in order to ensure good equilibrium. This is probably best accomplished by grinding the substances together and premelting before determining the melting point. The mixture should be well packed to a depth of about 2 mm in the bottom of the capillary in which the melting point is to be measured. The correct melting temperature is the temperature at which the last of the crystalline camphor skeleton dissolves in the liquid. When proper technique is employed, it is possible to obtain by this micromethod molecular weights within 5% of the theoretical value if the solute forms nearly ideal solutions with the solvent.

## 5  PURIFICATION OF SAMPLES BY RECRYSTALLIZATION

Reliable melting points for characterization or identification can be determined only on pure crystalline material. It is therefore essential that the experimenter use good crystallization techniques and procedures in preparing his substance for the melting-point determination.

### Solvent Recrystallization

If other methods for purification are available, recrystallization of the material should not be attempted until the substance is reasonably pure. When a binary mixture, for example, such as represented in Figs. 3.4 and 3.5, is dissolved in a solvent and cooled, its crystallization behavior can be represented by a ternary diagram [201]. From such diagrams it can be shown that the presence of large amounts of impurity reduces greatly the amount of pure crystals which can be obtained before the impurity also crystallizes out. Impurities, furthermore, frequently act as crystallization inhibitors, slowing down or even preventing crystallization in the temperature range where the material can be conveniently handled.

Unless the experimenter has previous knowledge of the best solvent to be used in a particular crystallization, the solvent must be determined by a trial-and-error process. The best solvent is usually one with a good temperature coefficient of solubility and one in which the impurities, if their nature is known, are very soluble. The solute should not be too soluble in the cold or the excessive losses encountered will require the use of special techniques. The usual procedure for selecting a solvent is to dissolve small amounts of the substance in small portions of different solvents by heating. These solutions, which are nearly saturated, are then allowed to cool slowly, note being made of the nature and amount of crystals which form. Several solvents with the desired characteristics may be found, in which case it is generally best to use the one which has a moderately high boiling point, since this reduces evaporation and allows the use of a larger temperature change during crystallization. High-boiling solvents have the disadvantage of being more difficult to remove completely from the final crystalline product.

The two most important steps in a good crystallization procedure are: first, the preparation of pure crystals (or as nearly pure crystals as it is possible to obtain); and, second, complete separation of the liquid from the solid. In general, the procedure for crystallization is to heat the mixture of solvent and solute until the latter dissolves, filter hot, reheat to dissolve any solid which may have formed, add a seed crystal, allow the mixture to cool slowly until crystals form in sufficient quantity, and finally separate the crystals from the mother liquor. The amount of solvent employed should usually be sufficient just to dissolve the material at a temperature fairly close to the boiling point of the solvent so that the maximum cooling effect can be obtained. To follow the progress of the purification, the melting point of these crystals should be taken. If it is higher than that of the original substance, the crystallization process should be repeated until no further rise is observed. Before determining the melting point, care should be taken to remove the last traces of solvent. This can best be done by heating *in vacuo* below the melting point in an Abderhalden drier.

An additional crop of crystals may be obtained from various mother liquors, either by cooling to a lower temperature or by partial evaporation, heating to dissolve any solid formed, and then cooling. The second crop of crystals is usually considerably less pure because the proportion of impurities is higher in the second mother liquor than in the first mother liquor.

Impurities may be carried down by occulsion in the crystals formed, especially if the rate of crystallization is too high. When crystals are formed rapidly, the concentration of impurity in the vicinity of the crystals is relatively great, owing to the removal of the main solute from solution. Further growth on these crystals will therefore be less pure. Slow cooling, in other words, slow crystal formation, permits the main solute to diffuse in the direction of the crystals and thus maintain a more uniform concentration throughout the solution. A low degree of supercooling, seeding, a slow rate of cooling, and gentle stirring—all aid greatly in obtaining pure crystals [202].

Evaporation of the solvent during crystallization should be avoided because evaporation from the surface of the liquid tends to leave a crust of solute and impurities on the edges or sides of the container, thus adding to the amount of impurity present in the final product. It is therefore advisable to carry out the crystallization in stoppered Erlenmeyer flasks or similar closed containers. Crystallization by evaporation of the solvent is usually considerably less efficient as a means of purification than is crystallization effected by temperature change.

Separation of the mother liquors from the crystals is the second important step in obtaining a pure product. The common method of decantation or filtration on a funnel leaves mother liquor upon the crystals even if the crystals are washed with pure solvent. When this liquid is evaporated off in drying the crystals, all the impurities present therein are deposited on the crystals. Experiments by Richards [203] demonstrated the superiority of centrifugation as a means of separating the liquid from the solid phase. Two crystallizations of sodium nitrate with nitric acid as an impurity removed only 90% of the impurity when the decantation method was used and 99.995% when centrifugation was used for separating the solid. If, however, the substance is almost pure, little may be gained for practical purposes by the use of the centrifugation method. The impurities remaining on the solid after decantation or filtration may often be removed to some extent by washing the wet crystals with more of the pure solvent, thus diluting the concentration of the impurities in the mother liquors remaining. This procedure has the disadvantage, however, of washing away part of the product and thus lowering the yield.

It is often difficult to find a solvent which has the correct solubility characteristics for a desired crystallization. In such cases, the method of mixed solvents may sometimes be successfully used. The material is first dissolved in solvent $A$, in which it is very soluble. The solution is then brought to an elevated temperature, and solvent $B$, in which the substance has a limited solubility, is added until turbidity is observed. The solution in the mixed solvents is then warmed a little until clear and allowed to cool slowly, permitting crystallization to take place. A less pure second crop of crystals may be obtained from the mother liquor by repeating with the addition of more solvent $B$ or merely by cooling the mother liquor to a lower temperature.

To crystallize small quantities successfully, mechanical loss must be avoided. This can be accomplished in many cases by means of a hot-extraction apparatus as described by Blount [204] or by means of a combination filter and crystallization vessel described by Craig [205]. In the Blount apparatus, the material is placed in a sintered-glass filter funnel or a small Soxhlet thimble suspended from a condenser and is dissolved by refluxing the hot solvent from a vessel directly below. When extraction is complete, the solution is allowed to cool and the substance crystallizes. The substance has thus been dissolved, filtered hot, and crystallized without handling, exposure, or removal from the apparatus.

An efficient method of crystallization in which separation is made by centrifugation instead of filtration, and which is applicable to crystallizations below room temperature, is afforded by the centrifugal filtration tube [206, 207] and its modifications [205].

A typical centrifugal filtration tube with a standard taper ground joint is illustrated in Fig. 3.28. The crystallization tube, $A$, containing the solute and sufficient solvent is placed in a bath at the desired temperature until solution is complete. The sides of the tube are then washed down with a

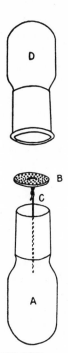

**Fig. 3.28** Centrifugal filtration tube [206, 207].

few drops of pure solvent so as to form a dilute layer on top of the solution. The porcelain filter, *B*, and the cap, *D*, are fitted into position as indicated, and the whole cooled slowly to the crystallizing temperature. The tube is then inverted and quickly spun in a centrifuge. The mother liquor is thus efficiently removed at the crystallizing temperature from the solid phase which is retained above the filter disk in a hard pack. After centrifuging, the cap is carefully removed from the tube, still in the inverted position to avoid contact between the liquid and the solid. If the disk remains seated in the cap, it is carefully removed with any loose crystals by means of the wire *C*. The mother liquor is removed from the cap, which is then washed with solvent for the second crystallization. The purified solid is kept in the crystallization tube and is redissolved in and recrystallized from a fresh batch of solvent as before. If necessary, the filter *B* may be covered by a piece of filter paper which has been slit where the wire is fastened to the disk. Glass tubes of various capacities made from tubing ranging from 6 mm for semimicrocrystallizations to 45 mm are available and can withstand centrifuge speeds corresponding to 300 times gravity. It is convenient to use tubular pieces of wood of various wall thicknesses to adapt the different sizes of tubes to the large centrifuge cups in a No. 2 International Centrifuge.

A metal centrifugal filtration tube has now been designed [208] which has a capacity of 250 ml and which can be centrifuged at much higher speeds. The centrifugal filtration tube is particularly applicable to the recrystallization of organic substances from solvents in which they have a high solubility and a high temperature coefficient of solubility. The sample is enclosed so that condensation of water vapor in the solution and evaporation of solvent are precluded. Thus 2,4-cholestadiene was prepared for the first time in highly pure form by a number of recrystallizations of 0°C from one-half its weight of ether [207]. Similarly a sample of lauric acid was subjected to six recrystallizations from one half or less its weight of acetone at 0°C with an overall yield of about 62% of highly pure product [208].

When it is desired to separate two or more substances or to purify a substance with maximum yield by recrystallization, a process of systematic fractional crystallization is employed. This involves separating the substance into a number of fractions of varied composition or purity and recombining crystals and mother liquors of like composition or purity according to a specified plan, thus avoiding unnecessary accumulation of small crops and mother liquors. This method has been used in the isolation of rare earth salts [209] and is described in many textbooks and manuals. It is concisely outlined by Cumming, Hopper, and Wheeler [210] as follows in connection with Fig. 3.29.

"The mixture is dissolved with the aid of heat in a solvent to give solution (1). From this solution on cooling, crystals separate which are filtered off, and solution (1) is thereby divided into crop (2) and mother liquor (3). Crop (2) is dissolved in the minimum quantity of hot solvent, and from the

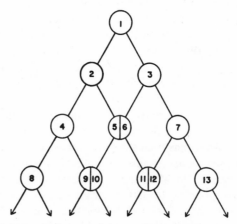

**Fig. 3.29**   Scheme for systematic fractional crystallization [209].

resulting solution, after cooling, crop (4) and mother liquor (5) are obtained. Mother liquor (3) is concentrated, and from the concentrated solution after cooling, crop (6) and mother liquor (7) are obtained. Crop (6) and mother liquor (5) are united to form a single fraction, and after being heated to dissolve, are subsequently cooled to give rise to crop (10) and mother liquor (11). Crop (4) is dissolved in a small portion of pure solvent by heating and after cooling is divided into crop (8) and mother liquor (9). Mother liquor (7) after concentration and cooling yields crop (12) and mother liquor (13). (9) and (10), likewise (11) and (12), are united to give single fractions. Proceeding in this way, the least soluble compound goes to the left in the diagram, while the most soluble goes to the right, and compounds of intermediate solubility lies between these extremes."

The crystalline crops obtained by this method should be tested for purity by observation under a lens or microscope or by melting-point determination. If the crop is found to be pure, it should be set aside. If the crystalline material consists of two different crystal fractions, the solution from which it was obtained was evidently saturated with respect to both. The solvent should be evaporated off in this case, and another solvent tried, one in which the substances have different solubilities.

The chance of contamination and the inevitable loss of material plus the tediousness involved in the filtration method makes it undesirable for use in systematic fraction crystallization. In their work on lead nitrate, for example, Richards and Hall [211] used decantation entirely. This is suitable in the case of an inorganic salt but is extremely inefficient when the crystals are less dense and when small amounts of volatile solvents are used as is often the case with organic compounds. The use of a series of centrifugal filtration tubes with interchangeable ground joints very effectively overcomes these difficulties.

## Fractional Freezing

A discussion of purification by crystallization would be incomplete without some mention of purification by slow fractional freezing. Two techniques have been described [202]. The first involves the slow lowering of a cylindrical cell, filled with the fused substance, through a heating coil in such a way that freezing begins at the bottom and progresses upward, as the cell emerges from the coil, until the whole mass is frozen. During the freezing, which may extend over a period of several hours, the liquid portion is constantly stirred. Finally the tube containing the solid material is cut into fractions as desired, the last portion freezing being discarded as the least pure.

In the second method, which is suitable for larger quantities, a spherical flask filled with the stirred liquid is cooled at a very low rate in such a way that freezing begins at the wall of the flask and proceeds inward until the desired fraction has solidified. The less pure liquid portion remaining is siphoned out of the flask. The crystallized material is partially melted and washed by insertion of a small radiant heater (such as an automobile light bulb) into the center of the flask to form some liquid which is also siphoned off. Schwab and Wichers estimated in their purification of benzoic acid by this second method that two fractional freezings were as efficient as eleven recrystallizations from benzene or twenty-five from water.

## Single-Crystal Formation

A third method, developed by Horton and Glasgow [212], is a highly refined version of the first method. Purification is accomplished by single-crystal formation under rigorously controlled conditions. Here the sealed tube containing the fused sample is lowered through the interface between two immiscible liquids. Advantage is taken of the remarkably low heat transfer across such an interface. Each of the liquid layers is heated separately through the walls of the outer tube so as to maintain two temperature zones above and below the melting point of the substance being crystallized. The rate of lowering of the growth tube and the temperature difference between the two layers (as much as 50°C for benzene purification) and so adjusted that the upper surface of the growing crystal is convex upward. This minimizes the effects of unwanted nucleation at the wall of the growth tube.

The "freezing staircase method" developed by Saylor [213] is another recent technique for preparing ultrapure monocrystalline substances. It depends on repeated controlled crystallizations with continuing removal of rejected impurities. An annular chamber, which may be either vertical or tilted but must not be horizontal, is slowly rotated clockwise in a space that is thermostated at a temperature a few degrees above the melting point of the substance being purified. The space inside the container has a rectangular cross section and is half to three-quarters filled with the substance being purified. Except where it is locally cooled, this material is liquid and flows to the low side of the cavity. In close promixity to the outside of the container are two sets of cooled surfaces maintained a few degrees lower than the melting point of the material. One set extends inwards from the rim and the other, at alternate positions, extends outward from the center. As the container rotates counterclockwise, the mound-shaped crystalline portions attached to the walls move with the chamber. The leading edges of each solid region move slowly away from the coolers, then melt. At the same time, the trailing edges move closer to the metallic cooler and the crystalline solid grows. The impurities join the flowing stream of liquid and move counterclockwise to the further end of the channel which must be maintained between the alternate-positioned opposing solid portions.

Provision is made for removing the impurity-rich liquid and for adding new material.

## Zone Melting

The principle of zone melting, developed originally by Pfann for the preparation of ultrapure germanium for transistors, has been applied to the purification of metals and organic compounds [214]. The solid in a vertical cylindrical tube is melted at the bottom by an annular heater which fits around the tube and moves upward allowing the liquid below to recrystallize so that the impurities are eventually carried with the mother liquor to the top of the tube.

The basic principles, apparatus, and applications of the above fractional solidification processes have been discussed in a number of recent reviews [215].

## 6   GUIDE FOR EVALUATION OF LITERATURE DATA

Melting and freezing temperatures published in the literature must be critically evaluated before being accepted as reliable physical constants. In judging the accuracy of such data a number of factors must be considered. No matter how accurate the method of measurement, the observed melting point will be incorrect if the specimen is impure and, conversely, no matter how pure the specimen, the expected accuracy is no greater than that attainable by the method of measurement. By special calorimetric procedures a correction can be calculated for small amounts of impurity.

From a chemical point of view the origin or method of preparation and purification of the compound are of primary importance. The known chemical reactivity of the compound may suggest special precautions which should have been taken when making the determination. From a physical point of view, the accuracy and precision of the method used have to be considered. The selection of standards for reference, in order that the results from different laboratories can be compared, is also important. Other factors to be considered are the reputation of the author publishing the data, whether he is a recognized specialist, and whether the determination of the melting or freezing temperature was one of the major purposes of the work reported or only incidental to it. Confirmatory data based on independent measurements by several investigators with different apparatuses and using specimens of different origins constitute the ideal proof of reliability.

In publishing melting-point and freezing-point data, all the above considerations should, of course, be kept in mind. Sufficient detailed information should be included to establish the probable degree of purity of the specimen used and the precision and probable accuracy of the physical measurement.

## 7 GUIDE FOR SELECTION OF METHODS

The various methods for determining melting and freezing temperatures have been classified in Table 3.7 with a list of the more important applications and the estimated maximum accuracy of each. The latter often varies with the characteristics of the substance under investigation. For example, a high accuracy is attainable by cooling curves only when the rate of crystallization is rapid enough to ensure true phase equilibrium. The accuracy also depends on the validity of the calibration and the particular modification of the technique or equipment used.

**Table 3.7** Classification of Melting- and Freezing-Point Methods

| Apparatus or Method | Special Applicabilities or Advantages | Approximate Maximum Accuracy (°C) |
|---|---|---|
| Pure or Nearly Pure Substances | | |
| Visual methods | | |
| Capillary, Thiele tube | Small sample, routine, convenient | 0.5–1.0 high |
| Capillary, copper block | Small sample, routine, convenient | 2 high |
| Capillary, Hershberg | Small sample, routine, convenient | 0.3–0.5 high |
| Dennis bar | Small sample, rapid, unstable and high-melting substances | ±3 |
| Kofler hot bar | Small sample, rapid, unstable substances | ±5 |
| Microscope hot stages | Microsample, −100 to +1500°C | ±0.2–10 |
| Sealed tube, static | 0.5-g sample, hygroscopic, volatile, transition points | ±0.1 |
| Microscope hot stage | Microsample | ±1 |
| Cooling-curve methods | | |
| Modified Beckmann | 2–5 g, convenient | ±0.1 |
| Lynn | 1 g, high-melting substances | ±0.2 |
| Skau | 0.5 g, −180° to +200°C, closed system, hygroscopic, also transition points | ±0.03 |
| Natl. Bur. of Standards | 50 ml, liquids, correction for impurities, high precision | ±0.001–0.003 |

(*continued*)

**Table 3.7** (*continued*)

| Apparatus or Method | Special Applicabilities or Advantages | Approximate Maximum Accuracy (°C) |
|---|---|---|
| Heating-curve methods | Substances with low rate of crystallization | |
| Lynn | 1 g, high-melting substances | ±0.2 |
| Skau | 0.5 g, −180° to +200°C, closed system, hygroscopic, also transition points | ±0.03–0.05 |
| Natl. Bur. of Standards | 50 ml, liquids | ±0.01 |
| Calorimetric methods | | |
| Heat-capacity calorimeter | Highest accuracy, static method, correction for impurities | ±0.001 |
| | **Binary Mixtures** | |
| Visual methods | | |
| Capillary, Rheinboldt | Small samples, also "thaw points" | ±2–5 |
| Sealed tube (static) | Most reliable, closed system, hygroscopic, volatile | ±0.1 |
| "Contact," Kofler microscopic | Microsamples, identification of type of binary system | qualitative |
| Grimm microscopic | Microsamples, also "thaw points" | ±2 |
| Cooling-curve methods | | |
| Modified Beckmann | Reliable only for dilute solutions | 0.2 to 10 low |
| Andrews, Kohman, Johnston | Easily crystallizable substances | ±0.1 |

It should also be noted that the reproducibility by a given method for determining melting and freezing temperatures is not a measure of accuracy. Thus, in the case of dynamic visual methods, the temperature measured is that of the bath or the environment and this must obviously be *above* the true melting temperature if it causes a change from the solid to the liquid phase in the sample.

In cooling- or heating-curve methods the temperature of the melt itself is measured, but we are still dealing with a dynamic system. For attaining high accuracy a static method is usually advantageous.

## References

1. S. Glasstone, *Textbook of Physical Chemistry*, 2nd ed., Van Nostrand, New York, 1946, pp. 465–466.
2. F. L. Swinton, *J. Chem. Educ.* **44**, 541 (1967).
3. P. W. Bridgman, *The Physics of High Pressure*, reprinted with supplement, 1949, Bell, London, 1952, pp. 189–256, 422–427.
4. P. W. Bridgman, *J. Chem. Phys.* **9**, 794 (1941).
5. H. T. Hall, *J. Phys. Chem.* **59**, 1144 (1955).
6. For complete derivation of this equation see S. Glasstone, *Textbook of Physical Chemistry*, 2nd ed., Van Nostrand, New York, 1946, p. 644.
7. J. H. Hildebrand, *Solubility of Nonelectrolytes*, 3rd ed., Reinhold, New York, 1950.
8. J. C. Smith, *J. Chem. Soc.* **1936**, 625.
9. T. W. Richards, E. K. Carver, and W. C. Schumb, *J. Amer. Chem. Soc.* **41**, 2019 (1919).
10. H. L. Ward, *J. Phys. Chem.* **30**, 1316 (1926).
11. E. L. Skau, *J. Amer. Chem. Soc.* **52**, 945 (1930).
12. *International Critical Tables*, Vol. 4, McGraw-Hill, New York, 1928, p. 120, System 612.
13. M. Gomberg and C. S. Schoepfle, *J. Amer. Chem. Soc.* **39**, 1664 (1917).
14. W. P. Jorissen and J. Rutten, *Chem. Weekblad* **6**, 261 (1909).
15. R. Kremann, *Monatsh.* **25**, 1252 (1904).
16. Data for Figs. 3.4–3.8 are from *International Critical Tables*, Vol. 4, pp. 180, 179, 136, 119, and 108, respectively.
17. Data from *International Critical Tables*, Vol. 4, p. 155 for Table 3.1, p. 154 for Table 3.2, and p. 123 for Table 3.3.
18. H. P. Cady, *J. Phys. Chem.* **3**, 127 (1899).
19. J. D. M. Ross and I. C. Somerville, *J. Chem. Soc.* **1926**, 2770.
20. H. Rheinboldt, *J. Prakt. Chem.* **111**, 242 (1925).
21. H. Rheinboldt, *Ber.* **74**, 756 (1941); H. Rheinboldt and M. Kircheisen, *J. Prakt. Chem.* **113**, 348 (1926); Houben-Weyl, *Die Methoden der organischen Chemie*, 4th ed., E. Müller, Ed., Vol. 2, Part 1, Thieme, Stuttgart, 1953, pp. 857–862.
22. E. L. Skau, *J. Chim. Phys.* **31**, 366 (1934); *Bull. Soc. Chim. Belges.* **43**, 287 (1934); *J. Amer. Chem. Soc.* **57**, 243 (1935).
23. H. C. Dickinson and N. S. Osborne, *J. Wash. Acad. Sci.* **5**, 338 (1915).
24. M. Centnerzwer, *Z. Physik. Chem.* **29**, 715 (1899).
25. A. Fredga, *Arkiv Kemi Mineral. Geol.* **18B**(4), (1944).
26. J. Timmermans, *Bull. Soc. Chim. Belges.* **39**, 239 (1930); **40**, 105, 689 (1931); **41**, 53, 399 (1932); **42**, 448 (1933); **48**, 33 (1939).
27. A. Fredga, "On the Use of Melting Point Curves for the Establishment of Steric Relationships between Optically Active Compounds," in *The Svedberg, 1884–1944*, A. Tiselius and K. O. Pedersen, Eds., Almqvist and Wiksells, Uppsala, 1945, pp. 261–273.

28. J. Timmermans, *Bull. Soc. Chim. Belges.* **28,** 392 (1919). See also W. E. Garner, F. C. Madden, and J. E. Rushbrooke, *J. Chem. Soc.* **1926,** 2498.
29. J. Timmermans, *Les Constantes Physiques des Composés Organiques Cristallisés,* Masson et Cie., Paris, 1953.
30. J. B. Austin, *J. Amer. Chem. Soc.* **52,** 1049 (1930).
31. J. Timmermans, *J. Chim. Phys.* **35,** 331 (1938).
32. A. C. Holler, *J. Org. Chem.* **13,** 70 (1948).
33. J. Timmermans, *Bull. Soc. Chim. Belges* **44,** 17 (1935); *J. Chim. Phys.* **35,** 331 (1938).
34. J. Timmermans, *J. Phys. Chem. Solids* **18,** 1 (1961).
35. E. F. Westrum and J. P. McCullough, "Thermodynamics of Crystals," in *Physics and Chemistry of the Organic Solid State,* D. Fox, M. M. Labes, and A. Weissberger, Eds., Interscience, New York, 1963, Vol. 1, pp. 89–102.
36. J. G. Aston, "Plastic Crystals," *ibid.* pp. 543–583.
37. E. L. Skau and R. E. Boucher, *J. Phys. Chem.* **58,** 460 (1954).
38. R. E. Boucher and E. L. Skau, "Solubility Charts for Homologous Long-Chain Compounds. A Comprehensive Graphical Correlation of Literature Data for 138 Systems Involving 11 Homologous Series and 17 Solvents," *Agr. Res. Serv.* **ARS-72-1,** U.S. Dept. Agr., 1954, 77 pp.
39. E. L. Skau and A. V. Bailey, *J. Phys. Chem.* **63,** 2047 (1959).
40. R. L. Harris, *Ind. Eng. Chem.* **24,** 455 (1932).
41. D. F. Othmer and M. S. Thakar, *Ind. Eng. Chem.* **44,** 1654 (1952).
42. A. I. Johnson, C.-J. Huang, and T.-K. Kwei, *Can. J. Technol.* **32,** 127 (1954).
43. A. V. Bailey, J. A. Harris, and E. L. Skau, *J. Amer. Oil Chemists' Soc.* **46,** 583 (1969).
44. E. L. Skau, *J. Amer. Oil Chemists' Soc.* **47,** 233 (1970).
45. G. Lynn, *J. Phys. Chem.* **31,** 1381 (1927).
46. E. L. Skau, *Proc. Amer. Acad. Arts Sci.* **67,** 551 (1933).
47. E. L. Skau, (a) *J. Phys. Chem.* **37,** 609 (1933); (b) *loc. cit.* ff 5.
48. E. L. Skau and H. F. Meier, *J. Amer. Chem. Soc.* **51,** 3517 (1929); E. L. Skau and R. McCullough, *ibid.* **57,** 2439 (1935).
49. E. L. Skau and H. F. Meier, *Trans. Faraday Soc.* **31,** 478 (1935).
50. W. M. Smit, *A Tentative Investigation of Fatty Acids and Fatty Acid Methyl Esters,* Thesis, Amsterdam, "De Mercuur", Hilversum, 1946, pp. 16, 26–30.
51. M. P. Mathieu, *Bull. Soc. Chim. Belges* **63,** 333 (1954).
52. W. J. Taylor and F. D. Rossini, *J. Res. Natl. Bur. Std.* **32,** 197 (1944).
53. S. R. Gunn, *Anal. Chem.* **34,** 1293 (1962).
54. W. M. Smit, *Rec. Trav. Chim.* **75,** 1309 (1956); W. M. Smit and G. Kateman, *Anal. Chim. Acta* **17,** 161 (1957).
55. L. T. Carlton, *Anal. Chem.* **27,** 845 (1955).
56. B. J. Mair, A. R. Glasgow, Jr., and F. D. Rossini, *J. Res. Natl. Bur. Std.* **26,** 591 (1941); A. J. Streiff and F. D. Rossini, *ibid.* **32,** 185 (1944); W. J. Taylor and F. D. Rossini, *ibid.* **32,** 197 (1944); A. R. Glasgow, Jr., A. J. Streiff, and F. D. Rossini, *ibid.* **35,** 355 (1945); A. R. Glasgow, Jr., N. C. Krouskop, J. Beadle, G. D. Axilrod, and F. D. Rossini, *Anal. Chem.* **20,** 410 (1948).
57. W. P. White, *J. Phys. Chem.* **24,** 393 (1920).

58. "Freezing Points of High-Purity Hydrocarbons," *ASTM D 1015-55* American Society for Testing and Materials, Philadelphia, 1955.
59. "Purity of Hydrocarbons from Freezing Points," *ASTM D 1016-55* American Society for Testing and Materials, Philadelphia, 1955.
60. A. R. Glasgow, Jr., N. C. Krouskop, and F. D. Rossini, *Anal. Chem.* 22, 1521 (1950).
61. A. R. Glasgow, Jr., and M. Tenenbaum, *Anal. Chem.* 28, 1907 (1956).
62. C. P. Saylor, *Anal. Chim. Acta* 17, 36 (1957); G. S. Ross and H. D. Dixon, *J. Res. Natl. Bur. Std.* 67A, 247 (1963); C. P. Saylor and G. S. Ross, *ibid.* 68C, 35 (1964); G. S. Ross and A. R. Glasgow, *Anal. Chem.* 36, 700 (1964). Also see K. L. Nelson, *Anal. Chem.* 29, 512 (1957); W. M. Smit and G. Kateman, *Anal. Chim. Acta* 17, 161 (1957); J. P. Hoare, *J. Chem. Educ.* 37, 146 (1960).
63. (*a*). P. Clechet and J. C. Merlin, *Bull. Soc. Chim. France 1964*, 2644; P. Clechet, Thesis, Univ. de Lyon, 1965. (*b*). B. Loiseleur, P. Clechet, and J. C. Merlin, *Bull. Soc. Chim. France 1967*, 677; M. Chavret, P. Clechet, and J. C. Merlin, *ibid. 1970*, 3745; Z. Cisse, B. Loiseleur, and P. Clechet, *ibid. 1971*, 335; and Z. Cisse, P. Clechet, M. Coten, J. Delafontaine, and H. Tachoire, *Thermochim. Acta*, in press.
64. D. R. Stull, *Anal. Chem.* 18, 234 (1946).
65. C. R. Witschonke, *Anal. Chem.* 24, 350 (1952).
66. E. F. G. Herington and R. Handley, *J. Sci. Instr.* 25, 434 (1948).
67. E. F. G. Herington and R. Handley, *J. Chem. Soc.* 1950, 199.
68. E. F. G. Herington, *Anal. Chim. Acta* 17, 15 (1957); also see R. Friedenberg and P. J. Jannke, *ibid.* 32, 589 (1965).
69. R. Handley, *Anal. Chim. Acta* 17, 115 (1957).
70. A. P. Gray, *Instr. News* 16 (3), 9, 15 (1966); E. S. Watson, M. J. O'Neill, J. Justin, and N. Brenner, *Anal. Chem.* 36, 1233 (1964); M. J. O'Neill, *ibid.* 36, 1238 (1964). Also see Thermal Analysis Newsletter, No. 6, Analytical Division, Perkin-Elmer Corp., Norwalk, Conn., 1966; and R. A. W. Hill and R. P. Slessor, *Trans. Faraday Soc.* 65, 340 (1969).
71. G. L. Driscoll, I. N. Duling, and F. Magnotta, "Purity Determinations Using a Differential Scanning Calorimeter" in *Analytical Calorimetry*, R. S. Porter and J. F. Johnson, Eds., Plenum Press, New York, 1968, pp. 271–278. Also see A. P. Gray and R. Scott, *Instr. News* 19(3), 1 (1969).
72. N. J. DeAngelis and G. J. Papariello, *J. Pharm. Sci.* 57, 1868 (1968).
73. C. Plato and A. R. Glasgow, Jr., *Anal. Chem.* 41, 330 (1969).
74. E. M. Barrall II and R. D. Diller, presented at the *159th Meeting of the American Chemical Society, Division of Analytical Chemistry*, Houston, Texas, February 20–27, 1970; see *Chem. Eng. News* 48 (10), 43 (1970).
75. F. W. Schwab and E. Wichers, in *Temperature, Its Measurement and Control in Science and Industry*, Reinhold, New York, 1941, pp. 256–264.
76. W. F. Giauque and R. A. Ruehrwein, *J. Amer. Chem. Soc.* 61, 2626 (1939); R. A. Ruehrwein and W. F. Giauque, *ibid.* 61, 2940 (1939); J. G. Aston and G. H. Messerly, *ibid.* 62, 1917 (1940); J. G. Aston, R. M. Kennedy, and S. C. Schumann, *ibid.* 62, 2059 (1940); R. A. Ruehrwein and H. M. Huffman, *J. Amer. Chem. Soc.* 65, 1620 (1943); G. D. Oliver, M. Eaton, and H. M. Huffman, *ibid.* 70, 1502 (1948); H. M. Huffman, *Chem. Rev.* 40, 1 (1947); S. S. Todd, G. D. Oliver, and H. M. Huffman, *J. Amer. Chem. Soc.* 69, 1519

(1947); D. D. Tunnicliff and H. Stone, *Anal. Chem.* **27**, 73 (1955). See also E. F. Westrum, Jr., "Determination of Purity and Phase Behavior by Adiabatic Calorimetry," in *Analytical Calorimetry*, R. S. Porter and J. F. Johnson, Eds., Plenum Press, New York, 1968, pp. 231–238.

77. M. R. Cines, "Solid-Liquid Equilibria of Hydrocarbons," in *Physical Chemistry of Hydrocarbons*, A. Farkas, Ed., Academic, New York, 1950, p. 315.

78. M.-P. Mathieu, *Acad. Roy. Belg., Classe Sci. Mem.* **28**, 10 (1953).

79. J. P. McCullough and G. Waddington, *Anal. Chim. Acta* **17**, 80 (1957); J. F. Messerly, S. S. Todd, G. B. Guthrie, and J. F. McCullough, "Study of the Calorimetric Method of Purity Measurement Using IUPAC Samples of Benzene," *Rept. Invest. No. 6273*, U.S. Bureau of Mines, Washington, D. C., 1963.

80. A. R. Glasgow, Jr., G. S. Ross, A. T. Horton, D. Enagonio, H. D. Dixon, C. P. Saylor, G. T. Furukawa, M. L. Reilly, and J. M. Henning, *Anal. Chim. Acta* **17**, 54 (1957).

81. E. Wichers, C. P. Saylor, and A. R. Glasgow, Jr. (National Bureau of Standards, Washington, D. C.), *Cooperative Determination of Purity by Thermal Methods*, report of the Organizing Committee, submitted to the IUPAC Commission on Physico-Chemical Data and Standards, July 14, 1961, Montreal, Canada.

82. D. Chapman, *J. Amer. Oil Chemists' Soc.* **37**, 243 (1960); *Nature* **183**, 44 (1959); W. P. Ferren, *Food Tech.* **17**, 1066 (1963); C. Y. Hopkins, *J. Amer. Oil Chemists' Soc.* **45**, 778 (1968); W. D. Pohle, *ibid.* **42**, 1075 (1965); W. D. Pohle and R. L. Gregory, *ibid.* **45**, 775 (1968); J. R. Taylor, *ibid.* **41**, 177 (1964); L. R. Wiedermann, *ibid.* **45**, 515A (1968).

83. L. J. Burnett and B. H. Muller, *Nature* **219**, 59 (1968).

84. E. F. G. Herington and I. J. Lawrenson, *Nature* **219**, 928 (1968); *J. Appl. Chem.* **19**, 337 (1969).

85. G. Karagounis, E. Papayannakis, and C. I. Stassinopoulous, *Nature* **221**, 655 (1969).

86. W. Swietoslawski, *Bull. Acad. Polon. Sci.* **Classe A**, 113 (1947); *Rocznicki Chem.* **21**, 94 (1947); *Przeglad Chem.* **6**, 249 (1948); *Bull. Intern. Acad. Pol. Sci. Lett., Cl. Sci. Math. Natur.* **10A** ,113 (1949); I. T. Plebanski, *Bull. Acad. Polon. Sci., Ser. Sci. Chim.* **8**, 23, 117, 239 (1960); W. Swietoslawski, *Zh. Fiz. Khim.* **36**, 2087 (1962). Also see A. P. Simonelli and T. Higuchi, *J. Pharm. Sci.* **50**, 1861 (1961).

87. T. Plebanski, *Pomiary Automat. Kontr.* **12**, 284 (1966).

88. G. S. Ross and L. J. Frolen, *J. Res. Natl. Bur. Std.* **67A**, 607 (1963).

89. A. Georg, *Helv. Chim. Acta* **15**, 924 (1932).

90. W. Dieckmann, *Ber.* **49**, 2204, 2213 (1916).

91. C. S. Glickman, *Ind. Eng. Chem., Anal. Ed.* **4**, 304 (1932); H. A. Jones, *ibid.* **13**, 819 (1941); J. R. Kach, G. J. Hable, and L. Wrangell, *ibid.* **10**, 166 (1938); H. A. Jones and J. W. Woods, *J. Amer. Chem. Soc.* **63**, 1760 (1941); L. B. Norton and R. Hansberry, *ibid.* **67**, 1609 (1945); L. Segal and D. J. Stanonis, *Anal. Chem.* **35**, 1750 (1963).

92. F. D. Snell, *Ind. Eng. Chem., Anal. Ed.* **2**, 287 (1930).

93. A. A. Morton, *Laboratory Technique in Organic Chemistry*, McGraw-Hill, New York, 1938.

94. G. F. Wright, *Can. J. Technol.* **34**, 89 (1956).
95. L. M. White, *Anal. Chem.* **19**, 432 (1947).
96. J. L. Hartwell, *Anal. Chem.* **20**, 374 (1948).
97. H. A. Bell, *Ind. Eng. Chem.* **15**, 375 (1923); L. M. Dennis, *ibid.* **12**, 366 (1920); F. C. Merriam, *Anal. Chem.* **20**, 1246 (1948); E. W. Blank, *Ind. Eng, Chem.*, *Anal. Ed.* **5**, 74 (1933); E. Conte, *ibid.* **2**, 200 (1930); E. Dowzard and M. J. Russo, *ibid.* **8**, 74 (1936); **15**, 219 (1943); M. M. Graff, *ibid.* **15**, 638 (1943); C. E. Sando, *ibid.* **3**, 65 (1931); M. S. Schechter and H. L. Haller, *ibid.* **10**, 392 (1938).
98. K. S. Markley, *Ind. Eng. Chem.*, *Anal. Ed.* **6**, 475 (1934).
99. E. B. Hershberg, *Ind. Eng. Chem.*, *Anal. Ed.* **8**, 312 (1936).
100. C.-L. Tseng, *J. Chinese Chem. Soc.* **1**, 143 (1933); *Sci. Quart.*, *Natl. Univ. Peking* **4**, 237, 283 (1934).
101. F. Francis and F. J. E. Collins, *J. Chem. Soc.* **1936**, 137.
102. F. Francis and S. H. Piper, *J. Amer. Chem. Soc.* **61**, 577 (1939).
103. H. Thiele, *Z. Angew. Chem.* **15**, 780 (1902).
104. E. Berl and A. Kullmann, *Ber.* **60**, 811 (1927).
105. W. L. Walsh, *Ind. Eng. Chem.*, *Anal. Ed.* **6**, 468 (1934); W. Friedel, *Biochem. Z.* **209**, 65 (1929); C. F. Lindström, *Chem. Fabrik* **7**, 270 (1934).
106. L. M. Dennis and R. S. Shelton, *J. Amer. Chem. Soc.* **52**, 3128 (1930).
107. L. Kofler and W. Kofler, *Mikrochemie* **34**, 374 (1949).
108. L. Kofler and H. Sitte, *Monatsh.* **81**, 619 (1950).
109. F. P. Zscheile and J. W. White, Jr., *Ind. Eng. Chem.*, *Anal. Ed.* **12**, 436 (1940). See also F. W. Matthews, *Anal. Chem.* **20**, 1112 (1948).
110. L. Kofler and A. Kofler, *Mikroskopische Methoden in der Mikrochemie*, Haim, Vienna, 1936.
111. W. C. McCrone, Jr., *Fusion Methods in Chemical Microscopy*, Interscience, New York, 1957.
112. R. V. Smith, *Amer. Lab.* **1969**, 1 (Sept.)
113. E. Kordes, *Z. Anorg. Allgem. Chem.* **154**, 93 (1926); **167**, 97 (1927); **168**, 177 (1927).
114. F. Francis, S. H. Piper, and T. Malkin, *Proc. Roy. Soc.* (*London*) **128A**, 214 (1930); A. C. Chibnall, S. H. Piper, and E. F. Williams, *Biochem. J.* **30**, 100 (1936).
115. J. H. Adriani, *Z. Physik. Chem.* (*Leipzig*) **33**, 453 (1900).
116. C. W. Gibby and W. A. Waters, *J. Chem. Soc.* **1931**, 2151.
117. G. Lock and G. Nottes, *Ber.* **68**, 1200 (1935). See also M. Brandstätter, *Mikrochemie* **32**, 33 (1944).
118. R. Hollmann, *Z. Physik. Chem.* (*Leipzig*) **43**, 129 (1903).
119. E. Fischer, *Ber.* **32**, 3641 (1899).
120. R. Kempf, *J. Prakt. Chem.* **78**, 242 (1908).
121. A. Michael, *Ber.* **28**, 1629 (1895).
122. I. J. Pisovschi, *Ber.* **41**, 1436 (1908).
123. R. Willstätter, *Ber.* **35**, 1375 (1902).
124. J. Thiele, *Ann.* **270**, 18 (1892).
125. C. Willgerodt, *Ber.* **26**, 358 (1893).

126. E. Noelting and K. Philipp, *Ber.* **41**, 584 (1908).
127. L. Kofler and A. Kofler, *Thermomikromethoden*, Universitäts Verlag-Wagner, Innsbruck, 1954.
128. A. Lacourt and N. Delande, *Microchem. J., Symp. Ser.* **2**, 259 (1962).
129. K. J. Frederick and J. H. Hildebrand, *J. Amer. Chem. Soc.* **61**, 1555 (1939).
130. A. Findlay, A. N. Campbell, and N. O. Smith, *The Phase Rule and Its Applications*, 9th ed., Dover, New York, 1951, pp. 64–69.
131. F. E. Wright, *J. Amer. Chem. Soc.* **39**, 1515 (1917).
132. F. K. Cameron, *J. Phys. Chem.* **2**, 409 (1898).
133. E. L. Skau and B. Saxton, *J. Phys. Chem.* **37**, 197 (1933).
134. A. Findlay, A. N. Campbell, and N. O. Smith, *op. cit.*, Chapter XII.
135. L. Knorr, O. Rothe, and H. Averbeck, *Ber.* **44**, 1138 (1911).
136. K. Meyer and P. Kappelmeier, *Ber.* **44**, 2718 (1911).
137. C. A. Rouiller, *Amer. Chem. J.* **49**, 301 (1913).
138. R. S. Morrell and E. K. Hanson, *Chem. News* **90**, 166 (1904); *J. Chem. Soc.* **85**, 1520 (1904).
139. E. L. Skau and B. Saxton, *J. Amer. Chem. Soc.* **52**, 335 (1930).
140. *International Critical Tables*, Vol. 1, McGraw-Hill, New York, 1926, pp. 314–320.
141. E. L. Skau and H. F. Meier, *Trans. Faraday Soc.* **31**, 478 (1935).
142. A. C. de Kock, *Z. Physik. Chem. (Leipzig)* **48**, 129 (1904).
143. G. Dupont and O. Lozac'h, *Compt. Rend.* **221**, 751 (1945); *Bull. Soc. Chim. France* **1946**, 525. See also R. Schenck, *Z. Physik. Chem.* **25**, 337 (1898); K. Auwers and E. Gierig, *ibid.* **42**, 631 (1902); and footnote 142.
144. E. D. Dietz, *Sci. Technol.* **83**, 10 (Nov. 1968).
145. J. Houben, *Die Methoden der organischen Chemie*, 3rd ed., Vol. 1, Thieme, Leipzig, 1925, pp. 812–820.
146. J. Piccard, *Ber.* **8**, 687 (1875). See also H. Landolt, *Z. Physik. Chem. (Leipzig)* **4**, 349 (1889).
147. B. S. Biggs, C. J. Frosch, and R. H. Erickson, *Ind. Eng. Chem.* **38**, 1016 (1946).
148. R. D. Evans, H. R. Mighton, and P. J. Flory, *J. Amer. Chem. Soc.* **72**, 2018 (1950).
149. P. J. Flory, L. Mandelkern, and H. K. Hall, *J. Amer. Chem. Soc.* **73**, 2532 (1951); L. Mandelkern, M. Tryon, and F. A. Quinn, Jr., *J. Polymer Sci.* **19**, 77 (1956).
150. B. Ke, "Differential Thermal Analysis," in *Newer Methods of Polymer Characterization*, B. Ke, Ed., Interscience, New York, 1964, pp. 347–419. Also see L. Mandelkern, *Crystallization of Polymers*, McGraw-Hill, New York, 1964, and B. Wunderlich, *Amer. Lab.*, June 1970, pp. 17, 18, 20, 22–27.
151. L. Kofler and A. Kofler, *Thermo-Mikro-Methoden zur Kennziechnung organischer Stoffe and Stoffgemische*, 3rd ed., Verlag Chem., Weinheim, 1954.
152. A. Kofler, *Naturwissenschaften* **31**, 553 (1943).
153. H. Rheinboldt, *Ber.* **74**, 756 (1941).
154. H. G. Grimm, M. Guenther, and H. Tittus, *Z. Physik. Chem.* **14B**, 169 (1931).
155. H. Lettré, H. Barnbeck, and W. Lege, *Ber.* **69**, 1151 (1936).
156. H. Lettré, H. Barnbeck, and H. Staunan, *Ber.* **69**, 1594 (1936).

157. M. Brandstätter, *Mikrochemie* **32**, 33 (1944); H. Rheinboldt, *et al.*, *Bol. Fac. Filosof. Cienc. Letras, Univ. Sao Paulo* **14**, Quím. No. 1, 3, 21 (1942); No. 2, 105, 110, 124, 139, 143 (1947); *J. Amer. Chem. Soc.* **68**, 973 (1946); H. Lettré et al. *Ber.* **70**, 1410 (1937); **71**, 1225 (1938).

158. A. Kofler, *Z. Physik. Chem.* (*Leipzig*) **A190**, 287 (1942).

159. E. L. Skau and B. Saxton, *J. Phys. Chem.* **37**, 183 (1933); see also E. L. Skau, *J. Amer. Chem Soc.* **52**, 945 (1930); E. L. Skau and L. F. Rowe, *ibid.* **57**, 2437 (1935).

160. D. H. Andrews, G. T. Kohman, and J. Johnston, *J. Phys. Chem.* **29**, 914 (1925).

161. I. Schröder, *Z. Physik. Chem.* (*Leipzig*) **11**, 449 (1893).

162. W. H. Walker, A. R. Collett, and C. L. Lazzell, *J. Phys. Chem.* **35**, 3259 (1931).

163. F. C. Magne and E. L. Skau, *J. Amer. Chem. Soc.* **74**, 2628 (1952).

164. R. E. Satterfield and M. Haulard, *J. Chem. Eng. Data* **10**, 396 (1965).

165. J. A. Harris, A. V. Bailey, and E. L. Skau, *J. Amer. Oil Chemists' Soc.* **45**, 639 (1968).

166. R. R. Mod and E. L. Skau, *J. Phys. Chem.* **56**, 1016 (1952); R. R. Mod, F. C. Magne, and E. L. Skau, *J. Amer. Oil Chemists' Soc.* **39**, 444 (1962); J. A. Harris, R. R. Mod, D. Mitcham, and E. L. Skau, *ibid.* **44**, 737 (1967).

167. E. L. Skau and L. F. Rowe, *Ind. Eng. Chem., Anal. Ed.* **3**, 147 (1931).

168. D. G. Grabar and W. C. McCrone, *J. Chem. Educ.* **27**, 649 (1950).

169. F. Henning and H. T. Wensel, *J. Res. Natl. Bur. Std.* **10**, 809 (1933).

170. J. F. Swindells, Ed., "Precision Measurement and Calibration—Temperature," *Natl. Bur. Std. Spec. Pub. 300*, USDC, Vol. 2, (Publ. 1968), 513 pp.

171. M. G. Lozinsky, *J. Roy. Microscop. Soc.* **86**(3), 211 (1967).

172. C. Solomons and G. J. Janz, *Rev. Sci. Instr.* **29**, 302 (1958); K. Motzfeldt, in *Physicochemical Measurements at High Temperatures*, J. O. Bockris, J. L. White, and J. D. Mackenzie, Eds., Butterworth, London, 1960; J. F. Schairer, *ibid.* p. 117.

173. S. J. Schneider, "Compilation of the Melting Points of the Metal Oxides," *Natl. Bur. Std. Monogr. 68*, 1963, 31 pp.

174. W. C. McCrone, *Mettler Tech. Inf. Bull. 3005*, 1968, 19 pp.

175. H. A. Liebhafsky, E. G. Rochow, and A. F. Winslow, *J. Amer. Chem. Soc.* **61**, 969 (1939).

176. H. J. Kostkowski and R. D. Lee, *Natl. Bur. Standards Monogr. 41*, 1962; E. A. Wynne and M. Zief, "Laboratory Scale Apparatus," in *Fractional Solidification*, Vol. 1, M. Zief and W. R. Wilcox, Eds., Marcel Dekker, New York, 1967, pp. 191–236; J. Reilly and W. N. Rae, *Physico-Chemical Methods*, 5th ed., Vol. 1, Van Nostrand, New York, 1954, pp. 333–363.

177. S. J. Schneider and C. L. McDaniel, *J. Res. Natl. Bur. Std.* **71A**, 317 (1967); S. J. Schneider, *Natl. Bur. Std. Spec. Publ. 303*, 1969, 19–39. See also IUPAC, Div. Inorg. Chem., Comm. on High Temperatures and Refractories, *Pure Appl. Chem. 21*, 115 (1970).

178. M. S. Van Dusen and A. I. Dahl, *J. Res. Natl. Bur. Std.* **39**, 291 (1947); W. F. Roeser, F. R. Caldwell, and H. T. Wensel, *J. Res. Natl. Bur. Std.* **6**, 1119 (1931).

179. W. H. Swanger and F. R. Caldwell, *J. Res. Natl. Bur. Std.* **6**, 1131 (1931); W. H. Swanger, ibid. **3**, 1029 [1929].

180. E. F. Westrum, Jr., and L. Eyring, *J. Amer. Chem. Soc.* **73**, 3399 (1951).

181. E. F. Westrum, Jr., "High Temperature Adiabatic Calorimetry," in *Advances in High Temperature Chemistry*, L. Eyring, Ed., Academic, New York, 1967.

182. S. J. Schneider and J. L. Waring, *J. Res. Natl. Bur. Std.* **67A**, 19 (1963). Also see Ref. 130, pp. 476–477.

183. J. F. Schairer, *J. Amer. Ceramic Soc.* **37**, 501 (1954); W. Schreyer and J. F. Schairer, *J. Petrology* **2**, 324 (1961); D. K. Bailey and J. F. Schairer, *J. Petrology* **7**, 114 (1966); J. F. Schairer, *J. Amer. Ceramic Soc.* **40**, 215 (1957).

184. G. W. Morey, *J. Amer. Ceramic Soc.* **36**, 279 (1953); J. Van den Heurk, *Bull. Geol. Soc. Am.* **64**, 993 (1963); O. F. Tuttle, *Amer. J. Sci.* **246**, 628 (1948).

185. F. Daniels, J. H. Mathews, J. W. Williams, P. Bender, and R. A. Alberti, *Experimental Physical Chemistry*, 5th ed., McGraw-Hill, New York, 1956, pp. 65–71.

186. A. Findlay, rev. by J. A. Kitchener, *Practical Physical Chemistry*, 8th ed., Longmans, Green, New York, 1953, pp. 107–112; see also 7th ed., pp. 125–136.

187. F. M. Batson and C. A. Kraus, *J. Amer. Chem. Soc.* **56**, 2017 (1934).

188. H. T. Rall and H. M. Smith, *Ind. Eng. Chem., Anal. Ed.* **11**, 387 (1939).

189. Data from M. Gomberg and F. W. Sullivan, *J. Amer. Chem. Soc.* **44**, 1810 (1922), for tetraphenyldi-$\beta$-naphthylethane; W. E. S. Turner and S. English, *J. Chem. Soc.* **105**, 1786 (1914), for phenol; E. R. Jones and C. R. Bury, *ibid.* **127**, 1949 (1925), tetrachloroethane; C. J. Peddle and W. E. S. Turner, *ibid.* **99**, 685 (1911), for propionic acid and aniline.

190. M. Ulmann, *Molekülgrössen-Bestimmungen Hochpolymerer Naturstoffe*, Steinkopff, Dresden, 1936, pp. 26–48.

191. M. L. Huggins, *Ind. Eng. Chem.* **35**, 980 (1943).

192. K. H. Meyer and H. Mark, *Ber.* **61**, 1939, 1946 (1928).

193. See, for example, H. Staudinger, M. Asano, H. F. Bondy, and R. Signer, *Ber.* **61**, 2575 (1928).

194. K. Freudenberg, *Tannin, Zellulose, Lignin*, 2nd ed. Springer, Berlin, 1933, p. 109. See also E. Paterno, *Atti Reale Accad. Naz. Lincei, Classe Sci. Fis. Mat. Nat.* **15**, 260 (1932).

195. K. Hess, *Die Chemie der Zellulose und Ihrer Begleiter*, Akadem. Verlagsgesellschaft, Leipzig, 1928, p. 590.

196. B. Hargitay, W. Kuhn, and H. Wirz, *Experientia* **7**, 276 (1951).

197. K. Rast, *Ber.* **55**, 1051, 3727 (1922).

198. F. Pregl, *Quantitative Organic Microanalysis*, 4th ed., J. Grant, Ed., Blakiston, Philadelphia, 1946, pp. 198–203.

199. N. D. Cheronis and J. B. Entrikin, *Semimicro Qualitative Organic Analysis*, 2nd ed., Interscience, New York, 1957, pp. 145–149.

200. W. W. Scott, *Standard Methods of Chemical Analysis*, 5th ed., Van Nostrand, Princeton, N. J., 1939, p. 2533; N. D. Cheronis, *Micro and Semimicro Methods*, Interscience, New York, 1954, pp 208–213.

201. H. W. B. Roozeboom, *Die heterogenen Gleichgewichte vom Standpunkte der Phasenlehre*, Vol. 3, Part 1, Vieweg, Braunschweig, 1911, p. 46.

202. F. W. Schwab and E. Wichers. *J. Res. Natl. Bur. Std.* **32**, 253 (1944).
203. T. W. Richards, *J. Amer. Chem. Soc.* **27**, 104 (1905). See also N. F. Hall, *J. Amer. Chem. Soc.* **39**, 1148 (1917).
204. B. K. Blount, *Mikrochemie* **19**, 162 (1936).
205. L. C. Craig, *Ind. Eng. Chem., Anal. Ed.* **12**, 773 (1940).
206. E. L. Skau, *J. Phys. Chem.* **33**, 951 (1929).
207. E. L. Skau and W. Bergmann, *J. Org. Chem.* **3**, 166 (1938).
208. E. L. Skau, unpublished work.
209. J. N. Friend, *Textbook of Inorganic Chemistry*, Vol. 4, Griffin, London, 1928.
210. W. M. Cumming, I. V. Hopper, and T. S. Wheeler, *Systematic Organic Chemistry*, 3rd ed. rev., Van Nostrand, Princeton, N. J., 1937, p. 15.
211. T. W. Richards and N. F. Hall, *J. Amer. Chem. Soc.* **39**, 531 (1917).
212. A. T. Horton, U.S. Pat. 2,754,180 (July 10, 1956); A. T. Horton and A. R. Glasgow, *J. Res. Natl. Bur. Std.* **69C**, 195 (1965).
213. C. P. Saylor, "The Freezing Staircase Method" in *Purification of Inorganic and Organic Materials*, M. Zief, Ed., Marcel Dekker, New York, 1969, pp. 125–138; *Natl. Bur. Stds. (U.S.), Tech. News Bull.* **49**, 210 (1965).
214. H. C. Wolf and H. P. Deutsch, *Naturwissenschaften* **41**, 425 (1954); H. Rock, *ibid.* **43**, 81 (1956); E. F. G. Herington, R. Handley, and A. J. Cook, *Chem. Ind. (London)* **1956**, 292. R. Handley and E. F. G. Herington, *Chem. Ind. (London)* **1956**, 304, W. R. Wilcox, R. Friedenberg, and N. Back, *Chem. Rev.* **64**, 187 (1964).
215. M. Zief and W. R. Wilcox, Eds., *Fractional Solidification*, Vol. 1, Marcel Dekker, New York, 1967; M. Zief, Ed., *Purification of Inorganic and Organic Materials*, Marcel Dekker, New York, 1969; G. F. Reynolds, "Crystal Growth," in *Physics and Chemistry of the Organic Solid State*, D. Fox, M. M. Labes, and A. Weissberger, Eds., Interscience, New York, 1963, pp. 223–286; E. A. D. White, "The Growth of Single Crystals from the Fluxed Melt," in *Technique of Inorganic Chemistry*, H. B. Jonassen and A. Weissberger, Eds., Interscience, New York, Vol. 4, 1965, pp. 31–64; R. G. Bautista and J. L. Margrave, "High Temperature Techniques," *ibid.* pp. 65–135. A. Lawley and D. R. Hay, "Zone Refining," in Kirk-Othmer, *Encyclopedia of Chemical Technology*, 2nd ed., Interscience, New York, Vol. 22, 1970, p. 680.

## General

Alper, A. M., Ed., *Phase Diagrams: Materials Science and Technology*, 3 vols., Academic, New York, 1970.
Bockris, J. O., J. L. White, and J. D. MacKenzie, *Physicochemical Measurements at High Temperatures*, Butterworth, London, 1960.
Bowden, S. T., *The Phase Rule and Phase Reactions*, Macmillan, London, 1945.
Clibbens, D. A., *The Principles of the Phase Theory*, Macmillan, London, 1920.
Deffet, L., *Repertoire des composes organiques polymorphes*, Desoer, Liege, 1942.
Findlay, A., A. N. Cambell, and N. O. Smith, *The Phase Rule and Its Applications*, 9th ed., Dover, New York, 1951.
Flory, P. J., *Principles of Polymer Chemistry*, Cornell Univ. Press, Ithaca, New York, 1953.

Glasgow, Jr., A. R., and G. S. Ross, "Cryoscopy," in *Treatise on Analytical Chemistry*, I. M. Kolthoff and P. J. Elving, Eds., Vol. 8, Part 1, Interscience, New York, 1968.

Gray, G. W., *Molecular Structure and the Properties of Liquid Crystals*, Academic, New York, 1962.

Hildebrand, J. H., *Solubility of Nonelectrolytes*, 3rd ed., Reinhold, New York, 1950.

Levin, E. M., and McMurdie, H. F., *Phase Diagrams for Ceramists*, Amer. Ceramic Soc., Columbus, Ohio, 1960.

Linke, W. F., *Solubilities of Organic and Metal Organic Compounds* (Revision of Seidell), 2 vols., Van Nostrand, New York, 1958–1965.

Masing, G., *Ternary Systems, Introduction to the Theory of Three-Component Systems*, Translation by B. A. Rogers, Reinhold, New York, 1944. (Reprinted by Dover.)

Reilly, J., and W. M. Rae, *Physico-Chemical Methods*, 5th ed., Vol. 1, Van Nostrand, Princeton, N. J., 1954.

Ricci, J. E., *The Phase Rule and Heterogeneous Equilibrium*, Van Nostrand, New York, 1951; corr. ed., Dover, New York.

Roozeboom, H. W. B., *Die heterogenen Gleichgewichte vom Standpunkte der Phasenlehre*, Vieweg, Braunschweig, 1901–1913.

Smit, W. M., Ed., *Purity Control by Thermal Analysis*, Proc. Intern. Symp., Amsterdam, 1957. Elsevier, New York, 1957.

Smothers, W. J., and Y. Chiang, *Differential Thermal Analysis*, Chemical Publishing Co., New York, 1958.

Stephen, H., and T. Stephen, *Solubilities of Inorganic and Organic Compounds*, 4 vols., Macmillan, New York, 1963–1964.

Tammann, G., *Kristallisieren and Schmelzen*, Barth, Leipzig, 1903.

Tammann, G., *The States of Aggregation*, translated by R. F. Mehl, Van Nostrand, Princeton, N. J., 1925.

Timmermans, J., *Physico-Chemical Constants of Binary Systems*, 4 vols., Interscience, New York, 1959–1960.

Timmermans, J., *Physico-Chemical Constants of Pure Organic Compounds*, Elsevier, New York, 2 vols., 1950–1965.

Timmermans, J., *Les constantes physiques des composés organiques cristallisés*, Masson et Cie, Paris, 1953.

Tipson, R. R., "Crystallization and Recrystallization" in *Technique of Organic Chemistry*, Vol. 3, Part 1, 2nd ed., Interscience, New York, 1956.

Utermark, W., and W. Schicke, *Melting Point Tables of Organic Compounds*, 2nd ed., Interscience, New York, 1963.

Zernike, Z., *Chemical Phase Theory*, Kluwer, Deventer (Netherlands), 1956.

Zief, M., Ed., *Purification of Inorganic and Organic Materials*, Dekker, New York, 1969.

Chapter **IV**

# DETERMINATION OF BOILING AND CONDENSATION TEMPERATURES

John R. Anderson

# 1  INTRODUCTION

The techniques of precise determinations of boiling temperatures and condensation temperatures by measurements on boiling systems and interpretation of such measurements in terms of thermodynamic law constitute the subject matter of ebulliometry. Ebulliometric measurements are carried out in ebulliometers.

The late Polish physical chemist, W. Swietoslawski systematically developed ebulliometry and the excellent ebulliometers which bear his name. Serious students of ebulliometry will, therefore, want to study his definitive works [1, 2] which summarize many dozens of papers by him and his collaborators [3]. However, it is relevant here, in a chapter in a general treatise on techniques of chemistry, to describe Swietoslawski's ebulliometers and ebulliometric techniques in simple terms, and to compare them in a fundamental way with certain other useful ebulliometric devices and techniques. It is also relevant to include an elementary discussion of (a) the basic principles of liquid–vapor equilibria, (b) the method of comparative measurements, (c) the use of water as primary tonometric and ebulliometric standard, and (d) the international temperature scale. Whereas an understanding of and appreciation for all of these are necessary for intelligent practice of ebulliometry, it is questionable whether one could develop meaningful proficiency in all of these adjunct subjects from a study of this chapter alone. However, if these elementary discussions will prompt the reader to look elsewhere for more detail, they will have accomplished much. At any rate, these are the objectives of including them in this chapter, which it is hoped may serve as a guide to understanding and hence mastering all ebulliometric techniques, and evaluating all ebulliometers and ebulliometric data as those techniques, devices, and data are detailed anywhere. The utility of precise ebulliometry reached its zenith even before the passing of Swietoslawski: many of its older analytical uses have been discontinued in favor of more convenient newer methods; even the normal boiling point of water is no longer required as a primary reference on international temperature scales and, aside from ebulliometric molecular-weight determinations, there is only infrequent need for precise work of any kind outside of national and international bureaus of standards. There are, however, many lessons to be learned about the systematic art of making precise measurements per se in studying Swietoslawski's brand of ebulliometry.

## Objectives and Scope of Ebulliometric Measurements

The objectives of ebulliometric measurements include the following: (a) determinations of the true thermodynamically significant equilibrium boiling points of pure substances, usually as a function of pressure; (b) determinations of steady-state boiling temperatures of solutions at constant pressure, usually as a function of concentration, for analytical uses based on Raoult's law; (c) estimations of the boiling points of zeotropic and azeotropic systems, usually as a function of concentration and often as a function of pressure, for engineering uses in separation and purification science; and (d) determinations of steady-state boiling temperatures *and* steady-state condensation temperatures, usually at constant pressure for empirical analysis of contaminated liquids and their vapors.

Ebulliometric measurements are useful both in science and in industry: in characterizing laboratory preparations and items of commerce; in calibrating thermometers; in determining pressure, molecular weight, solubility, the characteristics of azeotropic and zeotropic systems, equilibrium constants, impurity content, thermal stability at the boiling point, and the amount of vapors adsorbed by solid substances.

## The Major Problems in Ebulliometric Measurements

It is usually quite easy to attain high precision when making ebulliometric measurements; however, the exact significance, particularly the exact thermodynamic significance, of the result of the measurement often cannot be determined easily. The credibility of a single ebulliometric datum should be suspect until the extent by which the observed value may be changed by slight variation in relevant experimental parameters has been ascertained. This may involve varying the operating conditions of the experiment, or changing the apparatus, or redetermining the value under the same conditions in the same apparatus after additional purification of the sample, and so on, as will be described. Intelligent practice of ebulliometry is, thus, mostly a matter of evaluating observed data by determining the extent to which secondary phenomena have affected the result of the ebulliometric measurement or of carrying out the measurement in such a way that the errors caused by secondary phenomena are eliminated. Evaluation of ebulliometric data is usually made with the aid of additional ebulliometric data.

The second major problem in ebulliometric measurements arises because investigators have historically indulged themselves in faulty experimental apparatus and techniques, and imprecise definitions of terms. In using the ebulliometric data of others, therefore, the careful investigator must always evaluate ebulliometric data, usually listed in the literature as a "boiling point," in terms of the exact technique used to determine the data. Often such data are impossible to evaluate; indeed, if an appropriate sample is at hand, it is usually easier to do the work properly than to attempt to evaluate the literature result if an accurate value is required. Faulty definition of terms arises because most investigators, in assuming that they are dealing with a *pure* liquid when performing ebulliometric measurements, report boiling points (which imply thermodynamically significant data) while having simply determined condensation temperatures the thermodynamic significance of which is unknown until supplementary data are at hand.

### Boiling Phenomena

Boiling is a dynamic phenomenon characterized by a sustained generation of bubbles of vapor from a liquid through the action of heat. Boiling has an intermittent, sometimes violently intermittent, quality. The bubbles of vapor are generated in, rise through, and agitate the mass of the liquid; finally they break out of the liquid surface and escape into space previously occupied by other matter. Boiling requires a driving force; if a liquid actually boils, it is superheated with respect to its boiling point.

If a boiling liquid is thermally rectifiable, as is the case almost without exception in the experimental world, the composition of the vapor will be different from that of the boiling liquid; therefore, the boiling point of the boiling liquid is not the same as that of the original liquid before boiling commenced. Furthermore, if the vapor phase generated from a boiling, thermally rectifiable liquid is condensed, in a real (not hypothetical) experiment, the temperature at which it condenses will not be the same as the boiling point of the liquid before boiling commenced, or of the liquid from which the vapor phase is generated at the time of condensation.

A liquid exerts hydrostatic pressure as well as vapor pressure. Hence, at a given atmospheric pressure, the temperature at which liquid boils in the bulk-liquid phase varies according to the depth in the liquid where vapor is formed.

Precise determination of the boiling point of a substance or of a thermally rectifiable mixture of substances, by direct measurements made on a boiling bulk-liquid phase, is thus at best difficult for pure substances, and in the case of thermally rectifiable mixtures, impossible. Except in the case of pure substances the experimenter must always measure something other than the thermodynamically defined boiling point datum. He may, or may not, then wish to contrive to obtain the thermodynamic datum by supplemental measurements and calculations, depending on the situation.

## 2  DEFINITION OF TERMS

In contrast to the dynamic nature of the phenomenon of boiling, the term boiling point is given a purely static precise thermodynamic definition: The boiling point of a liquid at a specified pressure is the temperature at which the vapor pressure of the liquid is equal to the specified pressure. If the specified pressure is one standard atmosphere (1,013,250 dyne/cm$^2$, 760 mm Hg) this temperature is called the normal boiling point. We will define the condensation point of a vapor from a boiling liquid at a specified pressure as the temperature at which, in the absence of extraneous gases, the vapor would just entirely condense under that pressure. This definition follows from the obvious fact that if any liquid, at its boiling point but not boiling, is in thermodynamic equilibrium with an infinitesimally small proportion of vapor phase derived from the liquid phase, both phases are at the same temperature. For pure substances, of course, the relative proportions of the two phases involved is immaterial, but for thermally rectifiable liquids this point simply cannot be determined when the liquid is boiling and condensing, though it can be conjectured and it has known thermodynamic significance.

What then does one determine in ebulliometric measurement? One determines boiling temperatures and condensation temperatures—purely observational temperature data which may be given the following definitions, or descriptions:

Boiling temperature is the temperature established on a surface (usually the bulb of a thermometer) when the surface is in contact with a thin, moving film of liquid that has barely ceased to boil. It may or may not be equal to the boiling point. It usually is not. Condensation temperature identifies the temperature established on a surface (usually the bulb of a thermometer) on which a thin, moving film of liquid coexists with vapor from which the liquid has condensed, the vapor phase being replenished, at the moment of measurement, from a boiling liquid phase. It may or may not be equal to the condensation point. It usually is not. These temperatures may be measured in ebulliometers.

The thermodynamic significance of boiling temperature and condensation temperature determined on a boiling liquid is usually not known with a certainty at all comparable with the precision with which they can be determined. Each situation will depend on the composition of the boiling liquid, or its composition before boiling commenced, the size and kind of ebulliometer in which the liquid is being boiled, the operating conditions, and the supplemental data at hand. For example, for pure substances or other systems of more than one component which undergo no rectification upon boiling, both of these observational temperatures may be equivalent in value to the boiling point in appropriately conducted experiments, whereas for other solutions they may not.

# 3 BOILING POINT-COMPOSITION DIAGRAMS

In order to interpret boiling and condensation temperature data it is of great importance to understand what kinds of phenomena may occur when liquids consisting of one, two, or more components in various proportions are boiled and condensed.

## The Phase Rule

When studying the behavior of heterogeneous systems that are in thermo-dynamic equilibrium or that are in some kind of dynamic steady state that is influenced by a tendency to approach thermodynamic equilibrium, as in ebulliometric studies, it is helpful to be guided by the important generalization, now known as the phase rule, stated by J. Willard Gibbs in 1874.

$$F = C - P + 2.$$

This simply states that the number of independent variables $F$, such as temperature, pressure, and composition which must be specified in order to describe a system completely, is two greater than the difference between the number of individual independent chemical substances $C$, called components, which must be specified in order to describe the chemical nature of the system, and the number of separate physical states $P$, such as gas, liquid(s), solid(s), which exist in the system at equilibrium. The phase rule is a thermodynamic law relating $F$, $C$, and $P$ for systems in which it is assumed that temperature, pressure, and composition are the only determinative variables. Phase-rule systems may proceed from one equilibrium state to another by means of thermodynamic processes in which composition, temperature and pressure are state variables.

## Boiling-Point Curves of One-Component Systems

Strickly speaking, the concept of a one-component system presumes a sample of matter consisting of only one kind of molecule or atom. Actually such samples are exceeding difficult to prepare and keep, within the capabilities of the present state of the art of proving the presence of impurity by ebulliometric techniques.

Indeed as long ago as 1936, Smith and Wojciechowski [4], working at the National Bureau of Standards, determined the difference in boiling points of quite dilute aqueous solutions of duterium oxide. Although it was found that natural variations in the duterium oxide content of natural waters need not be considered to be of importance when using water as the primary standard in ebulliometric measurements, it is interesting that such variations can be detected cryometrically, and consequently that they should be taken into account when employing water as the primary standard in precise cryometric measurements [5].

In accordance with the phase rule, a one-component system may exist in equilibrium states involving one, two, or three phases, depending upon the conditions of temperature and pressure. These conditions have been determined experimentally for many purified substances, at the moderate temperatures and pressures germane here, and the pressure–temperature phase diagram for the water substance under these moderate conditions is indicated schematically in Fig. 4.1.

The diagram consists of three curves, a vapor-pressure curve, $AP$, a sublimation curve, $AS$, and a curve $AF$ which shows how the freezing temperature of the one-component system $H_2O$, changes as a function of pressure. At temperatures and pressures which can be represented along $AP$ both liquid and vapor coexist, along $AS$ both solid and vapor coexist, and along $AF$ both solid and liquid coexist. Only at the temperature and pressure represented by point $A$, the so-called triple point, can all three phases coexist; and at the temperatures and pressures lying within the fields, bound in part in Fig. 4.1 by the curves $FAP$, $FAS$, and $PAS$, only one phase, the phase specified in that field on the figure, can exist. Our primary interest here at

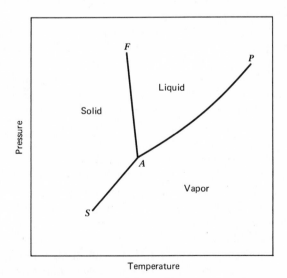

**Fig. 4.1**  Pressure–temperature diagram (schematic).

this juncture is in curve $AP$, the so-called "vapor-pressure curve." This curve $AP$ is also a boiling-point curve of the one component system since it relates, in accordance with our definition of boiling point, the temperatures at which water attains specified vapor pressures.

The shape and direction of the boiling-point curve $AP$ is given by the Clausius-Clapeyron equation

$$\left(\frac{dP}{dT}\right)_{l\to v} = \frac{\Delta H_{l\to v}}{T\left(V_v - V_l\right)},$$

in which $P$ is pressure, $T$ is absolute temperature, $\Delta H$ is the molar (or specific) heat absorbed in the vaporization of a molar (or specific) volume $V_l$ of the liquid to a molar (or specific) volume $V_v$ of vapor. Other equations relating vapor pressure and temperature (or boiling points with pressure) of substances are discussed in Chapter II.

## Boiling-Point Composition Diagrams of Two-Component Systems

Boiling point-composition curves of binary systems depict the boiling temperature of every conceivable mixture of the two substances, $A$ and $B$, from $100\%$ $A$, to $100\%$ $B$ at some specified pressure. Sometimes a family of boiling point-composition curves of binary systems at several pressures is available in the literature. Liquid mixtures of two components may consist of more than one liquid phase at the boiling points of the mixtures at specified pressures. If they do, the boiling point-composition curves of such mixtures at constant specified pressure will be considerably different than if only one liquid phase existed.

The liquid mixtures formed by two substances may be: (*a*) homozeotropic; (*b*) heterozeotropic; (*c*) homoazeotropic; and (*d*) heteroazeotropic. This nomenclature is used, first, to denote the fact that the liquid mixtures formed by two substances may consist of one liquid phase (homo-) or two liquid phases (hetero-) at the boiling point of the mixture at a specified pressure, and secondly, that the boiling point-composition diagrams of the liquid mixtures formed by two substances may (-azeotropic) or may not (-zeotropic) possess maxima or minima or both.

The vapor-pressure diagram, representing the dependence of vapor pressure $P$ (at a given temperature) on composition $C$ of a homozeotropic mixture, may be a straight line (Raoult's law) or it may deviate positively or negatively from a straight line (Fig. 4.2*a*). The corresponding boiling point-composition diagrams representing the dependence of boiling point $t$ (at a given pressure) on composition $C$ of heterozeotropic mixtures is given in Fig. 4.2*b*.

The boiling point-composition diagram of an ideal system (Raoult's law) is naturally concave downward, but its downward concavity may be augmented, nullified, or reversed by deviations from ideality. The degree of

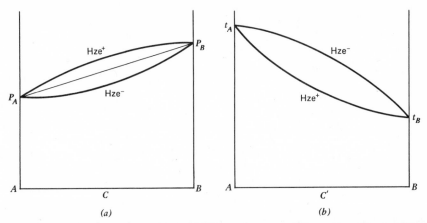

**Fig. 4.2** Vapor-pressure curves (*a*) and boiling point–composition curves (*b*) of positive and negative binary homozeotropic mixtures. Each schematic diagram shows more than one curve; each curve depicts the behavior of a different, hypothetical system.

concavity in an ideal system is a function of the difference in the boiling temperatures of the two components. A positive deviation in the isobar becomes a negative deviation from the straight line in the isotherm. When speaking of negative azeotropes or positive azeotropes subsequently we will use the convention that negative is used in the vapor pressure sense, that is, a negative azeotrope boils at a higher temperature than the boiling points of its components.

The boiling point-composition diagrams of binary heterozeotropic mixtures are composed of three portions—two curved lines and a straight line parallel to the axis of concentrations (Fig. 4.3). Heterozeotropy is rarely encountered.

The boiling-point curves of positive and negative homoazeotropic mixtures show a maximum (negative deviations) or a minimum (positive deviations) (Fig. 4.4). They may show both.

If a system of two components at the boiling temperature forms two liquid phases and the mixture boils at a temperature lower than that of the more volatile component, the mixture is heteroazeotropic. Only positive hetero-azeotropes are known. The boiling point-composition diagrams of binary heteroazeotropes are composed of three portions (Fig. 4.5): two curved lines, and a straight line parallel to the axis of concentrations. The straight line is called the heteroazeotropic line. All mixtures represented by any point on the heteroazeotropic line form two liquid phases, each of which boils at the same constant, minimum temperature. The heteroazeotropic point (Htaz, Fig. 4.5) lies on the heteroazeotropic line. This point represents the composi-

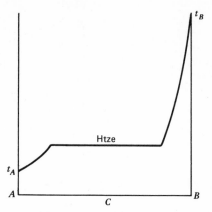

**Fig. 4.3**   Boiling point–composition curve of binary heterozeotropic mixture.

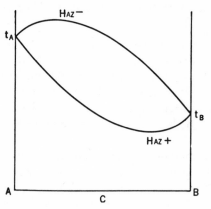

**Fig. 4.4**   Boiling point–composition curves of positive and negative binary homoazeotropic mixtures. Each schematic curve depicts the behavior of a different hypothetical binary system.

tion of the vapor in equilibrium with the two liquid phases at the minimum boiling temperature, as will be discussed shortly.

By changing the pressure, the heteroazeotropic point may be displaced toward one end or the other of the heteroazeotropic line (Fig. 4.6). If the pressure is further changed, the heteroazeotrope may be changed into a homoazeotrope—a transformation that demonstrates the close relation between homoazeotropy and heteroazeotropy. Likewise, homoazeotropic systems may be changed into homozeotropic systems by changing the pressure. There is a gradual change in one direction or the other in the azeotropic composition of a homoazeotrope until upon further changing the pressure

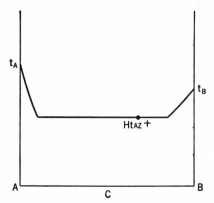

**Fig. 4.5**   Boiling-point curve of binary heteroazeotropic mixture.

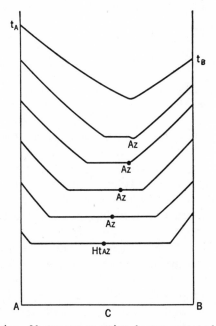

**Fig. 4.6**   Transformation of heteroazeotrope into homoazeotrope by increase of pressure.

the system becomes zeotropic. Similarly there is a gradual change in the shape of the isobars of various two-component systems composed of an azeotropic agent $A$, such as an alcohol, and the members of an homologous series of azeotropic agents $B$, $B_1 \ldots$ , such as a homologous series of $n$-paraffins (Fig. 4.7). This phenomenon is not unrelated to transformation by

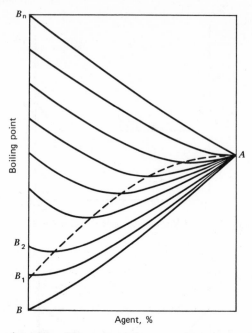

**Fig. 4.7**  Boiling point–composition curves of an azeotropic agent with a homologous series of azeotropic agents. The dashed line passes through the azeotropic compositions.

change of pressure since such change simply alters the difference between the boiling points of the components of the system. Isobars possessed by the various two-component homoazeotropic systems composed of an azeotropic agent and the members of a homologous series of azeotropic agents, illustrated in Fig. 4.7, show that if the boiling point of the homolog is too high or too low, as compared to the boiling point of the common azeotropic agent, no azeotropic phenomena will be observed.

Between these two extremes, there is found a family of isobars which illustrates how the azeotropic compositions, the azeotropic boiling points, and the differences between the boiling points of the azeotropic composition and the low-boiling component gradually change. In addition to those isobars for the clearly defined azeotropic and zeotropic systems, Fig. 4.7 shows two isobars that have in common the property that, over a range of concentration, they are nearly tangential to a horizontal line that may be drawn through the point representing the boiling point of the low-boiling component. The difference between the highest and the lowest boiling temperatures forming tangential isobars in arrays such as this is called the azeotropic range of the common azeotropic agent vis a vis that homologous series. The azeotropic range of an azeotropic agent is a fundamental property of the agent and of the class of substances for which it is an agent since each substance that can enter into azeotropic behavior with another substance is an azeotropic agent. Some agents have wide azeotropic ranges with hydrocarbons, for example, while others have narrow ranges. Hydrocarbons may form azeotropes with hydrocarbons. A homologous series of azeotropic agents such as *n*-paraffins, or aromatic hydrocarbons, is often employed to compare the azeotropic ranges of polar azeotropic agents such as alcohols, ketones, and the like.

## Liquid-Vapor Equilibrium Diagrams of Two-Component Systems

Since the act of determining the boiling temperature of a boiling mixture creates a vapor phase whose composition may be different than that of the boiling liquid, and since the boiling point of the liquid must be estimated from such boiling temperature data, or the experiment must be performed so as to circumvent this complication, it is of great importance that those who carry out ebulliometric measurements also understand and have a real appreciation for the kinds of condensation point-composition curves that are possible in liquid systems of two or more components. It should be recalled that the definition of the thermodynamically significant condensation point (but not the actually determined condensation temperature) equates this point to the boiling point of the liquid or liquids with which the vapor can be in thermodynamic equilibrium. Condensation point-composition diagrams, thus become most meaningful when presented as one of two loci of points joining equilibrium compositions of the vapor and liquid(s) phases as a function of temperature. Such diagrams are most commonly called liquid–vapor equilibrium diagrams and have for the most part been determined in so-called equilibrium stills. They may however, also be studied in ebulliometers [2]. Equilibrium stills are usually not designed for precise equilibrium temperature determination in either phase, and ebulliometers are usually not designed for precise determination of either compositions or temperatures of phases in thermodynamic equilibrium. The two devices should thus be employed to complement each other in elucidation of liquid–vapor equilibrium diagrams.

Equilibrium boiling point diagrams of homozeotropic liquid mixtures of two components may be concave upward or concave downward from the ideal curve (normally concave downward), depending on the direction and extent of the deviation of the behavior of the system from Raoult's law, as already discussed. If they are abnormally concave downward, as in Fig. 4.8, and there is extreme deviation bordering on azeotropy (Fig. 4.7), the vapor and liquid curves of the liquid–vapor equilibrium diagram may coalesce over a rather wide range of compositions at a temperature near the boiling temperature of the pure low-boiling component (Fig. 4.8a). If concave up-

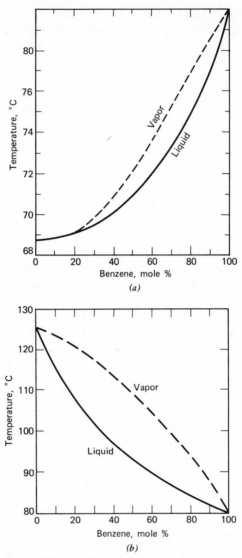

**Fig. 4.8** Liquid-vapor equilibrium diagrams of homoazeotropic systems. (*a*) is for the system *n*-hexane and benzene [6], and (*b*) for the systems *n*-octane and benzene [6].

ward and the deviations are extreme, the curves may coalesce over a rather wide range of compositions near the boiling temperature of the high-boiling component. Furthermore, the separation of the curves may be extremely wide as well as extremely narrow near either end and/or in all intermediate ranges of the composition axis, depending on the system (Fig. 4.8*b*).

These variations in types of liquid–vapor diagrams are manifestations of interactions of the components of the system as a function of concentration and temperature. Correlation of the subtle differences in the behavior of boiling binary homogeneous solutions with chemical structure of the components and deducing possible interactions constitutes a fertile research area that could employ both ebulliometric and equilibrium-still measurements, and that could further contribute substantially to solution theory.

The liquid–vapor equilibrium diagram of a heterozeotrope is depicted in Fig. 4.9. As mentioned before, heterozeotropy is but rarely encountered. Any zeotropic mixture boils at a temperature higher than the temperature of the more volatile component, and lower than that of the less volatile component. The boiling temperature of a homozeotropic system increases gradually as a function of the proportion of higher boiling component, and this same phenomenon occurs in the case of a heterozeotropic mixture, but only within the concentration limits in which the boiling mixture forms a single phase. At the extreme compositions the second liquid phase appears, and, in the whole range of compositions that two liquid phases exist, all such mixtures boil at a constant temperature which is somewhat above the boiling temperature of the more volatile component, and both liquid phases are in equilibrium with the same vapor phase. The composition of this vapor phase however lies outside the miscibility gap in heterozeotropic systems and within it in heteroazeotropic systems, as will be described.

The liquid–vapor equilibrium diagrams of homoazeotropic systems either possess a maximum or a minimum or, in at least one remarkable example, both a maximum and a minimum [7]. The diagram for a minimum boiling (positive) homoazeotrope is depicted in Fig. 4.10a, and that for a binary system possessing both a maximum and minimum is depicted in Fig. 4.10b.

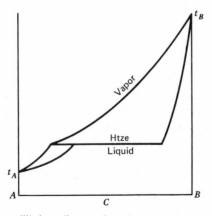

**Fig. 4.9**   Liquid-vapor equilibrium diagram for a heterozeotrope.

Note that the liquid and vapor curves of the diagram for cyclohexane–benzene are also coalesced over a wide range of concentrations (Fig. 4.10a) that boil near the composition of the azeotrope, if indeed this azeotrope can be said to have an exact composition except in terms of an extremely precisely determined boiling point. Systems such as depicted in Fig. 4.8a have been called nearly tangential systems. Nearly tangential systems were once thought to be characteristic of so-called nonideal zeotropic systems only, but it is now generally recognized that homoazeotropes, too, generally form nearly tangential isobars, at temperatures near the azeotropic boiling point, although they may be depicted otherwise in the schematic x-y equilibrium diagrams in some textbooks for chemical engineers. The occurrence of nearly tangential isobars is responsible for many of the conflicting data found in the literature on the composition of azeotropes (Fig. 4.10a) and, indeed, on the actual existence of azeotropes (Fig. 4.8a).

The components of homoazeotropic mixtures cannot be separated by ordinary distillation when working at the pressure at which a minimum or maximum was found when the isobar was established. For instance, benzene and cyclohexane (Fig. 4.10a), form a positive homoazeotrope at atmospheric pressure, and mixtures of these two substances cannot be separated by simple distillation at atmospheric pressure. They are, however, easily separated by azeotropic distillation with a selective polar azeotropic agent, such as acetone or the nonselective azeotropic agent methanol, whereas methyl-cyclohexane, another hydrocarbon, boiling some 20° higher than benzene and forming a nearly tangential coalesced liquid–vapor equilibrium system with benzene in the range of very high proportions of the low-boiling component, cannot easily be separated with polar azeotropic agents [8].

The most interesting and by far the most important liquid–vapor equilibrium behavior, in so far as ebulliometric measurements on highly purified organic preparations are concerned, is that of systems that form heteroazeotropes. Typical behavior is shown in Fig. 4.11. This behavior is important because water forms a heteroazeotrope with many liquids of ebulliometric interest, and the removal of traces of water from such liquids is both extremely important and extremely easy, in appropriately designed apparatus and using an intelligent approach easily deduced from Figure 4.11. Furthermore, incredibly small proportions of water can be detected in such systems ebulliometrically. Indeed before the advent of the Karl Fischer method, water was often determined ebulliometrically. It is remarkable therefore, that so much ebulliometric data have been reported on liquids that form heteroazeotropes with water without the investigator determining by auxiliary ebulliometric measurements if water were or were not present in the liquids in ebulliometrically significant but actually very small proportions.

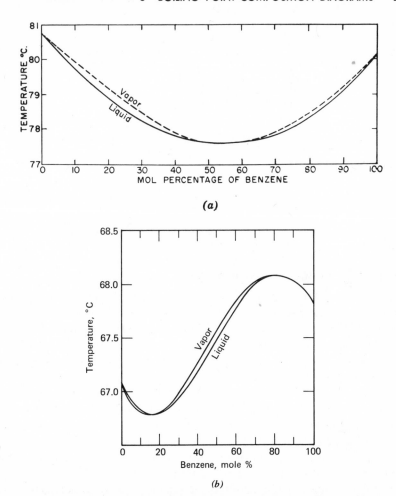

**Fig. 4.10**  Liquid–vapor equilibrium diagram (*a*) for the minimum boiling (positive) homo-azeotropic system of cyclohexane-benzene [6] and (*b*) for the system benzene–hexafluoro-benzene at a pressure of 5000 mm Hg[7].

Investigators need be cautious in retrieving data on so-called heteroazeo-tropes from the literature. Often a binary system (or system of higher order) may be homozeotropic at certain boiling temperatures resulting from arbitrarily selected pressures, but form a system of two (or more) liquid phases when not boiling at room temperature. Such systems are often in-correctly reported in industrial literature as heteroazeotropes at the selected pressure because the observation of two phases in the receiver was made at atmospheric temperature!

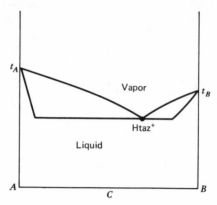

**Fig. 4.11** Liquid–vapor equilibrium diagram for a binary heteroazeotropic mixture (schematic).

## Equilibria in Multicomponent Systems

Ebulliometric data may become significantly more difficult to interpret and graphical representation may be formidable when multicomponent systems are dealt with. They are however of great scientific interest and technical importance. For instance, if a high- or low-temperature coal tar is submitted to a fractional distillation, these mixtures, containing at least six series of homologs, form dozens of azeotropes of various kinds most of which have been discovered recently. A number of ternary and quaternary positive and positive–negative azeotropes has been discovered, and precise methods for elucidating their equilibrium diagrams have been described in a series of papers published by Swietoslawski and by his collaborators [2]. These investigators have shown that most of the organic raw materials of commerce are polyazeotropic mixtures [10].

Though presentation of diagrams depicting the behavior of polycomponent systems of more than two components is beyond the scope of this chapter, two points of extreme importance will be mentioned. They are: (*a*) A system consisting of *n* homologs, each nonazeotropic with the others, does not necessarily form a system of only *n* azeotropes with an azeotropic agent that forms binary azeotropes with each of the homologs. Polyazeotropes of more than two components may be involved but are unlikely except at carefully selected pressures. (*b*) The compositions of homoazeotropes are not always determined easily by fractionation of seemingly appropriate mixtures, even in extremely efficient rectifying columns. The vapor and liquid curves of the equilibrium diagram of the system may coalesce over a very wide range of composition at approximately the azeotropic temperature. Thus without a knowledge of the shape of the isobars through a wide range of concentrations, it is impossible to draw conclusions about the composition of azeotropes from a knowledge of the composition of the main fraction collected in a single distillation. The latter composition may correspond to that of the azeotrope, or it may be very different. In examining such systems the azeotropic composition should of course be approached from all directions, with very careful ebulliometric measurements of both boiling and condensation temperatures on several compositions.

## 4 EBULLIOMETERS

Anyone wishing to embark upon ebulliometric studies is faced with the problems of choosing and obtaining suitable apparatus. Ebulliometric studies are usually made in specially designed devices called ebulliometers. Several kinds of ebulliometers are required to perform all kinds of ebulliometric measurements. Most ebulliometers are custom-made. Their specific design depends upon the size and kind of samples available for study, the kind of thermometer available, the accuracy desired, and often on the investigator's preferences in method of approach. The special glassblowing techniques required for constructing satisfactory ebulliometers of diverse types have been described in detail [11].

In the earliest edition of this volume and elsewhere [11], emphasis was placed on ebulliometers designed to hold Beckmann thermometers of either the long- or short-stem type. Though inexpensive, the Beckmann thermometer is an adequate instrument for most comparative ebulliometric work. But, in recent years, the art of Beckmann thermometry has died out, particularly in the United States where more convenient absolute certified equipment, including recording bridges for certified resistance thermometers, has become rather commonplace. The component parts used for constructing diverse ebulliometers illustrated in other editions of this volume are for the most part scaled to meet the size requirements of platinum resistance thermometers of conventional standard design [12], and these same components will be emphasized in the present revision. However, if other thermometers are chosen or the size of sample is the dominant consideration, the apparatus may be made larger or smaller than shown here, by changing the dimensions of the boiler and of the thermometer wells. Different ebulliometers, each for use in any kind of ebulliometric measurement, can be assembled readily by combining the various components illustrated in this chapter. These are based upon but are not identical to original Swietoslawski designs. Thermometry and thermometers will not be treated here.

All ebulliometers, no matter by whom designed, may be classified into four groups:

1. In the first group, in which is included all simple ebulliometers, only the boiling temperature or the condensation temperature may be measured.

2. To the second group belong those types of differential apparatus in which neither the boiling temperature nor the condensation temperature can be measured; only their precisely measured difference as reflected on some arbitrary scale is the focus of attention.

3. To the third group belong all of those differential ebulliometers in which both the boiling temperature of a liquid and the condensation temperature of the vapors from the boiling liquid can be measured precisely and simultaneously.

4. Finally, to the fourth group, belong all of those types of differential apparatus in which two or more boiling temperatures, or two or more condensation temperatures, or a boiling temperature and two condensation temperatures, or other combinations may be measured precisely and simultaneously.

## Simple Condensation Temperature Ebulliometers

The best known but by far the crudest apparatus which might be called an ebulliometer is that used in the distillation test familiar to every scientifically oriented schoolboy. It is depicted schematically in Fig. 4.12. Items of commerce, such as petroleum fuels, aromatic hydrocarbons, alcohols, and other chemical products, are frequently subjected to distillation tests in such apparatus. These involve determining condensation temperatures in the sense of the definition given in this chapter, but they are often reported in terms of various so-called boiling points, or boiling-point curves. For a pure substance, the condensation temperature at the beginning and the end of a distillation should be identical and equal to the boiling point of the substance at constant pressure; for mixtures, the condensation temperature range and shape of the distillation curve (temperature as a function of volume distilled) is related to the composition of the sample being distilled.

Organic chemists also frequently use such apparatus, or more expensive but no more complicated versions of it, both to distill a preparation and thereby obtain a presumably most pure, heart cut of the preparation, and simultaneously determine the condensation temperature or condensation temperature range of the heart cut which data they report in the literature as the boiling point. The boiling points of the vast majority of organic substances which have been reported have been determined in ebulliometric devices similar to this, or in modifications of such devices arranged for distillation at lower than ambient pressure. Often a more elaborate, but fre-

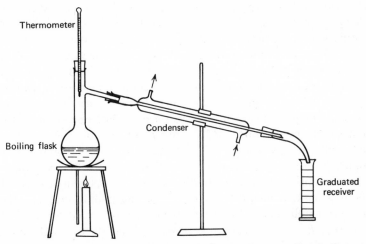

**Fig. 4.12**  Apparatus for distillation test—the simplest and most familiar ebulliometer.

quently not much more effective, distillation column than the simple open tube shown in Fig. 4.12, is interposed between the boiling liquid and the thermometer.

When examining items of commerce, the sample is distilled under prescribed conditions which have usually been proven to be appropriate to its nature [13]. Systematic observations of thermometer readings and distillate volumes are made, and from these data evaluation of the sample with respect to a standard specification is achieved.

The thermometer reading which is observed at the instant the first drop of condensate falls from the lower end of the condenser tube has been called the "initial boiling point"; the maximum thermometer reading obtained during the test has been called the "final boiling point"; and the locus of points representing temperature and volume of distillate plotted on rectangular coordinates has been called the "boiling-point curve," or more often, perhaps, the "distillation curve" [13]. Each of these terms represents certain empirical temperatures that reflect on the compositional characteristics of the sample; these temperatures for a given mixture are affected by the operating conditions, the dimension and arrangement of the apparatus, and other factors which are prescribed in the standardized tests. The exact thermodynamic significance of each temperature is unknown but each temperature and, in particular, the condensation-temperature curve may nonetheless be quite useful and adequate for establishing that certain items of commerce that have been prepared by established processes and that have been adequately studied previously, do or do not have certain desirable compositional characteristics.

The American Society for Testing and Materials and the American Petroleum Institute have adopted the same standard method [13] of test for the distillation of appropriate liquids. The operating conditions, apparatus, thermometers, and vocabulary for use in reporting the results are all defined for several classes of liquids, arranged according to boiling points.

Many devices are suitable for realizing precise condensation temperatures during a distillation or under total reflux. The condensation head depicted in Fig. 4.13a is quite adequate for this purpose. For operation under total reflux it should be equipped with a condenser system, Fig. 4.13b, and a suitable boiler. It may be used as a head for a distillation column, or it may be employed as part of a differential ebulliometer, as will be shown.

Its most important parts are a thermometer well, B, and a drop counter, F. The boiling flask, a distillation column, or the boiling unit of a boiling-temperature ebulliometer, to be described, may be sealed on at E. Vapors rise, through E, around thermometer well B, and thence to the condenser sealed on at D. Some condensed liquid moves down the thermometer well, and, under total reflux, a very constant temperature may be established inside the lower part of the thermometer well owing to heat exchange and mass exchange between the flowing liquid and the vapors that rise from the

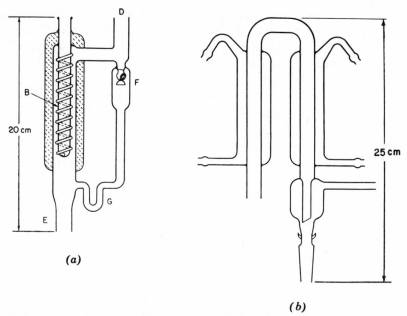

(a)

(b)

Fig. 4.13   (a) Head for condensation temperature measurements. (b) Condenser system for distillation or total reflux. When equipped with an appropriate boiler, this head and condenser system form a condensation temperature ebulliometer, see text.

boiling liquid; see a definition of the term condensation temperature on page 203. The thermometer is placed inside the thermometer well. Liquid formed in the condenser returns to the boiling liquid through drop-counter $F$ and U-tube $G$. The device may be fitted with a stopcock at the base of the condensate line leading to $G$. It may then be used as a still head. The reflux may be heated appropriately, if required, at U-tube $G$, (p. 232). The function of the drop-counter is described in the next section.

### Simple Boiling-Temperature Ebulliometers

Early ebullioscopists were concerned with boiling-point apparatus primarily for determining the boiling temperatures of dilute solutions of essentially nonvolatile solutes in various solvents. For this purpose, the simple condensation-temperature apparatuses just described here were, of course, worthless. These investigators were thus forced to make measurements on the boiling liquid phase itself, and they started by concentrating their attention on the elimination of superheat because this was at first their most distressing problem. They placed the thermometer in the boiling liquid phase and used glass beads, Prince Rupert's tears, garnets, platinum tetrahedra, indirect heating, electrical heating, stirring, and a variety of other methods in attempts to minimize the errors caused by superheat; and, under the best conditions that were devised, the errors that could be ascribed to superheat were probably reduced to a few hundredths of a degree. But should these errors have been eliminated entirely this approach could not have adequately coped with the requirements of precise measurements on dilute solutions, since observational errors attributable to variations in hydrostatic pressure were also quite high. As a rule, the boiling point of a liquid is about 0.1°C lower at the surface of the boiling liquid than at a depth of 3–4 cm below the surface. Hence, even if superheat could be eliminated entirely, the temperature read on a thermometer immersed in a boiling liquid would be difficult to interpret. First, there is the problem of a temperature gradient. Also, the liquid level surrounding the thermometer bulb oscillates, as does the mean density of the liquid, and these changes depend upon the number and size of the vapor bubbles. Though outmoded, ebulliometers in which the thermometer is placed in the boiling liquid phase are still used by some microchemists. The only redeeming feature of such ebulliometers is the small volume of liquid which they require. They certainly cannot produce precise results.

About 1892, the idea was conceived that for molecular-weight work the solution should be heated to the boiling point by passing in vapor from boiling solvent contained in a separate vessel. The apparatus was more or less perfected by McCoy in 1900 and has plagued students trying to determine molecular weights ebullioscopically in elementary laboratory courses in physical chemistry ever since. Condensation of vapor, in order to release the latent heat required to bring the solution to the equilibrium point (and to compensate for heat loss), produces a solution of unknown and constantly changing composition; and, since the thermometer is immersed in the solution, the basic problem of a correction for hydrostatic pressure of course remains. Truly though, McCoy's device was a step forward.

Over a hundred publications on the attempted determination of boiling point were rendered obsolete when Cottrell [14], in 1919, revealed that in 1910 he had discovered the utility of the vapor-lift pump for determination of the boiling temperature of solutions. The vapor-lift pump, patented [15] for coffee brewing in 1869, is an adaptation of the air-lift pump. No matter how constructed, the lift pump is essentially a tube standing in a liquid, with provision for introducing a gas into the tube below the liquid level. Adjustment of the gas pressure brings about automatic formation of alternate layers of liquid and gas which will rise in the tube above the level of the liquid surrounding the tube. In the vapor-lift device used in early ebulliometers, the tube is usually a small, inverted long-stemmed funnel, and the gas is formed by boiling the liquid which is to be pumped (Fig. 4.14).

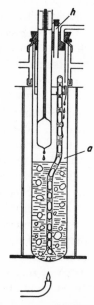

**Fig. 4.14**  Cottrell's boiling temperature ebulliometer. The liquid is boiled in *a;* the inverted funnel lift pump carries a broken stream of liquid to the upper part of the apparatus; a thin film of liquid drains down the thermometer. Solute may be introduced through tube *h*.

Cottrell placed the thermometer bulb in the gas phase, above the boiling liquid, and directed the broken stream of liquid from the pump, first to a chamber in the upper part of the apparatus and thence, in a thin film, down the thermometer (Fig. 4.14). In Cottrell's words [14]. "The liquid in its long path through the pump tube, the chamber, and down the thermometer stem continually in contact with vapor, has a good opportunity of coming to temperature equilibrium before reaching the bulb of the thermometer." Thus in one inspired stroke Cottrell essentially eliminated the two most annoying problems, superheat and hydrostatic pressure, but did so at the expense of further complicating the errors caused by rectification of the solution, especially the solution actually in contact with the thermometer. The liquid in its long path is continually evaporating (to lose superheat) and hence preferentially losing solvent to the vapor phase, so that, on reaching the thermometer bulb, its composition may be quite different from that of the boiling liquid phase or of the system as a whole before boiling commenced. The matter is further complicated by the fact that the vapors surrounding the film at the thermometer bulb are derived, for the most part, from the boiling, bulk-liquid phase. The temperature established on the bulb, though remarkably constant, must therefore represent some "steady-state" condition that is dependent upon a very large number of factors, such as the size and shape of the apparatus, the concentration of the solution before boiling commenced, the amount of the solution, the boiling rate, the heat losses inherent in the apparatus excluding the condenser, and the temperature, rate of flow, and specific heat of the condenser fluid, and so on.

One does not have to turn the page of the journal in which Cottrell's contribution was published in order to find the article by Washburn and Read [16] announcing the first modification of Cottrell's ebulliometer; and modifications have been forthcoming ever since.

The ebulliometer of Washburn and Read, depicted in Fig. 4.15, does not differ in principle from Cottrell's. Improvements include all-glass construction, a sheath which "serves to protect the thermometer bulb from the influence of cold liquid running down from the condenser" [16], and a side arm through which samples from the boiling solution (but not from the liquid film on the thermometer) may be withdrawn for analysis. Despite scores of publications on such apparatus, nothing differing in principle from Cottrell's device, except in one minor respect to be mentioned later, has appeared. Thus the boiling point of a solution cannot yet be measured directly, with high accuracy, when the solution is boiling. In fact, no one has yet suggested how to determine the composition of the solution that is in direct contact with the thermometer, probably because serious workers in the field realize that the situation would be little improved if that information were available. One has, therefore, in attempting to determine the

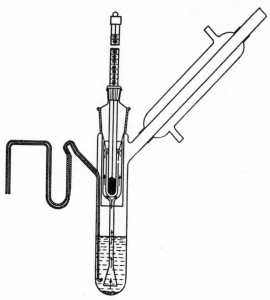

**Fig. 4.15**   Washburn and Read ebulliometer.

boiling point of a solution by boiling the solution, a choice either of measuring the highly variable temperature of a superheated liquid phase under hydrostatic pressure or of measuring the temperature of a film whose composition and state of equilibrium are both matters for conjecture. The latter alternative is preferable, because the advantage of high precision in measuring the difference between two boiling temperatures, though the absolute significance of each is unknown, overshadows all other considerations. In properly executed experiments, the difference between the two boiling temperatures will essentially be equal to the difference between the boiling points of the two solutions under study. However, two definitions are nevertheless required—the thermodynamically significant boiling point which, for a solution, cannot be measured directly by measurements made on a boiling solution, and the ebulliometric boiling temperature which, though a value of relative significance, can conveniently be measured with high precision and can be put to diverse uses provided that one possesses the requisite supplemental information for proper interpretation of the result of the measurement. The supplemental information usually includes a knowledge of the difference between the boiling temperatures of at least two solutions, and often the difference between the boiling temperature and the condensation temperature of at least one solution.

Cottrell's ebulliometer has been equipped with precisely controllable electric heaters so that, under defined operating conditions, the deficiency of solvent in the film on the thermometer may be assumed to be essentially comparable from one measurement to the next. A given apparatus may then be calibrated for molecular-weight determinations by employing substances of known molecular weight. By using this calibration technique in conjunction with a 10-junction thermoelement between two Cottrell ebulliometers, one containing solvent (so as to be able to compensate for the effect of variations in atmospheric pressure) and the other various solutions, Mair performed, in 1935, some of the most systematic and hence most satisfactory molecular-weight measurements. By employing the principles of comparative measurements, he was able to show, with this twin apparatus, the dependence of apparent molecular weight upon concentration with an accuracy well within 1%. Most of the molecular-weight work reported in the literature today, though done in more sophisticated apparatus does not come up to the high standard set by the systematic investigation of Mair. This is because of the appearance, in 1921, of a more convenient but eminently less satisfactory device invented by Menzies and Wright [17], described on p. 227. Though always more satisfactory than the device of Menzies and Wright, any arrangement of two Cottrell ebulliometers can be made even more useful if differences in condensation temperatures, as well as differences in boiling temperatures, can be measured, as in the differential apparatus, invented even before Mair's work, yet to be described.

An exceedingly simple and convenient ebulliometer for measuring boiling temperatures was constructed in 1924 by Swietoslawski and Romer and has been used for nearly half a century by Swietoslawski and his co-workers. The device, essentially but not exactly as originally built is depicted (but without a condenser) in Fig. 4.16. The liquid, which just fills the activated, electrically heated insulated vessel $A$, is brought to a steady boil. The vapor phase leaves $A$ carrying boiling liquid (cf. the Cottrell pump) through tube $I$ (7-mm outside diameter, standard Pyrex tubing) and ejects it, through $C$ onto the surface of the well $B$ in which the thermometer is situated. The well is sealed into the apparatus to prevent contact of the liquid and vapor with possible extraneous contaminants. The vapor then passes to a condenser (Fig 4.13$b$) sealed on at $D$, and the liquid phase formed in the condenser is recycled to $A$ through $D$ and drop-counter $F$.

Experiments have shown that the liquid passing through $C$ is always superheated. However, this superheating may ordinarily be made not to exceed 0.030–0.050°C in apparatus with adequately activated boiling vessels, and hence equilibrium can be established very rapidly (i.e., before the liquid reaches that part of the thermometer-well which surrounds the bulb of the thermometer), and with a very small degree of rectification during passage down the thermometer-well. Moreover, vapor from the boiling bulk phase cannot reach $B$ except through tube $C$, an improvement in principle over Cottrell's ebulliometer. To control superheat, heating is regulated in accordance with the indication of the drop-counter at $F$. The number of drops, $n$, passing $F$/sec depends upon the intensity of heating. For proper functioning of the apparatus, the number of drops/sec must be neither too small nor too large. There is a range in which the temperature in $B$ does not change with a change in the number of drops/sec. The intensity of heating should be maintained so as to have the number of drops within this range. For instance, for water, the number of drops may vary from 8 to 25/min. If the number is less than 8/min, the temperature at the thermometer is too low; if the number is too large, the temperature at the thermometer is too high. With increased intensity of heating, superheating of the liquid becomes too great to permit sufficiently rapid attainment of equilibrium between the liquid and gaseous phases. Water possesses a very high heat of condensation, and hence a rela-

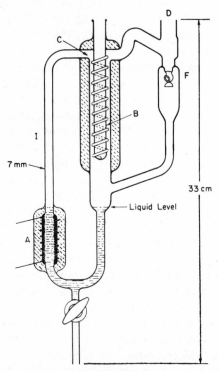

**Fig. 4.16** Boiler unit for boiling-temperature measurements, according to Swietoslawski. When a condenser system is attached at $D$, the unit becomes a simple boiling-temperature ebulliometer.

tively small number of drops of water/min is required for heating an insulated apparatus adequately; with other liquids, the number of drops flowing through the drop counter per unit time should be greater.

## Differential Ebulliometers of the First Kind

Differential ebulliometers provide information about the difference between the boiling temperature of a boiling solution and the condensation temperature of vapors generated from the boiling solution. Minimal or substantial rectifying capability may be incorporated into differential ebulliometers. Differential ebulliometers that deal only with one boiling temperature and one condensation temperature are of two kinds. In the first kind, neither the boiling temperature nor the condensation temperature is measured; a change in the difference between these temperatures is the only datum. Consequently there is no way of determining which temperature changed or whether both temperatures changed.

Menzies and Wright [17] designed the first differential apparatus for ebulliometric molecular-weight determinations; their apparatus is of the first kind and is illustrated in Fig. 4.17. Menzies and Wright in effect adapted

**Fig. 4.17** Menzies-Wright differential ebulliometer of the first kind [17]. The liquid is boiled in the lower part of the apparatus; the inverted funnel carries a broken stream of liquid and ejects it onto the lower bulb of a Menzies differential thermometer. Solute is introduced through the condenser.

Cottrell's lift pump to the Menzies differential thermometer [18]. But neither the boiling temperature of the solution nor the condensation temperature of the vapor may be measured directly in this apparatus. A change in the difference between these temperatures will cause a correlative change of level of the liquid in the two limbs of the thermometer—a change whose magnitude may be used, with the aid of an appropriate table, as a measure of some factor that caused it, such as a change in the molar concentration of a non-volatile solute in the boiling solution. In practice, in molecular-weight determinations the apparatus including the thermometer is calibrated directly with substances of known molecular weight. Detailed instructions for calibration have been published [19].

The Menzies-Wright apparatus is quite convenient and popular for accumulating data on a routine basis, but it cannot always be recommended because the data obtained with the differential thermometer cannot be evaluated thoroughly. All the details of construction and manipulation of a particularly well-designed modification of the Menzies-Wright apparatus are available [11, 19]. Menzies thermometers have, over the years, come under scrutiny, with the object of increasing their sensitivity and range, and Cottrell ebulliometers have been modified in many ways, presumably to improve their usefulness. However, no amount of improvement in detail can overcome their defect in principle, as compared to differential apparatus of the second kind which will now be described.

### Differential Ebulliometers of the Second Kind

In differential apparatus of the second kind (first built by Swietoslawski in 1925), both the boiling temperature of a liquid and the condensation temperature of a vapor may be measured precisely and simultaneously. Though somewhat more tedious in manipulation, ebulliometers of the second kind enable the experimenter to conduct all the supplemental measurements incident to thorough interpretation of ebulliometric data.

A differential ebulliometer of the second kind for molecular-weight studies is depicted in Fig. 4.18 and for degree-of-purity studies in Fig. 4.19. The device for molecular-weight work consists of a boiler unit (Fig. 4.16), a condensation unit (Fig. 4.13a), and a condenser system (Fig. 4.13b) all sealed together. The boiler unit may be equipped with a side arm for introduction of solid samples (see p. 251). The importance of the double condenser system will be discussed subsequently (see p. 243). What Swietoslawski did in constructing a differential ebulliometer of the second kind was to recognize the great advantage of differential devices (cf. Menzies and Wright, 1921) but to use two independent thermometers and thus determine both temperatures simultaneously. He then had only to learn to use a differential device and a simple ebulliometer together, and hence be able to measure three

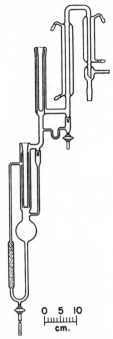

0 5 10

cm.

**Fig. 4.18** Differential ebulliometer for molecular-weight studies, according to Swieto-slawski.

temperatures simultaneously, to perfect in principle ebulliometric molecular-weight determinations.

The ebulliometer depicted in Fig. 4.18 and in Figs. 4.19 and 4.20 are sized to accommodate long-stemmed Beckmann thermometers. They may, of course, be made to accommodate platinum resistance thermometers by combining units like those depicted in Figs. 4.13a and b and 4.15.

The differential ebulliometer for degree-of-purity studies differs from the molecular-weight apparatus shown in Fig. 4.18 in that a short distilling column to provide some enhanced degree of rectification of vapor is located between the thermometer wells. This column contains only a spiral packing so as to minimize pressure drop. Since the distance between the two levels at which the thermometers are located can be measured, a correction may be made for the difference in temperatures caused by the slight difference in pressure at these levels. This ebulliometer is used for the determination of the ebulliometric degree of purity in connection with estimating boiling points of purified preparations and for many other purposes (see pp. 245, 246, and 249). Use of it is essential for serious, precise boiling-point studies of substances (p. 245).

**Fig. 4.19** Differential ebulliometer for degree-of-purity studies and boiling-point determinations according to Swietoslawski.

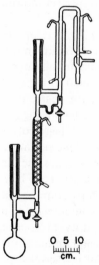

**Fig. 4.20** Differential ebulliometer for measuring two condensation temperatures [1]

## Miscellaneous Differential Ebulliometers

Still a different kind of differential ebulliometer is depicted in Fig. 4.20. It permits determination of two condensation temperatures, measured at two levels, one before and one after the vapors pass through the column. It is convenient for the study of the ebulliometric degree of purity of high-boiling liquids and of liquids contaminated with water, and so forth. Many other differential and multistage differential ebulliometers have been suggested for special ebulliometric studies [1, 2]. A description of these is beyond the scope of this chapter but they should be studied by specialists working in the field of ebulliometry.

# 5 CONSTRUCTION OF EBULLIOMETERS

## Glassblowing

A detailed description of the glassblowing techniques for the construction of thermometer wells, drop counters, and activated boilers has been published [11]. Activated boilers are made by fusing small glass particles (50–60 mesh) to the boiler surfaces. It is important to avoid overheating the pulverized glass during this operation. Production of the thermometer wells and drop counters requires special techniques; it is important that the bottom surface of the drop counters be uniform, and that the spiral of the thermometer wells be true and not deformed. Although it was earlier recommended that the boiler element, the condensation element, and the condenser system be equipped with ground joints of standard taper, it should be pointed out that, for very precise work, the use of ground joints other than the one indicated for removal of distillate from the double condenser system, Fig. 4.13b, should be abandoned. Standardized elements without ground joints should, however, be retained during construction and shipment; various ebulliometers may then be assembled by sealing the elements together in the laboratory. The ground joint sealed to the condenser system is designed to accommodate a receiver for collecting wet distillates during azeotropic drying of samples (p. 245).

## Insulation

Careful attention should be given to the adequate insulation of ebulliometers. Asbestos-paper tape and 0.125-in. asbestos cord have been found by the author to be convenient insulating materials. After having been soaked with water, the cord is wound on; this is followed by alternate layers of tape and cord until a layer at least a half-inch thick has been built up. There are certainly other perhaps even easier methods of applying insulation.

## Electric Heaters

Electric heaters should be wound around the cylindrical boiler of the boiler unit and, if desired, around the left half of the U-tube of the condensation unit. Nichrome wire having a resistance of about 10 $\Omega$/ft is most convenient for this purpose. Five to 10 ft is usually adequate for the boilers and about 1 ft for the U-tube. Both heaters should be metered, provided with fuses, and controlled by autotransformers. The resistance wire is insulated with approximately 0.06-in. asbestos tubing, is wound onto the boiler or U-tube, and is then covered with several layers of other insulation. Nichrome ribbon has also on occasion been wound into the insulation on the thermometer wells to introduce heat during measurements of condensation temperature on high-boiling liquids.

## 6   MANIPULATION OF APPARATUS

The manipulation of all types of ebulliometer is practically the same. Heating of the boiler is regulated in accordance with the indications at the drop-counter. In differential and multistage ebulliometers, the U-tube connecting the upper drop-counter with the lower part of the ebulliometer may be heated so that the same number of drops flows through each drop-counter.

If the boiling temperature of the liquid under examination does not exceed 100–110°C, bulbs of the thermometers in the thermometer wells may be surrounded with mercury and the latter covered with mineral oil. If higher-boiling liquids are examined, high-boiling mineral oil may be used exclusively, but more time must then be allowed for heat transfer between the surface of the thermometer well and the bulb of the thermometer.

Electrical resistance thermometers, thermocouples, mercury thermometers, and so on, may be used for temperature measurement (see Chapter I).

If only long-stemmed Beckmann thermometers are available, the wells of the apparatus should be scaled to accommodate the long tube that connects the bulb and fine capillary tube in front of the thermometer scale. The components of various ebulliometers of this type (Figs. 4.18–4.20) have been described in the previous edition of this volume and elsewhere [1, 2, 11].

## 7   EBULLIOMETRIC MEASUREMENTS

### Principle of Comparative Measurements

Precise ebulliometric measurements require specialized techniques that are soundly based in the principle of comparative measurements. The advantages in simplicity, convenience, and precision of the method of comparative measurements cannot be emphasized too strongly.

All physicochemical measurements may be divided into two groups: absolute and comparative [20]. When making absolute measurements, all values are expressed in absolute units and all corrections are introduced, see for example the remarkable paper of Stimson quoted presently in this chapter. Such measurements, which determine the physicochemical properties of standards as accurately as they can possibly be determined by some dedicated person at a given time, should be carried out by specialists who usually work in national or international bureaus of standards, or other laboratories especially equipped for such measurements. In other situations, the use of comparative measurements based on these standards wherever possible, or on secondary standards, is to be recommended.

A comparative measurement consists in the direct comparison of the properties of the substance under examination with those of a standard substance. Comparative measurements are made in such a manner that most or all of the corrections, calculated from estimates of various secondary effects, are unnecessary. By means of comparative measurements, the accuracy of results can be increased many times over absolute measurements made under similar conditions. For instance, if water is used as primary standard, or if other liquids are used as auxiliary reference substances, a relatively unskilled worker can determine the boiling temperature of a particular liquid with an accuracy of 0.005°C or better, and the coefficient $dt/dp$ with an accuracy of 0.1% or better, even when using uncertified thermometers or without absolute control of pressure (p. 249). Such accuracy has never been achieved where ordinary laboratory apparatus and instruments have been used for making measurement in the absolute manner. Certainly the experimenter should make full intelligent use of certified thermometric equipment and of precisely controlled monostats if they are available to him, but, even so, the measurements should be carried out systematically, employing the principles of the comparative method of which more will be said subsequently.

### International Practical Temperature Scale

Those reporting precise ebulliometric data in the modern scientific literature or retrieving it from the older scientific literature must face the problems created by the need to adjust occasionally the values of certain fixed points in practical thermometry and indeed, of the definition and name of the intervals of practical and thermodynamic temperature scales. Current practice in ebulliometry, when employing the principle of comparative thermometry, often succeeds in measuring temperatures with a precision approaching 0.0001°C; consequently, to realize the full capability of the results of precise ebulliometry there is need for comparable definition of the temperature scale that is employed.

Temperature is a physical quantity just as mass, length, and time are physical quantities, but the impracticality of measuring temperature routinely with high precision in absolute terms, makes it necessary to measure it in terms of scales that are defined on the basis of certain physical properties of standard substances, and certain equilibrium states of standard substances, and, until recently, of a certain mixture of standard substances. Such scales are studied in national and international laboratories and are adopted, from time to time, by international agreement.

Prior to 1927 there was no international agreement with respect to practical temperature scales but the intervals of the most important scales of temperatures probably were those intervals based on certain equilibrium states of water. These states were the normal boiling point of water and the normal freezing point of water. Both of these can be found on the pressure–temperature diagram of water, see Fig. 4.1. Thus zero exactly was usually defined in the so-called Centigrade scale as the temperature at which water and ice are in equilibrium under a pressure of one standard atmosphere, but not contaminated with air. The other fixed point, the normal boiling point of water, was assigned the value 100 exactly on the Centigrade scale.

The first International Practical Temperature Scale was introduced in 1928. Water and other substances were chosen as primary standards, and various practical scales covering various regions of the measurable magnitude of the physical quantity temperature were defined in terms of selected physical properties of those standards. An attempt was made to assign temperature values to certain equilibrium states of standard substances as closely as possible to the true thermodynamic temperatures which were measured laboriously by gas thermometry and which at the time, were usually expressed using as the intervals, the centigrade degree.

It is important to point out that the International Temperature Scale of 1927 did not define the zero point of the Centigrade scale in terms of the ice–water equilibrium of the water substance itself at one standard atmosphere, but instead in terms of the ice–solution equilibrium of the water substance saturated with air at one standard atmosphere. The temperature difference that can be measured for these two equilibrium states is about 0.0028° Centigrade and hence the intervals of these two so-called "centigrade" scales are not exactly the same. But the official Centigrade degree was at least finally defined in 1927.

The International Temperature Scale of 1927 was modified slightly by international agreement in 1948, and by 1954 agreement was reached that the temperature of the ice–solution equilibrium of the water substance saturated with air was equal to exactly 273.15° Kelvin; that is to say agreement was reached in 1954 that absolute zero was 273.15° Centigrade below the ice–solution equilibrium of the water substance saturated with air, at one standard atmosphere, and measured in terms of the interval defined by 1/100 of the temperature difference between the normal boiling point of water and the ice–solution equilibrium of the water substance saturated with air at one atmosphere.

This agreement of 1954 thus provided a sound basis for: (a) arbitrarily defining once and for all the fundamental unit of the physical quantity temperature and, in doing so, (b) to make the fundamental unit of the physical quantity temperature as compatible as practicable with the interval of a particular practical temperature scale. The Centigrade degree, however, was not chosen as that interval. Rather, another agreement was reached that the triple point of water (Fig. 4.1), was exactly 0.01° Kelvin above the ice–solution equilibrium temperature of the water substance saturated with air at one atmosphere, and the unit of the physical quantity temperature was thus defined by assigning the triple point of water the temperature of 273.16° Kelvin exactly. Further, the unit of the International Practical Temperature Scale then became the degree Celsius which was defined as exactly 1/99.99 of the difference between the temperatures of the normal boiling point of water and the triple point of water. Thus there was established in 1954 a new interval for a practical temperature scale that must differ in value in absolutely precise terms from both the interval of the Centigrade scale and that for the new thermodynamic temperature scale. Actually, the triple point of water probably lies nearer +0.0098° Centigrade than exactly 0.01° Centigrade above the ice–solution equilibrium temperature of the water substance saturated with air at one standard atmosphere. But the ice–solution equilibrium cannot be realized routinely with high precision whereas the triple point can.

However, these apparent discrepancies may be made philosophically compatible perhaps, by the agreements reached in 1968 for establishing the International Practical Temperature Scale of 1968, and adopting a new set of conventions for expressing the physical quantity temperature, on the one hand, and measuring it with practical temperature scales on the other.

The unit of the physical quantity temperature is defined by dividing the magnitude of the physical quantity temperature of the triple point of water by 217.16. The unit is called the kelvin (sic) symbol K (sic) and the term degree Kelvin and symbol °K is by international agreement to be discontinued. Kelvin temperature ($T$) and Celsius temperature ($t$) are to be employed as needed to express temperatures, and are expressed in kelvins ($K$) and degrees Celsius (°C), respectively, but temperature differences are expressed always in kelvins.

A summary of the definition of the International Practical Temperature Scale of 1968 has been given by Barber [21] and is reproduced here.

(1) Defining fixed points (see Table 1).

Table 1.    FIXED POINTS

| | $T_{68}$ K | $t_{68}$ °C |
|---|---|---|
| Triple point of equilibrium hydrogen | 13·81 | −259·34 |
| Temperature of equilibrium hydrogen when its | | |
| vapour pressure is 25/76 standard atmosphere | 17·042 | −256·108 |
| Boiling point of equilibrium hydrogen | 20·28 | −252·87 |
| Boiling point of neon | 27·102 | −246·048 |
| Triple point of oxygen | 54·361 | −218·789 |
| Boiling point of oxygen | 90·188 | −182·962 |
| Triple point of water | 273·16 | 0·01 |
| Boiling point of water* | 373·15 | 100 |
| Freezing point of zinc | 692·73 | 419·58 |
| Freezing point of silver | 1235·08 | 961·93 |
| Freezing point of gold | 1337·58 | 1064·43 |

* The freezing point of tin has the assigned value of $t_{68} = 231·9681$ °C and may be used as an alternative to the boiling point of water.

(2) Interpolation instruments (see Table 2).

Table 2.    INTERPOLATION INSTRUMENTS

| Interpolation | Instrument |
|---|---|
| 13·81 K to 903·89 K (630·74 °C) | Platinum resistance thermometer $R_{100}/R_0 > 1·3925$ |
| 630·74 °C to 1064·43 °C (903·89 K to 1337·58 K) | Platinum—10% rhodium/platinum thermocouple |
| 1337·58 K upwards | Radiation pyrometer using Planck law of radiation |

(3) Definition of IPTS-68 in different temperature ranges. (a) From 13·81 to 273·15 K, the temperature $T_{68}$ is defined by the relation

$$W(T_{68}) = W_{CCT-68}(T_{68}) + \Delta W(T_{68}) \tag{5}$$

where $W(T_{68})$ is the resistance ratio $\dfrac{R(T_{68})}{R(273 \cdot 15 \text{ K})}$ of the platinum resistance thermometer and $W_{CCT-68}$ is the resistance ratio as given by a standard reference function*. $\Delta W(T_{68})$ is the deviation of the $W$ value of the resistance thermometer from the reference function and is determined from several polynomials as follows:

For 13·81 K to 20·28 K

$$\Delta W(T_{68}) = A_1 + B_1 T_{68} + C_1 T_{68}^2 + D_1 T_{68}^3 \tag{6}$$

The constants are determined by the measured deviations at the triple point of equilibrium hydrogen, the temperature of 17·042 K and the boiling point of equilibrium hydrogen and by the derivative of the deviation function at the boiling point of equilibrium hydrogen as derived from equation (7).

For 20·28 K to 54·361 K

$$\Delta W(T_{68}) = A_2 + B_2 T_{68} + C_2 T_{68}^2 + D_2 T_{68}^3 \tag{7}$$

* $T_{68} = [A_0 + \Sigma_{i=1}^{20} A_i (\ln W_{CCT-68}(T_{68}))^i]$K.

The constants are determined by the measured deviations at the boiling point of equilibrium hydrogen, the boiling point of neon and the triple point of oxygen and by the derivative of the deviation function at the triple point of oxygen as derived from equation (8). For $54 \cdot 361$ K to $90 \cdot 188$ K

$$\Delta W(T_{68}) = A_3 + B_3 T_{68} + C_3 T_{68}^2 \tag{8}$$

The constants are determined by the measured deviations at the triple point and boiling point of oxygen and by the derivative of the deviation function at the boiling point of oxygen as derived from equation (9). For $90 \cdot 188$ K to $273 \cdot 15$ K

$$\Delta W(T_{68}) = A_4 t_{68} + C_4 t_{68}^3 (t_{68} - 100 \ ^\circ C) \tag{9}$$

where $t_{68} = T_{68} - 273 \cdot 15$ K. The constants are determined by the measured deviations at the boiling point of oxygen and the boiling point of water.

(b) From 0 °C ($273 \cdot 15$ K) to $630 \cdot 74$ °C, $t_{68}$ is defined by

$$t_{68} = t^1 + 0 \cdot 045 \left(\frac{t^1}{100 \ ^\circ C}\right)\left(\frac{t^1}{100^\circ \ C} - 1\right)\left(\frac{t^1}{419 \cdot 58 \ ^\circ C} - 1\right)$$
$$\left(\frac{t^1}{630 \cdot 74 \ ^\circ C} - 1\right) \ ^\circ C \tag{10}$$

where $t^1$ is defined by the equation

$$t^1 = \frac{1}{\alpha} [W(t^1) - 1] + \delta \left(\frac{t^1}{100 \ ^\circ C}\right)\left(\frac{t^1}{100 \ ^\circ C} - 1\right) \tag{10a}$$

where $W(t^1) = \dfrac{R(t^1)}{R(0 \ ^\circ C)}$. The constants $R(0 \ ^\circ C)$, $\alpha$ and $\delta$ are determined by measurement of the resistance at the triple point of water, the boiling point of water (or the freezing point of tin, see footnote, Table 1) and the freezing point of zinc.

(c) From $630 \cdot 74$ °C to $1064 \cdot 43$ °C, $t_{68}$ is defined by the equation

$$E(t_{68}) = a + b t_{68} + c t_{68}^2 \tag{11}$$

where $E(t_{68})$ is the electromotive force of a standard thermocouple of rhodium-platinum alloy and platinum, when one junction is at a temperature $t_{68} = 0$ °C and the other is at a temperature $t_{68}$. The constants $a$, $b$ and $c$ are calculated from the value of $E$ at $630 \cdot 74$ °C $\pm 0 \cdot 2$ °C, as determined by a platinum resistance thermometer, and at the freezing points of silver and gold.

(d) Above $1337 \cdot 58$ K ($1064 \cdot 43$ °C) the temperature $T_{68}$ is defined by the equation

$$\frac{L_\lambda(T_{68})}{L_\lambda(T_{68}(Au))} = \frac{\exp\left[\dfrac{c_2}{\lambda T_{68}(Au)}\right] - 1}{\exp\left[\dfrac{c_2}{\lambda T_{68}}\right] - 1} \tag{12}$$

in which $L_\lambda(T_{68})$ and $L_\lambda(T_{68}(Au))$ are the spectral concentrations at temperature $T_{68}$ and at the freezing point of gold, $T_{68}(Au)$, of the radiance of a black body at the wavelength $\lambda$; $c_2 = 0 \cdot 014 \ 388$ metre kelvin.

The numerical values of temperatures defined by the International Practical Temperature Scale of 1968 (IPTS-68) differ from those given by the International Practical Temperature Scale of 1948 (IPTS-48) by the amounts shown in Table 4.1 below, which is also taken directly from Barber's summary.

Table 4.1 (IPTS-68)—(IPTS-48) in Kelvins

| $t_{68}$ °C | 0 | -10 | -20 | -30 | -40 | -50 | -60 | -70 | -80 | -90 | -100 |
|---|---|---|---|---|---|---|---|---|---|---|---|
| -100 | 0.022 | 0.013 | 0.003 | -0.006 | -0.013 | -0.013 | -0.005 | 0.007 | 0.012 | (0.008 at O$_2$ point) | 0.022 |
| -0 | 0.000 | 0.006 | 0.012 | 0.018 | 0.024 | 0.029 | 0.032 | 0.034 | 0.033 | 0.029 | 0.022 |

| $t_{68}$ °C | 0 | 10 | 20 | 30 | 40 | 50 | 60 | 70 | 80 | 90 | 100 |
|---|---|---|---|---|---|---|---|---|---|---|---|
| 0 | 0.000 | -0.004 | -0.007 | -0.009 | -0.010 | -0.010 | -0.010 | -0.008 | -0.006 | -0.003 | 0.000 |
| 100 | 0.000 | 0.004 | 0.007 | 0.012 | 0.016 | 0.020 | 0.025 | 0.029 | 0.034 | 0.038 | 0.043 |
| 200 | 0.043 | 0.047 | 0.051 | 0.054 | 0.058 | 0.061 | 0.064 | 0.067 | 0.069 | 0.071 | 0.073 |
| 300 | 0.073 | 0.074 | 0.075 | 0.076 | 0.077 | 0.077 | 0.077 | 0.077 | 0.077 | 0.076 | 0.076 |
| 400 | 0.076 | 0.075 | 0.075 | 0.075 | 0.074 | 0.074 | 0.074 | 0.075 | 0.076 | 0.077 | 0.079 |
| 500 | 0.079 | 0.082 | 0.085 | 0.089 | 0.094 | 0.100 | 0.108 | 0.116 | 0.126 | 0.137 | 0.150 |
| 600 | 0.150 | 0.165 | 0.182 | 0.200 | 0.23 | 0.25 | 0.28 | 0.31 | 0.34 | 0.36 | 0.39 |
| 700 | 0.39 | 0.42 | 0.45 | 0.47 | 0.50 | 0.53 | 0.56 | 0.58 | 0.61 | 0.64 | 0.67 |
| 800 | 0.67 | 0.70 | 0.72 | 0.75 | 0.78 | 0.81 | 0.84 | 0.87 | 0.89 | 0.92 | 0.95 |
| 900 | 0.95 | 0.98 | 1.01 | 1.04 | 1.07 | 1.10 | 1.12 | 1.15 | 1.18 | 1.21 | 1.24 |
| 1000 | 1.24 | 1.27 | 1.30 | 1.33 | 1.36 | 1.39 | 1.42 | 1.44 | | | |

| $t_{68}$ °C | 0 | 100 | 200 | 300 | 400 | 500 | 600 | 700 | 800 | 900 | 1000 |
|---|---|---|---|---|---|---|---|---|---|---|---|
| 1000 | | 1.5 | 1.7 | 1.8 | 2.0 | 2.2 | 2.4 | 2.6 | 2.8 | 3.0 | 3.2 |
| 2000 | 3.2 | 3.5 | 3.7 | 4.0 | 4.2 | 4.5 | 4.8 | 5.0 | 5.3 | 5.6 | 5.9 |
| 3000 | 5.9 | 6.2 | 6.5 | 6.9 | 7.2 | 7.5 | 7.9 | 8.2 | 8.6 | 9.0 | 9.3 |

Barber says, "since the IPTS-68 values are in agreement with the thermodynamic values, the differences" [shown in the table] "also represent the divergences of the IPTS-48 from thermodynamic temperatures [21]".

## Water as Primary Ebulliometric Standard

In 1938 the International Union of Chemistry accepted the proposal of the Committee on Physicochemical Data that water be used as the primary standard substance in ebulliometry and tonometry [22]. Therefore, in all cases in which the boiling point or vapor pressure or the coefficient $dt/dp$ or $dp/dt$ of a substance is measured, water should be used as the primary reference substance.

A large number of very careful investigations have been carried out for the purpose of establishing the temperature–pressure relation for water precisely. In 1934, Osborne and Meyers [23] at the National Bureau of Standards, developed a formulation which, by a single equation, represented an adjusted composite appraisal of all available data in the temperature range between the freezing point and the critical region. Subsequently, in 1939, the Committee on Physicochemical Data of the International Union of Chemistry examined all existing data, accepted the data of Table 4.2 for use in ebulliometry and tonometry, and recommended the following equations for determining $p$ and $t$ in degrees Centigrade in the range from 680 to 860 mm Hg.

$$t = 100 + 0.0368578(p - 760) - 0.000020159(p - 760)^2$$
$$+ 0.00000001621(p - 760)^3$$

$$p = 760 + 27.1313(t - 100) + 0.40083(t - 100)^2 + 0.003192(t - 100)^3.$$

**Table 4.2**   Boiling Points of Water under Different Pressures[a]

| Pressure (mm Hg) | Temperature (°Centigrade) |
|:---:|:---:|
| 660 | 96.096 |
| 680 | 96.914 |
| 700 | 97.712 |
| 720 | 98.492 |
| 740 | 99.255 |
| 760 | 100.000 |
| 780 | 100.729 |
| 800 | 101.443 |
| 820 | 102.142 |
| 840 | 102.828 |
| 860 | 103.500 |

[a] In the original table, figures are given to ten-thousandths of a degree.

Still later, Zmaczynski and Moser [24], working at the Physikalischtechnische Reichanstalt, Berlin, employed two quite different methods but with each investigator simultaneously employing the same manostat. These data were in almost exact agreement and moreover were in almost exact agreement with those recommended by the International Union of Chemistry. The equations given by Zmaczynski and Moser [24] for calculating $p$ and $t$ in degrees Centrigrade in the range 300–2000 mm Hg are as follows:

$$p - 760 = 27.12912(t - 100) + 0.400793(t - 100)^2 +$$
$$3.04131 \times 10^{-3}(t - 100)^3 + 1.12141 \times 10^{-5}(t - 100)^4$$

$$t - 100 = 3.68608 \times 10^{-2}(p - 760) - 2.0073 \times 10^{-5}(p - 760)^2 +$$
$$1.625 \times 10^{-8}(p - 760)^3 - 1.61 \times 10^{-11}(p - 760)^4.$$

Later, in 1942, Stimson and Cragoe working at the National Bureau of Standards made some measurements which indicated the desirability of minor revisions in the Osborne-Meyers formulation—a formulation that in 1934 was suspected of being somewhat defective below 100°C. These data, referenced as private communications, were incorporated into a new formulation, by Goff and Gratch [25], which extends only to 100°C, and they have been used, together with the original Osborne-Meyers formulation, as the standard of reference for determining vapor pressures of hydrocarbons by the American Petroleum Institute [26].

These corrections now nearly 25 years old were finally published in 1967 by Gibson and Bruges [27], but in the meantime the quest for still more reliable data had already been taken up at the National Bureau of Standards, this time in 1947 by Wilson, Cross, and Stimson. The full story of the events of both 1942 and 1947 were only recently recorded by Stimson himself [28], long-since retired from the Bureau. As already indicated, even the temperature scale itself had changed, and hence Stimson duly recorded the results of this work in terms of both the International Temperature Scale of 1948, Table 4.3, and the International Practical Temperature Scale of 1968, Table 4.4. In Stimson's own words:

"Table [4.3] gives the combined results from the 1942 and the 1947–1949 periods. In column 4 are the estimated standard deviations of pressure in dynes per square centimeter. Using values for $dp/dt$ gives the corresponding deviations of temperature in the next column. The last column gives the values in newtons per square meter. From the estimates of the standard deviations in [the] table it appears that (except for the 25°C measurement) the pressure measurements were consistent with each other to one part in 50,000. It also appears that temperatures in this range were measured with a standard deviation estimated to be 0.00040 degree.

The International Practical Temperature Scale of 1968 (IPTS-68) now replaces the International Temperature Scale of 1948, which has been used for all foregoing temperature computations. In the range between 0 and 100°C the values of temperature on the new scale are smaller than on the 1948 scale. This makes the values of pressure larger, when adjusted to the corresponding even temperatures. Table [4.4] gives the even temperatures on the 1968 scale in the first column, the temperature correction in the second column, and the values of pressure adjusted to the even temperatures in the third column."

**Table 4.3**  Summary of Measured Values for the Vapor Pressure of Water

| Temperature ITS-48 (°C) | Year | Pressure (dyne/cm²) | Standard Deviation (dyne/cm²) | Standard Deviation (°) | Pressure (N/m²) |
|---|---|---|---|---|---|
| 25 | 1942 | 31,670. | 1.4 | 0.00075 | 3,167.0 |
| 40 | 1948 | 73,772.7 | 1.5 | 0.00041 | 7,377.3 |
| 50 | 1942, 1947, 1948 | 123,383.0 | 1.8 | 0.00019 | 12,338.3 |
| 60 | 1942, 1949 | 199,242.2 | 1.6 | 0.00017 | 19,924.2 |
| 70 | 1949 | 311,661.5 | 2.2 | 0.00016 | 31,166.2 |
| 80 | 1942 | 473,639. | 5.2 | 0.00027 | 47,363.9 |
| 100 | 1947–1949 | 1013,250. | 15.9 | 0.00044 | 101,325.0 |

**Table 4.4**  Values of Pressure at Temperatures on the IPTS-68

| Temperature IPTS-68 (°C) | Correction $t_{68} - t_{48}$ (°) | Pressure (N/m²) |
|---|---|---|
| 25 | −0.00854 | 3,168.6 |
| 40 | −0.01034 | 7,381.3 |
| 50 | −0.01037 | 12,344.6 |
| 60 | −0.00957 | 19,933.0 |
| 70 | −0.00805 | 31,177.0 |
| 80 | −0.00589 | 47,375.2 |
| 100 | 0.00000 | 101,325.0 |

Someone will certainly derive a new formulation from Stimson's tables.

For pressures outside Stimson's tables, use may be made of the Osborne-Meyers [23] formulation. If other pure substances are employed as reference liquids, use may be made of data collected by Zmaczynski [29]. He examined ethyl bromide, carbon disulfide, acetone, chloroform, carbon tetrachloride, benzene, toluene, chlorobenzene, and bromobenzene.

Also the data on pure hydrocarbons published by the American Petroleum Institute [26, 30] and other data scattered about the literature may be useful if the investigator possesses adequately pure preparations. Certainly, the standard of reference, together with the ebulliometric characteristics of the standard, should be stated when publishing boiling-point data.

## Ebulliometric Degree of Purity of Liquids

Simultaneous determination of the boiling temperature of a liquid substance and the condensation temperature of its vapor in a differential ebulliometer of the type shown in Fig. 4.19 makes it possible to determine the ebulliometric degree of purity of that liquid. Such a determination ought to be made in conjunction with all measurements of boiling point, $dt/dp$, and $dp/dt$, as it serves to define the ebulliometric quality of the material submitted to those measurements.

From thermodynamic considerations, these temperatures, when corrected for the difference in pressure at the two levels in the ebulliometer, should be identical for a pure substance, or for certain mixtures of substances forming azeotropes if these substances are present in the mixture in their azeotropic proportions. In practice, with the exception of water, it is exceedingly difficult to prepare a liquid whose boiling and condensation temperatures, as measured in a differential ebulliometer, are exactly the same to 0.001°C. With organic liquids, the boiling temperature is often higher than the condensation temperature because of the presence of small proportions of volatile impurities, especially moisture which concentrates in the vapor phase because of azeotropic (often heteroazeotropic) phenomena.

If the boiling temperature and condensation temperature of a system are equal, after correction has been made for the difference in pressure at the two levels in the ebulliometer, both are equal to the boiling point of the liquid in the ebulliometer. If they are not equal, often neither is equal to the boiling point, regardless of the amount of auxiliary data the experimenter has accumulated to substantiate high purity for his material. For example, although cryometric determinations of purity of hydrocarbons are relatively insensitive to the presence of traces of moisture, ebulliometric determinations of purity of hydrocarbons are particularly sensitive to this impurity.

An essential part of each determination of the purity of a liquid consists of the removal of moisture or other volatile impurities which often contaminate otherwise pure liquids. Some moisture may be present in the ebulliometer despite careful drying. The following procedure is recommended. First of all, the ebulliometer must be cleaned, dried, and washed with a sample of the liquid under examination. Then it is filled and the liquid brought to a boil. Another differential ebulliometer, containing a reference material, is similarly prepared and is placed by the side of the differential apparatus. An adequate heating rate is established for each ebulliometer, by intercomparing the various temperatures as a function of heating rate (p. 226). The rates may require minor readjustment after removal of moisture. After the boiling and condensation temperatures of the sample have been compared and their difference noted, the coolant is removed from both sides of the condenser assembly (Fig. 4.19), and, of the total amount of 50 ml of liquid, 2 ml is distilled off through the condensers and collected in a receiver connected to the condenser system by a ground joint. Then coolant is returned to the condensers and a second determination of the difference in the boiling and condensation temperatures is made. These operations are repeated until four determinations of the differences have been made: $\Delta t_1$, $\Delta t_2$, $\Delta t_3$, and $\Delta t_4$. Comparison of these values gives an idea of the kinds of impurity present in the liquid. For instance, if $\Delta t_1 > \Delta t_2$, and the differences, $\Delta t_3$ and $\Delta t_4$, are equal to or slightly smaller than $\Delta t_2$, it can be concluded that moisture and other quite volatile impurities were removed by the first distillation. If, however, $\Delta t_1$, $\Delta t_2$, $\Delta t_3$, $\Delta t_4$ are all relatively large, and decrease only slightly after each distillation, it is to be concluded that the impurities are close boiling and are not removable by this simple operation. The sample should then be repurified by some other means and submitted again to ebulliometric study.

If the reference substance is not water, the same procedure should be used to remove moisture from the reference substance, and the boiling and condensation temperatures of the substance under investigation compared with the boiling and condensation temperatures of the standard. In this way not only can the liquid under examination be characterized, but also any contamination of the reference material can be detected.

An arbitrary scale of purity, known as the Swietoslawski scale, is shown in table 4.5.

Experiments have revealed that the differences between the boiling temperatures of several different samples of a liquid, each of them characterized by the degree of purity V, did not exceed 0.008°C and ordinarily were within

**Table 4.5** Swietoslawski's Scale of Purity

| Degree of Purity | Difference Between Boiling and Condensation Temperatures (°C) |
|:---:|:---:|
| I | 1.00 –0.10 |
| II | 0.10 –0.050 |
| III | 0.050–0.020 |
| IV | 0.020–0.005 |
| V | 0.005–0.000 |

0.001–0.003° [4, 31, 32]. In most cases, a liquid with purity V does not contain more than 0.005% of impurities. This value, however, cannot be accepted as an infallible rule; the magnitude of the difference between the boiling and the condensation temperatures depends to a large extent upon the nature of the impurities. In spite of the limitations of the ebulliometric method in this case, one should not underestimate the usefulness of ebulliometric degree-of-purity data; when the boiling point of a liquid is reported to better than tenths of a degree, even after systematic purification, the data can be assumed to have little significance unless it has been established that the liquid was of degree II of purity or better. This statement is particularly true when the "boiling point" is reported as the result of condensation-temperature measurements, and when it is not established that water has been completely removed from the area in the apparatus where the temperature was measured.

In table 4.6 are shown the results obtained in relevant tests on isopropyl alcohol. In following the suggestion that the second difference between the boiling and the condensation temperatures be accepted for the classification of the purity of a liquid, the isopropyl alcohol examined was found to be of degree of purity I.

**Table 4.6**    Determination of the Degree of Purity of Isopropyl Alcohol [33], Secondary Standard Substance: Benzene

| Liquid Examined | ΔBoiling and Condensation Temperature (°C) | ΔBoiling Temperature of Liquid and Benzene (°C) |
|---|---|---|
| Isopropyl alcohol | $\Delta t_1 = 0.196$ | $\Delta T_1 = 2.060$ |
| After removal of 2 ml | $\Delta t_2 = 0.195$ | $\Delta T_2 = 2.098$ |
| After second removal of 2 ml | $\Delta t_3 = 0.144$ | $\Delta T_3 = 2.108$ |
| After third removal of 2 ml | $\Delta t_4 = 0.083$ | $\Delta T_4 = 2.124$ |

After careful rectification of the same isopropyl alcohol in an efficient distillation apparatus, the data presented in Table 4.7 were collected; the sample was then characterized by the degree of purity IV. The determination of the difference between the boiling temperature of isopropyl alcohol and benzene was made by transferring the same Beckmann thermometer from one ebulliometer to the other.

It is perhaps unfortunate that Swietoslawski used the term degree of purity in connection with the difference in boiling and condensation temperatures as determined in a differential ebulliometer. The greatest value of these measurements lies in their usefulness in telling the experimenter when his

**Table 4.7**  Determination of the Purity of Rectified Isopropyl Alcohol [33]

| ΔBoiling and Condensation Temperature (°C) | ΔBoiling Temperature of Liquid and Benzene (°C) |
|---|---|
| $\Delta t_1 = 0.010$ | $\Delta T_1 = 2.168$ |
| $\Delta t_2 = 0.009$ | $\Delta T_2 = 2.169$ |
| $\Delta t_3 = 0.009$ | $\Delta T_3 = 2.170$ |
| $\Delta t_4 = 0.008$ | $\Delta T_4 = 2.170$ |

preparation is free from moisture, and/or in indicating the advisability of careful repurification.

The mechanical difficulty of removing moisture from substances that form heteroazeotropes with water is very often underestimated. A system of two condensers is highly recommended for this operation. It can be manipulated to preclude the possibility of trapping water in the apparatus (see p. 243). It has been proposed [34] that a differential ebulliometer of standardized dimensions be used in determinations of the ebulliometric degree of purity, but for exceptionally pure materials a distilling column more effective than the one originally proposed may be used between the two temperature wells.

### Boiling Points of Reasonably Pure Preparations

A precise determination of the boiling point of a preparation should be made only after the determination of the degree of purity of the preparation. In no other way can the experimenter be sure that the preparation is in appropriate condition to warrant acceptance of precise boiling temperature measurements. To determine the boiling point of a preparation from acceptable boiling temperature data obtained in the degree-of-purity test, one needs only to know the pressure under which the two systems, standard and preparation (Tables 4.6 and 4.7) are boiling, and the boiling point of the standard substance.

For the purpose of comparing the boiling point of a preparation and of a reference liquid at a specified pressure, precise determinations of pressure are not required. If, for example, the normal boiling point of the reference liquid has been determined by a specialist, and if the worker of less skill has found that the difference between the boiling temperature of this reference substance and that of the substance under examination is 2.632 ± 0.002° at 760.0 ± 0.1 mm Hg, and both substances are of degree of purity IV or better, then the difference between the normal boiling points (at exactly 1 atm pressure) will be 2.632 ± 0.002°.

Should it be desired to measure the boiling point of a liquid with an accuracy of ±0.002° without the use of a reference substance, the pressure must be known with an accuracy which exceeds that which usually can be established in ordinary laboratories, and the temperature-measuring equipment must, of course, be correct within 0.002°. This is one of the principal reasons why very precise measurements of the boiling points of pure liquids as a function of pressure should be carried out by specialists in national and international bureaus of standards. Given a suitable number of adequately characterized auxiliary reference substances, it is doubtful whether certified thermometric equipment would be required in most industrial distillation laboratories. The accuracy of the determination of the boiling temperature of liquids in these laboratories should, however, always be 0.05°C or better.

Considerable effort has been expended in past years toward designing and building precision manostats. It should be emphasized that if in principle a manostat can be calibrated by measuring the boiling temperature of water, then the boiling temperature of water can be measured simultaneously with the measurement of the boiling temperature of the substance under investigation.

If the boiling temperature of the substance under examination is compared with that of the reference substance under a pressure considerably different from normal, the difference between the normal boiling temperatures of the substance and the standard can, of course, be calculated from a knowledge of the coefficient $dt/dp$ for the substance and for the standard, respectively. Should $dt/dp$ for the substance under examination not be known, advantage may be taken of the fact that, in general, substances normally boiling at about the same temperature usually have coefficients $dt/dp$ which do not differ very much. If it is suspected that this assumption is not valid within the accuracy desired, the measurements should be made at substantially normal pressure, or $dt/dp$ for the substance under examination should be determined as described presently.

### Condensation Temperatures of Liquids during Rectification

In the precise fractionation of reasonably pure liquids, it is a common practice to determine the condensation temperature as a function of the volume of the distillate removed from the distilling head. These data are often recorded erroneously as boiling points. In the usual case the experimenter is interested in determining precisely how the condensation temperature changes rather than in determining the exact temperature at any stage in the distillation.

If it is desired to determine temperature differences with a precision of a few thousandths of a degree and with a minimum of effort, a simple or, preferably, a differential ebulliometer of the type shown in Fig. 4.19, containing a sample of the substance which is to be distilled, should be placed alongside the head of the distilling column. The condensation temperatures measured in the distilling head should then be compared with the boiling temperature of the sample in the ebulliometer. By this artifice variations in the condensation temperature occasioned by fluctuations in atmospheric pressure cannot cause inaccurate results, although barometric pressure is not measured at any time during the distillation, and thermometric corrections need not be made. The same thermometer may successively be transferred from the distilling head to the ebulliometer and then back to the distilling head, and, if the substance being purified is already reasonably pure, there will not be more than a few tenths of a degree difference between its boiling temperature and any of the various condensation temperatures determined at the head of the column. When a differential ebulliometer is employed in such measurements, the thermometer located in the condensation-temperature well of that device can be used to detect and correct for the most minute of pressure changes occurring while the other thermometer is successively transferred.

Figure 4.21 illustrates the results obtained when 3 liters of already reasonably pure benzene was fractionated in a column of about 50 theoretical plates. The curve, $T$, shows the condensation temperatures of the distillate as compared, on the scale of the same Beckmann thermometer, with the boiling temperature, $X$, of a sample of the same benzene. About 150 comparative temperature readings were made during the distillation, and not more than 15 of these, after correcting for changes in atmospheric pressure, deviated more than $\pm 0.002°$ from the best curve drawn through all the values.

For comparing purities, the solid–liquid equilibrium temperatures of the fractions ($FT$, Fig. 4.21), with $5\%$ of the sample in the solid phase, were

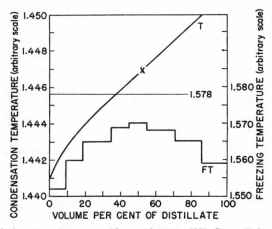

**Fig. 4.21** Distillation curve for reasonably pure benzene [37]. Curve $T$ shows condensation temperatures. $FT$ is a graph of the freezing temperatures of the fractions.

compared with the solid–liquid equilibrium temperature of the best preparation of benzene then obtainable. The latter temperature, as read on the scale of the same Beckmann thermometer, is represented by the horizontal line, 1.578, in Fig. 4-21.

Should the absolute values of the condensation temperatures be required, the ebulliometer should be filled with a properly chosen reference substance. Better still, a second ebulliometer should be filled with the reference substance, and that ebulliometer should be operated for short periods during the distillation. In this way it can be ascertained whether or not the sample in the first reference ebulliometer has undergone changes due to prolonged boiling. For example, in the experiment described above, a second ebulliometer, containing the purest sample of benzene then obtainable was operated during the first hour, the ninth hour, and the seventeenth hour of the distillation, and it was found that the difference in boiling temperatures of the samples in the two ebulliometers was constant.

Such precision, though very easy to obtain with ordinary temperature-measuring equipment, is not required in everyday laboratory technique and is not at present obtainable with small amounts of material. However, the corrections for change of pressure, and the direct comparison of the condensation temperature with the boiling temperature of a standard, should always be made. The difference found should be published together with the other data characterizing the liquid.

This technique may be appropriate to studies of homoazeotropy. Here the condensation temperature of the azeotrope, measured at the head of a distilling column is compared with the boiling temperature of the lowest-boiling component, measured in an ebulliometer. However additional data are needed to establish azeotropic composition (see p. 214).

In the distillation of reasonably pure liquids, attention should be paid to changes in the condensation temperature at the beginning of the distillation to ascertain whether or not a water azeotrope contaminates the foreruns. Often, when only a few hundredths of a per cent of moisture contaminates the sample, an unexpectedly low condensation temperature is noticed. Should the substance form a heteroazeotrope with water, the first fraction should be removed through the condenser system as explained on p. 243.

Should a differential ebulliometer be unavailable, advantage may still be taken of the principle of comparative measurements in distillation studies of reasonably pure liquids or of azeotropes. A still head, identical with that employed on the column, may be equipped with a boiling flask, and the condensation temperatures of a reference substance in that device may be compared with the temperatures observed at the top of the distilling column. If a liquid other than the material under study is used as reference substance and if the absolute value of the condensation temperature is required, correction must be made for the difference in pressure exerted by the columns of vapor that exists between the thermometer wells and condensers (see p. 229).

## Determination of $dt/dp$

It very often occurs that boiling points are determined at the pressure which happens to prevail in the laboratory, and that these temperatures are corrected to the normal values. To calculate the normal boiling point of a substance from data obtained at a pressure other than normal, $dt/dp$ over the desired range must be known. These data are rather easily determined when employing the method of comparative measurements.

The experiment consists in measuring the ratio of the change in the boiling temperatures of the liquid and of water, both produced by the same change in pressure:

$$(dt/dp)_s/(dt/dp)_w = a.$$

The absolute value of $(dt/dp)_s$ is then calculated from a knowledge of $(dt/dp)_w$. With this method, precise measurements of pressure are not required.

## Vapor Pressure of "Pure" Preparations by the Ebulliometric Method

Although the determination of vapor pressure is the subject of Chapter II, the comparative ebulliometric method will be discussed here.

In the comparative ebulliometric method, successive or, preferably, simultaneous measurements are made of the boiling point of the substance under investigation and the boiling point of water, or of an adequately characterized secondary standard substance. The ebulliometers are joined to the same manostat by means of which the pressure can be varied over the desired range. Thus the boiling point of the substance, $t_s$, at a given pressure, and the pressure under which that substance is boiling, may be ascertained precisely by temperature measurements alone; the pressure is determined by measuring the boiling point of water or the secondary standard.

Thus the relations $t_s = f(p)$, $p_v = f(t_s)$, where $p_v$ is the vapor pressure of the liquid, may be established directly. Similarly, the relation $t_s = f(t_w)$ between the boiling point of the substance, $t_s$, and that of water, $t_w$, boiling under the same pressures may be found. This relation may be expressed:

$$t_s = A + B(t_w) + C(t_w)^2,$$

in which $A$, $B$, and $C$ are constants. In addition,

$$(dt/dp)_s/(dt/dp)_w = a,$$

the ratio of the change in the boiling temperature of the substance and of water, both produced by the same change in pressure, may be established with a very high accuracy.

The apparatus consists of a simple ebulliometer filled with water (or a differential ebulliometer filled with an auxiliary reference substance) and a differential ebulliometer filled with the substance under examination. Both ebulliometers are connected to the same manostat. To avoid oxidation of the liquid and its vapors, air may be replaced by nitrogen.

For protection of the substances in the ebulliometers against the mutual penetration of vapors, the tubes leading to the manostat are provided with stopcocks and side tubes so that, after completion of the experiments, the inner spaces of the ebulliometers may be separated from the rest of the apparatus. Small amounts of the liquid, which may contain volatile impurities, may be removed from the ebulliometers (see p. 243). The ebulliometer containing the substance under examination should be of the differential type so that periodical determination of the differences in boiling and condensation temperatures may be made. A change in this difference may indicate the penetration of moisture into the differential ebulliometer, or decomposition of the sample under investigation. Indeed, such measurements may be made for the purpose of determining under similar but not identical experimental conditions, the thermal stability of an organic substance submitted to prolonged boiling under various pressures. Should a standard substance other than water be employed, it too should be contained in a differential ebulliometer.

From the description given above it might be concluded that precise electrical thermometers are required for precise comparative vapor-pressure determinations. On the contrary, it should be emphasized that, if the difference between the boiling points of two substances is known with high accuracy at an approximately known pressure, this difference will not change appreciably if the pressure is changed by a few tenths of a millimeter. For example, the differences in the boiling points of water and of benzene are 20.108, 20.029, and 19.948°C at 725.77, 739.35, and 753.08 mm Hg, respectively. This means that in this range the difference in the boiling points is changing by only 0.006°C/mm. Therefore if we can establish the temperature difference to an accuracy of only ±0.002°C. at, say, 740 mm, it is immaterial whether we establish the pressure with an accuracy of ±0.001, ±0.01, or ±0.1 mm Hg. In any case we may subtract the temperature difference from the boiling point of water at exactly 740 mm and obtain the boiling point of benzene, to an accuracy of ±0.002°C, at exactly 740 mm. Thus precise vapor-pressure determinations may be made without elaborate thermometric equipment and without precise pressure determination or regulation. Indeed, the above method is to be recommended even though certified equipment is available.

Beckmann thermometers are often employed for these measurements. It should be remembered that the divisions of the scale do not represent exact differences in degrees Centigrade and that calibration corrections must be introduced.

## Molecular Weight by the Ebulliometric Method

Though based on Raoult's law, ebulliometric molecular-weight studies are comparative studies involving purely observational data of unknown thermodynamic significance. Emphasis is placed on reducing secondary effects to a minimum and on accommodating, as much as possible, the vagaries of the solutions under study. A maximum of information and the minimum of assumption is requisite for acceptable molecular-weight work.

The easiest, but generally the most unsatisfactory, way of determining molecular weight by the ebulliometric method is to use a differential apparatus of the first kind. In such apparatus one measures neither the boiling temperature nor the condensation temperature; only the difference between these temperatures, as a function of concentration, is the focus of attention. If the solute is completely nonvolatile, if neither it nor the solvent from which it has been crystallized contains volatile impurities such as moisture, and if no large changes in atmospheric pressure occur during the measurements, it is permissible to employ differential apparatus of the first kind. The Menzies-Wright apparatus (Fig. 4.17) is the best known apparatus of this type, but a differential apparatus like that depicted in Fig. 4.18 should be classified as belonging to the first kind if differential thermocouples, differential resistance thermometers, or similar devices are used instead of two thermometers.

The next easiest, but still far from satisfactory, technique involves the use of one differential ebulliometer of the second kind. The ebulliometer depicted in Fig. 4.18 may be used in this way for molecular-weight work if a side arm is sealed into the spherical reservoir so that solute can be introduced at the appropriate time. The determination consists first in establishing the difference, $\Delta t$, between the boiling and condensation temperatures of the solvent. If two Beckmann thermometers are used, their readings should be successively compared in the same thermometer well, while a third thermometer is employed in the other well. Then $\Delta t$ is determined by employing the thermometers which have been compared. Second, a pellet of solute is introduced through the side arm and the increase in boiling temperature, $\Delta T$, is measured. It is assumed that any correction of the boiling temperature of the solution resulting from a change in atmospheric pressure is determined by observing a corresponding change in the condensation temperature of the solvent; but volatility of the solute and the presence of volatile impurities in the solute, of course, make this assumption invalid.

A more satisfactory approach involves the simultaneous determination of the boiling temperatures of solvent and solution. Two of the devices shown in

Fig. 4.16 are suitable. Both units are equipped with condensers. One is equipped with a side arm for introduction of the solute. Solvent is placed in both ebulliometers, brought to the boil, and the boiling temperature of the solvent is read on both thermometers. Solute is then added to the contents of one ebulliometer, through the side arm, and the boiling temperatures are again compared. A correction of the boiling temperature of the solution, occasioned by any fluctuation in atmospheric pressure, is determined by observing the change in boiling temperature of the pure solvent. The solution is then removed, fresh pure solvent is introduced into this ebulliometer, and a new determination is made, using a standard substance as solute. In principle, this method entailing the use of two simple ebulliometers does not differ from the method usually employed when using two of Cottrell's ebullioscopes or two sets of Washburn-Read boiling-point apparatus. In all of these, one is simply determining the change in boiling temperature occasioned by the addition of solute to the solvent, and changes that may occur in the composition of the vapor, occasioned by volatile or impure solutes, are not recognized.

If, instead of using two thermometers, one measures the difference in the two boiling temperatures directly (e.g., with differential thermocouples or differential thermohms), one encounters the added disadvantage of being unable to assess the changes in boiling point occasioned by changes in atmospheric pressure. These changes may be larger than the elevation of the boiling point occasioned by addition of the solute, and under these conditions the experimental errors may become greater than permissible. This is one reason the literature contains so much inadequate molecular-weight data.

Despite their shortcomings, the methods thus far described are all capable of producing highly precise results; their respective popularity seems to increase in proportion to their simplicity of manipulation, rather than to their potentiality for supplying information.

The most convenient method for taking all the variables under control is to use two ebulliometers—one simple and one differential of the second kind—in the molecular-weight determination. The experiment starts with a direct comparison of the boiling temperature of the solvent in the simple ebulliometer with the boiling and the condensation temperatures of the same solvent in the differential ebulliometer. At the same time the thermometers are compared by interchange between the thermometer wells. Afterward, in two successive experiments, almost equimolecular solutions of the substance whose molecular weight is being determined and that used as standard are examined in the differential apparatus. The simple ebulliometer is used exclusively for ascertaining the temperature corrections (due to pressure fluctuations) which must be applied to the boiling and condensation temperatures observed in the differential ebulliometer. Before the data are accepted as adequate, the following considerations should be given attention. The relative volatility of the substance is calculated by

$$V = \Delta t'/\Delta t'_s,$$

where $V$ is the relative volatility, $\Delta t'_s$ is the corrected change in condensation temperature occasioned by the addition of the standard solute, and $\Delta t'$ is the corrected change in condensation temperature occasioned by the addition of the substance under examination. Obviously $\Delta t'$ and $\Delta t'_s$ should both be small, their signs should be the same, and their ratio should be equal, or nearly equal, to unity. If both differences are large and positive, in other words, the condensation temperature is higher after adding solute, a solvent with a lower boiling temperature should, if possible, be employed. If both differences are relatively large and positive, and there is little likelihood of selecting a lower boiling solvent, the ratio $V$ should be adjusted as near unity as practicable by choosing if necessary a different standard substance whose volatility is either greater than or smaller than the standard previously employed. It should be emphasized that, when examining mixtures of hydrocarbons boiling about 250°C as solutes, and employing acetone as solvent, increases in the condensation temperature are noticeable. In this particular case it may be found necessary to employ two reference substances—one slightly more, and one slightly less volatile than the material whose average molecular weight is being determined.

As stated before, noticeable volatility of the solute increases the condensation temperature of the vapor, but volatile impurities such as moisture in the solute may decrease the condensation temperature of the vapor. The two effects may cancel each other. Hence, when the condensation temperature is inordinately changed by either the substance under examination or the standard substance and yet both are known to have substantially the same boiling temperature, it is to be suspected that moisture or some other volatile impurity may contaminate one of the solutes.

Small amounts of moisture are particularly annoying with solvents (e.g., benzene) forming a heteroazeotrope with water. For this reason it is often informative to determine the relative volatility of the solutes, standard and unknown, when using two different solvents, one forming a homoazeotrope, the other a heteroazeotrope, with water. It is also informative to distill some of the vapor through the condensers of the differential ebulliometer, both before and after addition of solute as previously described, and so to ascertain how the condensation temperature changes as a function of the volume of distillate collected. Volatile impurities, especially water as a heteroazeotrope, are effectively removed by this simple distillation, and the change in condensation temperature will indicate their removal. Obviously, if volatile impurities are present after addition of the solutes, further purification of the solute(s) so contaminated is indicated.

Having made all the adjustments, and having chosen an adequate solvent and standard solute, one may expect that the molecular weight of a substance or the average molecular weight of a mixture can be compared with the molecular weight of a standard with an accuracy of 0.5–0.05%. As far as is known, the use of a simple and a differential ebulliometer as here described is the only way of bringing all of the variables under control. The refinement here described may sometimes be advisable, for instance, when average molecular weights are used to characterize mixtures of substances. Regardless of the apparatus used, the molecular weight of the substance under investigation is usually calculated from the equation:

$$M = M_s \frac{G_s}{G} \frac{\Delta T_s}{\Delta T} \frac{a}{a_s},$$

where subscript $s$ refers to the standard substance, $M$ is the molecular weight, $G$ is the weight of solvent in grams, and $\Delta T$ is the (corrected) change in boiling temperature occasioned by the introduction of $a$ grams of solute. In practice, $G_s$ is usually equal to $G$ and the ratio $a/a_s$ may be so chosen in a check determination that $\Delta T_s/\Delta T$ is almost equal to unity. Calibration of the thermometers is then unnecessary provided that the atmospheric pressure has not changed greatly during the execution of the series of measurements.

Comparative ebulliometric measurements may be made by successively introducing given amounts of the reference substance and then of the substance under examination into the same ebulliometer. The equation used for calculation of the molecular weight of the unknown substance then becomes

$$M = M_s \, (\Delta T_s a / \Delta T a_s).$$

It should be emphasized, however, that deviations from Raoult's law are not infrequent. Therefore it is rather exceptional if $\Delta T$ plotted against molar concentration gives a straight line. In addition, in using the method of successive comparative measurements the thermometers should be calibrated.

### References

1. W. Swietoslawski, *Ebulliometric Measurements*, Reinhold Publishing, New York, 1945.
2. W. Swietoslawski, *Azeotropy and Polyazeotropy*, Macmillan, New York, 1963.
3. W. Malesinski, *Azeotropy and Other Theoretical Problems*, Interscience, London, 1965.
4. E. R. Smith and M. Wojciechowski, *J. Res. Natl. Bur. Std.* **17,** 841 (1936).

5. H. F. Stimson, *J. Res. Natl. Bur. Std.* **65**, 139 (1961).
6. L. Sieg, *Chem. Eng. Tech.* **22**, 322 (1950).
7. W. J. Gaw and F. L. Swinton, *Nature* **212**, 284 (1966).
8. J. R. Anderson, U.S. Pats. 2,581,344 and 2,618,591 (1952).
9. W. Swietoslawski, *Bull. Acad. Polon. Sci., Cl. III* **5**, 1141 (1957); A. Galska, *ibid.* **3**, 479 (1955); K. Zieborak and A. Galska, *ibid.* **3**, 383 (1955) (in English), and so forth by the same authors through 1969.
10. W. Swietoslawski, *Bull. Acad. Polon. Sci., Cl. III*, **1**, 201 (1953); *Physical Chemistry of the Coal Tar*, Warsaw; Ger. trans., in print; *Rev. Pol. Acad. Sci.* **1**, 21 (1956) (in Eng.).
11. W. E. Barr and V. J. Anhorn, *Scientific and Industrial Glassblowing and Laboratory Techniques*, Instruments Publishing, Pittsburgh, Pennsylvania (1949).
12. Leeds and Northrup Co., Philadelphia, Pennsylvania.
13. "Method of Test for Distillation of Petroleum Products," *ASTM Std.* **D86-62**, American Society for Testing and Materials, Philadelphia, 1964.
14. F. G. Cottrell, *J. Amer. Chem. Soc.* **41**, 721 (1919).
15. O. F. Stedman, U.S. Pat. 94,787 (Sept. 14, 1869).
16. E. W. Washburn and J. W. Read, *J. Amer. Chem. Soc.* **41**, 729 (1919).
17. A. W. C. Menzies and S. L. Wright, *J. Amer. Chem. Soc.* **43**, 2314 (1921).
18. A. W. C. Menzies, *J. Amer. Chem. Soc.* **43**, 2309 (1921).
19. W. E. Barr and V. J. Anhorn, *Instr.* **20**, 342 (1947).
20. W. Swietoslawski, *Compt. Rend. Conf. Union Intern. Chim.*, 12th Conf. Lucerne, 1936.
21. C. R. Barber, *Nature* **222**, 929 (1969).
22. W. Swietoslawski, *Compt. Rend. Conf. Union Intern. Chim.* 13th Conf. Rome, 1938.
23. N. A. Osborne and C. H. Meyers, *J. Res. Natl. Bur. Std.* **13**, R.P. 691 (1934).
24. A. Zmaczynski and H. Moser, *Physik. Z.* **40**, 221 (1939).
25. J. A. Goff and S. Gratch, *Trans. Amer. Soc. Heating Ventilating Eng.* **52**, 95 (1946).
26. C. B. Willingham, W. J. Taylor, J. M. Pignocco, and F. D. Rossini, *J. Res. Natl. Bur. Std.* **35**, 219 (1945).
27. M. R. Gibson and E. A. Bruges, *J. Eng. Sci.* **9**, 24 (1967).
28. H. F. Stimson, *Bur. Std. J. Res.* **73A**, 493 (1969).
29. A. Zmaczynski, *J. Chim. Phys.* **27**, 496 (1930).
30. F. D. Rossini, R. L. Pitzer, R. L. Arnett, R. M. Baun, and G. C. Pimentel, *Selected Values of Physical and Thermodynamic Properties of Hydrocarbons and Related Compounds*, Carnegie Press, Pittsburgh, Pennsylvania, 1953.
31. E. R. Smith and M. Wojciechowski, *Res. Natl. Bur. Std.* **17**, 84 (1936); M. Wojciechowski, *ibid.* **17**, 453, 721 (1936).
32. J. Timmermans and L. Gillo, *Roczniki Chem.* **18**, 812 (1938) (in French).
33. W. Swietoslawski, *J. Phys. Chem.* **38**, 1169 (1934).
34. W. Swietoslawski and W. Romer, *Bull. Intern. Polon. Sci., Cl. A*, **1924**, 59; W. Swietoslawski, *Ebulliometric Measurements*, Reinhold, New York, 1945, pp. 85, 87.
35. J. R. Anderson, unpublished data.

Chapter **V**

# DETERMINATION OF SOLUBILITY

William J. Mader and Lee T. Grady

---

## 1  INTRODUCTION TO SOLUBILITY

The solubility of compounds is a fundamental property with a variety of applications: appropriate solvents furnish media for reactions; differential solubility relations underlie several methods for the isolation, purification, and determination of substances; solubility analysis reveals the purity of substances; comparison of the solutions one compound forms with others provides information about its molecular structure and the nature and extent of intermolecular forces; solubility tests are one of the bases of systematic qualitative analysis. It is fortunate that so useful a property as solubility is so often easy to measure with adequate precision. This chapter is intended to serve as a working guide for making determinations of solubility and interpreting the results.

## Definition of Solubility

Solubility may be defined as the capacity of two or more substances to form spontaneously, one with the other, without chemical reaction, a homogeneous molecular, or colloidal dispersion. First note that the state of the dispersion is not limited; it may be gaseous, liquid, crystalline, mesomorphic, or amorphous. A swollen polymer such as rubber in benzene is as much a solution as is alcohol in water. A colloidal dispersion of a water-insoluble dye in an aqueous soap solution may be considered a category of solution. It is required that the dispersion be formed spontaneously, so that the formation of the solution is accompanied by a decrease in free energy of the system. Free energy changes due to chemical reaction, as in the "solution" of base metals in acid, are ruled out.

The three phases, gas, solid, and liquid give rise to nine classes of binary solutions. Three classes of solution are predominant in the interest of organic chemists and are discussed in further detail, solids in liquids, liquids in liquids, and gases in liquids. The subject of micelles is not treated in this chapter.

## Expression of Solubility

The capacity of any system of substances to form a solution has definite limits and these limits find concise expression in the phase rule of Gibbs:

$$F = C + 2 - P,$$

where $F$ is the number of degrees of freedom (temperature, pressure, compositions) in a system of $C$ components with $P$ phases. At constant temperature and pressure this simplifies to $F^0 = C - P$, where $F^0 = F - 2$. Thus for two components and two phases (solid and liquid, two liquids, or two solids) under the pressure of their own vapor and at constant temperature, $F^0$ equals zero, or in other words, composition is not free to vary. If one of the phases consists solely of one component (i.e., is a pure substance), the quantitative measure of solubility is a single number, namely, the amount of the substance (solute) which is contained in saturated solution in a unit amount of the other component (solvent). Therefore, whenever $F^0$ is zero a definite, reproducible solubility equilibrium can be reached. The expression of solubility in this general case is the set of numbers giving the composition of each of the phases at the given temperature. Complete representation of the solubility relations requires determination of the phase diagram, which gives the number, composition, and relative amounts of each phase present at any temperature in a system containing the components in any specified proportion. A saturated solution is one in which no more solute will dissolve at a given temperature, or, in which the chemical potential (partial pressure) of the undissolved solute equals that of the dissolved solute.

Solubilities of solid or liquid compounds normally are expressed as weight of solute per weight or volume of solvent or solution. See Table 5.1.

**Table 5.1**   Solubility of Various Compounds Expressed in Different Units [1]

| Compound | Solvent | Temperature (°C) | g/100 g Solvent | g/100 ml Solvent | g/100 g Solution | g/100 ml Solution |
|---|---|---|---|---|---|---|
| | | | | *Solubility* | | |
| Tartaric acid | 50% by wt ethyl alcohol water | 25 | 88.6 | 80.9 | 47.0 | 55.7 |
| o-Nitrophenol | Ethyl alcohol | 30.2 | 60.6 | 47.3 | 41.0 | 44.5 |
| Sodium chloride | Water | 20 | 35.8 | 35.8 | 26.4 | 31.6 |

Interconversion of these expressions is simple if the respective densities are known. Solubility in g/100 ml solvent or solution is equal to solubility in g/100 g solvent or solution multiplied by the density of the solvent or the solution, respectively.

Other terms such as molarity, molality, normality, and mole fraction are commonly used as solubility expressions. Solutions by weight or volume commonly are expressed by percentage. In any event, the expression of solubility includes temperature and, where applicable, partial pressures of the system.

Solubilities of gases find some rather different expressions. Mole fraction, $X$, is used commonly, additionally in the form $- R \ln X_2$ as by Hildebrand [2]. The classical expression for the solubility of a gas in a liquid is the Bunsen coefficient, $\alpha$ [3]. The Bunsen coefficient or absorption coefficient is defined as the volume of gas $V_0$, reduced to 0°C and 1 atm, dissolved by unit volume of solvent $V_p$, at the temperature of the experiment under a partial pressure, $p$, of the gas of 1 atm.

$$\alpha = \frac{V_0}{V_p}.$$

Bunsen used the ideal gas law to reduce the gas volume to standard conditions. Since this law is not exact, the coefficients found by different methods, namely, physical and chemical, can be expected to differ. Many of the past workers have not controlled the total pressure carefully, or have neglected solvent partial pressure. Some have corrected results at other partial pressures to 1 atm using Henry's law which describes a direct, linear relation between solubility and partial pressure: $P_2/X_2 = K_2 = P_2^0/X_2^0$.

If gas solubility is calculated according to the Bunsen coefficient, except that the amount of the solvent is 1 g, the result is known as the Kuenen coefficient. Markham and Kobe [4] express the volume of gas, reduced to standard conditions, dissolved by the quantity of solution containing 1 g of solvent; this is designated by $S$ and is proportional to gas molality. If the solubility is calculated as grams of gas dissolved/100 ml of solvent at the temperature of the experiment and a partial gas pressure of 760 mm Hg, the result is known as the Raoult absorption coefficient. The Ostwald coefficient of solubility, $\beta$, is defined as the volume of gas, measured under the temperature and pressure at which the gas dissolves, taken up by unit volume of the liquid. Thus, the Ostwald and Bunsen coefficients are related by $\beta = \alpha(T/273)$.

## Measurements of Solubility

Generally the methods used in the determination of solubility are classified as "synthetic" or "analytic," the terms applied originally by Alexejew [5] to the type of determination. Synthetic refers to those methods applied to an arbitrary system of solute and solvent in which the temperature or the pressure or both are varied until the solute just dissolves. An analytical method refers to a method in which the composition of the solution phase is determined by analysis in a system containing excess solute at a given temperature and pressure.

Methods for determining solubility can be classified on the basis of the phase rule. Thus, Hill [6] would group all the existing methods of solubility determination according to a constant factor, that is, thermostatic, plethostatic (constant composition), and barostatic. Williams [7] called those methods in which sampling is not required "isosystic."

In general, the analytical methods consist of obtaining a saturated solution at equilibrium and analyzing the resulting solution by some suitable physical or chemical method. Equilibration may be obtained by intimately mixing the solute and solvent, percolation of the solvent through the solute, and convection of the solvent through the solute. When solvent and solute are intimately mixed, separation of the phases can be accomplished by decantation, possibly after centrifugation, or filtration. Alternatively, undissolved solute is measured in some procedures.

## Factors Affecting Solubility Determinations

The solubility of a substance in a solvent at equilibrium is a function of temperature, pressure, and purity of solute and solvent. Equilibration, in turn, is dependent on many factors.

From a practical standpoint, one cannot generally calculate with confidence the solubility of a solid in a liquid, and even a rough experimental measurement is probably more accurate than calculated values. Moreover, calculation requires data not likely to be available for compounds whose solubility, at least in common solvents, has not been determined. Hildebrand [2] has found that the most significant parameter in calculating solubilities of non-electrolytes in normal liquids is the energy of vaporization of the solute at its boiling point, this being most indicative of entropy. The natural solubility of an ideal solute also can be related to the heat of fusion and the melting point of the solute. If the melting point of the solute is high, the natural solubility is small; if the melting point is low, the natural solubility is large; if the solute is liquid, the natural solubility is infinite. Abnormally high solu-bilities (negative partial-pressure curves) can be explained by supposing that some kind of combination occurs between the components. Abnormally low solubilities (positive partial-pressure curves) can be shown to be given by two components which are approaching the temperature at which they will separate into layers.

### Temperature

The solubility diagram records the change in equilibrium concentration with temperature. In general, the solubility of a solid increases with increasing temperature (Fig. 5.1, curve $D$). However, there are available examples that illustrate little or no change (curves $B$ and $C$), and a decrease (curve $E$) with increase of temperature. The solubility curve may exhibit a maximum (curve $A$), or a minimum (curve $B$).

As temperature has such a great effect on solubility, a good thermostatic system and an accurate temperature-measuring device are prerequisites for precise solubility determinations.

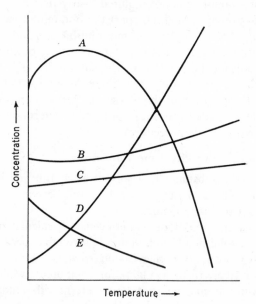

**Fig. 5.1**  Temperature–concentration curves.

## Purity

Solubility measurements are of little value unless the nature and amount of impurities in both solute and solvent are considered, an impurity being any substance other than the solute and solvent of the system under consideration. The importance that an impurity will assume in a solubility measurement is dependent on the amount present, the number of impurities present, the component that is being determined, and the affinity or reactivity that the impurity has for the solute or solvent. There is considerable leeway for the exercise of the experimenter's judgment as to how rigorously purification must be carried out. The following examples illustrate the considerations governing minimum purity acceptable.

1. In measuring the solubility of anhydrous sodium stearate in alcohol, both solute and solvent must be very dry, since in the presence of even small amounts of water the equilibrium phase is not anhydrous sodium stearate but a stoichiometric hydrate [8]. If 1 g of anhydrous sodium stearate is equilibrated with 1000 g of ethyl alcohol, the latter must contain less than about 0.07% moisture if the equilibrium solid is not to be entirely converted to $C_{17}H_{35}COONa \cdot {}^1/_8H_2O$, the lowest reported hydrate.

2. The solubility of succinic acid in absolute acetone [9] at 40°C is 7 g/100 g solvent, whereas for acetone containing 5% water it is 13 g/100 g solvent. For acetone adulterated with 5% carbon tetrachloride or methyl, ethyl, or propyl alcohol, the solubility of succinic acid is 6.2, 9.3, 8.0, and 7.5 g/100 g, respectively. From these figures one can readily calculate (assuming a linear mixture rule) how pure the acetone must be if the measured solubility is to be correct within any given precision. The differences just cited are not extraordinarily large, although the effect of chemically similar impurities, as toluene in benzene, would be expected to be small. Thus the solubility of *m*-nitroaniline in benzene is diminished by only 0.6% for a contamination of benzene with 5% toluene.

3. Similarly the solubility of naphthalene in benzene, toluene, and ethylbenzene exhibits the same mole fraction singly or in binary solvent mixtures [10]. However other normal solvents give different equilibrium solubilities, and prediction of solubility behavior in ternary mixtures can only be made empirically.

Conversely, the effect of solute impurities on solubility measurements is the basis for the determination of the purity of the solute as discussed in later sections.

## Equilibration

The third factor and the one which often causes the most difficulty is the attainment of equilibrium. The chemical and physical nature of the solute and solvent have a great effect and consequently the equilibration time varies from system to system.

A reliable method used to determine whether an equilibrium condition has been obtained is one in which equilibration is approached from both directions (i.e., undersaturation and supersaturation). When using this method; heat one of two identical samples to a temperature well above the equilibration temperature so that the equilibrium solubility is exceeded. Then both these two samples are placed in a thermostatic bath to equilibrate. If identical analytical results on aliquots of these two samples are obtained after a period of time, adequate equilibration has been reached.

The attainment of equilibrium within a system also may be revealed by periodic sampling: when the analytical values so obtained establish a plateau, adequate equilibration has been reached. One must be careful to interpret sample data only where the elapsed times are meaningful, in other words, analyses of samples taken a few minutes apart cannot define the equilibrium condition of slowly dissolving solutes.

To decrease the time necessary to obtain an equilibrated system, some means of continual agitation is usually necessary. A rotating wheel which has clamps attached to it is the best agitation apparatus. This rotating device is placed in a constant-temperature bath in such a manner that the sample containers clamped on to the wheel are always completely submerged. A less reliable, but somewhat faster method uses rapid vibration equipment and in this case best results are obtained by attaching the sample containers directly to the vibrator.

Chugaev and Khlopin [11] suggested that a solution be equilibrated at the boiling point of the saturated solution, thus eliminating the need for a constant-temperature bath. The desired temperature in such a system is obtained by varying the pressure. Thorough mixing is ensured by vigorous boiling. The apparatus used is simply a wide-mouthed flask, equipped with a sensitive thermometer and coupled to a vacuum pump, in which the solution is boiled in the presence of excess solute. In the upper part of the flask is suspended a weighing tube which can be filled with the saturated solution by means of a syphon reaching to the bottom of the flask.

According to the theory of Noyes and Whitney [12–14] solubility or dissolution is a diffusion process (see Section 7 for a more thorough treatment). A thin layer of solvent surrounding the solute crystal first becomes saturated, and material is transferred from this layer to the solvent by diffusion. The higher the viscosity of the solvent, the slower is the diffusion; the higher the temperature, the faster is the diffusion. The higher the solubility, the faster equilibrium is established, since diffusion velocity is proportional to the concentration gradient. The particle size of a solute may affect the rate of equilibration, and the mathematical relationship has an area term. The solubility one obtains for a solute in a solvent *at equilibrium* is independent of particle size (colloidal suspensions excepted).

## 2 APPLICATIONS OF SOLUBILITY IN CHEMISTRY

The uses of solubility are many and varied and it is not the intention of the authors to consider all of them. However, a number of examples will be considered to manifest what can be done with quantitative or semiquantitative knowledge of this fundamental physical property, solubility.

### Isolation and Identification of Organic Compounds

Differential solubility relationships have long constituted the major means of separating, and thereby distinguishing, one organic compound from another. Differential solubilities of solutes, or partitioning, between two liquid phases are outstanding in significance and form the basis of extraction, countercurrent distribution, and partition chromatography techniques. Similarly, gas chromatography depends on partition behavior. These important techniques for isolation and identification are discussed elsewhere in these volumes.

### Recrystallization

Differences in the solubility of two compounds in a given solvent as a function of temperature can be used to effect a more or less complete separation in which the less soluble compound is readily produced in pure form. This kind of separation is made evident by Fig. 5.2, which shows two hypothetical solubility curves. Suppose, for example, that it is desired to separate compound I from compound II, which is contained in I to the extent of about 20%. Fifty grams of the mixture, containing 40 g of I and 10 g of II, will dissolve completely in 100 g of solvent when heated to 45°C. Upon cooling to room temperature (20°C), II will remain entirely in solution, while 31 g of I will crystallize out, since its solubility at 20°C is only 9 g/100 g solvent. Thus, pure I is obtained in a 77.5% yield by one recrystallization. In this elementary example, complete independence of the two curves has been assumed, the possibility of coprecipitation is neglected, and the problem of washing the crystals of pure I free from the solution containing a considerable amount of II is ignored. For details on fractional crystallization, see Tipson [15]. The solubility curve for compound I is not smooth: the inflection at point *A* indicates either that there are two allotropic forms of I between which a transition occurs at 39.5°C or that I forms one or more solvates with this solvent. It would be necessary to ascertain whether or not the solid formed at room temperature is pure compound I.

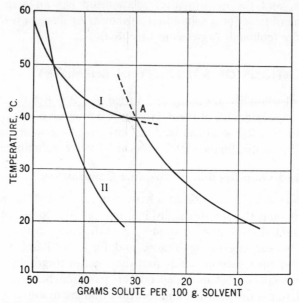

**Fig. 5.2** Typical solubility curves.

## Salting-Out

The solubility of solutes may be decreased by changes in the solvent. The separation of many biochemical preparations, particularly proteins and protein complexes, is carried out by the addition of successive amounts of some salt, for example, ammonium sulfate or potassium citrate, to a buffered solution of a protein mixture in aqueous solution. This technique also indicates the number of components present in a complex mixture. Standard fractionation procedures for plasma proteins are described in biochemistry texts.

Solutes of low polarity show reduced solubility in polar solvents as the ionic strength increases. Gordon and Thorne [16] studied the activity coefficient of naphthalene in aqueous electrolyte solutions composed of mixtures of two salts and, from Setschenow plots, determined that inorganic ion salting out effects were additive; however, organic salts were not additive with the inorganic, tending towards greater naphthalene solubility.

Organic solvents are used frequently to suppress the solubility of slightly soluble salts. The selection of the organic solvent and its most effective concentration has usually been determined by trial and error. The common ion effect may be employed simultaneously. Jentoft and Robinson have presented an objective method which is based upon mathematical derivation and involves the graphical analysis of solubility data to determine the ratio of organic solvent to water for the most complete separation [17]: tangents are drawn to the plotted solubility curve; the point giving the lowest tangent intercept represents the most efficient solvent composition.

## Identification Schemes

The solubility of a pure compound at any given temperature is a characteristic of the compound just as are other physical properties such as the melting point or the boiling point. Reeve and Adams [18] suggest that when conventional melting points and mixed melting points fail to prove the identity of two materials or to serve as criterion of purity, owing to decomposition, a procedure based on the temperature of complete solution often can be used. When thermal analysis of binary mixtures by means of a melting-point curve is not feasible the substitution of a similar type of curve based on the temperature of complete solution as determined by a solubility procedure should be attempted. This section later describes a convenient method of determining the temperature at which a solute will dissolve in a fixed amount of solvent and demonstrates how this temperature, designated as the "solubility temperature," can be used in the characterization of compounds and in the analysis of mixtures.

The qualitative solubility behavior of organic compounds toward a selected group of solvents is the basis of one of four general procedures used for the systematic identification of unknown compounds [19]. The solvents employed for such general classifications are usually water and ether, as representatives of polar and nonpolar solvents, dilute solutions of acid and base, and, finally, more concentrated acids. More elaborate systems based on the use of a sequence of organic solvents are employed only in special cases, for example in the partition pairs and equipment worked out by Beroza and Bowman [20] for pesticides and residues. In general, the solubility relations of substances are too specific to permit a division of compounds into mutually exclusive small classes.

High polymers [21] show a behavior toward solvents quite different from that of substances with smaller molecules. Some, notably chain polymers, swell continuously in certain solvents until finally a homogeneous solution is formed. There are no saturation equilibria and no numerical expressions of solubility. Rather, at a given temperature, liquids may be classed informally as solvents or nonsolvents for the particular polymer. For some solvents, a critical temperature exists above which "solubility" of the polymer is unlimited and below which it is negligible. Other polymers, notably space and net polymers, swell to a limited extent in various solvents, but no molecular dispersion forms in the equilibrated solvent. The behavior of a given polymer toward various standard solvents is a means of identification of unknown samples.

The amounts of nonsolvent required to precipitate a polymer from its solution in a miscible solvent, at varying polymer concentrations, can be used to estimate the molecular weight of the polymers [22]. The separation of a given polymer into fractions of varying molecular weight by the progressive addition of a nonsolvent, or precipitant, to a solution of the polymer has become of very great practical importance. The separation is based on the generalization that, for chains of like structure, those of high molecular weight are precipitated by smaller amounts of nonsolvent, although examples are known of the reverse relation for relatively short chains where the contribution of the terminal groups to solubility is not negligible [23].

## Purity Determination for Compounds

### Phase Solubility Analysis

Phase solubility analysis is the application of precise solubility measurements to the determination of the purity of a substance. This method is applicable to all species of molecules. A knowledge of the nature of the impurities is not required. The solubility method has been established on the sound theoretical principle of Gibbs' phase rule. Temperature and pressure are held constant so that $C = P + F^0$, and the degrees of freedom are expressed only in composition. For a pure solid in solution, one phase is present and one degree of freedom is possible as the concentration varies from zero to a saturated solution.

Mader has published a comprehensive treatment and methodology of phase solubility analysis [24, 25]. Consequently, the authors will not treat this subject extensively here but merely include the following illustrative example. An aqueous solution of pure DL-isoleucine in water will give the solubility curve $ABC$ in Fig. 5.3, while a mixture 'of 85% DL-isoleucine and 15% L(+)-glutamic acid in aqueous solution will yield the solubility curve $ADEF$. From this solubility curve, $ADEF$, the following information can be extracted:

1. The solute is composed of two components.
2. The principle component of the solute is present to the extent of 85% (100% − slope $GDE$).
3. The impurity is present to the extent of 15% (slope $GDE$).
4. The solubility of the principle component is 21.2 mg/ml of water (point $G$).
5. The solubility of the impurity is 10.3 mg/ml of water (point $H$ − point $G$).

Tarpley and Yudis [26] discovered that the classical methods of purity determination, namely, freezing and melting curves, were not suitable for steroids, the steroids being both heat labile and prone to undergo polymorphic

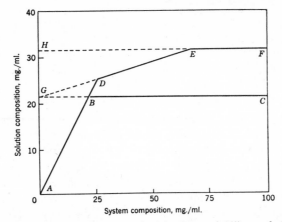

**Fig. 5.3**  Examples of solubility curves obtained in phase-solubility analysis.

modifications. They were, however, able to apply the phase-solubility method of purity determination to these systems with great success. The purity of a slightly soluble salt can be determined by use of another solubility procedure in which the conductance of a saturated solution, in the presence of a solid solute, is determined. Then the supernatant liquid is poured off and fresh solvent added to the solute. The conductance of this second saturated solution should be the same as the first if the solid is pure. However, this procedure is of no value if the impurities in question are nonconducting.

### Solubility Temperature

Solubility temperature supplies an alternate procedure [24] for purity determinations. The temperature at which the last crystal of solute dissolves in a given volume of solvent, at constant pressure, is plotted against added levels of impurities. A disadvantage of this method is that the kinds of impurities should not vary and that the solubility temperature of pure compound be known. Where a pure sample is not available, phase solubility analyses of samples having several levels of impurities may be used to construct the calibration curve. The main advantage of solubility temperature, and a significant one, is relative speed which may be used to advantage in routine process control after initial definition by phase solubility analysis. Satterfield and Haulard [27] have described the magnitude of overshoot of the true solubility temperature caused by various rates of temperature increase and recommend three hours agitation at the solubility temperature. Heric [28] has used their data to develop a rapid dynamic method based on extrapolation of data from varying heating rates, say 1–10°/min, to zero rate, finding the temperature so obtained agreed with the value obtained statically.

## Study of Intermolecular Forces

The mutual solubility of two substances has long been used as a qualitative measure of the extent of the interaction between their molecules, varying from simple departure from the laws of ideal solution (Raoult's law, Henry's law) to a genuine formation of a compound between the solvent and the solute.

The question of the existence of solution complexes between aromatic hydrocarbons and aluminum halides has, because of its importance to a knowledge of the mechanism of Friedel-Crafts reactions, attracted the attention of a number of workers. The use of solubility data in this investigation serves as an excellent example of the application of solubility data to physical chemical problems. The existence of a chemical solvent–solute interaction, in a strictly binary aluminum chloride–hydrocarbon system was not readily established. The reason for this was that the low solubility of aluminum chloride in hydrocarbon solutions containing neither hydrogen chloride nor moisture made it impractical to study a possible complex formation by vapor-pressure lowering or the formation of univariant two-component systems. Such a system can be studied, however, by a solubility method, because complex formation in solution is reflected in the temperature dependence of solubility. Fairbrother, Scott, and Prophet [29] determined the solubility of aluminum chloride in a number of hydrocarbons under rigorously anhydrous conditions, from 20 to 70°C. Their solubility results gave evidence of weak complex formation at room temperature in aromatic hydrocarbons. This complex formation was found to become greater as the electron-donating character of the hydrocarbon increased and less as the temperature increased.

Another example which illustrates the use of solubility data is the study of Gill et al. [30], in which diketopiperazine was used as a model compound in an investigation of hydrogen-bond interactions between the peptide bond and urea. They determined the solubility of diketopiperazine in aqueous urea solutions as a function of temperature and urea concentrations. The enthalpy of solution of diketopiperazine was estimated and consequently provided thermodynamic information on the interaction effects of diketopiperazine and urea in aqueous solutions.

Solubility measurements can be used to determine the stability constant and the stoichiometric ratio of a complex or chelate. Higuchi and Lach used a solubility method to investigate the complexation of p-aminobenzoic acid, (PABA), by caffeine [31], Fig. 5.4. Obviously, the concentration of PABA at the initial point of the curve, that is, at the vertical axis intersection, is the solubility of PABA in water. At low caffeine concentration a straight-line relationship is obtained between the concentration of PABA and caffeine in solution. This increase in solubility of PABA is due to complex formation

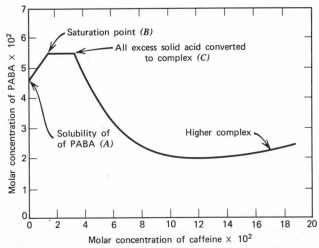

**Fig. 5.4**   Phase diagram of PABA–caffeine system in water at 30°C. (Reproduced from T. Higuchi and J. L. Lach, *JAPhA Sci. Ed.* **43**, 349 (1954) with permission of the copyright owner.)

and continues until the solubility limit of the complex is reached. The saturation point of the solution with respect to the complex is noted by a sharp break in the curve, point *B*. On further addition of caffeine the complex continues to form but precipitates out of the already saturated system. The total concentration of PABA in solution is constant in this region as shown by the flat plateau and is independent of the amount of caffeine, because any acid precipitated in the form of the complex is replaced by further dissolution of the excess solid PABA. The next break in the curve, point *C*, is noted when there is no more excess solid PABA. That is, all the PABA has either passed into solution or been converted to the complex. Further addition of caffeine results in depletion of PABA in solution. Although the solid PABA is exhausted and the solution is no longer saturated, some of the PABA remains uncomplexed in solution, and it combines further with caffeine to form higher complexes, as shown by the curve at the right of the diagram. The stoichiometric ratio of the components of the PABA–caffeine complex which was formed in the plateau region can be calculated from the phase diagram. Once the stoichiometry has been established the stability constant, *K*, also can be calculated.

## 3   DETERMINATION OF THE SOLUBILITY OF SOLIDS IN LIQUIDS

Methods of determining the solubility of solids in solvents vary in the degree of accuracy and precision obtainable and in the type of systems to which they are best suited. The first few methods considered here will be methods which are applicable to common systems which have no unusual restraints for which special treatment must be given. After these methods of general applicability are described some specialized methods will be treated.

## General Solubility Method

Reilly and Rae [32] have described a method of general applicability with medium accuracy which requires a method of analysis for the solute. Weigh out, crudely, four samples of the solid of a size which will not dissolve completely at the given temperature in the volume of solvent contemplated. The solid should consist partly of fine and partly of coarse crystals. Place the sample in each of four flasks or cylinders, and pour in the determined volume of solvent so that the flasks are about two-thirds full. Stopper tightly. Heat two of the samples well above the final temperature of measurement in order to approach equilibrium from the side of supersaturation, as most substances have a positive temperature coefficient for solubility. Place all four flasks on a suitable agitation device in such a way that they are completely submerged in the thermostat. After equilibrium seems likely to have been reached, discontinue agitation and set the flasks upright in the thermostat to facilitate settling of the excess solid to the bottom of the cylinder. If the flask has a long neck and the bath temperature is not very different from room temperature, the stopper may be allowed to project above the bath to facilitate removal of the saturated solution. Weigh samples of the solution which had a known volume at the temperature of the experiment. This procedure provides data for interconversion of units. The removal of a sample from the flasks is usually done with a pipet to which there is attached a filtering device of some sort. The filtering device may be made from sintered glass, cotton, or filter paper held to the pipet by rubber bands. Analyze the solution for solute content. All four results should agree within the precision called for in the determination, and there must be no significant difference between the pair originally supersaturated and the pair originally undersaturated.

To ensure that a change in the solid phase has not occurred during equilibration, obtain a sample of the solid by filtering the residue, and examine a crystal under a hand magnifier or a low-power microscope. If any doubt exists about its identity with the original, free a sample from the excess solution by rapid washing on a filter with fresh solvent or a more volatile solvent miscible with the first, and analyze the residue for solute content, or take a melting point of it alone and in admixture with the original.

## Rapid Methods

Some of the more popular rapid methods are described below. These procedures do not ordinarily require any special equipment or manipulations. One usually sacrifices accuracy and precision in gaining time.

Ward suggested the following method [33]. A test tube containing 3 ml of solvent is immersed in a beaker of water at 10–20°C above the temperature at which the solubility is to be determined. Solid is added in divided portions until a portion remains undissolved. Then the test tube is transferred to another beaker of water at a temperature at which the data is required and held for 10 min with occasional shaking. Push a small thimble made of folded filter paper, similar to a miniature Soxhlet thimble, down into the liquid and allow filtration to proceed from the exterior to the interior. Pipet a definite volume of the clear solution to a tared container. Evaporate the solvent and weigh the residue. Individual results can be obtained within an hour.

A rapid solubility approximation was presented by Pastac and Lecrivain [35]. Weigh out $2A$ g of sample, where $A$ represents more solute than can be dissolved in 10 ml of solvent. Then this $2A$ g of sample is divided into portions weighing $A$, $0.5A$, $0.25A$, $0.125A$, etc. Each portion is placed separately into a test tube containing 10 ml of solvent. The tubes are heated to 70°C and then cooled to the desired temperature. The last tube that shows undissolved sample is noted.

Another method which has been proposed for the rapid estimation of solubility is based on the rate of solution [35]. This method is particularly useful in measuring the slight solubility of some electrolytes in concentrated solutions of others, the solubility in systems that tend toward supersaturation, and the solubility of some electrolytes in supersaturated unstable solutions of other electrolytes. An analytical method is not required. This method is based on the relation:

$$V = K(C - C_x) = \frac{\rho^{2/3}(P_0^{1/3} - P^{1/3})}{2\tau},$$

where $V$ is the rate of solution at concentration $C_x$, which is close to the true solubility, $C$; $\rho$ is the specific gravity of the solid solute; $P_0$ is the initial weight of the dissolving crystals; $P$ is the final weight of the dissolving crystals; $\tau$ is the time; and $K$ is the coefficient of solution rate. Values of $V$ are determined experimentally, and $C_x = C$ (when $V = 0$) is determined graphically or by the equation. This equation is for cubic, crystalline solutes and factors other than "2" are required for other crystal geometries.

The residue-volume method of solubility determination is a rapid method with reasonable precision and can be used with pure substances, compound mixtures, and substances containing insoluble impurities [36]. Some of the advantages and disadvantages of this method are illustrated in Table 5.2. The principle of this method rests on the fact that when the solubility limit of a solute in a solvent has been reached, further additions of solute result in a proportional increase in the amount of undissolved residue. The volume of these residues can be measured for a series of solute–solvent ratios. These volumes are plotted as a straight line and extrapolated to zero residue volume

**Table 5.2**   Residue-Volume Method of Solubility Determination

| Advantages | Disadvantages |
| --- | --- |
| Accuracy sufficient for most commercial applications. | Method not applicable to solutes of density lower than solvent nor effectively to solutes having very slow rates of solution. |
| Quickly gives preliminary view of solubility limits in many systems. | |
| Solubility found from number of individual determinations which operate to check on one another and to define precision of determination. | The particle size of residue must be fine enough to give a compact, reproducible residue in capillary and must be coarse enough to be readily precipitated. |
| No analysis needed. | |
| Gives good results even when insoluble impurity present. | The time and force of centrifuging must be constant. |
| Difficulties with supersaturation not likely to be met. | Temperature control is difficult since centrifuging causes a small rise in temperature of the solutions. |

as the solubility limit. When insoluble impurities are present it will be found that the plotted points may be connected by two straight lines. Below the solubility limit, the points will yield a line of very low slope corresponding mainly to traces of impurities. Above the solubility limit a line of very steep slope will be obtained. The intersection of these two lines is taken as the solubility limit.

### Apparatus for Solubility of Solids in Liquids

One of the most efficient solubility devices was originally designed by Campbell [37] and recently modified [38] as is shown in Fig. 5.5. In using this apparatus, bottle $A$ is charged with the solvent and excess solute and rotated until equilibrium is reached. The glass jacket then is inverted while still in the thermostatic bath. The saturated solution is permitted to filter through the tube containing glass wool from $A$ to $B$, the air being displaced simultaneously from $B$ to $A$ through the capillary. The filtration thus occurs at exactly the same temperature as that at which the saturated solution is prepared.

Lombardo [39] has devised a two-chamber automatic-filtration cell (Fig. 5.6), using readily available glassware, and which is operated by pulses of pressure delivered from a plastic bellows. The whole unit can be submerged in a bath and the apparatus therefore can be operated at otherwise inconvenient temperatures. Solute and solvent are placed in side $A$ which was prepared from a filter tube. The bellows is connected by tubing to side $A$ and operated at about two impulse cycles per minute.

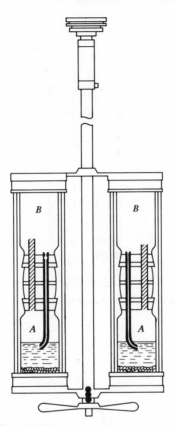

**Fig. 5.5** Campbell solubility apparatus.

## Determination in Volatile Solvents

The procedure of Reilly and Rae, previously described, may be used for solvents of moderate volatility, in other words, 50–100 mm Hg at the temperature of the thermostat, if time is allowed for the air space in the flasks to become saturated with the vapor before stoppering. When the vapor pressure of the solvent is high, that is, 1 atm or more, special equipment is required.

One such piece of equipment is a modified solubility tube which has been used to determine the solubility of solids in volatile liquids by a synthetic method [40]. In this procedure known weights of the solute and solvent are contained in a sealed glass apparatus. The amount of solvent admitted to the tube containing the solute may be varied at will without opening the apparatus. The solvent is kept in the graduated reservoir tube, *D* (Fig. 5.7). The solvent is distilled from *D* to *A*, the solubility vessel containing the solid

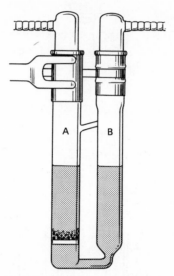

Fig. 5.6 Lombardo apparatus. (Reproduced with permission from J. B. Lombardo, *J. Chem. Ed.* **44**, 600 (1967).)

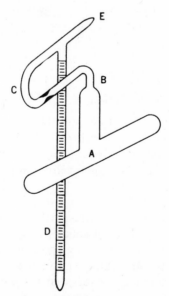

Fig. 5.7 Apparatus for determination of solubility of solids in volatile liquids according to Menzies.

under investigation, through trap *C*. Trap *C* is closed during determinations by melting a pellet of silver iodide–silver chloride mixture above the constriction in the U–tube and *C*, and allowing it to run into the constriction. This method permits the solubility of nonvolatile solids in volatile solvents to be measured with an accuracy of greater than 1% over a wide temperature range.

### Slightly Soluble Solutes in Liquids

Gross [41] made use of the interferometer to determine the solubility of slightly soluble liquids in water. Adams earlier stated that the interferometer is well adapted to determine a single varying component in a transparent mixture, whether that component be solute or solvent, electrolyte or non-electrolyte [42].

Mitchell extended the use of the interferometric method by developing a procedure that could be used in the determination of the solubility of any sparingly soluble solute, including liquids, in any solvent [43]. A double cell was constructed through which liquid could be circulated, and readings were made with pure solvent circulating through one side of the cell and the solution through the other. The equilibrium-saturated solution was prepared and interference compensation data for serial dilutions of this solution were plotted against % saturation. In all but one case, that being benzene, a linear relationship was obtained. In view of this one noted exception, it is always necessary as a preliminary to a solubility determination by this method to investigate the relationship between compensator readings and % saturation. A change in solute concentration of 2 ppm will be detected by this method.

Strongly absorbing substances are measured easily by ultraviolet spectrophotometry, so this analytical method is very good for determining the solubility of slightly soluble substances. A good example of this technique is the determination of aromatic hydrocarbons in water by Bohon and Claussen [44]. Fluorescence and phosphorescence methods also may find application in the determination of slightly soluble solutes, particularly in the presence of other absorbing solutes.

Weyl devised an apparatus for the conductometric determination of the solubility of slightly soluble electrolytes in a closed system [45]. The unique internal pumping system consists of a column of mercury that changes position when the apparatus is gyrated mechanically. This process of equilibration is continuously monitored by measuring the conductance in a cell in the line of the liquid flow. Figure 5.8 is a perspective drawing of the solubility apparatus which consists of two conductance cells, a bulb filled with the electrolyte under investigation, and a pump for circulating the solution. The entire apparatus is made of glass and can be completely filled with the solution under investigation. The mineral bulb is terminated by two coarse

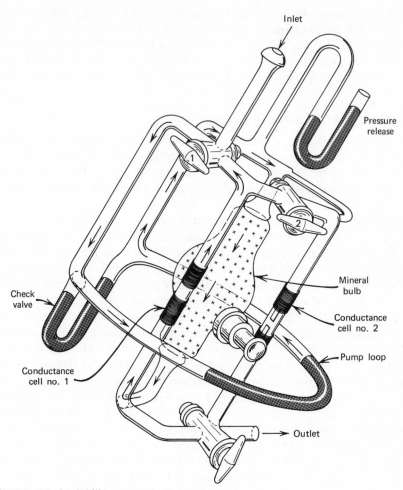

Inlet

Pressure release

Mineral bulb

Conductance cell no. 2

Pump loop

Check valve

Conductance cell no. 1

Outlet

**Fig. 5.8** Weyl solubility apparatus for determination of slightly soluble solids. (Reproduced from P. K. Weyl, *Rev. Sci. Instr.* **28,** 722 (1957).)

glass frits to retain the material under investigation. Filling and emptying is accomplished through a side port, normally closed by a glass stopper. The mounting for this apparatus is so constructed that the entire apparatus can be immersed in a thermostatic bath. By replacing the conductance cell by a suitable counter, this apparatus could be used to measure the solubility of radioactive solids.

Methods have been developed for determining the solubility of slightly soluble materials by use of radioactive indicators [46, 47]. In the procedure proposed by Jordan [46] the tracer is added to a saturated solution of the substance at room temperature. The mixture is warmed to the temperature of the solubility determination. At first, the radioisotope is evenly distributed in solution but on cooling it is distributed both in solution and in the solid. The activity in the solid and liquid phases is determined and the solubility calculated by the ratio.

Aqueous solubilities of fatty acids and alcohols have been determined by a film-balance technique. Robb [48] measured interfacial tensions and related these to concentration. He found that the logarithm of the solubility varied linearly with chain length, with some deviation due to fatty acid ionization.

### Apparatus for High-Temperature Solubility

Marshall obtained the solubility data in the dilute region for the system, uranium trioxide–sulfuric acid–water, at elevated temperatures and, in so doing, devised a method for sampling an equilibrated solution at elevated temperatures [49]. In this method the pressure bomb, in which is found the solution plus solid, contains a length of thin-walled capillary tubing attached at one end to a pressure valve equipped with a sampling tip (Fig. 5.9). Part of the capillary passes through a wet ice bath so that the solution rapidly is

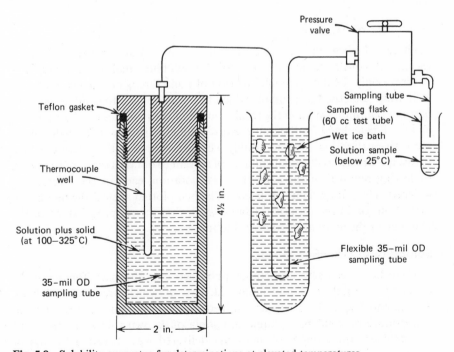

**Fig. 5.9**  Solubility apparatus for determinations at elevated temperatures.

cooled during the sampling process, thereby eliminating a separate cooling or isolation chamber and facilitating the procurement of many samples per run. With this direct sampling technique there is neither a distillation problem nor a need for correction for loss of solvent or other components to the vapor phase at the equilibration temperature, as the liquid phase alone is sampled at the elevated temperature and can be analyzed for all components. However, with systems having a positive temperature coefficient of solubility, the precipitation characteristics and solubility range of the individual system would have to be considered and evaluated. The solubility relations for the system that Marshall investigated fell within the most sensitive range of pH variation with mole ratio and thus the solubility of the uranium trioxide was determined by comparing the pH of the isolated sample at 25°C with control pH data. For the application of this method to other systems the sensitivity of pH in the particular solubility range must be considered. There are, however, possible modifications of the analytical method.

To determine the solubility of salts at high temperatures, a method suggested by d'Ans is used, which involves the determination of the temperature at which a known amount of salt just dissolves [50]. This determination is made in small tubes placed in an air thermostat and observed through a glass window.

### Solubility of Mixtures

Phase solubility analysis and solubility temperature, as outlined above (Section 2) constitute the primary tools for solubility analysis of mixtures.

Where only the more soluble fraction of a mixture is of interest, and time is limiting, a method is available which has been applied to soluble fractions of wax [51]. In this method the total amount of the more soluble fraction of a mixture is calculated from the amounts of material dissolved in each of several successive extractions. An accurately weighed sample of the mixture is successively extracted with equal volumes of solvent to obtain a series of fractions. Following removal of the solvent by evaporation, each of the fractions is weighed. The total weight extracted, $w$, is then plotted against the number of extractions, $n$, in the form $n/w$ against $n$, so that the reciprocal slope of the graph gives the amount of soluble fraction.

### Solubility in Liquefied Gases

There have been several methods published over the years for the determination of solids in liquid ammonia, two of which are discussed here. Schenk and Tulhoff [52] have determined solubilities of salts as a function of temperature by condensing liquid ammonia on known solute samples until all dissolved; the amount of ammonia delivered was measured by loss of pressure in an ammonia storage vessel. Such a procedure would be generally applicable to solutes and gases.

The 25°C solubility data for a number of inorganic halide salts in liquid ammonia and liquid sulfur dioxide have been determined from weight measurements or chemical analysis obtained during the course of a process based on effecting solution in a glass filter tube, separating the saturated solution from excess solid by centrifugation, and subsequent removal of the solvent [53]. The exact procedure of these authors is as follows. A weighed sample of solid, approximately 0.1 g, is introduced into one end of a glass filter tube (fritted-glass disk at midpoint). This end of the tube is sealed and the tube and contents are weighed. The tube is flushed out with anhydrous gas, after which the open end of the tube is attached to the source of the gas. The end of the tube containing the sample is immersed in a dry ice–acetone bath ($-75$°C) and a suitable quantity of solvent, 0.8–1.0 g, is condensed on the solid sample. The open end of the tube is sealed off under conditions which permit the determination of the weight of glass removed in making the seal. The sealed tube is allowed to warm to room temperature and then weighed and agitated in a thermostat (25 ± 1°C) for 48 hr. The tube is removed and centrifuged, thus effecting separation of the saturated solution and the excess undissolved solid. The end containing the saturated solution is cooled to $-75$°C and the other end is drawn out to a fine capillary through which solvent is allowed to escape. Thereafter the tube is maintained at $10^{-3}$ mm Hg for 1–2 hr. The weight of solute in the saturated solution is determined by removal of that portion of the tube which contains the sample of the solid that was dissolved, weighing it before and after removal of the solid (recommended procedure), or a similar determination of the weight of the excess solid and the weight of the dissolved portion is obtained by difference. The weight of the solvent may be measured by a determination of the weight of the tube assembly before and after introduction of solvent and application of the correction for the weight of the glass removed.

## 4   DETERMINATION OF THE SOLUBILITY OF LIQUIDS IN LIQUIDS

### General Methods

There are a number of methods, both analytical and synthetic, which have broad applicability to solubility determinations in liquid–liquid systems.

When an analytical procedure is used there are essentially two operations, namely, securing equilibrium between the two liquid phases during a sufficient equilibration time, with the customary precaution of approach to the final state from both sides (i.e., higher and lower temperature) and obtaining samples of each layer at the temperature of equilibration.

The following general analytical method is suggested. Measure out the desired volumes of the two liquids into two suitable containers which can be tightly stoppered. If either liquid has an appreciable vapor pressure, allow the vessels to stand uncovered for several minutes before stoppering, or reduce the pressure inside after stoppering by means of a suitable pump connected through the stopper and a stopcock. Place one vessel in a beaker of cold water and agitate thoroughly. Place the second in hot water and agitate. Then submerge the two in a thermostat at the desired temperature and rock mechanically for a suitable time interval. With adequate mixing, if the liquids are not too viscous, equilibrium should be reached in a few minutes. However, if the liquids tend to emulsify, the only permissible agitation is a gentle rotation which renews the liquid near the surface without rupturing the meniscus separating the two liquids. In such a case, several hours may be required to establish equilibrium, depending on the ratio of interfacial area to volumes of the phases and in this case long, narrow tubes are to be preferred. To withdraw samples, allow the vessels to come to rest in an upright position with their stoppers barely projecting above the surface of the bath. If distillation occurs because of the temperature gradient thus established along the neck of the flask, leave the vessels completely submerged but so arranged that they can be raised without disturbing the liquid layers. The samples desired can be taken from the top layer simply by pipet. A sample of the bottom layer can also be secured by maintaining a slight air pressure on the pipet to prevent contamination as it is lowered through the top layer into the bottom. Alternatively, a graduated pipet with the tip sealed by a thin-blown glass membrane can be used to sample the lower layer, the membrane being broken against the bottom of the vessel. Weigh known volumes of the two layers to obtain the data needed for expressing the results in any units desired. Use any convenient procedure to determine the composition of the layers. If the necessary calibrations have been made, physical methods such as measurements of refraction or density may be adequate. When separation occurs readily into two liquid layers, each of moderate volume, the authors prefer the method described above.

However, synthetic methods are also generally useful. In one such synthetic method, the "cloud-point" method, the temperature of incipient separation into two phases is determined as the isotropic, single-phase liquid solution is cooled. Minute droplets of the second phase begin to form throughout the formerly homogeneous system and give it a cloudy appearance. The temperatures of appearance of the cloud on cooling, and of its disappearance or heating, are usually the same within experimental error, $\pm 0.1°C$. However, instances of real discrepancies of up to 3° have been reported [54]. A single experiment gives no information about the composition of the second phase. To determine the composition of both phases at any given temperature, a series of experiments must be performed so that the complete solubility curve can be constructed.

Sometimes no readily visible cloud is formed. If a dye can be found whose color differs in the two phases, a minute amount dissolved in the solution may change color sharply when the second phase separates out. Klobbie [55] used this effect successfully in studying the system diethyl ether–water. To increase the visibility of the cloud one can place the sample in a heavy-walled tube held in a horizontal position. A wire network is viewed through the tube, which acts as a lens. Distortion or disappearance of the image shows when a second phase begins to form. In the special cases in which the separating phase is liquid crystalline rather than liquid, observation through crossed polaroid plates renders its appearance conspicuous.

The thermostatic method of Hill is a procedure of wide application in liquid–liquid systems [6]. This involves the phase-rule principle that two-component systems consisting of two liquid phases under their vapor pressure are univariant, that is, the fixing of one condition determines the system in all other conditions. Thus if the temperature is fixed, the composition of the liquid phase as well as the vapor pressures will be fixed at equilibrium. The two liquids are mixed in two different ratios by weight in two separate experiments at the same temperature, using the flasks shown in Fig. 5.10. The $m$ and $m'$ represent the weights of the first component used in the two experiments; $x$ represents the first component's concentration in g/ml at equilibrium in the upper phase in both experiments, since by the phase rule the concentration at saturation cannot vary. Similarly, $y$ represents the concentration in the lower phase in both experiments. If the measured volumes of the upper phase are $a$ and $a'$ and the measured volumes of the lower phase are $b$ and $b'$, then $ax + by = m$, and $a'x + b'y = m'$.

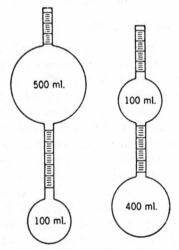

500 ml.

100 ml.

100 ml.

400 ml.

**Fig. 10** Solubility flasks used in the Hill solubility method.

Solving the equations for $x$ and $y$ will give the concentration of the first component in each phase. If the equations are again solved, substituting the weights $n$ and $n'$ of the second component in place of $m$ and $m'$, then the second component's concentration in each phase becomes known. By adding together the weight of each component present in 1 ml of a given phase, the weight/ml or density is obtained and the % composition by weight thus follows.

Evans [56] suggested the use of an oil centrifuge bottle instead of Hill's original apparatus. He looked into the question of what volume ratio gives maximum accuracy in this method and determined graphically that $a$ and $b$ should be as small as possible.

## Rapid Methods

Herz [57] described an approximate procedure for the solubility of one liquid in another at room temperature. One liquid is added to the other dropwise from a buret, with continuous agitation. The first excess gives the whole liquid in the flask a cloudy appearance which may be amplified by the modifications mentioned above. This procedure also is applicable to ternary liquid systems. A refinement of the above method is reported by Sobotka and Kahn [58], who used minute, jagged crystals insoluble in the first liquid as indicator particles. The slightest excess of the second liquid rapidly dissolved the indicator particles, converting them from jagged crystals into rounded droplets of a different color. They determined the water solubility of the ethyl esters of the homologous series from propionic to capric acid and from malonic to sebacic acid using Sudan IV as the dye. An accuracy of $\pm 0.001\%$ or 0.01 ml in 1 liter was obtained.

## Slightly Soluble Liquids in Liquids

Analytical difficulties have often characterized determinations of the slightly soluble liquids. Physical methods often are inapplicable. Gas chromatography is most useful, particularly where the slightly soluble compound is the more volatile. Some specific applications of other methods are listed here. Methods of choice in the determination of water in organic solvents are gas chromatography, Karl Fischer titrimetry [59], and infrared.

Gross [41] used an interferometer to determine the solubility of slightly soluble liquids in water. The interferometer scale is calibrated by measuring synthetic solutions. The estimated errors of the solubility values obtained in this manner are 1–2% which represents an error of only 0.01% in the total composition of the solution when the solubility of the solute is only 1%.

Hayashi and Sasaki [60] determined the solubility of sparingly soluble organic liquids in water by turbidity tritrations using Tween 80. There is a linear relationship between the measured turbidity and the solute concentration in water.

## 5   MICROMETHODS FOR SOLUBILITY DETERMINATION

Although many solubility methods can be modified to determine the solubility of microsamples, very few are designed specifically for such use. Yet it is quite important that one have available a solubility determination method which is generally applicable when only a limited amount of sample is available.

Such a general procedure was devised by L. K. Nash [61]. In this procedure one determines the solubility through the measurement of the vapor-pressure lowering in a saturated solution. The accuracy of the measurements depend on the value of $P_0$ (the vapor pressure of pure solvent at the temperature of the measurement) and on the intrinsic solubility of the solute. However, an accuracy of at least 5% can generally be maintained when dealing with moderately soluble substances which obey Raoult's law. Only enough sample to saturate approximately 0.1 ml of solvent is required and no weighings are involved. However, Raoult's law must apply, and molecular weights must be known or determined. The apparatus that is required for this submerged bulblet method is rather elaborate. Figure 5.11 is a diagram of the equipment used in this method of solubility determination. However, if microscale determinations are required in a laboratory with any frequency at all, it might be advisable to construct this apparatus. This same apparatus can be used for vapor pressure, decomposition pressure, phase study, purity of solute or solvent and molecular weight as a function of concentration measurements.

The procedure used in determining solubility is as follows. A quantity of solute estimated to be sufficient to saturate 0.1–0.2 ml of the solvent at the maximum temperature of the equipment is introduced into the bulblet, $L$, and 0.2–0.3 ml of the solvent is added. The vapor jacket, $H$, is charged with the same solvent, the apparatus is assembled as shown in Fig. 5.11. The pressure in the jacket is set at an appropriate low value. The air is swept from the bulblet, and after a pause to establish equilibrium, the pressure in the jacket, $P_0$, is read from barometer $A$ and the value of $\Delta p$ is taken from manometer $B$ or $C$. The temperature of the experiment is fixed by the value of $P_0$, and if $X$ is taken as the mole fraction of the solute, then from Raoult's law:

$$\frac{1}{X} = \frac{P_0}{\Delta p}.$$

If the molecular weight of the solvent and solute is known, the mole fraction can readily be converted to a solubility value.

The attainment of equilibrium is checked by successive approaches from under- and oversaturation. The assumption of Raoult's law can be checked with the same sample by a determination of the apparent molecular weight, even if association or dissociation occurs, as a function of concentration. If the results are independent of concentration, then that value (which may, of

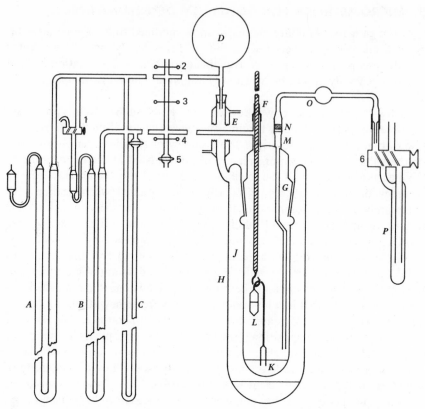

**Fig. 5.11** Submerged-bulblet apparatus for solubility determination of microsamples.

course, be only an apparent molecular weight of a substance either associated or dissociated) can be used in converting the mole fractions to gram solubilities. If the apparent molecular weight varies with concentration this method is not valid.

Accurate measurements of the solubilities of various compounds of the synthetic elements have been obtained on a sample as small as a few hundredths of a microgram [62]. Such measurements are facilitated by radiometric assay of the solution in equilibrium with the solid phase. For example, in the case of $^{239}$Pu, as little as 0.01 $\mu$g may be determined with an error of less than 2%. Solubility measurement of a nonradioactive material usually requires several micrograms of sample, as precise chemical analysis is quite difficult with smaller samples. In this procedure, the solid and solvent are placed in a glass capillary, which is sealed and attached to the periphery of a notched wheel that is rotated by means of a motor. This rotation produces a continuous mixing action within a specially designed liquid thermostat. After equilibration, the solid phase is centrifuged to the bottom of the tube and the sample withdrawn. The sample is then spread on a thin plate and dried for radiometric or chemical assay.

# 6 DETERMINATION OF THE SOLUBILITY OF GASES IN LIQUIDS

Determination of gas solubilities in liquids may be categorized by the type of measurement made. Physical methods are based on the gas law and measurements commonly are manometric within constant volume, isothermal systems. Physical methods may be classed further as saturation or extraction methods. A saturation method measures the amount of gas required to saturate the previously degassed solvent. An extraction method measures the gas extracted or liberated from a saturated solution of that gas in the solvent. Several systems of general usefulness are considered here.

The generally applied methods are physical ones, mostly of the saturation type. Precision varies according to the apparatus used and the specific gas–liquid pairs. Accuracy depends heavily on the approach to ideality or the validity of correction for nonideality. In all physical methods, the liquid must be gas-free at the start and this is a major source of error if not carried out scrupulously. This condition is usually obtained by boiling, followed by vacuum cooling [3] or vacuum sublimation [63]. See Chapter II for pressure–volume determinations.

Chemical methods are those which measure a property of the gas molecule other than gas law relations, whether that measurement be spectrophotometric, radiometric, or by any other "physical" determination or by chemical reaction. As such methods are specific to the gas–liquid pair, extensive treatment is not required in this chapter.

## Rapid Methods

Two generally applicable methods are available which attain equilibrium rapidly.

### Dymond and Hildebrand Apparatus

These authors [64] recently described a glass apparatus for the accurate and rapid determination of gas solubility in liquids (Fig. 5.12). Bulb $A$ contains a known volume of liquid and opens into bulb $B$ into which a known amount of gas is introduced. The solvent is *pumped* into the upper bulb by means of a sealed magnetic pump in the side arm between bulbs $A$ and $B$. The liquid then runs down the walls of bulb $B$ and in this way fresh interface is continually exposed to the gas. Equilibrium is determined when consecutive manometric readings on the undissolved gas remain constant, typically 1–3 hr. Measurements made by reequilibrating at several temperatures permit calcu-

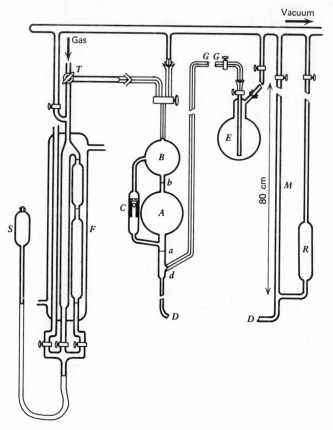

**Figs. 5.12** Apparatus for determining solubility of gases. [Reproduced from J. Dymond and J. H. Hildebrand, *I&EC Fund*, **6**, 130 (1967)].

lation of the entropy of solution. Equilibrium for nonviscous solvents was obtained in 1–3 hr, and accuracy was reported as better than 1%.

To operate, the whole apparatus is evacuated; mercury from reservoir $S$ is admitted to the gas buret until the meniscus is level with the lower mark. Purified gas is admitted to the buret via the three-way stopcock, $T$, until the pressure is about 1 atm. The mercury height in the central column is adjusted until the pressure equals 1 atm. Mercury from reservoir $R$ is admitted to $D$ until the meniscus is just below $d$. Purified solvent in $E$ is degassed scrupulously by a combination of pumping on the frozen solvent and boiling away a portion of it under vacuum. The degassed solvent is then impelled under its own vapor pressure via $G$ into $A$ until the bulb is nearly full. In the case of a liquid of insufficient vapor pressure, helium at a pressure of about 50 mm of mercury is used to force the liquid over. After the liquid has attained the temperature of the bath, the mercury level in $D$ is raised to cut off the capillary side arm and to bring the level of the liquid to $b$. The distance from the mercury meniscus to $a$ is measured and the volume of solvent calculated exactly.

Gas from the buret is admitted slowly to $B$ while the amount of mercury from reservoir $R$ is increased, so as to keep the mercury meniscus just below $a$. The gas is shut off as the pressure in $B$ approaches 1 atm, as shown on manometer, $M$. The mercury height in $F$ is adjusted until the pressure of the gas remaining is 1 atm and the amount of gas introduced into $B$ is calculated from the difference between the initial and final buret readings. The motor and attached eccentric operating the magnet are turned on and slugs of liquid at the rate of about 1/sec are pumped into $B$, where they dissolve gas as they run down the inside of the bulb. Under these conditions, no bubbles of gas are carried into $A$. The pressure in $B$ becomes constant in from 1 to 3 hr, depending on the system studied. More gas is added, if necessary, to raise the equilibrium pressure to at least 300 mm of mercury. The mercury level in $D$ is then adjusted to bring the liquid meniscus to $b$, and the pumping continued for a short time. The number of moles of gas undissolved in $B$ is calculated from the observed pressure less the vapor pressure of the solvent and the pressure of the head of liquid. More gas is added and equilibrated as a check on the attainment of equilibrium and, in the case of a very soluble gas, to test the applicability of Henry's law. To obtain values for heats or entropies of solution, solubilities at a series of temperatures are measured just by changing the temperature and reequilibrating the system.

This apparatus has been used to study the solubility of fluorocarbon gases in cyclohexane [65], as well as inert gases and lower alkanes [64]. The solubility of inert gases was interpreted by Miller and Hildebrand [66].

### Loprest Apparatus

Loprest [67] developed an apparatus (Fig. 5.13) for a physical saturation method employing manometric measurements to calculate solute distribution in the system.

Prior to carrying out a determination, the volume of flask $H$ is calibrated from mark $h$ up to and including the bore of stopcock 2, about 100 ml, and the volume of flask $I$, whose weight is already known, is calibrated to mark $i$ and to the mercury surface in U-tube $J$, about 120 ml. Initially flask $I$ is empty. With stopcock 1 open, the level of mercury in $H$ is set at the mark $h$ using leveling bulb $B$. That stopcock is closed and the system is evacuated through 2, 4, 5, and 9. Stopcocks 5 and 9 are closed and solute gas is introduced from $P$, a gas cylinder, through 8 until atmospheric pressure is reached and the gas is allowed to bubble into the atmosphere through $Q$, a mercury

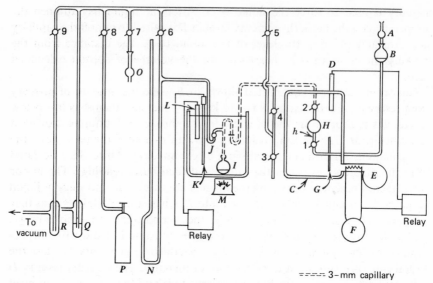

**Fig. 5.13**  Apparatus for the rapid determination of gas in liquid solubility. [Reproduced from F. J. Loprest, *J. Phys. Chem.* **61**, 1128 (1957)].

bubbler. Stopcock 5 is then opened slowly so that the bubbling continues through $Q$. The valve of tank $P$ and 8 are then closed. Stopcock 9 is then opened slightly and some solute gas is pumped out of the system until the pressure falls to some value below atmospheric where the gas-measuring operation is to be carried out. The introduction of solute gas into the system is carried out in this way to avoid possible contamination of the gas which might occur if the valve on the gas tank $P$ was opened to the vacuum manifold. The temperature in the room is noted and the temperature of the air thermostat, $C$, is set only a few degrees higher. (Temperature control is obtained in this Lucite box, $C$, by means of a thermoregulator, $D$, and a hair dryer, $E$, with its heating element connected to an electronic relay through a Variac, $F$.) This reduces the amount of heat transfer and permits accurate control of the temperature of the airbath ($\pm 0.05°$). After sufficient time is allowed for thermal equilibrium to occur, the pressure of the gas is measured on the manometer, $N$, using a cathetometer. Stopcock 2 is then closed rapidly and the number of moles of gas in $H$ can then be calculated.

The system, exclusive of $H$, is brought to atmospheric pressure with air or nitrogen by opening 7. Flask $I$ is removed and an appropriate amount of solvent and the magnetic stirring bar are added. The solvent is degassed in the following manner. First it is frozen by placing a low-temperature bath around flask $I$. The system is then evacuated through stopcocks 4 and 5. The solvent is allowed to thaw and a portion of it is "boiled off." The boiling-off operation can be eliminated if solvent must be conserved by performing a series of freeze–evacuate–thaw cycles.

With the constant-temperature bath $K$ in place, the vapor pressure of the solvent is measured. This is compared with the literature value or, if no literature value is available, the degassing cycle is repeated until a constant vapor pressure is obtained. The pressure measurement is made after closing stopcocks 4 and 5, opening 6 and 7, thus allowing air or nitrogen to build up the pressure to the point where the levels of the mercury in both arms of $J$ (a mercury-filled U-tube containing fritted-glass disks which do not permit the passage of mercury) are equal. The pressure is then read on the manometer $N$ with a cathetometer. If the surfaces in $J$ are not level, the difference is determined with the cathetometer and appropriate corrections to the pressure reading and the volume of $I$ are made. With stopcock 4 closed, 1 and 2 are opened and mercury is allowed into $H$ and into the capillary lines to the mark, $i$. The solvent is stirred magnetically, and the system is allowed to come to equilibrium. The pressure is again measured with sufficient time allowed between the leveling of the surface in $J$ and the final pressure measurement with $N$ to ensure complete equilibration.

The number of moles of solute gas remaining in the gas phase is calculated and the amount dissolved determined by difference. The temperature of bath $K$ is either raised or lowered, and the system again equilibrated. It is, therefore, possible to obtain solubility data over a wide temperature range with a single charging of flask $I$. The weight of the solvent is determined at the conclusion of the experiment.

The equation used to obtain the solubility data is

$$ n = \frac{76}{WR} \left[ \left( \frac{P_1 \pm V_1}{(P_2 - P_v)T_1} \right) - \left( \frac{V_2 - V_0}{T_2} \right) \right], $$

where

$n$ = solubility in moles of gas/g of solvent at $T_2°K$, and at a partial pressure of gas of 1 atm

$W$ = weight of solvent in grams

$R$ = the gas constant in (cm)(ml)/(moles)(°K)

$P_1$ = pressure of initial quantity of gas in $H$ at $T_1$ in cm

$P_2$ = total pressure at equilibrium in the solubility vessel $I$ at $T_2$ in cm

$P_v$ = vapor pressure of the solvent at $T_2$ in cm

$V_1$ = volume of $H$ in ml

$V_2$ = volume of $I$ in ml

$V_0$ = volume of solvent at $T_2$ in ml

$T_1$ = temperature of air-bath $C$ and gas in flask $H$ in °K

$T_2$ = temperature of bath $K$ and contents of $I$ at which equilibration is carried out, in °K

This equation assumes that the ideal gas laws are obeyed, that the vapor pressure if the solvent in the saturated solution is the same as the pure solvent, and, that Henry's law is obeyed up to a pressure of 1 atm. These assumptions produce deviations well within experimental error under the conditions employed in the experiments. It is believed that a precision of $\pm 0.5\%$ can be obtained easily with this apparatus when $n$ is of the order of $10^{-5}$. The precision is poorer for solubilities of a lower order of magnitude.

The Loprest method has several advantages over other apparatus previously used. A liquid bath whose temperature is easily controlled at a constant value is used and the need for elaborate air thermostats is eliminated. Solvent does not come in contact with mercury surfaces. The necessity of reading a gas buret with the attendant error also is eliminated. In this apparatus the liquid is degassed easily. The vapor pressure of the solvent may be determined. The trend to equilibrium from above and below saturation can be followed easily, and equilibrium is attained rapidly (often less than 20 min). The solubility may be measured at various partial pressures; however, if Henry's law is assumed to be applicable, the solubility can be determined on a single sample of solvent and an entire gas–liquid system can be characterized in several hours.

## Classical Saturation Apparatus

In general, the kinds of gas saturation apparatus in use today are modifications of the apparatus and technique used by Henry in 1803, by Bunsen in 1855, and by Ostwald in 1890. However, the Ostwald method and its various modifications have largely displaced the others.

### Markham-Kobe Modification, Ostwald

The Markham and Kobe modification of the Ostwald apparatus and technique will be used to indicate the general nature of the Ostwald method [4, 68]. In this method a measured volume of gas is brought in contact with a measured quantity of gas-free liquid, equilibrium is established by agitation; the volume of gas remaining is measured and the change in volume gives the amount dissolved by the liquid. Thus the preceding methods also are generally related.

The Markham and Kobe apparatus (Fig. 5.14) differed from others in the method of providing a gas–liquid interface and in the provision for agitating the absorption flask. The buret, $A$, is connected at the bottom by a T-tube to the mercury leveling bulb, $B$, and the manometer tube, $C$, open at the top. The cock, $D$, is between the buret and the manometer tube. At the top, the buret is connected by the ground-glass joint, $E$, to the T-tube, $F$. One branch of this T-tube connects to the vacuum through the stopcock, $G$. The other ends in a straight tube at $H$. The absorption flask consists of two bulbs, $J$ and

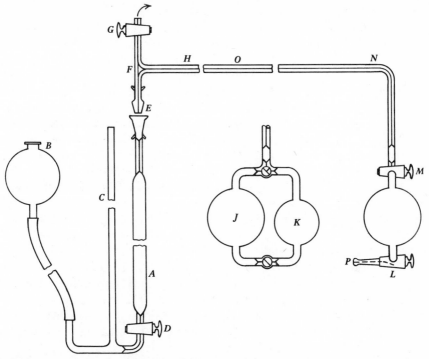

**Fig. 5.14** Markham and Kobe modification of Ostwald apparatus. [Reproduced from A. F. Markham and K. A. Kobe, *Chem. Rev.* **28**, 519 (1941)].

*K*, one having twice the volume of the other. They are connected at the bottom through the three-way cock, *L*, and at the top through the three-way cock, *M*. A capillary tube leads from *M* to *O*.

During a run the absorption flask is in a thermostat with a water level nearly up to *N*. The buret and manometer tube, *C*, are jacketed in a large glass tube through which water flows from the thermostat. The tube from *E* to *H* is capillary, so only a very small volume of gas is outside the thermostat. A framework of metal supports the absorption flask. It connects to a motor which oscillates the flask through an arc of about 10°, with *O–N* as an axis, at 160 oscillations/min. The buret and the bulbs of the absorption flask are calibrated by filling with mercury.

These steps are used to determine the solubility of a gas in a liquid. Boil the solvent under reflux to free it from dissolved gas, seal the condenser and cool the solvent under vacuum to the temperature of the thermostat. The liquid is then withdrawn from the reflux flask by means of a withdrawal tube which dips to the bottom of the flask and is connected to the absorption flask at *P*. Cocks *L* and *M* are so arranged that there is a passage through one bulb from $P$ to $O$. Suction is applied at $O$ and the bulb filled with the solvent. After one bulb is filled with liquid, the cocks are turned to give free passage through the other bulb from $P$ to $O$, thereby isolating the full bulb. Several drops are then drawn into the empty bulb and the refluxing flask and vacuum disconnected. The buret is filled with mercury by raising the leveling bulb and the cock $D$ is closed. The absorption flask is put into the thermostat, connected to the gas supply at $P$ with $O$ and $H$ connected by tubing. The gas from the cylinder goes first through a saturator filled with the solvent under study at the thermostat temperature. The several drops of solvent previously drawn into the bulb ensure saturation. With the gas and suction connected at $P$ and $G$, respectively, the bulb is alternately evacuated and filled four times with gas by manipulation of the cocks $L$ and $G$. Then cock $D$ is opened, the leveling bulb lowered and the buret filled with gas. Cock $L$ is shut and the gas supply disconnected. The leveling bulb is adjusted to such a level that the partial pressure of the gas is exactly 760 mm Hg, allowance being made for the density of the mercury, capillary effects, barometric pressure, and the vapor pressure of the solution. The buret level at this time is recorded. Cock $L$ is opened between the two bulbs, and cock $M$ is opened between the bulbs and to the buret. The entire apparatus is then shaken. The leveling bulb is raised as solution proceeds to maintain pressure relationships. After there is no perceptible change in the buret levels, shaking is stopped. The leveling bulb is adjusted as before, and the final buret reading is recorded.

This apparatus gives a precision of 0.2% and is simple in construction. The possible sources of error in using this method have been considered by Markham and Kobe. According to their findings, supersaturation is not a problem. A small error may reflect solvent expansion due to the dissolved solute gas which thereby affects the observed volume of gas.

### Cook and Hansen

It should be noted in the above procedure that the gas was saturated with liquid vapor before filling the buret. Other investigators have kept the gas in the buret dry. If the gas in the buret is saturated, the vapor pressure of the solvent is of little consequence. If the gas is dry, however, the vapor pressure must be known accurately, since all gas coming into the free space above the liquid in the absorption vessel picks up vapor, increasing its volume to an extent determined by the vapor pressure. On the other hand, if the gas in the buret is saturated, any part of the apparatus that is not in the thermostat may collect condensed solvent if the thermostat is above room temperature. The capillary between the buret and the absorption vessel is usually out of the thermostat. Drops of liquid in this capillary would make the pressure adjustment in the buret uncertain.

If the gas in the buret is dry, the temperature of the whole apparatus can be changed and thus a range of temperatures can be covered with one filling which is a valuable feature where the thermodynamics of solution are of interest. The Ostwald-type apparatus involves a mercury surface contact with the gas and sometimes with the solvent as well. This is a serious drawback when dealing with a system that reacts with or is soluble in mercury.

Cook and Hansen [69] have given an excellent discussion of the problems involved in saturation methods of solubility determinations. It struck these authors that deviations in gas solubility data of greater than 1%, common in the literature, are not in accord with the accuracy obtainable in the physical measurements involved, for surely an accuracy of 0.1% in the measurement of a pressure, a volume, or a mass can be obtained with minimal care. They reasoned that if the purity of the materials is satisfactory and the measurements of temperature, volume, and pressure are sufficiently accurate, the discrepancies in the solubility values must result from other sources of error. These errors can be one or more of the following; failure to reach equilibrium, failure to degas the solvent completely, failure to ascertain the true amount of gas dissolved, and failure to ensure that the transfer of gas from a primary container to the apparatus does not involve air contamination. Cook and Hansen proceeded to design an apparatus which would eliminate the above errors. This apparatus is shown in Fig. 5.15.

## Classical Extraction Apparatus

### Van Slyke

One of the best and the most popular types of apparatus for solubility determination by the extraction method is that of Van Slyke and Neill [70, 71]. This manometric apparatus is based on the principle of extraction of the gas from the liquid and subsequent measurement of the pressure of the liberated gas. The apparatus is shown in Fig. 5.16. The 50-ml short pipet A has several graduations on it and a corresponds to 2 ml. The pipet is connected to the manometer and to the mercury leveling bulb. The sample of a solution of the gas is introduced through stopcock b by a special pipet in such a way that the solution does not come in contact with the air. The gas solution is evacuated by lowering the leveling bulb, and the pipet is shaken for 2–3 min to assist in liberating the gas. The liberated gas is compressed into the volume a and the pressure read on the manometer.

A later modification of the Van Slyke-Neill apparatus eliminates the transfer step [70]. Modified Van Slyke designs have been used extensively in biochemical studies. Orcutt and Seevers [72] determined blood levels of the anesthetic gases, cyclopropane, ethylene, nitrous oxygen, and carbon dioxide.

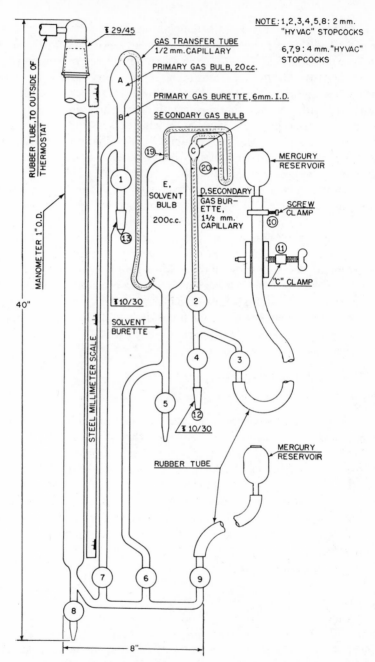

**Fig. 5.15** Cook and Hansen apparatus for determination of the solubility of a gas in liquid. [Reproduced from M. W. Cook and D. N. Hansen, *J. Chem. Phys.* **28**, 370 (1957)].

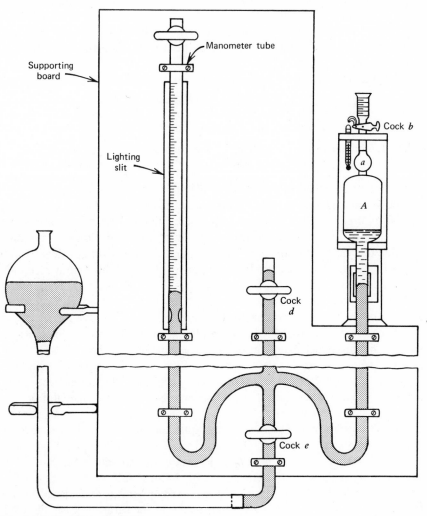

**Fig. 5.16**  Van Slyke-Neill Apparatus.

The general equation for calculating the total gas content of a solution from the amount of gas extracted in an evacuated chamber of definite volume is

$$V^0 = \frac{V_t(P_b - P_v)}{760(1 + 0.000367t)}\left(1 + \alpha'\left(\frac{S}{A - S}\right)\right),$$

where

$V^0$ = volume of gas, measured at $0°$, 1 atm in the solution analyzed,

$V_t$ = volume of gas at $t$,

$P_b - P_v$ = barometric pressure in mm Hg corrected for the vapor pressure of the liquid at the temperature of the experiment,

$t$   = temperature in °C,

$\alpha'$ = the Ostwald distribution coefficient of the gas between gas and liquid phases, that is, $\alpha' = (t/273)$,

$A$   = volume of extraction chamber,

$S$   = volume of solution in extraction chamber.

The factor in the above equation which corrects for the gas which remains unextracted is $[1 + S\alpha'/(A - S)]$.

## Baldwin-Daniel Apparatus for Viscous Liquids

Baldwin and Daniel designed [73] an extraction apparatus to measure the solubility of gases in liquids, especially viscous liquids, which consists essentially of three operations: deaeration of the liquid, saturation of the liquid with the test gas at atmospheric pressure, and determination of the amount of gas liberated under vacuum from a known volume of the saturated liquid.

The apparatus used for the first two steps is shown in Fig. 5.17. The solvent is placed in funnel $F$ and allowed to drip slowly into vessel $A$ which is evacuated. It might be expected that this procedure would remove all dissolved gas, but tests showed that 2–3% of the orginal gas still remained in solution. Hence, the gas under examination is passed into liquid through tap $T_1$ for 2–3 hr. The gas supply is then stopped, and the liquid is allowed to drip into vessel $B$, which is evacuated continuously. The liquid finally is free of the dissolved gas initially present (usually air) and contains only a small fraction of the gas under examination.

For the saturation and storage of the liquid solvent before test, a specially-designed displacement buret, $D$, is used. The evacuated liquid from $B$ is introduced into $D$ through taps $T_2$ and $T_3$, tap $T_4$ of the buret being open to the vacuum pump. When the displacement buret is full, taps $T_3$ and $T_4$ are closed, and the buret is disconnected from the apparatus and placed in the thermostat. Taps $T_3$ and $T_4$ are then opened and the saturating gas bubbled into the liquid through tap $T_3$ until saturation is completed. When no gas bubbles can be detected in the bulk of the liquid, the buret tube above $T_3$ is filled with mercury to displace the gas from the buret. When all the gas has been expelled, tap $T_4$ is closed, and $T_3$ is left open so that there is a head of some 100 mm Hg in the buret tube, which suffices to keep the gas in solution before test begins.

The apparatus used for determining the amount of gas is shown in Fig. 5.18. It consists essentially of a vessel, $V$, with a side arm, taps, and so on, for

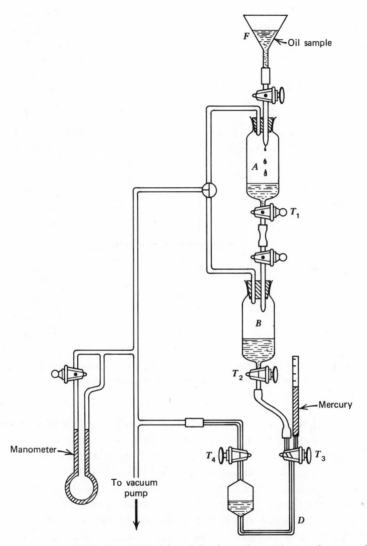

**Fig. 5.17** Apparatus of Baldwin and Daniel used for deaeration and saturation procedure in their solubility method. [Reproduced from *J. Appl. Chem.* **2,** 161 (1952)].

connection to the displacement buret *D*, and connected via a ground-glass joint at the top to a Töpler pump, *P*, manometer, *M*, and calibrated vessel, *S*. The whole apparatus is initially evacuated, and the gas released from the liquid is transferred by the Töpler pump to the calibrated vessel, *S*, the pressure it exerts in this volume being measured by the manometer, *M*.

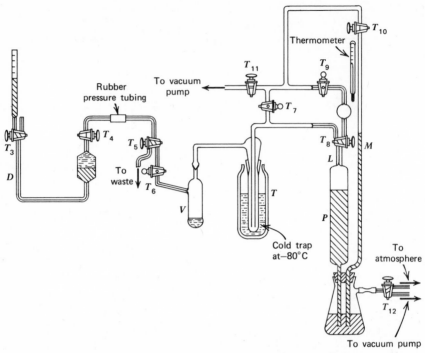

**Fig. 5.18**   Apparatus of Baldwin and Daniel used for the determination of the amount of gas liberated from the liquid in their solubility method. (Reproduced from *J. Appl. Chem.* **2**, 161 (1952)].

$T_5$ is closed, the apparatus is evacuated through $T_{11}$ by a high-vacuum pump and $T_7$ is then closed to prevent loss of gas to the vacuum pump. The displacement buret is then connected to the solubility apparatus by a short length of pressure tubing. The space between $T_4$ and $T_5$ is filled with liquid by displacing, with mercury, a little of the liquid to waste through $T_5$. The evacuated space between $T_5$ and $T_6$ is then filled. The mercury reading in the buret tube is noted. $T_6$ is opened and liquid is allowed to run into vessel $V$ which is immersed in bath at 150°C to assist gas volution. During this time, the Töpler pump is connected to the vessel, $V$, through $T_8$ and the mercury is maintained at the bottom of its stroke. When about 10 ml of liquid has been introduced into $V$, $T_6$ is closed, and the change in mercury level in the buret tube noted. In this manner, the volume of oil introduced can be measured to ±0.02 ml. $T_8$ is reversed, and the collected gas forced into $S$. $T_8$ is again reversed so as to connect to vessel $V$ and the mercury is drawn to the bottom of the Töpler pump, the process being repeated until all liberated gas has been collected.

To measure the final amount of gas collected, the Töpler pump is brought to a fixed mark on the neck of the pump at L by adjustment of the pressure at the base of the pump. The readings of the manometer, $M$, and the thermometer are then noted. To obtain the zero of the manometer, the volume, $S$, and the Töpler pump are evacuated through $T_9$ and the manometer reading is again noted when the mercury level in the Töpler pump is again brought to $L$. The difference in the two manometer readings gives the pressure of the gas, and the volume of the gas is known from the calibration of $S$ and the additional volume between $T_8$ and the mark $L$.

The volume of gas collected is corrected to standard conditions by the gas laws. This method has an accuracy of at least $\pm 1\%$.

### Special Applications

The following techniques serve to illustrate the range of solubility problems encountered with gases.

Durrill and Griskey [74] have modified the pressure vessels of Newitt and Weale [75], and Lundberg and coworkers [76], which allow determination of both the solubility and diffusivity of simple gases in molten or thermally softened polymers. In these, the unusual observation was made that Henry's law held up to 20 atm.

Bott and Schulz [77] studied the solubility of gaseous chlorine in brine solutions as a function of pH in order to improve the operation of electrolytic cells. The study was complicated by reaction of chlorine with water to yield hydrochloric and hypochlorous acids and with chloride ion to form a complex, which reactions in turn were influenced by pH and salt concentration. The complexity of problems illustrates the need to understand the overall chemistry of any system for which solubility data are desired.

The physical solubility of gases at 1 atm in fused silica was investigated by Doremus [78]. He found that, in marked contrast to most gas solubility situations, gas molecular solubilities in fused silica were relatively insensitive to temperature, pressure, and molecular size. These observations are consistent with an inert matrix with some free volume, and it is this free volume alone which is available for gas solutions.

## 7  DETERMINATION OF DISSOLUTION RATE

Dissolution-rate studies are rapidly gaining the attention of organic and, especially, pharmaceutical chemists. Such rate studies are important for the understanding and the design of experiments in partition, adsorption, and dialysis involving the transfer of solids to solutions. Dissolution-rate data can guide the pharmaceutical scientist in designing the solid oral dosage forms which represent the majority of drugs on the market. Orally administered drugs must pass through the gastrointestinal barrier and to accomplish this the drug first must be in solution. For some drugs it is desirable that absorption be rapid and for others that absorption should be delayed.

Dissolution rate can be expressed by the equation advanced by Noyes and Whitney [12, 13]: $dC/dt = k(C_s - C)$, where $C$ is the concentration of solute in the solvent at time $t$, $C_s$ is the solubility of the solute in the solvent and $k$ is a constant with the dimension time$^{-1}$. They developed the relationship from studies on benzoic acid and lead chloride of essentially constant surface area prepared by depositing molten solute on a glass core and noted that rate studies previously were lacking because of the constant surface problem. Subsequently [14], the surface area of the solute $S$ was incorporated in equation form. The modern equation is:

$$dW/dt = k_1 S(C_s - C)$$

where $dW$ is the mass of solute entering solution, in which case the first-order rate constant $k_1$ has the dimensions mass/(area)(time).

Dissolution rate is viewed thus as a simple transfer process involving a concentration gradient between a thin film of solvent surrounding the solute, and saturated by it, and the bulk of the solution. Broadly, the dissolution rate is directly proportional to the equilibrium solubility. The dissolution rate constant, $k$, may be interpreted in terms of diffusivity (see Chapter VI, Part IV) and, therefore, solution viscosity. Similarly, as $k$ depends on the thickness of saturated film, it can be viewed in terms of speed of convection, or turbulence, of solvent in the vicinity of undissolved solute. The thermodynamics of the transfer of a solute molecule across the interface of the solid and liquid phases is not a component of $k$, as this transfer is a feature of the equilibrium solubility, $C_s$.

Dissolution methods can be classified into two general categories—intrinsic and apparent methods. Intrinsic methods determine the rate in terms of mass/(area)(time) [mg/(cm$^2$)(hr)] under conditions of known or controlled surface area. Apparent methods measure the total mass dissolved per unit of time. Dissolution methods are further classified as "sink" or "nonsink" methods. A broad interpretation of a sink method is one in which the solute is removed from the media as dissolution proceeds.

As there are almost as many methods for determining dissolution as there are workers, the authors have taken the liberty of selecting examples of only some of the methods now in general use.

## Determination of Intrinsic Dissolution Rate

### Fine Particle Method

Edmundson and Lees [79] developed a procedure for dissolution rates of fine particles in which the changing solute particle-size distribution is followed in a stirred solvent by means of a particle-size counter. The dissolution rate is related to loss of particle diameter per unit of time. They obtained dissolution rate data on crystalline hydrocortisone acetate by this method. The major diasdvantages of this method are the requirement for an expensive particle-size distribution-counter, poor precision, and difficult calibration and control of stirring dynamics.

## Rotating Disk Method

Levy and Sahli [80] proposed a procedure wherein the solute is compressed by a hydraulic press into a plane-faced disk which is placed in an acrylic holder which in turn mounts at the end of a stirrer shaft. The shaft extends into a 500-ml round-bottom flask (three-neck) containing 200 ml of dissolution medium. The shaft is rotated at the desired speed and samples withdrawn through the other necks. Observed dissolution rates were highly sensitive to variations in rotation speed, so Levy and Tanski [81] developed a power unit from commercial pump motors and speedometer cable which allowed constancy of rotation over a 3 to 400-rpm range. Another modification [82] featured a compression die which was used directly as the disk holder, eliminating several manipulations, and ensuring that a single planar surface was exposed to the medium.

Control and constancy of rotation rates, and absence of shaft wobble are critical to this method. The method is usable only where direct compression of the solute is feasible and where the compressed disk does not disintegrate in the presence of dissolution medium.

## Hanging Pellet Method

Another compressed-disk procedure [83] involves mounting the disk on a strip of aluminum with wax so as to expose just one surface of the disk. The strip is then suspended in the dissolution medium from the arm of a suitable balance. The loss of weight of the disk is observed with time. This is a static method and depends, therefore, on solvent convection and viscosity.

## Determination of Apparent Dissolution Rate

General methods are available to determine apparent or total dissolution rate, mass per unit time. Methods for apparent rates need not control surface area.

## Rotating Basket Method

The rotating basket was first proposed by Pernarowski [84] and coworkers and was adopted with modifications by the United States Pharmacopeia [85] and the National Formulary [86] as an official method for determining the apparent dissolution rate of solid oral-dosage forms (Fig. 5.19). The dosage form is placed in the stainless steel, 40-mesh screen basket, $A$, which clamps

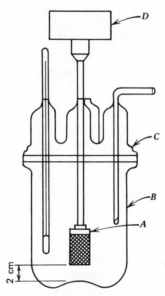

**Fig. 5.19** Apparatus for Dissolution Test, *USP/NF*.

onto the end of the stainless steel shaft. This shaft is rotated, by suitable means such as a variable-speed stirrer, *D*, at a carefully controlled speed between 50 and 150 rpm. The basket is immersed in 900 ml. of the selected dissolution medium which is maintained at 37°C by locating the resin flask, *B*, in a constant temperature bath. Samples are removed at periodic intervals through the necks,*C*, of the resin flask and analyzed by a specified procedure. Figure 5.20 illustrates the type of data one obtains using commercial prednisone tablets produced by eight different manufacturers.

The rotating basket assembly is commercially available and its inherent versatility permits a much broader application than just for standardization of pharmaceuticals.

### Beaker Method

This method, widely employed in pharmaceutical research, was introduced by Parrott, Wurster, and Higuchi [87] in a study of benzoic acid spheres. Levy and Hayes [88] modified this for drugs in tablet form and generally the procedure is as follows: 250 ml of dissolution medium at 37°C is placed in a 400-ml beaker and a 5-cm three-bladed polyethylene stirrer is immersed at the exact center of the beaker to a depth of 2.7 cm. The stirrer usually is rotated at about 60 rpm. The tablet is dropped down the side of the beaker to the bottom where it disintegrates into granules covering a small (but uncontrollable) area. Samples are withdrawn using a sintered-glass immersion filter.

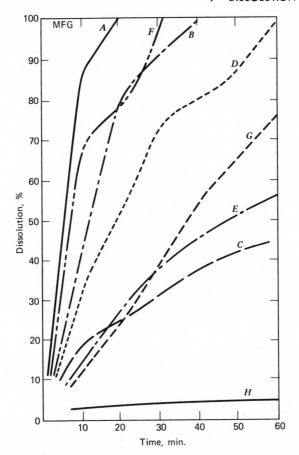

**Fig. 5.20** Time-dissolution curve in water at 37.5°C, 50 rpm.

A variety of modifications of this basic beaker method have been reported. The advantages of the method are simplicity and ready availability of components. The disadvantages are poor reproducibility due to variable geometry between solute and stirrer, and restriction to solutes more dense than the medium (capsules, for example, float). Further modifications have included supernatant, immiscible layers [89] to supply a partition effect. Immiscible layers and adsorbents such as charcoal have been used to maintain "sink" conditions.

### Solvometer

The apparatus developed by Klein [90] features a boat or pan suspended from a float which in turn is suspended from the arm of a balance or some form of graduated scale. The pan and float are submerged in the medium and loss of weight of solute is observed. Widespread disintegration of solute or carrying-off of particles by convection is a serious problem and places limits on the agitation which can be given to the medium. The principle and equipment is simple and this method would be particularly convenient for moderately dense solids, polymer, or mineral samples.

### Adhesive Tape Method for Particles

In this procedure, particles of solute are pressed onto 3/4-in.-wide pressure-sensitive tape and the tape placed in a frame-like holder. The beaker containing dissolution medium is fitted with runners into which the frame containing the solute tape is slipped and held firmly. A variable speed stirrer also projects into the medium. The method is suitable to multiparticulate systems and was used originally [91] to determine dissolution rates of benzoic acid and salicylamide.

## References

1. A. Seidell, *Solubilities of Inorganic and Organic Compounds*, Vol. 2, 3rd ed., Van Nostrand, Princeton, N. J., 1941; *ibid.* with W. F. Linke, *Supplement* 1952; *ibid.* Vol. 2, 4th ed., American Chemical Society, 1965.
2. J. H. Hildebrand, *Proc. Natl. Acad. Sci.* **57**, 542 (1967).
3. Bunsen, *Phil. Mag.* **9**, 116 181 (1855).
4. A. E. Markham and K. A. Kobe, *J. Amer. Chem. Soc.* **63**, 449 (1941).
5. W. Alexejew, *J. Prakt. Chem.* **25**, 518 (1882).
6. A. E. Hill, *J. Amer. Chem. Soc.* **45**, 1143 (1923).
7. V. C. Williams, *J. Phys. Colloid Chem.* **52**, 1263 (1948).
8. M. J. Buerger, L. B. Smith, A. de Boetteville, Jr., and F. V. Ryer, *Proc. Natl. Acad. Sci.* **28**, 526 (1942).
9. W. D. Bancroft and F. J. C. Butler, *J. Phys. Chem.* **36**, 2515 (1932).
10. E. L. Heric and C. D. Posey, *J. Chem. Eng. Data* **9**, 35 (1964).
11. L. A. Chugaev and V. G. Khlopin, *J. Russ. Phys. Chem. Soc.* **46**, 1659 (1914).
12. A. A. Noyes and W. R. Whitney, *Z. Physik. Chem. (Leipzig)* **23**, 689 (1897).
13. A. A. Noyes and W. R. Whitney, *J. Amer. Chem. Soc.* **19**, 930 (1897).
14. L. Bruner and St. Tolloczko, *Z. Physik. Chem. (Leipzig)* **35**, 283 (1900); F. Brunner, *Z. Physik. Chem. (Leipzig)* **47**, 56 (1904).
15. R. S. Tipson, in *Physical Methods of Chemistry*, Vol. 3, 2nd ed., John Wiley, New York, Part 1, 1956.
16. J. E. Gordon and R. L. Thorne, *J. Phys. Chem.* **71**, 4390 (1967).
17. R. E. Jentoft and R. J. Robinson, *Anal. Chem.* **26**, 1156 (1954).
18. W. Reeve and R. Adams, *Anal. Chem.* **22**, 755 (1950).
19. R. L. Shriner, R. C. Fuson, and D. Y. Curtin, *Systemic Identification of Organic Compounds, a Laboratory Manual*, 5th ed., John Wiley, New York, 1964, pp. 67 ff.
20. M. Beroza and M. C. Bowman, *Anal. Chem.* **38**, 837 (1966).

21. K. H. Meyer, *Natural and Synthetic High Polymers*, Interscience, New York, 1942, pp. 9, 25, 565–598.
22. G. V. Schulz and B. Jirgensons, *Z. Physik. Chem. (Leipzig)* **B46**, 105, 137 (1940).
23. D. R. Morey and J. W. Tamblyn, *J. Phys. Colloid Chem.* **51**, 721 (1947).
24. W. J. Mader, "Phase Solubility Analysis," in *Organic Analysis*, Vol. 2, Interscience, New York, 1954, pp. 253–275.
25. W. J. Mader, *Critical Reviews in Analytical Chemistry* **1**, 193 [1970].
26. W. Tarpley and M. Yudis, *Anal. Chem.* **25**, 121 (1953).
27. R. G. Satterfield and M. Haulard, *J. Chem. Eng. Data* **10**, 397 (1965).
28. E. L. Heric, *J. Chem. Eng. Data* **12**, 71 (1967).
29. F. Fairbrother, N. Scott, and H. Prophet, *J. Chem. Soc. (London)* **1956**, 1164.
30. S. J. Gill, J. Hutson, J. R. Clopten, and M. Downing, *J. Phys. Chem.* **65**, 1432 (1961).
31. T. Higuchi and J. L. Lach, *J. Amer. Pharm. Assoc., Sci. Ed.* **43**, 349 (1954).
32. J. Reilly and W. N. Rae, *Physico-Chemical Methods*, Vol. 1, 3rd ed., Van Nostrand, Princeton, N. J., 1939, p. 589.
33. T. J. Ward, *Analyst* **44**, 137 (1919).
34. I. A. Pastac and R. Lecrivain, *Chim. Anal.* **30**, 28 (1948).
35. A. B. Zdanovskii, *Zh. Neorg. Khim.* **1**, 1279 (1956) in Israel Program for Scientific Translations, *J. Inorg. Chem.* **1**(6), 164 (1956).
36. T. H. Vaughn and E. G. Nutting, Jr., *Ind. Eng. Chem., Anal. Ed.* **14**, 454 (1942).
37. A. N. Campbell, *J. Chem. Soc. (London)* **1930**, 179.
38. G. Aravamudan and K. R. Krishnaswami, *Current Sci. (India)*, **25**, 287 (1956).
39. J. B. Lombardo, *J. Chem. Educ.* **44**, 600 (1967).
40. A. W. C. Menzies, *J. Amer. Chem. Soc.* **58**, 934 (1936).
41. P. Gross, *J. Amer. Chem. Soc.* **51**, 2362 (1929).
42. L. H. Adams, *J. Amer. Chem. Soc.* **37**, 1181 (1915).
43. S. Mitchell, *J. Chem. Soc. (London)* **1926**, 1333.
44. R. L. Bohon and W. F. Claussen, *J. Amer. Chem. Soc.* **73**, 1571 (1951).
45. P. K. Weyl, *Rev. Sci. Instr.* **28**, 722 (1957).
46. P. Jordan, *Z. Physik Chem. Neue Folge* **9**, 187 (1956).
47. N. B. Mikheev, *Intern. J. Appl. Radiation Isotopes* **5**, 32 (1959).
48. I. D. Robb, *Aust. J. Chem.* **19**, 2281 (1966).
49. W. L. Marshall, *Anal. Chem.* **27**, 1923 (1955).
50. J. d'Ans, *J. Chem. Appar.* **28**, 197 (1941).
51. W. B. Bunger, *J. Amer. Oil Chemists' Soc.* **36**, 466 (1959).
52. W. Schenk and H. Tulhoff, *Ber. Bunsengesell. Physik. Chem.* **71**, 206 (1967).
53. G. W. Watt, W. A. Jenkins, and C. V. Robertson, *Anal. Chem.* **22**, 330 (1950)
54. H. S. Davis, *J. Amer. Chem. Soc.* **38**, 1166 (1916).
55. E. A. Klobbie, *Z. Physik. Chem.* **24**, 615 (1897).
56. T. W. Evans, *Ind. Eng. Chem., Anal. Ed.* **8**, 206 (1936).
57. W. Herz, *Ber.* **31**, 2669 (1898).
58. H. Sobotka and J. Kahn, *J. Amer. Chem. Soc.* **53**, 2935 (1931).
59. M. M. Acker and H. A. Frediani, Jr., *Ind. Eng. Chem., Anal. Ed.* **17**, 793 (1945).
60. M. Hayashi and T. Sasaki, *Bull. Chem. Soc. Japan* **29**, 857 (1956).
61. L. K. Nash, *Anal. Chem.* **21**, 1405 (1949).

62. B. B. Cunningham, *Nucleonics* **5**(5), 62 (1949).
63. J. H. Hibben, *J. Res. Natl. Bur. Std.* **3**, 97 (1929).
64. J. Dymond and J. H. Hildebrand, *Ind. Eng. Chem. Fund.* **6**, 130 (1967).
65. K. W. Miller, *J. Phys. Chem.* **72**, 2248 (1968).
66. K. W. Miller and J. H. Hildebrand, *J. Amer. Chem. Soc.* **90**, 3001 (1968).
67. F. J. Loprest, *J. Phys. Chem.* **61**, 1128 (1957).
68. A. E. Markham and K. A. Kobe, *Chem. Rev.* **28**, 519 (1941).
69. M. W. Cook and D. N. Hansen, *Rev. Sci. Instr.* **28**, 370 (1957).
70. D. D. Van Slyke, *J. Biol. Chem.* **130**, 545 (1939).
71. D. D. Van Slyke and J. M. Neill, *J. Biol. Chem.* **61**, 523 (1924).
72. F. S. Orcutt and M. H. Seevers, *J. Biol. Chem.* **117**, 501, 509 (1937).
73. R. R. Baldwin and S. G. Daniel, *J. Appl. Chem.* (*London*) **2**, 161 (1952).
74. P. L. Durrill and R. G. Griskey, *Amer. Inst. Chem. Eng. J.* **12**, 1147 (1966).
75. D. M. Newitt and K. E. Weale, *J. Chem. Soc.* (*London*) **1948**, 1541.
76. J. L. Lundberg, M. B. Wilk, and M. J. Huyett, *Ind. Eng. Chem., Fund.* **2** (1), 37 (1963).
77. T. R. Bott and S. Schulz, *J. Appl. Chem.* (*London*) **17**, 356 (1967).
78. R. H. Doremus, *J. Amer. Ceram. Soc.* **49**, 461 (1966).
79. I. C. Edmundson and K. A. Lees, *J. Pharm. Pharmacol.* **17**, 193 (1965).
80. G. Levy and B. A. Sahli, *J. Pharm. Sci.* **51**, 58 (1962).
81. G. Levy and W. Tanski, Jr., *J. Pharm. Sci.* **53**, 679 (1964).
82. J. H. Wood, J. E. Syarto, and H. Letterman, *J. Pharm. Sci.* **54**, 1068 (1965).
83. E. Nelson, *J. Pharm. Sci.* **47**, 297 (1958).
84. M. Pernarowski, W. Woo, and R. O. Searl, *J. Pharm. Sci.* **57**, 1419 (1968).
85. *The Pharmacopeia of the United States of America, Eighteenth Revision*, U.S. Pharmacopeial Convention, Bethesda, Md., General Tests, 1970.
86. *The National Formulary, Thirteenth Edition*, American Pharmaceutical Association, Washington, D. C., 1970, p. **802**.
87. E. L. Parrott, D. E. Wurster, and T. Higuchi, *J. Amer. Pharm. Assoc., Sci. Ed.* **44**, 270 (1955).
88. G. Levy and B. A. Hayes, *New Engl. J. Med.* **262**, 1053 (1960).
89. B. Gibaldi and S. Feldman, *J. Pharm. Sci.* **56**, 1238 (1967).
90. L. Klein, *Bull. biolistes. pharmaciens.* **1932**, 273; as reprinted by G. H. Elliott, *Pharm. J.* **131**, 514 (1933).
91. A. H. Goldberg, M. Gibaldi, J. L. Kanig, and J. Shanker, *J. Pharm. Sci.* **54**, 1722 (1965).

Chapter **VI**

# DETERMINATION OF OSMOTIC PRESSURE

J. R. Overton

# 1 INTRODUCTION

The objective of this chapter is to acquaint the reader with the theory and practice of osmometry as a laboratory tool for the characterization of macromolecules and the study of solute–solvent interactions in dilute solutions. Both the theory and applications of osmometry to aqueous and nonaqueous solutions are presented. However, the author's experience, and therefore prejudice, is confined to the area of nonaqueous media and synthetic polymers, and may not be as complete for aqueous and biological systems. The reader is urged to consider this background and temper any comments accordingly.

If a solution is separated from pure solvent by some medium which can transmit solvent, but not solute, there will occur a net flow of solvent into the solution. Transport of solvent under these conditions is known as osmosis. The osmotic pressure of a solution is conventionally defined as that pressure required, under isothermal conditions, to prevent the net flow of solvent into the solution when solvent and solution are separated by a perfectly semipermeable membrane. A semipermeable membrane allows the free passage of solvent molecules and totally prevents permeation by solute molecules. Perfect semipermeability is not always achieved in practice and therein lies one of the major difficulties in osmometry. This problem will be discussed in greater detail later in the chapter.

The phenomenon of osmosis was probably the first experimental observation of the effect of a solute on the thermodynamic state of the solvent. Such effects are of vital importance in the biological sciences as they control numerous life processes. The generation of an osmotic pressure belongs to the class of solution properties designated as colligative properties. These include vapor-pressure lowering, boiling-point elevation, and freezing-point lowering. These properties are useful in the physical sciences as they allow experimental determination of solvent activity in a solution, from which one can calculate the solute activity. At the limit of infinite dilution the solute activity is equal to solute concentration in moles/cc. One in effect "counts" the number of solute molecules in solution, via solution behavior, from a given mass of material, thus the determination of molecular weight. Of the colligative properties, osmotic pressure is by far the most sensitive. The sensitivity of osmotic pressure is some $10^3$ times that of ebullioscopic and cryoscopic techniques [1].

The molecular weight range over which osmometry is applicable is generally considered to be from $10^4$ to $10^6$. The limits, of course, are only approximations as they depend on a number of factors in practice. The lower limit is determined by the porosity of the membrane, in other words, the solute must be sufficiently large so as not to penetrate the membrane. The upper limit is determined by the sensitivity of the osmometer used to small osmotic pressures. The solute must generate a detectable osmotic pressure at concentrations where dilute solution behavior is applicable.

Osmometry is thus applicable to the characterization of macromolecules—polymers. For most polymers, except some of biological origin, the term molecular weight is ambiguous in that a distribution of species is always present. Therefore an average molecular weight must be defined to render the term useful. Osmotic-pressure measurements yield the number-average molecular weight, $\bar{M}_n$, as do all colligative measurements. This average is synonymous with the classical definition of molecular weight in that it is the total weight of a sample divided by the number of moles present,

$$\bar{M}_n = \frac{\Sigma N_i M_i}{\Sigma N_i} = \frac{\Sigma W_i}{\Sigma N_i},$$

(6.1)

where $N_i$ is the number of moles of the $i$th species whose molecular weight is $M_i$, and $W_i$ is the total weight of species $i$ in the sample.

Number-average molecular weights are important parameters in polymer science. These data can be used to calibrate nonabsolute methods such as solution or melt viscosity and gel-permeation chromatography. Molecular-weight data are essential as they are directly relatable to polymer physical properties and the processability of polymer melts and dopes (concentrated solutions usually containing 25–35% polymer by weight). Molecular-weight data are also important in mechanistic and kinetic studies of polymerization and degradation. In addition to the determination of molecular weights, the extent and nature of solute–solute and solute–solvent interaction can be determined from the concentration dependence of the reduced osmotic pressure. These concepts will be developed in the theoretical section which follows.

## 2 THEORY

Osmometry as a science dates to 1887 when Van't Hoff satisfactorily expressed the relationship between the experimentally observed osmotic pressure and the molecular weight of the solute. Thus the technique can truly be considered a classical one. The original equation, the Van't Hoff relationship, was arrived at by analogy to the behavior of ideal gases:

$$\pi = \frac{n_2 RT}{V} = \frac{CRT}{M},$$

(6.2)

where $\pi$ is the osmotic pressure, $V$ the volume of solution containing $n_2$ moles of solute, $T$ the absolute temperature, and $R$ the usual ideal gas constant. The same relationship can be derived from strictly thermodynamic considerations.

If two portions of pure solvent are separated by a membrane through which the solvent molecules can readily permeate, the liquid level on the two sides will be equal at equilibrium and the hydrostatic pressure on side one will equal that on side two. Equilibrium will be maintained and we may write

$$a_1{}^0 \text{ (left)} = a_1{}^0 \text{ (right)}, \tag{6.3}$$

where $a_1{}^0$ equals the activity of pure solvent.

If we now add solute to the right side which cannot permeate the membrane, equilibrium will be destroyed and solvent will pass through the membrane from left to right so as to dilute the solution. Now

$$a_1{}^0 \text{ (left)} > a_1 \text{ (right)}. \tag{6.4}$$

Equilibrium can be reestablished if sufficient external pressure is applied to the solution side so that

$$a_1{}^0 \text{ (left)} = a_1 \text{ (right)} \tag{6.5}$$

$$a_1{}^0 \text{ (left, } P = P^0) = a_1 \text{ (right, } P = P^0 + \pi). \tag{6.6}$$

At constant temperature, the solvent activity is a function of solute concentration and applied pressure.

$$a_1 = f(N_2, P), \tag{6.7}$$

where $N_2$ represents solute mole fraction. Therefore with complete generality we may write

$$d \ln a_1 = \left[ \left( \frac{\partial \ln a_1}{\partial P} \right)_{T,N_2} dP \right] + \left[ \left( \frac{\partial \ln a_1}{\partial N_2} \right)_{P,T} dN_2 \right]. \tag{6.8}$$

Now if equilibrium is established so that

$$a_1{}^0 \text{ (left)} = a_1 \text{ (right)} \tag{6.9}$$

then

$$d \ln a_1 = 0 \tag{6.10}$$

and

$$\left[ \left( \frac{\partial \ln a_1}{\partial P} \right)_{T,N_2} dP \right] = - \left( \frac{\partial \ln a_1}{\partial N_2} \right)_{P,T} dN_2. \tag{6.11}$$

It can be shown rigorously that

$$\left( \frac{\partial \ln a_1}{\partial P} \right)_{T,N_2} = \frac{\overline{V}_1}{RT}, \tag{6.12}$$

where $\overline{V}_1$ is equal to the molar volume of the solvent. We now assume the validity of Raoult's law for the solvent, that is,

$$a_1 = a_1{}^0 N_1 \tag{6.13}$$

$$a_1 = a_1^0(1 - N_2), \tag{6.14}$$

which in logarithmic form is

$$\ln a_1 = \ln a_1^0 + \ln (1 - N_2), \tag{6.15}$$

and differentiating at constant pressure and temperature one gets

$$d \ln a_1 = d \ln (1 - N_2) = \frac{-dN_2}{1 - N_2} \tag{6.16}$$

in the limit of infinite dilution $(1 - N_2)$ approaches unity and therefore

$$\left( \frac{\partial \ln a_1}{\partial N_2} \right)_{T,P} = -1. \tag{6.17}$$

Therefore substituting (6.17) and (6.12) into (6.11) gives

$$\frac{\overline{V}_1}{RT} dP = dN_2, \tag{6.18}$$

which when integrated from zero to some low concentration $N_2$, gives

$$P - P_0 = \frac{RT}{\overline{V}_1} N_2, \tag{6.19}$$

which can be written as

$$\pi = n_2 \frac{RT}{\overline{V}} = \frac{CRT}{M}. \tag{6.20}$$

This derivation assumes isothermal conditions, and the validity of Raoult's law, and is therefore strictly applicable only in the limit of infinite dilution. One should also note that a perfectly semipermeable membrane must be assumed. The theoretical osmotic pressure, $\pi$, is, however, independent of membrane structure and the mechanism by which solvent is transported through it. Therefore, gaseous and liquid membrane behavior as well as that of conventional solid membranes is explained equally well by the preceding relationships.

Thus from the Van't Hoff relationship and an experimental determination of the osmotic pressure one can calculate the number-average molecular weight of the solute. Since the relationship is valid only at infinite dilution, however, one must make measurements as a function of concentration and extrapolate to zero concentration. The concentration dependence of the osmotic pressure is conventionally expressed in virial form similar to the ideal gas law.

$$\pi = RT[A_1C + A_2C^2 + A_3C^3 + \ldots], \tag{6.21}$$

where $A_1, A_2, \ldots$, are the virial coefficients.

$A_1$ is equal to the reciprocal of the solute number-average molecular weight,

$$A_1 = \frac{1}{\overline{M}_n}. \tag{6.22}$$

Equation 6.21 is more conveniently expressed as

$$\frac{\pi}{C} = RT\left[\frac{1}{\overline{M}_n} + A_2C + A_3C^2 + \ldots\right]. \tag{6.23}$$

The obvious technique then is to determine $\pi$ at a series of concentrations and plot $\pi/C$ versus $C$ and extrapolate to zero concentration. The intercept is then equal to $RT/\overline{M}_n$ and the slope given by $A_2RT$.

The validity of the virial expansion to the osmotic pressure relationship has been rigorously demonstrated [2, 3]. Just as the second virial coefficient of a real gas is relatable to a volume parameter [4], so is the osmotic second virial coefficient. In the case of real gases the entire volume of the system is not available to the center of gravity of each molecule in that the molecules themselves occupy a finite volume. This is the so-called forbidden volume [5], the result of which is deviation from ideal behavior.

A quantitative interpretation of the osmotic second virial coefficient has been carried out by Flory and Krigbaum [3]. They evaluated the partition function for a dilute solution by considering the volume available to successive solute molecules as they are added to the system. The procedure must obviously consider the finite volume of the solute molecules already added. In solution theory this parameter is the so-called excluded volume, $U_{ij}$, and is considered to be the effective volume of some molecule $i$ with respect to molecule $j$ (for the details of this derivation the reader is referred to the original literature).

The partition function $Q$ is given by

$$\ln Q = N \ln V - \frac{1}{2} \frac{\Sigma_i\Sigma_jN_iN_jU_{ij}}{V}, \tag{6.24}$$

where $N_i$ and $N_j$ are the number of solute molecules of molecular weight $M_i$ and $M_j$, $V$ the total volume of the system, and $N$ the total number of molecules.

The osmotic pressure is given by [6]

$$\pi = kT\frac{\partial \ln Q}{\partial V}, \tag{6.25}$$

and therefore one has

$$\pi = kT\frac{N}{V} + \frac{kT}{2V^2}\Sigma_i\Sigma_jN_iN_jU_{ij}. \tag{6.26}$$

Now since $N/V = CN/\overline{M}_n$ and $\overline{M}_n = N\Sigma_i m_i N_i/N$ where $m_i$ is the mass of species $i$, comparison of (6.23) and (6.26) gives

$$A_2 = \frac{N}{2N^2\overline{M}_n{}^2} \Sigma_i\Sigma_j N_i N_j U_{ij} .$$  (6.27)

Introducing the definition of $U_{ij}$ derived by Flory [7] gives

$$A_2 = \frac{V_1{}^2}{V_2} (\Psi_1 - \kappa_1)F(X).$$  (6.28)

The function $F(X)$ includes the effect of the discontinuous nature of dilute polymer solutions which had been omitted in earlier solution theories and is therefore a function of molecular weight. Osmotic-pressure studies carried out above about $1-2\%$ where the polymer segment distribution is uniform throughout need not consider this function. The parameter $\Psi_1$ is the entropy-of-mixing parameter characteristic of a given solute segment–solvent pair, and is equal to $^1/_2$ for an ideal solution, in other words, the entropy of dilution is given by

$$\Delta\overline{S} = R\Psi V_2{}^2.$$  (6.29)

The parameter, $\kappa$, is the heat term such that the partial molar heat of dilution is given by

$$\Delta\overline{H}_1 = RT\kappa_1 V_2{}^2.$$  (6.30)

From (6.28) it is apparent that when $\Psi_1$ equals $\kappa_1$ the osmotic second virial coefficient is equal to zero and the reduced osmotic pressure $\pi/C$ will be independent of concentration, in other words, the Van't Hoff equation is applicable over all concentrations. Under these conditions the excluded volume also reduces to zero and solute molecules freely penetrate each other. It is apparent of course that solute molecules must in reality occupy a finite volume. Therefore the conditions of ideality are not those which might be superficially assumed of no interaction between the solvent and solute. There is in fact a net repulsion between solute and solvent so that solute molecules slightly prefer their own environment to that of the solvent—thus the net effect is as if the volume occupied by the solute were zero.

In general, thermodynamic solvent power is a function of temperature. The temperature at which $A_2$ reduces to zero for a given system is termed the Flory theta temperature and is defined by

$$\Theta = \frac{\kappa_1 T}{\Psi_1} ,$$  (6.31)

thus giving an alternative expression for $A_2$ as

$$A_2 = \frac{\bar{V}^2}{V_1} \Psi_1(1 - \Theta/T)F(X).$$

The $\Theta$ temperature is also termed the critical miscibility temperature, in other words, the temperature at which phase separation would occur for a solute of infinite molecular weight. In general the heat of mixing and entropy of mixing are both positive and $(\kappa_1 - \Psi_1)$, and thus solvent power, increases with temperature. Numerous exceptions to this behavior have been noted however where solvent power actually goes through a maximum, decreases, and phase separation occurs at some higher temperature, designated the lower critical solution temperature [8–10].

It has in fact been suggested that the phenomenon is general and will occur in all polymer–solvent systems below the vapor–liquid critical temperature of the solvent [11]. The phenomenon is not generally observed since it usually lies above the normal boiling point of the solvent.

For systems in which $A_2$ is reasonably high the plot of $\pi/C$ versus $C$ will exhibit upward curvature reflecting the importance of the third virial coefficient $A_3$. The third coefficient is related to interactions involving three molecules. Flory [7] using a hard-sphere approximation derives the following relationship between $A_2$ and $A_3$.

$$A_3 = \frac{5}{8} \frac{A_2^2}{A_1}. \tag{6.32}$$

Therefore

$$\frac{\pi}{C} = RT\left[\frac{1}{\bar{M}_n} + A_2C + (\tfrac{5}{8})A_2^2\bar{M}_nC^2 + \ldots\right].$$

Perhaps a more convenient form of the virial equation is

$$\frac{\pi}{C} = \left(\frac{\pi}{C_0}\right)(1 + \Gamma_2C + \Gamma_3C^2), \tag{6.33}$$

where $\Gamma_i = A_i\bar{M}_n$.

The Flory hard-sphere approximation gives the relationship

$$\Gamma_3 = (\tfrac{5}{8})\Gamma_2^2 \tag{6.34}$$

and therefore

$$\left(\frac{\pi}{C}\right) = \left(\frac{\pi}{C_0}\right)(1 + \Gamma_2C + \tfrac{5}{8}\Gamma_2^2C^2 + \ldots). \tag{6.35}$$

For the purpose of facilitating the extrapolation of $\pi/C$ to zero concentration under conditions where curvature exists, (6.33) can be forced to a linear form by assuming that

$$\Gamma_3 = 0.25\Gamma_2^2. \tag{6.36}$$

Substituting (6.36) into (6.33) and taking the square root gives

$$\left(\frac{\pi}{C}\right)^{1/2} = \left(\frac{\pi}{C_0}\right)^{1/2}(1 + 0.50\Gamma_2 C). \tag{6.37}$$

The assumption leading to (6.36) is found to be valid in numerous cases. Stockmayer and Cassassa have shown theoretically that the value of $\frac{5}{8}$ deduced from the hard-sphere approximation is too high for real polymer coils [12]. They find that the ratio of $\Gamma_3/\Gamma_2^2$ depends on $\alpha$, the molecular expansion coefficient, and that for normal values of $\alpha$, the relationship

$$\Gamma_3 = 0.25\Gamma_2^2 \tag{6.38}$$

is a reasonable approximation.

The solution theories within whose framework the preceding relationships were developed, namely those of Flory, Huggins, and Flory and Krigbaum, are all statistical in nature. Each approach suffers from a number of deficiencies. The Flory-Huggins theory is applicable only at concentrations above about 2% where polymer segment distribution is uniform throughout the system. The development by Flory and Krigbaum using different interaction parameters is only applicable in very dilute solutions.

More recently a solution theory has been proposed by Maron [13] which to a large extent eliminates these difficulties. This theory is nonstatistical in nature and is derived based on the effective volume occupied by the solute and the interaction between solute and solvent. Experimental evaluations have shown that the relationships developed are applicable over extended ranges of concentration and solvent power [14–18].

The osmotic pressure relationship derived by Maron is

$$\frac{\pi}{C} = \frac{RT}{M_n}(\epsilon/\epsilon_0) + B_1[\tfrac{1}{2} - (\mu - \sigma v_1)]C + B_2 C^2 + B_3 C^3 + \ldots, \tag{6.39}$$

Where $B_1$ is defined by

$$B_1 = \frac{RT}{V_1^0 \rho_2^2}, \tag{6.40}$$

and $B_n$ $(n > 1)$ is

$$B_n = \frac{RT}{[(n + 1)V_1^0 \rho_2^{(n+1)}]}, \tag{6.41}$$

where $V_1^0$ is the molar volume of the solvent and $v_1$ is its volume fraction, $\rho_2$, is the solute density, and $\mu$ is the interaction parameter. Sigma, $\sigma$, is given by

$$\sigma = \left(\frac{\partial \mu}{\partial V_2}\right)_{P,T}$$

where $V_2$ is the volume fraction of the solute, $\epsilon$ the effective volume of the solute at a given concentration, and $\epsilon_0 = \epsilon$ at infinite dilution.

The Maron theory has recently been extended to include ternary systems and in addition the correction term $\epsilon/\epsilon_0$ has been eliminated [19]. It is perhaps not within the scope of this chapter to delve deeper into the field of solution theory and the reader is referred to the original literature.

## 3   OSMOMETRY WITH PARTIALLY PERMEATING SOLUTES

The choice of a proper membrane is of utmost importance in osmometry studies [20]. Data obtained using a membrane which permits solute permeation may be more characteristic of the membrane than of the macromolecular solute under investigation. The problem is intensified when studying unfractionated polymers in which significant quantities of low-molecular-weight material are present. Solute diffusion is exemplified by a gradual decrease in apparent osmotic pressure with time. This behavior indicates the need for a membrane with lower permeability.

In the real case where solute diffusion is present one may resort to two experimental techniques: (a) the determination of the osmotic pressure at equilibrium, in other words, when the diffusing species has distributed itself equally between the two chambers, and (b) the determination of the osmotic pressure at time zero when no solute has diffused from the solution chamber.

In the first case one should obtain the theoretical osmotic pressure for that portion of the solute which cannot penetrate the membrane. Thus if the concentration change due to diffusion is determined it is possible to calculate the molecular weight of the nondiffusing species. In practice this turns out to be unsatisfactory [21]. The decrease in osmotic pressure with time becomes immeasurably small after reasonably short times (several hours), however, the osmotic pressure will continue to decrease slowly. In the limit of $t = \infty$, of course, true equilibrium will be established and the theoretical osmotic pressure will be obtained.

The time required for a diffusing solute to equilibrate is excessively long. In addition the contribution to the osmotic pressure of a solute molecule in the solution chamber which can diffuse through the membrane is a function of its permeability [22]. Those molecules which can penetrate the membrane, but have not at some time prior to equilibrium, contribute less than the theoretical amount to the osmotic pressure. Therefore in the long time experiments equilibrium is achieved only after excessive times and at any point prior to equilibrium the osmotic pressure is lower than the theoretical value. Molecular weights determined by this technique will be too high.

In short time experiments, it is customary to measure the apparent osmotic pressure as a function of time and extrapolate to time zero where no solute diffusion has occurred. Under these conditions the extrapolated value of the osmotic pressure, $\pi_0$, is related to the theoretical value by

$$\pi_0 = s\pi_{th},$$

(6.42)

where $s$ is the Staverman reflection coefficient and is equal to the difference in the relative permeabilities of the solute and solvent [22]. In the case where solute permeation occurs the true molecular weight can be obtained only if $s$ is known. For polydisperse solutes one has a distribution of $s$ values and calculating a true molecular weight requires knowledge of the molecular-weight distribution and the variation of $s$ with molecular weight. In the absence of this information the molecular weight determined from $\pi_0$ is designated by Staverman as a "reflection-average" molecular weight [22]. It is apparent that this value can be a stronger function of the membrane employed than of the solute of interest.

Solute permeation is not always undesirable. Low-molecular-weight additives such as plasticizers and stabilizers have little influence on the number-average molecular weight determined by osmometry in the long time experiments. The presence of such compounds completely invalidate molecular-weight data obtained by other colligative techniques, for example, boiling-point elevation.

Another application of the use of partially permeable membranes first suggested by Staverman [23] is the study of the polydispersity of the polymer itself (osmodialysis). Successful applications of this technique have been reported by Hoffman [24] and Hudson [25]. This approach to the determination of molecular-weight distribution requires extensive calibration for a given osmometer, solvent, membrane, and temperature. Calibration as a function of molecular weight is necessary in order to determine the reflection coefficients, $s$, and the time constants, $\alpha$. The molecular-weight distribution can be estimated from analysis of the osmotic pressure–time curve represented by the following relationship [20]:

$$\pi = RT\Sigma_i \frac{s_i C_i}{M_i} \exp\left(-\alpha_i t\right). \tag{6.43}$$

Although the technique is of interest it is probably of little practical value due to the extensive calibration and approximations required.

## 4  OSMOMETRY OF POLYELECTROLYTES

### Donnan Membrane Equilibria

Osmotic pressure measurements of polyelectrolytes in solutions containing diffusible electrolytes, for example, $Na^+R^-$ in aqueous $Na^+Cl^-$ are complicated by the unequal distribution of $Na^+Cl^-$ across the membrane. This is the so-called Donnan equilibrium effect first predicted by Gibbs [26] and Donnan [27] and subsequently experimentally verified by Donnan [28]. The presence of the low-molecular-weight electrolyte is conventional in order to reduce the virial expansion for the polyelectrolyte to integral powers of the concentration [20].

Donnan equilibrium is demonstrated by the following example. If a solution of $C_3$ moles/liter of polyelectrolyte, $Na^+R^-$, is formed from solvent containing $C_2$ moles/liter of sodium chloride, $Na^+Cl^-$ and the solution and solvent are separated by a membrane through which $R^-$ cannot permeate, the condition illustrated below will exist at equilibrium.

Membrane

| Solution Side | Solvent Side |
| --- | --- |
| $[R^-] = C_3$ | $[Na^+] = C_2 + \delta$ |
| $[Na^+] = C_3 + C_2 - \delta$ | $[Cl^-] = C_2 + \delta$ |
| $[Cl^-] = C_2 - \delta$ | |

The net effect being depletion of $\delta$ moles/liter of $Na^+Cl^-$ from the solution chamber with an equivalent increase in the solvent chamber. The conditions for equilibrium are that electrical neutrality be maintained on both sides of the membrane and that the chemical potential of the material on both sides of the membrane, in other words, $Na^+Cl^-$, must be equal.

$$\mu_{NaCl(1)} = \mu_{NaCl(2)} \tag{6.44}$$

where the subscripts (1) and (2) refer to the two sides of the osmometer. The chemical potential of an electrolyte is equal to the sum of the potentials of its respective ions and since

$$\mu = \mu_0 + RT \ln a, \tag{6.45}$$

equilibrium is characterized by

$$\mu^0_{Na^+} + RT \ln a_{Na^+(1)} + \mu^0_{Cl^-} + RT \ln a_{Cl^-(1)} =$$
$$\mu^0_{Na^+} + RT \ln a_{Na^+(2)} + \mu^0_{Cl^-} + RT \ln a_{Cl^-(2)}, \tag{6.46}$$

which reduces to

$$a_{Na^+(1)} \cdot a_{Cl^-(1)} = a_{Na^+(2)} \cdot a_{Cl^-(2)}. \tag{6.47}$$

If one assumes sufficient dilution such that the activity can be replaced by the concentration, in other words, the activity coefficient equals unity, and substituting for $C$ from the equilibrium state the relationship below is obtained.

$$(C_3 + C_2 - \delta)(C_2 - \delta) = (C_2 + \delta)(C_2 + \delta). \tag{6.48}$$

Solving for $\delta$ one gets

$$\delta = \frac{C_2 C_3}{4C_2 + C_3}. \tag{6.49}$$

Assuming the validity of the Van't Hoff equation,

$$\pi = RTC, \tag{6.50}$$

where $C$ has the units of moles/liter, and since the observed osmotic pressure will be the difference between the theoretical osmotic pressure on the solution side and that on the solvent side the equation below follows.

$$\pi_{obs} = \pi_1 - \pi_2 \tag{6.51}$$

$$= RT[C_3 + (C_3 + C_2 - \delta) + (C_2 - \delta)]$$
$$- RT[(C_2 + \delta) + (C_2 + \delta)], \tag{6.52}$$

which reduces to

$$= RT[2C_3 - 4\delta]. \tag{6.53}$$

Since $RTC_3$ equals the osmotic pressure due to the polyelectrolyte anion and is the quantity of interest, this is separated to give

$$\pi_{obs} = \pi_{polymer} + RT[C_3 - 4\delta]. \tag{6.54}$$

Substituting from (6.49) into (6.54) gives

$$\pi_{obs} = \pi_{polymer} + RT\left[\frac{C_3^2}{4C_2 + C_3}\right]. \tag{6.55}$$

Since the molar concentration of the polymer will invariably be significantly lower than that of the salt, (6.55) can be sufficiently approximated by

$$\pi_{obs} = \pi_{polymer} + RT\left[\frac{C_3^2}{4C_2}\right]. \tag{6.56}$$

The preceding derivation is for the simplest case where activity coefficients are equal to unity and the nondiffusible species contains a single charge. The extension to multicharged polyelectrolytes requires the obvious modification of the concentration terms.

$$\pi_{obs} = \pi_{polymer} + RT\left[\frac{Z_3^2 C_3^2}{4C_2}\right]. \tag{6.57}$$

$Z_3$ represents the effective net charge/molecule of the nondiffusible species. The Donnan pressure is minimized in practice by increasing the ionic strength of the solvent, decreasing the polyelectrolyte concentration and decreasing the charge/molecule.

In the case of ampholytes such as proteins the effective charge/molecule is controlled by adjusting the pH of the solvent. It is impractical however to reduce the effective charge to zero by using isoionic protein in salt-free water. Scatchard [29] found interpretation of osmotic pressure data from bovine albumin-water to be complicated by the sensitivity of the system to trace quantities of acid or alkali. To overcome the Donnan effect some salt invariably is used.

The most extensive study of the osmometry of proteins, namely, albumins, has been conducted by Scatchard and coworkers [29, 30]. He pointed out that the osmotic pressure of most protein solutions could be represented by only two terms in the virial expansion.

$$\pi = AW_2 + ABW_2{}^2. \tag{6.58}$$

$W$ has units of g salt/kg $H_2O$, $A$ is $RT/(V^0{}_m M_2)$, and $V^0{}_m$ is the total volume containing 1 kg salt. The coefficient $B$ is the interaction term related to the effective volume of the solute [31]. Scatchard's relationships lead to the following:

$$\pi = \frac{RT}{V^0{}_m M_2}\omega_2 + \frac{RT}{2V^0{}_m M_2{}^2}\left[\frac{Z_2{}^2}{2m_3} + \frac{\partial \ln \gamma_2}{\partial m_2} - \frac{\left(\dfrac{\partial \ln \gamma_2}{\partial m_3}\right)^2 m_3}{2 + \dfrac{2\partial \ln \gamma_3 m_3}{\partial m_3}}\right]\omega_2{}^2. \tag{6.59}$$

(The two above equations are given in Scatchard's notation where concentrations are in molalities, $m$ and the $\gamma$'s are the respective activity coefficients. Subscript 3 refers to salt and subscript 2 to protein in this case.)

Comparison of (6.58 and 6.59) reveals the interaction term $B$ to be the product of $1/(2M_2)$ and the bracketed terms. The first term in the bracket is the Donnan term, the second reflects protein–protein interaction and is a function of the size and shape of the molecule. The third term includes protein–salt and salt–salt interactions. For further discussion of the osmometry of proteins see Kupke [32]. Doty and Edsall [33], and Edsall [34].

## 5  EXPERIMENTAL METHODS

### The Static-Elevation Method

The static-elevation method is the classical technique for the determination of osmotic pressure and is that most commonly used with conventional osmometers. The method consists simply of allowing the solvent to pass through the membrane from either direction until equilibrium is achieved. The volume change in the solution side of the chamber will cause the liquid in the attached capillary tube to rise or fall until the hydrostatic pressure exerted by the liquid head reduces the net solvent flux to zero. The pressure exerted by the excess liquid head under conditions of equilibrium is by definition the osmotic pressure. The major advantages of the technique are its simplicity, and that it requires little or no attention once the osmometer is assembled and placed in a thermostated bath. When using the static-elevation method it is desirable to run several experiments simultaneously due to the long equilibration times required. This is facilitated by the relatively low cost and compactness of such osmometers.

A unique advantage of equilibrium osmometry is that the presence of low-molecular-weight additives, such as plasticizers, or stabilizers, have little or no effect on osmotic pressure in that these materials distribute themselves, to a first approximation, evenly between the two chambers. The same phenomenon, solute permeation, is a disadvantage when studying polydisperse solutes where the molecular size range extends below the region of membrane rejection.

The time required to reach equilibrium can vary from several hours to several days depending on the experimental arrangement and the system being investigated. It is quite essential in this technique that the liquid level in the capillary be initially adjusted to within a few millimeters of the estimated equilibrium value. Variables which contribute to the rate of equilibration are: (a) the ratio of membrane surface area to solution volume; (b) membrane thickness and porosity; (c) the diameter of the capillary; and (d) the viscosity of the solvent. Each of these factors should be considered and adjusted accordingly for each situation.

The obvious disadvantage of the static-elevation or equilibrium method is the long time required to reach equilibrium. This in turn amplifies the normal problems associated with osmometry in general. For example, the diffusion of solute through the membrane, and precise temperature control is required over a much longer time scale. The technique is obviously not applicable for solutes which are not stable for extended periods. The deficiencies of the static-elevation method have led to the development of more rapid techniques for the determination of osmotic pressure.

Experimental techniques discussed in the next section are grouped under the heading of Dynamic Methods more for the reason that they all differ in principle from the static-elevation technique than for their mutual similarity. Dynamic techniques offer two potential advantages over the equilibrium approach. They are more rapid, and some of the dynamic methods allow the determination of the osmotic pressure at time zero, $\pi_s$, prior to the permeation of any solute. As already discussed the osmotic pressure at zero time, for cases where solute permeation can occur, will differ from the theoretical value by the Staverman coefficient.

# 6  DYNAMIC METHODS

## The Half-Sum Method of Fuoss and Mead

This approach consists of determining the asymptote to a plot of pressure versus time (or capillary-liquid height versus time). The asymptote is deter-

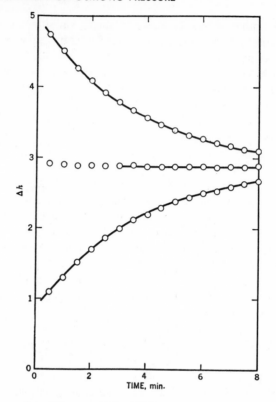

**Fig. 6.1**  Half-sum plot of Fuoss and Mead [35].

mined using the half-sum method first described in 1943 [35]. After the osmometer is assembled and thermal equilibrium is established the liquid level in the capillary is adjusted to some value 1–2 cm above the estimated equilibrium height. Then the height is recorded at 30-sec intervals until the meniscus is moving quite slowly. This usually requires about 10 min. The same procedure is then repeated with the liquid level adjusted below the equilibrium value. One then plots the average (one half the sum) of the two readings at equal time intervals. This value should become constant after only a few minutes and is equal to the equilibrium osmotic pressure (the asymptote of the pressure time plot). Example plots of descending and ascending curves as well as the half-sum values are shown in Fig. 6.1. Such plots are typical only under ideal conditions where the membrane is rigid and semipermeable.

The technique is quite rapid compared to the static-elevation method. An osmotic-pressure determination can be made in about 30 min. It does however require the operator's total attention during that time period. Results obtained compare favorably with those obtained by equilibrium osmometry.

The method should not be used if the ascending and descending branches of the curves are not symmetrical [36, 37]. This will be the case if (a) solute permeation occurs; (b) diffusional layers are present; and (c) asymmetrical membrane ballooning occurs.

### The Zero-Flow Method

The zero-flow method consists of automatic compensation for the tendency of the solvent to flow through the membrane (in either direction depending on the initial pressure) by adjusting the pressure applied externally to either of the osmometer chambers. This technique, first described in 1934 [38, 39] is used by two of the commercially available versions of the automatic osmometer. Earlier versions adjusted an applied gas pressure in order to maintain zero net flow. The technique is obviously quite rapid compared to the static-elevation method as no solvent is required to pass through the membrane.

An interesting modification of the technique is to determine the time required to reach maximum pressure using different starting pressures. The technique is useful for those cases where some solute permeation occurs. The lower the starting pressure the lower the maximum pressure will be since in the longer time period required for the measurement more solute will have diffused through the membrane. The maximum pressure is plotted versus the time required, $t_{max}$ and the data extrapolated to $t_{max=0}$ [40]. Fresh solvent and solution must be used for each determination. The molecular weight obtained for a polydisperse solute using this technique is the "reflection average" and will be higher than the number average, the degree of the discrepancy depending on the porosity of the membrane. The technique is not applicable for nonpermeating solutes.

### The Method of Elias

This method is based on the relationship between the velocity of solvent permeation at a given time and the pressure in excess (positive or negative) of the equilibrium value. The relationship below has been experimentally verified [41].

$$\frac{dH}{dt} = K(\pi_{s,t} - P_t) \tag{6.60}$$

or

$$\pi_{s,t} = P_t + K_e \frac{dH}{dt}. \tag{6.61}$$

$\pi_{s,t}$ is the theoretical Staverman pressure at time $t$, related to the theoretical osmotic pressure by

$$\pi_s = s\pi_{th}, \tag{6.62}$$

where $s$ has its usual significance. In the case of solute permeation $\pi_s$ is a function of time.

$(dH/dt)$ represents the rate of change of liquid level in the capillary, $P_t$ the real pressure at $t$ and $K_e$ the permeation constant. The permeation constant is characteristic of the system and ideally independent of time. The value of $K_e$ is determined with solvent in both osmometer chambers and is given by [40]

$$\frac{dH}{dt} = -\frac{1}{K_e} h, \qquad (6.63)$$

where $h$ is the excess liquid height in the capillary.

For the case of a semipermeable membrane it is thus only necessary to determine the rate of change of pressure at a given pressure. The equilibrium value can then be calculated from the predetermined value of the permeation constant. For cases where solute permeation occurs it is necessary to calculate $\pi_{s,t}$ as a function of time and extrapolate to zero time. The success of this technique is dependent on the constancy of the permeation constant, $K_e$. According to Elias [40] a constant value of $K_e$ is contingent upon the following conditions: (a) absence of any balloon effect; (b) absence of capillary-drainage effects and temperature gradient in the osmometer; (c) stability of the membrane; (d) tightness of the measuring system. Bruss and Stross [41] find that the most serious contribution to variations of $K_e$ is the blockage of membrane pores by solute molecules which are unable to completely penetrate. This effect is of course a function of the molecular-weight distribution of the solute. To ensure that $K_e$ has not changed during the course of the measurement it should be checked before and after each experiment. Results from this technique agree well with those of other dynamic methods when $K_e$ is constant. It is essential that the initial value of the capillary height be not too different from the equilibrium value such that the rate is sufficiently slow so as to be determined with precision [42].

Another dynamic technique consists of measuring the rate of change of pressure or liquid level in the capillary as a function of pressure and extrapolating the data to zero rate. The general approach was first introduced by Hepp [43] where externally-generated gas pressure was used. It was later modified by Bruss and Stross so as to be useful with nonaqueous systems using a small-volume high-speed osmometer [44, 45]. The hydrostatic head was preset at various positions above and below the equilibrium value by adjusting the osmometer push rod. The initial rate of solvent permeation was determined and the rate–pressure data extrapolated to zero rate. The extrapolated value gives $\pi_s$ near zero time. (Bruss and Stross determined the rates within 5 min from the time the osmometer was filled.) This technique is probably the most reliable of the dynamic methods as applied to partially permeating solutes [42].

### Phillip's Method

The principle underlying this technique, as for all dynamic methods, is that the rate of solvent permeation through the membrane is proportional to the excess liquid head, $h$, in the capillary [46]. In effect:

$$\frac{dH}{dt} = -Kh \qquad (6.64)$$

where $h = H - \pi$. $H$ is the capillary height in cm and $\pi$ the equilibrium capillary height. Restated, the pressure decays exponentially with time.

$$\frac{h}{h_0} = e^{-kt}. \qquad (6.65)$$

Thus

$$\frac{h_2}{h_1} = \exp\left[-k(t_2 - t_1)\right] \qquad (6.66)$$

and

$$\frac{h_3}{h_2} = \exp\left[-k(t_3 - t_2)\right]. \qquad (6.67)$$

Therefore in the dynamic mode one can record the data at equal time intervals such that

$$(t_2 - t_1) = (t_3 - t_2) \qquad (6.68)$$

and

$$\frac{h_2}{h_1} = \frac{h_3}{h_2}. \qquad (6.69)$$

Substitution of (6.64a) into (6.69) gives

$$\frac{H_2 - \pi}{H_1 - \pi} = \frac{H_3 - \pi}{H_2 - \pi} \qquad (6.70)$$

and consequently

$$\pi = \frac{H_1 H_3 - H_2^2}{H_1 + H_3 - 2H_2}. \qquad (6.71)$$

Thus the osmotic pressure can be computed from three equally spaced observations of capillary height. The Phillip's method is not applicable for partially permeating solutes. In this case the fundamental assumption of exponential pressure decay is invalid.

All of the dynamic methods discussed require essentially the same experimental data. In effect one records capillary height, or some other pressure response, as a function of time. The methods differ only in the way the data are treated. The availability of modern computing systems allows one to treat raw data by several techniques with essentially no additional effort. It seems obvious that this should be common practice, particularly for partially-permeating solutes. When using the semiautomatic, high-speed osmometer it seems reasonable to employ automatic data acquisition with a small laboratory computer and generate $\pi_s$ values by one or more dynamic computational schemes as well as determining the equilibrium value. (The response time of the osmometer must be fast compared to that of the membrane.) Once these values are determined as a function of concentration for a given solute, it remains to determine the reduced osmotic pressure at infinite dilution $(\pi/c)_0$, by an appropriate extrapolation scheme. This is subject matter discussed in the next section.

## 7  OSMOTIC PRESSURE AT INFINITE DILUTION

The inapplicability of the Van't Hoff equation at finite concentrations requires that experimentally determined values of the reduced osmotic pressure, $\pi/c$, be corrected for concentration dependence. The problem reduces to determining the osmotic pressure at a series of concentrations to establish concentration dependence and from this functionality determine, graphically or computationally, the reduced osmotic pressure at infinite dilution. The fundamental relationship between osmotic pressure and concentration is the virial expansion of the Van't Hoff equation.

$$\frac{\pi}{c} = RT\left[\frac{1}{\overline{M}_n} + A_2 c + A_3 c^2 + \ldots\right] \tag{6.72}$$

or

$$\frac{\pi}{c} = \frac{\pi}{c_0}[1 + \Gamma_2 c + \Gamma_3 c^2 + \ldots]. \tag{6.73}$$

The theoretical significance of the virial coefficients has been discussed in Section 2. The problem of evaluating experimental data is a matter of deciding which terms in the expansion must be included to yield a reliable extrapolation.

For a large number of cases (for reasonably low molecular weights, low concentrations, and poor solvents) a plot of $\pi/c$ versus $c$ will be linear with a slope equal to the second osmotic virial coefficient, $A_2$, and intercept equal to $RT/\overline{M}_n$. Should this plot be nonlinear, indicating the importance of the third virial coefficient, extrapolation becomes hazardous. For these cases an alternative form of (6.73) can be used.

$$\frac{\pi}{c} = \left(\frac{\pi}{c}\right)_0 (1 + \Gamma_2 c + g\Gamma_2^2 c^2).$$ (6.74)

The value of $g$ most convenient to use is 0.25 as this renders (6.74) a perfect square. Taking the square root therefore gives

$$\left(\frac{\pi}{c}\right)^{1/2} = \left(\frac{\pi}{c}\right)_0^{1/2} \cdot (1 + 0.50\ \Gamma_2 c).$$ (6.75)

Thus if the third virial coefficient, but not the fourth, is important a plot of $(\pi/c)^{1/2}$ versus $c$ will be linear and yield a reliable extrapolation.

An alternative technique, due to Bawn [47] is based on the derivative of (6.72).

$$\frac{d(\pi/c)}{dc} = RTA_2 + 2RTA_3 c.$$ (6.76)

Using data point pairs, Bawn plots $[(\pi/c_1) - (\pi/c_2)]/(c_1 - c_2)$ versus $(c_1 + c_2)$ obtaining a linear relationship with a slope of $RTA_3$ and an intercept of $RTA_2$. Having obtained $A_3$ (6.72) is rearranged to give

$$\frac{\pi}{c} - RT \cdot A_3 c^2 = \frac{RT}{\overline{M}_n} + RTA_2 c$$ (6.77)

and accordingly $[(\pi/c) - RT \cdot A_3 c^2]$ is plotted versus $c$ yielding an intercept of $RT/\overline{M}_n$. The technique does not apply of course if $A_4$ is important.

Arnett and Gregg [48] have suggested the use of a two-parameter function which is essentially a virial expansion in closed form.

$$\frac{\pi}{c} = \frac{RT}{\overline{M}_n} f(c),$$ (6.78)

where

$$f(c) = \exp \frac{\Gamma c}{1 + g'\Gamma c}$$ (6.79)

and

$$g' = \frac{1}{2} - g.$$ (6.80)

Equation 6.78 expands to the virial form for the case where $(g'\Gamma c)^2 < 1$.

According to Arnett and Gregg this relationship is found to fit experimental data quite well over extended concentration requirements and yields reliable extrapolations to infinite dilution. An obvious advantage is that it does not require that one assume a value of $g$.

With the automatic osmometer it is customary to determine an initial pressure, $P_0$, with solvent on both sides of the membrane. This value is then substracted from each pressure, $P_c$, for the concentration series. It has been noted [48, 49] that this procedure gives undue statistical weight to the value of $P_0$.

Valtasaari [49] suggests that the $P_0$ value be computed as one of the parameters in the virial equation,

$$P = P_0 + RTA_1c + RTA_2c^2,$$

and that the coefficients of the parabolic function be evaluated by least squares analysis of the experimental data.

Arnett and Gregg have shown that using this procedure, that is, calculating a value for $P_0$, decreases the standard deviation of the parameters in (6.79) and that the procedure gives improved accuracy as well as improved precision.

Certainly in the majority of laboratories where osmometry is used, calculations are carried out on some type of computer. It is therefore no longer necessary to decide a priori which computational scheme is best suited for the problem at hand. It is highly desirable that the experimental data, which can sometimes be difficult to come by, be routinely treated by several approaches and then based on the statistics of the outcome choose the appropriate analysis. Once a versatile computer program is established the additional computational time takes only a few seconds.

## 8 MEMBRANES

The soul of osmometry as a science lies in the transport properties of the membrane. The use of the right membrane is imperative if absolute molecular weights are to be accurately determined. The importance of membrane selection cannot be overemphasized. Fundamental requirements are chemical and dimensional stability under the conditions of the experiment. Structural homogeneity is essential, in other words, the absence of pin holes and the presence of a uniform pore-size distribution. If these conditions are met, success is further governed by the relative retention power of the membrane for the solvent and solute. Ideally the membrane should totally reject solute while allowing the solvent to pass freely. For the case of polydisperse solutes, the former condition is often not realized in practice. It is therefore important that the experimentalist recognize conditions of solute permeation (often referred to as a leaking membrane) and deal with the situation within the framework of the experimental methods and computational schemes available. The failure of early investigators to appreciate the significance of this problem has caused osmotic-pressure data to be viewed with skepticism on more than a few occasions [50].

Practitioners of osmometry now have at their disposal a variety of membranes available from commercial sources. Commercial preparation and distribution has contributed significantly to overall membrane quality. The requirements for conventional osmotic-pressure measurements with high polymers (molecular weights > 10,000) have to a large extent been satisfied by cellulose-based membranes. Cellulose membranes are available in the form of gel cellophane, deacetylated cellulose acetate, and bacterial cellulose. The most commonly used membranes are the gel-cellophane series, availably dry (designated D) and wet (designated W); the latter never having been allowed to dry since preparation. The gel cellophanes are not recommended for aqueous systems. The dry membranes, after reconditioning to the solvent of interest, are reported to have a smoother surface and be less susceptible to adsorption phenomena [51]. The deacetylated cellulose acetate membranes (S & S07, S & S08) are thinner than the gel-cellophane membranes and therefore reach equilibrium more rapidly. Bactiflex (B19, B20) membranes are recommended for use with aqueous as well as organic media. Table 6.1. gives the molecular-weight range and response time for cellulose membranes These membranes are available from several sources [52].

**Table 6.1**  Response Time and Molecular Weight Range Limitations of Commercially Available Membranes for High-Speed Osmometers

| Type of Membrane | Response Time (min) | Molecular Weight Range |
|---|---|---|
| Gell Cellophane 600-W | 10 to 20 | 15,000 and up |
| Gel Cellophane 450-W | 7 to 10 | 30,000 and up |
| Gel Cellophane 600-D | 15 to 45 | 10,000 and up |
| Gel Cellophane 450-D | 10 to 20 | 15,000 and up |
| Gel Cellophane 300-D | 5 to 12 | 100,000 and up |
| S & S 08 | 7 to 10 | 20,000 and up |
| S & S 07 | 7 to 10 | 80,000 and up |
| S & S B-19 | 7 to 15 | 20,000 and up |
| S & S B-20 | 10 to 20 | 7,000 and up |

Special conditions such as high temperatures and particular solvents require special membranes. The problem of what type of membrane to use when cellulose is not applicable is by no means solved. The life of gel cellophane can vary from a few days to several months at elevated temperatures depending on the solvent [53]. Immergut [54] has reported the successful use of poly(vinyl butyral) and trifluorochloroethylene (Kel-F) membranes with cuprammonium hydroxide and cupriethylene diamine, which are cellulose solvents. The Kel-F membrane, although difficult to prepare free of pin holes was reported to have an indefinite life at high temperatures.

Other approaches to the problems of high temperature and special solvents include the use of glass membranes [55], metal membranes [56], and graphite oxide membranes [57]. None of these materials are completely satisfactory. Glass membranes are too slow, primarily because they must be thicker than conventional membranes. Sintered-metal membranes are too porous, only retaining molecular weights greater than about 500,000. The graphite oxide membranes are unstable above 70°C. Kel-F holds perhaps the greatest promise for a universal membrane. The author is unaware of a commercial source for Kel-F membranes but preparation procedures have been published [58].

## Membrane Conditioning

Membranes that are shipped wet are usually in a mixture of ethanol and water. It is important that these membranes be properly conditioned to the solvent which is to be used in the osmometer. The standard technique is to condition the membrane from water to absolute ethanol to the solvent of interest. An accepted procedure is given below [59].

1. Wash in distilled water for 2 hr (2 changes).
2. Wash in 25% ethanol/water for ½ hr.
3. Wash in 50% ethanol/water for ½ hr.
4. Wash in 75% ethanol/water for ½ hr.
5. Wash in pure ethanol ½ hr.
6. Wash in absolute ethanol ½ hr.
7. Wash in 25% solvent/ethanol ½ hr.
8. Wash in 50% solvent/ethanol ½ hr.
9. Wash in 75% solvent/ethanol ½ hr.
10. Wash in pure solvent for 1 hr (1 change).

Membranes that are shipped dry can be used directly with the solvent but should be presoaked before installing in the osmometer.

## 9 INSTRUMENTATION

The objective of this section is to describe in some detail the more conventional types of osmometers which are now being used in the laboratory. Being a rather simple device, at least in principle, a very large number of variations on basic designs have been created in order to meet specific requirements. In the majority of cases, however, the classical concepts have been maintained, in other words, thermally insulated chambers containing solvent and solution, respectively, are separated by a rigidly supported semipermeable membrane. The osmotic pressure is determined by the excess liquid head on the solution side required to yield net zero solvent flux. In general these instruments lend themselves to either dynamic or equilibrium measurements. For special designs the reader is referred to the original literature.

The discussion which follows will be limited to instruments which employ flat membranes and use pressure or liquid level as the primary response. We therefore omit discussion of osmometers using pouch-shaped membranes [60, 61] and also the osmotic balance of Jullander [62, 63].

Certainly, the most significant advance in osmometry hardware in the last ten years is the commercialization of the high-speed or automatic osmometers. Three such devices will be discussed in detail in the pages which follow. First, however, three of the classical devices will be briefly described, both for historical and practical reasons in that such instruments are still being used in many laboratories.

### The Schulz Osmometer

This design first described by Schulz [64] and later modified by Wagner [65], Yanko [66], and others is simple to construct, easy to assemble, and is apparently quite reliable. A detailed drawing of the component parts of the Schulz design as modified by Yanko is shown in Fig. 6.2.

The osmometer consists of a stainless-steel cell with a conical-shaped chamber, an attachable capillary tube assembly and a membrane support which is attached to the bottom of the cell. The membrane also serves as a gasket in sealing the solution chamber from the solvent reservoir. It is essential that the bottom of the cell and the membrane support have perfectly smooth surfaces in order to provide a perfect seal. This is best achieved via hand lapping.

The potential problem of minute leakage at the connection between the capillary tube assembly and the osmometer cell has been successfully eliminated by Yanko. The capillary tube (1-mm-bore, pyrex), is sealed to a small length of $1/_8$-in. Kovar metal tubing. The Kovar tube is soldered into one end of a brass coupler, the other end of which contains $8/_{32}$-in. threads. The osmometer cell is also tapped with $8/_{32}$-in. threads as shown in Fig. 6.2. The coupler threads are then filled with solder and then smoothed and cleaned. The coupler is then rethreaded two turns with an $8/_{32}$-in. die. It can then easily be screwed into the osmometer cell after filling and the solder-filled threads are self-sealing. In Wagner's all-glass modification a specially-ground-glass joint (7/25) was found to be satisfactory if properly sealed. This probably requires a good deal of care and some experience on the part of the operator.

The Schulz osmometer is assembled by first placing the membrane on the membrane support. It is perhaps essential to point out that care must be taken that the membrane surface remains wet with solvent throughout the assembly

Detailed drawing of osmometer

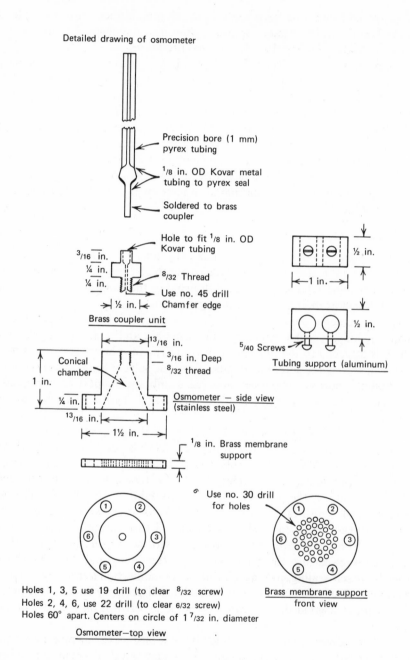

Precision bore (1 mm) pyrex tubing

$1/8$ in. OD Kovar metal tubing to pyrex seal

Soldered to brass coupler

Hole to fit $1/8$ in. OD Kovar tubing

$3/16$ in.
$1/4$ in.
$1/4$ in.
$1/2$ in.

$8/32$ Thread

Use no. 45 drill
Chamfer edge

Brass coupler unit

$1/2$ in.
1 in.

$1/2$ in.

$5/40$ Screws

Tubing support (aluminum)

$13/16$ in.

Conical chamber

$3/16$ in. Deep
$8/32$ thread

1 in.

$1/4$ in.

Osmometer – side view
(stainless steel)

$13/16$ in.

$1\frac{1}{2}$ in.

$1/8$ in. Brass membrane support

Use no. 30 drill
for holes

Holes 1, 3, 5 use 19 drill (to clear $8/32$ screw)
Holes 2, 4, 6, use 22 drill (to clear 6/32 screw)
Holes 60° apart. Centers on circle of $1\,7/32$ in. diameter

Osmometer—top view

Brass membrane support
front view

**Fig. 6.2**  Components of the Schulz osmometer [66].

334

of any osmometer. If the membrane is allowed to dry its permeability may be significantly reduced and it will become brittle. Wagner [65] suggests that the membrane rest on several thicknesses of ash-free filter paper to ensure a rigid support. If the membrane is cut so that it overlaps the assembly screw holes in the membrane support, these holes should be cut into the membrane and filter paper (if used). This prevents tearing the membrane when the assembly screws are inserted. After properly placing the membrane, the cell is fastened to the membrane support and the assembly screws tightened just beyond finger tight.

The interior of the assembled cell is then rinsed several times with the solution whose osmotic pressure is to be determined. Then the cell is filled, taking care to expel any air bubbles by lightly tapping with a solid object. The capillary tube is then filled to the appropriate height by drawing the solution up through the brass coupler, and is then screwed into the cell block, again exercising care not to introduce or entrap air bubbles. The exterior of the assembled cell is then rinsed several times with solvent to remove residual solution. For the Schulz osmometer it is customary to use a 500-ml graduated cylinder or some container of similar dimensions as a solvent reservoir. The osmometer may be suspended from the top by means of a small hook or as Yanko [66] has done, the assembly screws can serve as a stand if made sufficiently long. It is important that the apparatus be vertical in its final position.

As the osmometer is placed in the solvent reservoir air will be trapped in the holes beneath the membrane support plate. This is removed by tilting the reservoir to 45° and moving the osmometer rapidly in an up and down motion. During this operation the membrane should be protected from pressure fluctuations by placing a finger over the capillary tube. The entire assembly can then be placed in a thermostated water bath for temperature control.

Osmotic equilibrium is achieved when a sufficient quantity of solvent passes through the membrane so that the hydrostatic head in the capillary tube is sufficient to prevent further flux. In determining the osmotic pressure, it is of course necessary to correct for capillarity due to surface tension of the liquid. This can be done as a separate experiment using the same capillary or simultaneously using a matched capillary of identical bore. The latter method is used by Yanko and the matched capillary is mounted parallel to the osmometer capillary.

The Schulz design is sufficiently compact and easy to assemble that it readily lends itself to multiple installations. By using, for example, five assemblies simultaneously an entire concentration series for a given sample can be run in about 24 hr. This approaches the rapidity of a single, modern, high-speed membrane osmometer.

## The Zimm-Myerson Osmometer

The Zimm-Myerson osmometer, first described in 1946, has simplicity of design in common with the Schulz osmometer. It also has several features which can be advantageous. The components of this design are shown in Fig. 6.3. The capillary tube as well as a reference capillary are an integral part of the osmometer cell, eliminating the possibility of a leak at a connection. In addition the osmometer employs two membranes, one on either side of the body of the cell thus increasing the ratio of membrane surface area to solution volume. This is a desirable feature of course and proportionally reduces the time required to achieve equilibrium. The metal membrane supports can be drilled or cut in a variety of patterns to expose the membrane to the solvent chamber. Some difficulty has been experienced achieving a satisfactory seal routinely using the conventional screw arrangement. This problem has been circumvented by using a modified C-clamp to hold the membrane support plates to the cell body [68]. With this arrangement the pressure is applied evenly to every point on the circumference of the cell.

After filling, the cell is sealed by placing the metal rod in the filling tube. A more effective seal can be achieved by adding a drop of mercury at the top. The same type of solvent chamber is described for the Schulz osmometer is useful for the Zimm-Myerson and again its simplicity suggests the simultaneous employment of several such devices.

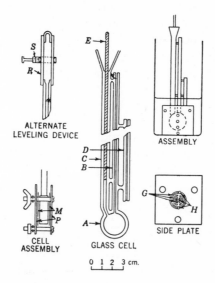

Fig. 6.3   The Zimm-Myerson osmometer [67].

## The Fuoss-Mead Osmometer

The Fuoss-Mead design differs fundamentally from other classical osmometers. This instrument consists of symmetrical half-cells with the membrane oriented in the vertical plane. Each of the half-cells is constructed of stainless steel and is held together about the membrane by machine screws around the periphery. The assembled cell is shown in Fig. 6.4. The face of the cell proper is machined to give a ½-in. flat ring, 5-in. outside diameter. This serves as the outer dimension of the cell. Inside the ring a set of concentric grooves 2-mm wide and 2-mm deep are cut which are connected by a vertical cut running from the inlet to the outlet of each half-cell. The machined face of the half-cell is shown in Fig. 6.5. Alignment pins on each face ensure perfect registration of the two half-cells. After assembly the osmometer is thermostated using dry insulation such as cotton waste.

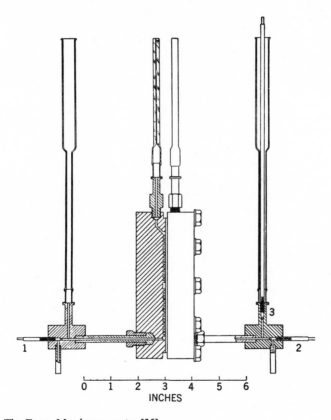

**Fig. 6.4** The Fuoss-Mead osmometer [35].

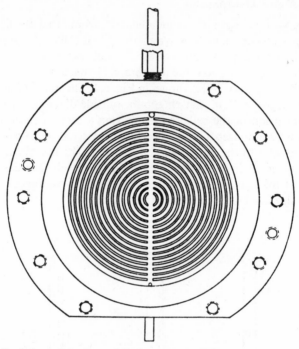

**Fig. 6.5**　Half-cell face of the Fuoss-Mead osmometer [35].

The static-elevation or equilibrium technique can be used with this osmometer, however, it was designed for the dynamic technique (half-sum method). The valves and capillary tubes are arranged so that the liquid level in the left capillary (Fig. 6.3) can be adjusted to the desired position and the liquid level in the right capillary noted as a function of time using a cathetometer.

### The Automatic Osmometer

The advent of commercially-available, high-speed membrane osmometers has significantly enhanced the use of osmometry as a tool for routine molecular-weight determinations. Such devices allow osmotic-pressure determinations to be done in a matter of minutes (10–60) rather than hours. In addition to the obvious advantages of rapid data acquisition, such measurements are less subject to error due to solute permeation through the membrane. There are at present three commercial designs available. Each of these will be discussed individually.

The principle of high-speed osmometry is that solvent flux across the membrane is detected by some sensing device and a pressure change is automatically induced to reduce the net flux to zero. Consequently the condition of

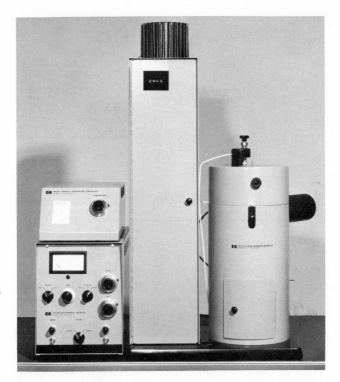

**Fig. 6.6** Mechrolab instrument.

zero net flux is achieved quite rapidly and the osmotic pressure is determined by the induced pressure necessary to achieve such a state. The commercial instruments vary primarily in the methods employed to detect solvent flux and to change the pressure applied to the system.

### The Mechrolab Design

This instrument is manufactured by the Hewlett-Packard Company* and was the first such device to be made available commercially. A photograph of the instrument is shown in Fig. 6.6 and a schematic in Fig. 6.7.

The Mechrolab design employs an optical system to detect solvent flow through the membrane and adjusts the pressure to achieve zero flow through an electromechanical servo system. The solvent side of the osmometer is in communication with a solvent reservoir through a glass capillary tube. The solvent reservoir, the position of which controls the pressure on the solvent side of the chamber, is fixed to the servo-screw. Prior to clamping the membrane in the instrument, the solvent system is filled and a small air bubble is

---

* Hewlett-Packard Company, Avondale, Pa.

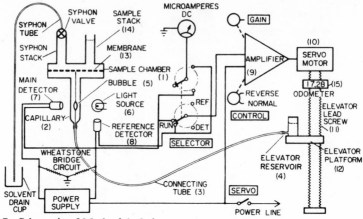

**Fig. 6.7**  Schematic of Mechrolab design.

injected into the capillary tube by means of a Hamilton microsyringe. The bubble passes through the capillary to a bubble trap and then the membrane is clamped into position. The solvent reservoir is then run to the top of the servo-screw by means of a switching device thus placing a positive pressure on the solvent side of the osmometer. This causes the bubble to rise into the capillary. Ideally the bubble should be about ½-in. long in organic solvents. The position of the upper meniscus of the bubble is detected by a light source and a photodetector oriented in a plane perpendicular to the capillary. In the normal mode of operation the servo-screw automatically positions the solvent reservoir so as to maintain the bubble meniscus in a constant position. This of course represents zero flow across the membrane.

The position of the solvent reservoir with solvent on both sides of the membrane is taken as the reference point, in other words, zero osmotic pressure. The reservoir position is displayed digitally to the nearest 0.01 cm. The difference between this value and the position with solution in the solution chamber is taken as the osmotic pressure in centimeters of solvent. The output from the servo system can also be displayed on a strip-chart recorder. This is very convenient for detecting equilibrium and noting the variation in osmotic pressure with time due to solute diffusion, thus extrapolation back to zero time is greatly facilitated.

The cell volume on the solvent side of the osmometer is only 0.02 ml, the solution side is approximately 0.3 ml. This volume ratio allows ready detection of diffusion through the membrane. A distinct disadvantage in the design, however, is that the solvent chamber cannot be flushed with fresh solvent. In the event the solvent becomes contaminated due to solute diffusion or solvent degradation, the osmometer must be completely disassembled, the solvent chamber flushed, a new bubble injected, a new membrane installed, and the instrument reassembled.

The instrument is available in three models which vary primarily in their operating temperature ranges. The pressure range for each is approximately 20 cm of solvent. Model 501 operates from ambient to 65°C, Model 502 from ambient to 130°C, and Model 503 from 5–65°C. The latter device is designed for aqueous and biological applications. The inside of the chambers and the capillary bore have been treated to reduce the interfacial surface tension, allowing the bubble to move up and down the capillary more freely. In general, however, the device is not very satisfactory for aqueous work [51].

### The Shell Design*

This design is manufactured under license by Hallikainen Instruments.† A photograph is shown in Fig. 6.8. The instrument is considerably more compact than the Mechrolab design and has a built-in strip-chart recorder.

The Shell design employs a horizontal membrane, as do the other two commercial versions. It is the only one of the three, however, with the solution on the lower side of the membrane. The sensing element in this instrument is located in the solution side of the half-cell and consists of a pressure-sensing diaphragm. The diaphragm is constructed of 0.0015-in.-thick beryllium copper spaced 0.002 in. from a stationary electrode by means of synthetic rubies. Displacement of the diaphragm by solvent flux is detected as an electrical capacity change by the electrode which is in a $R$-$F$ oscillator circuit. The capacity-sensing oscillator drives a servo-motor which, through a sprocket-and-chain device, adjusts the height of a plummet in a manometer tube thus adjusting the solvent head so as to restore the diaphragm to its original position. The instrument has a digital output of solvent head, a range of 10 cm, and a mechanical reproducibility of 0.02 cm.

In this instrument the solution side of the chamber is a closed system, accomplished by means of leakproof valves at the inlet and outlet. For this reason precise temperature control is critical to successful operation; more so than in the Mechrolab version. The instrument is provided with an electronic thermoregulator which controls the rate of change of temperature to less than 0.001°C/min.

The Shell design requires about 8 ml of solution for a measurement, 6 ml of which is used to flush out the previous sample. The Mechrolab design requires only 1 ml for the total operation. In the Shell design, however, the solvent side is readily flushed with clean solvent by means of a syringe.

---

* Shell Development Company.
† Hallikainen Instruments, Richmond, Calif.

**Fig. 6.8** Hallikainen instrument.

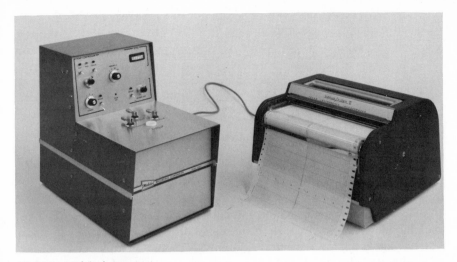

**Fig. 6.9**  Melabs instrument.

The standard model is capable of operating at any temperature between 35–135°C. At temperatures above 65°C an auxiliary sample funnel heater is required.

Some users will find the instrument unsatisfactory as the beryllium–copper diaphragm is slowly attacked by chlorinated solvents at elevated temperatures [51]. In some cases it has been found that elevated temperatures induce a permanent change in the diaphragm and it must be replaced for subsequent operation at a lower temperature. As with the Mechrolab design operation with aqueous systems is considerably less satisfactory than with organic solvents.

### The Melabs Osmometer*

This instrument is the latest version of the high-speed osmometers to be made available commercially. The employment of solid state electronics provides considerably greater compactness than the other models discussed. A photograph of the instrument is shown in Fig. 6.9 and Schematic in Fig. 6.10.

In this design the horizontal membrane is again employed with the solvent on the lower side. The solvent side is closed to the atmosphere and is in contact with a stainless-steel diaphragm connected to a strain gage. Solvent flux through the membrane is accompanied by a deflection of the diaphragm and the output from the strain gage is recorded directly on a 1-mv potentiometric strip-chart recorder. The design is therefore unencumbered by a servo system. Equilibrium is achieved when the pressure change is the solvent side, due to

* Manufactured under license to Theodore Reiff, M.D., by Melabs, Inc., Palo Alto, Calif.

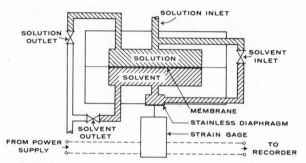

**Fig. 6.10**   Schematic of Melabs instrument.

solvent flux from the closed system, is equal to the osmotic pressure. Unlike the other devices however there is no digital display of the signal and since a strain gage is employed it must be calibrated. This is accomplished by means of a calibrated glass stack and setting full-scale deflection of the recorder for the solvent head in the stack. The hydrostatic pressure for full-scale reflection can be any value less than 100 cm $H_2O$. This provides a wide range of operating pressures particularly useful in biomedical applications.

Each chamber of the osmometer has a volume of 0.5 ml and can be readily flushed with new solution and solvent. Two models are available whose operating temperatures are 40–130°C, and 5–130°C, respectively. This instrument is more suited to aqueous applications than either the Mechrolab or Shell designs.

One of the added advantages to the use of commercial osmometers is that the manufacturer provides a detailed description of how the instrument should be operated to give maximum reliability. With conventional osmometers a considerable amount of laboratory experience and expertise are required in order to obtain reliable osmotic pressure data on a day to day basis.

### References

1. Paul J. Flory, *Principles of Polymer Chemistry*, Cornell University Press, Ithaca, New York, 1953, p. 272.
2. W. G. McMillan and J. F. Mayer, *J. Chem. Phys.* **13,** 276 (1945).
3. Flory and Krigbaum, *J. Chem. Phys.* **18** (8), 1086 (1950).
4. E. A. Moelwyn-Hughes, *Physical Chemistry*, American Elsevier Publishing, New York, 1967, p. 204.
5. T. L. Hill, *J. Chem. Phys.* **30,** 93 (1959).
6. M. L. Huggins, *J. Chem. Phys.* **9,** 440 (1941); *J. Amer. Chem. Soc.* **64,** 1712 (1942).
7. P. J. Flory, *J. Chem. Phys.* **9,** 660 (1941); **10,** 51 (1942).

8. G. Delmos, D. Patterson, and T. Somcynsky, *J. Poly. Sci.* **57,** 79 (1962); *Trans. Faraday Soc.* **58,** 2116, 2117 (1962).
9. C. H. Baker, W. B. Brown, G. Gee, J. S. Rowlinson, D. Stickley, and R. E. Yeadon, *Polymer* **3,** 215 (1962).
10. A. A. Tager, *Kolloid. Zh.* **15,** 6910 (1953).
11. D. Patterson and A. A. Tager, *Vysokomol. Soyed.* A9 (8), 1814–1825 (1967).
12. W. H. Stockmayer and E. F. Cassassa, *J. Chem. Phys.* **20,** 1560 (1952).
13. S. H. Maron, *J. Poly. Sci.* **38,** 329 (1959).
14. S. H. Maron and N. Nakajima, *J. Poly. Sci.* **42,** 327 (1960).
15. S. H. Maron and N. Nakajima, *J. Poly. Sci.* **40,** 59 (1959).
16. S. H. Maron, M. Wang, and N. Nakajima, *J. Poly. Sci.* **46,** 333 (1960).
17. S. H. Maron and N. Nakajima, *J. Poly. Sci.* **47,** 157 (1960).
18. L. C. Cerny, T. E. Helminiak, and J. F. Meir, *J. Poly. Sci.* **44,** 539 (1960).
19. L. C. Cerny and D. M. Stasiw, *J. Macromol. Sci.-Phys.* **B3,** (2) 293 (1969).
20. Hans Coll and F. H. Stross, *Conference on the Characterization of Macromolecular Structure,* Warrenton, Va., April, 1967.
21. A. J. Staverman, D. T. F. Pals, and C. A. Kruissink, *J. Poly. Sci.* **23,** 57 (1957).
22. A. J. Staverman, *Rec. Trav. Chim.* **70,** 345 (1951); **71,** 673 (1952).
23. A. J. Staverman, *Ind. Chim. Belge* **18,** 235 (1953).
24. M. Hoffman and M. Unbehend, *Makromol. Chem.* **88,** 256 (1965).
25. B. E. Hudson, presented at 152nd Meeting ACS Division of Polymer Chemistry, September 1966.
26. J. W. Gibbs, *Trans. Conn. Acad. Sci.* **3,** 228 (1876).
27. F. G. Donnan, *Z. Electrochem.* **17,** 572 (1911).
28. F. G. Donnan, *Chem. Rev.* **1,** 73 (1924).
29. G. Scatchard, A. C. Batchelder, and A. Brown, *J. Amer. Chem. Soc.* **68,** 2320 (1946).
30. G. Scatchard, A. C. Batchelder, A. Brown, and M. Zora, *J. Amer. Chem. Soc.* **68,** 210 (1946); G. Scatchard, A. Gee, and J. Weeks, *J. Phys. Chem.* **58,** 783 (1954).
31. G. Scatchard, Y. V Wu, and A. L. Shen, *J. Amer. Chem. Soc.* **81,** 6104 (1959).
32. D. W. Kupke, *Adv. Protein Chem.* **15,** 57 (1920).
33. P. Doty and J. T. Edsall, *Adv. Protein Chem.* **6,** 35 (1951).
34. J. T. Edsall, *The Proteins,* H. Neurath and K. Bailey, Eds., Vol. 1, Academic Press, New York, 1953, p. 549.
35. R. M. Fuoss and D. J. Mead, *J. Phys. Chem.* **47,** 59 (1943).
36. H-G. Elias, presented at *Conference on Macromolecular Structure,* Warrenton, Va., April, 1967.
37. C. R. Masson and H. W. Melville, *J. Poly. Sci.* **4,** 323 (1949).
38. F. Urban, *Rev. Sci. Instr.* **5,** 375 (1934).
39. M. Sumwalt and E. M. Landis, *J. Kal. Clim. Med.* **22,** 402 (1936).
40. H.-G. Elias, *Chem. Ing. Tech.* **33,** 359 (1961).
41. D. B. Bruss and F. H. Stross, *J. Poly. Sci.* **55,** 381 (1961).
42. D. B. Bruss and F. H. Stross, *J. Poly. Sci.* **A1,** 2439 (1963).
43. O. Hepp, *Z. Ges. Exp. Med.* **99,** 709 (1936).
44. D. B. Bruss and F. H. Stross, *J. Poly. Sci.* **55,** 381 (1961).

45. D. B. Bruss and F. H. Stross, *Anal. Chem.* **32**, 1456 (1960).
46. H. J. Phillip, *J. Poly. Sci.* **6**, 371 (1951).
47. C. E. H. Bawn, R. F. J. Freeman, and A. R. Kamaliddin, *Trans. Faraday Soc.* **46**, 862 (1950).
48. R. L. Arnett and R. Q. Gregg, *J. Phys. Chem.* **74**, 1593 (1970).
49. L. Valtasaari, *Makromol. Chem.* **131**, 313 (1970).
50. H. J. Philipp and C. F. Byosk, *J. Poly. Sci.* **6**, 383 (1951).
51. Gerald M. Armstrong, "Critical Evaluation of Commercially Available Hi-Speed Membrane Osmometers," presented at International Symposium on Polymer Characterization, Battelle Memorial Institute, Columbus, Ohio, November, 1967.
52. ArRo Laboratories, Inc., Joliet, Illinois; Carl Schleicher and Schuell, Keene, N. H.; J. V. Stabin, 84-21 Midland Parkway, Jamaica, Queens, N. Y.; Utopia Instrument Co., Box 27, Mokena, Ill.
53. W. Schneider, *Kunststoffe* **50**, 166 (1960).
54. E. H. Immergut, S. Rollin, A. Salkind, and H. F. Mark, *J. Poly. Sci.* **12**, 439 (1954).
55. H.-G. Elias and T. A. Ritscher, *J. Poly. Sci.* **28**, 648 (1948).
56. T. A. Barr, *J. Chem. Phys.* **25**, 669 (1956).
57. K. H. Hellwege, W. Knappe, and G. Muh, *Kolloid-Z.* **174**, 46 (1964).
58. E. H. Immergat, B. G. Ranby, and H. F. Mark, *Ind. Eng. Chem.* **45**, 2483 (1953).
59. *Instruction Manual*, Hallikainen Automatic Osmometer.
60. N. F. Burk and D. M. Greenberg, *J. Biol. Chem.* **87**, 197 (1930).
61. H. B. Oakley, *Biochem. J.* **30**, 668 (1936).
62. I. Jullander, *Arkiv Kemi, Mineral. Geol.* **A21** (8), (1945).
63. B. Enoksson, *J. Poly. Sci.* **3**, 314 (1948).
64. G. V. Schulz, *Z. physik. Chem.* **A176**, 317 (1936).
65. R. H. Wagner, *Ind. Eng. Chem., Anal. Ed.* **16**, 520 (1944).
66. J. A. Yanko, *J. Poly. Sci.* **19**, 437 (1956).
67. B. H. Zimm and I. Myerson, *J. Amer. Chem. Soc.* **68**, 911 (1946).
68. E. Riesel and A. Berger, *J. Poly. Sci.* **37**, 337 (1959).

Chapter **VII**

# CALORIMETRY

Julian M. Sturtevant

# 1 INTRODUCTION

## Scope

Calorimetry as applied to chemistry is the measurement of the energy changes taking place in chemical processes. The purpose of this chapter is to indicate important applications of calorimetry in the various areas of chemistry and to outline briefly the experimental methods involved. Since numerous types of chemical processes have been studied from the point of view of energetics, it is to be expected that a very wide range of calorimetric applications and methods has been developed. It is thus not possible to present here a complete discussion of all of these, and the selection of topics for treatment will undoubtedly reflect to some extent the interests and prejudices of the author.

It is fortunately possible to adopt here a rather different approach from that employed in the discussions of calorimetry in the previous editions of this series. There have appeared in recent years a number of excellent comprehensive reviews of various aspects of calorimetry, so that our discussion here can in many instances be limited to indicating significant areas of application and giving references to appropriate detailed presentations of methodology. These reviews are listed in the General References at the end of this chapter.

## First Law of Thermodynamics

All chemical and physical processes involve energy changes. The direct evaluation of these changes is the aim of the methods of calorimetry. Those energy changes which accompany changes of state and chemical reactions form the subject matter of thermochemistry.

According to the law of the conservation of energy, or the First Law of Thermodynamics, the energy of a system is a state function, that is, it is a single-valued function of the state of the system. Therefore the energy change associated with the change of the system from one state to another is dependent only on the initial and final states, and not on the path followed by the system in going from the initial to the final state. Thus we may evaluate the energy change associated with the process $A \rightarrow C$ by summing the changes corresponding to the processes $A \rightarrow B$ and $B \rightarrow C$; or, if it is more convenient, those corresponding to the processes $A \rightarrow B'$ and $B' \rightarrow C$. This statement of the First Law of Thermodynamics, in its applications to calorimetry, is frequently called the law of Hess [1].

It is not possible to define the total absolute energy of a system; we are concerned only with changes in energy content. Thus when we speak of the energy content of a system in a certain state, we mean the energy change associated with bringing the system to that state from some reference state in which the energy content is arbitrarily set equal to zero.

Since we shall be interested in changes of energy accompanying changes in state, it is at once obvious that the greatest care should be exercised in defining the initial and final states under consideration with sufficient precision to confer on the corresponding energy change a definiteness consistent with the experimental accuracy achieved in measuring it. All too frequently in the past, otherwise precise thermochemical work has been marred by the failure to describe with appropriate detail the states or system involved.

If, in changing from an initial to a final state, a system performs work against mechanical or other forces, the heat effect accompanying the change is not equal to the total energy change. Since calorimetric measurements are directly concerned only with heat changes, it is essential to have a clear understanding of this point. Suppose the quantity of heat absorbed by the system in some process is $Q$, and the quantity of work performed by the system is $W$. Then the increase in energy accompanying the process is

$$\Delta E \equiv E_2 - E_1 = Q - W, \tag{7.1}$$

where $E_1$ and $E_2$ are the energy contents, referred to some arbitrary zero, of the initial and final states, respectively. We shall deal only with processes in which the work terms arise from expansion or contraction of the system under a finite pressure; for a reversible process the term describing the work of expansion or contraction is

$$W = \int_{V_1}^{V_2} P dV, \tag{7.2}$$

where $P$ and $V$ are the pressure and volume of the system, respectively. In the important special case of a reversible process taking place at constant pressure, (7.2) becomes

$$W = PV_2 - PV_1 = P\Delta V. \tag{7.3}$$

Thus the heat absorbed by a system during such a process is

$$Q = \Delta E + P\Delta V = (E_2 + PV_2) - (E_1 + PV_1). \tag{7.4}$$

It is convenient to represent the property $E + PV$ by $H$; this quantity is called the *enthalpy*.

In view of the fact that the change in energy or enthalpy associated with a given process is dependent only on the initial and final states, it is clear that it is permissible to speak of the change of enthalpy ($\Delta H$), for example, when a system changes from an initial state at one pressure to a final state at some other pressure. In other words, changes in the energy content $E$ and the enthalpy $H$ are not necessarily restricted to constant volume and constant pressure processes. For the important special case in which the initial and final volumes are equal, it is customary to speak of the heat absorbed "at constant volume."

$$Q_V = E_2 - E_1 = \Delta E \tag{7.5}$$

even though the volume may undergo some changes during the process. Likewise, if the initial and final pressures are the same, we speak of the heat absorbed "at constant pressure:"

$$Q_P = H_2 - H_1 = \Delta H = \Delta E + P\Delta V \tag{7.6}$$

without in any way implying that the pressure remains constant at all times during the process. For more complete discussions of the First Law, see the standard texts on thermodynamics.

## Thermochemistry

Interest in thermochemistry was stimulated in the middle of the last century by the belief that the heat change during a reaction is a measure of the affinity of the reacting substances for each other, and that a process will take place spontaneously only if it is accompanied by evolution of heat. The extensive thermochemical investigations of Thomsen and of Berthelot were undertaken primarily from this point of view. As a result of the development of the law of mass action by Guldberg and Waage, and the extremely important theoretical deductions of Gibbs and of Helmholtz, it became evident that the requirement for the spontaneity of a change in a system is that it be accompanied by a decrease in *free energy* rather than in enthalpy. Since there is no general parallelism between enthalpies and free energies, so that the former are of no direct value in predicting the direction of spontaneous change, the subject of thermochemistry suffered neglect for an extended period of time.

Interest in thermochemistry was revived after the formulation of the theorem of Nernst (1906), which indicated how thermochemical measurements, of enthalpies (Section V, p. 386), and heat capacities (Section IV, p. 371), could be employed to evaluate free energies, and thus to accomplish the purposes which Thomsen and Berthelot had hoped to accomplish by observations of enthalpy changes alone. A more recent factor in renewed interest in thermochemistry is the fact that changes in enthalpy are more closely related to the energy values deduced from the majority of theoretical considerations of molecular structure than are changes in free energy, since the latter contain entropic contributions concerning which the theoretical treatments give no information.

## Indirect Evaluation of Thermochemical Quantities

Many of the changes in enthalpy accompanying chemical processes which are directly measured by the methods of calorimetry may also be evaluated from other types of measurements. Thus the molar heat of vaporization, $\Delta H_{vap}$, of a substance is related to its vapor pressure, $P$, by the Clapeyron equation,

$$\frac{dP}{dT} = \frac{\Delta H_{vap}}{T\Delta V}, \tag{7.7}$$

where $\Delta V$ is the increase in volume when one mole of substance is vaporized, and $T$ is the absolute temperature. This relation, or others derived from it, can evidently be utilized to evaluate $\Delta H_{\text{vap}}$ from measurements of $P$ over a range of temperatures.

The variation with temperature of the equilibrium constant of a chemical reaction gives, by means of the van't Hoff equation,

$$\left(\frac{\partial \ln K}{\partial (1/T)}\right)_P = -\frac{\Delta H^0}{R} \tag{7.8}$$

a value for the standard enthalpy change, $\Delta H^0$, accompanying the reaction. This is the enthalpy change when all reactants and products are in their standard states. $\Delta H^0$ will be given in cal/mole if the gas constant, $R$, is expressed in cal/(mole)(deg). The equilibrium constant is related to the standard free energy change for a reaction by the equation

$$\Delta G^0 = -RT \ln K. \tag{7.9}$$

It is important to note that while both $\Delta G^0$ and the standard *entropy* change for a reaction,

$$\Delta S^0 = \frac{\Delta H^0 - \Delta G^0}{T}, \tag{7.10}$$

in general have values which depend on the choice of standard states, the value of $\Delta H^0$ is independent of this choice and furthermore is very nearly equal to the $\Delta H$ observed calorimetrically in a dilute system. If we employ the following general representation for a chemical reaction,

$$-\nu_A A - \nu_B B - \ldots = \nu_P P + \nu_Q Q + \ldots, \tag{7.11}$$

then

$$\Delta G = \Delta G^0 + RT\Sigma\nu_i \ln a_i \tag{7.12}$$

$$\Delta S = -\partial(\Delta G)/\partial T = \Delta S^0 - R\Sigma\nu_i \ln a_i - RT\Sigma\nu_i\partial(\ln a_i)/\partial T \tag{7.13}$$

$$\Delta H = \Delta H^0 - RT^2\Sigma\nu_i\partial(\ln a_i)/\partial T. \tag{7.14}$$

In these expressions the mole numbers, $\nu_i$, are negative quantities for reactants and positive quantities for products, and the $a_i$ are the activities of the various species. The summations are over all the reactants and products. If, as is usually the case, the standard states are chosen so that the activity of each species is equal to its concentration, $c_i$, at infinite dilution, and in the case of solution reactions the activity of the solvent is equal to its concentration at infinite solute dilution, then the free energy and enthalpy changes at high dilution are

$$\Delta G_\infty = \Delta G^0 + RT\Sigma\nu_i \ln c_i \tag{7.15}$$

$$\Delta H_\infty = \Delta H^0 - RT^2\Sigma\nu_i\partial(\ln c_i)\partial T. \tag{7.16}$$

When the concentrations are expressed in weight units, the last term on the right side of (7.16) vanishes and

$$\Delta H_\infty = \Delta H^0. \qquad (7.17)$$

This is very nearly the case when concentrations are expressed in volume units. For any given set of concentrations $\Delta G$ has a definite value; it is thus evident that in any case where $\Sigma \nu_i \neq 0$, the value of $\Delta G^0$ must depend on the concentration units employed. This is also true of $\Delta S^0$ but not of $\Delta H^0$.

Equations 7.8 and 7.9 can be combined in the form of the Gibbs-Helmholtz equation:

$$\left( \frac{\partial (\Delta G^0/T)}{\partial T} \right)_P = \frac{\Delta H^0}{T^2} \qquad (7.18)$$

The applicability of this equation is broader than indicated by (7.18) in that it is not restricted to standard state quantities.

When the van't Hoff equation is applied to an equilibrium involving small molecules, there is no uncertainty as to the amount of material to which the enthalpy change $\Delta H^0$ applies. This is not in general the case with macromolecular systems. It is evident from the form of the van't Hoff equation that the value obtained for $\Delta H^0$ is completely independent of the concentration units used in expressing $K$. Indeed, the equilibrium constant can be expressed in terms of changes in absorbance or other physical properties the relation of which to concentration may be completely unknown. Thus there is no information concerning the actual size of the mole given by the equilibrium experiment, and additional nonthermodynamic information is required to permit specification of the number of g/mole. Such information can be obtained in many instances from calorimetric experiments in which enthalpy changes/unit mass are determined.

A variety of indirect methods for evaluating enthalpy changes can be devised which are essentially based on the van't Hoff equation. For example, if an equilibrium is found experimentally to depend on pH, it follows that protons are liberated or absorbed during the reactions [2, 3]. Consider the simplest possible case,

$$A \leftrightarrows B + nH^+, \qquad (7.19)$$

where $n$ may be of either sign and nonintegral. If at every pH the temperature is selected so that $a_A = a_B$, that is, $T = T_{1/2}$, then

$$\left( \frac{d \ln K}{d \ln a_H} \right)_{T = T_{1/2}} = n, \qquad (7.20)$$

where $a_H$ is the activity of hydrogen ions. Differentiation with respect to $T_{1/2}$ gives

$$\frac{d \ln K}{dT_{1/2}} \cdot \frac{dT_{1/2}}{d \ln a_H} = \frac{\Delta H^0}{RT_{1/2}} \cdot \frac{dT_{1/2}}{d \ln a_H} = n. \tag{7.21}$$

Thus measurements of $T_{1/2}$ as a function of pH $= -0.434 \ln a_H$ lead to an estimate of the enthalpy change for (7.19).

## Definition of Units

The modern calorimetric method consists essentially in measuring the amount of electrical energy necessary to duplicate (or nullify, in the case of an endothermic process) the thermal effect accompanying a physical or chemical process. Modern calorimetric data are therefore primarily expressed in units of electrical energy. Since potential differences are usually measured in absolute volts and currents in absolute amperes, the rational unit of energy is the absolute joule, the product of absolute volts by absolute amperes by seconds.* However, chemists are so accustomed to thinking of thermal energies in terms of the energy required to increase the temperature of a mass of water by a given number of degrees that the calorie has been retained as a unit of energy. In order to free this unit from any dependence on past or future determinations of the specific heat of water, it has been redefined [4] in terms of the absolute joule, 1 cal being set equal to 4.1840 absolute joules (one kcal equals 1000 cal).

Although the calorie continues in wide use, it is probable that its use will decline and that calorimetric and thermodynamic data will increasingly be presented in terms of the absolute joule, or more briefly, the joule. A resolution was adopted in 1948 by the 9th International General Conference on Weights and Measures [5] recommending the joule as the unit of heat and requesting that the results of calorimetric experiments be expressed in joules when possible.

Unless otherwise specified, we use the terms joule for the absolute joule, and calorie for the defined calorie. The temperature at which ice is in equilibrium with air-saturated water under a pressure of 1 atm is taken as 273.16°K, frequently abbreviated 273°K.

## Temperature Measurement and Control

Calorimetric measurements almost universally involve temperature measurements, and in many cases also problems of temperature control. These topics are discussed in detail in Chapter I.

---

* The International Committee on Weights and Measures, Paris, October, 1946, adopted the equivalence: 1 international joule = 1.000165 absolute joule.

## 2   MEASUREMENT OF ELECTRICAL ENERGY [6]

Measurements of electrical energy can be carried out with an exceptional degree of accuracy using readily available instruments. All modern calorimetric measurements are, therefore, in principle if not in fact, referred to measurements of electrical energy, the calorimeter itself serving as an instrument for comparing the chemical or physical energy change in the system under investigation with electrical energy. The energy dissipated in a resistor in an ac circuit cannot be determined with as great accuracy as that in a dc circuit, so that, almost without exception [7], dc circuits are employed in calorimetry. Aside from this consideration, the use of ac would have some advantages; in particular, electrical leakage to the very low voltage dc circuits frequently used in temperature measurement would give less trouble.

The fundamental quantities in measuring electrical energy are potential difference, resistance, and time. If a potential difference, $V$, measured in absolute volts, is impressed on a resistance, $R$, measured in absolute ohms, for a period of $t$ sec, the energy dissipated is $V^2t/R$ absolute joules, or $V^2t/4.1840R$ cal. We shall consider first the methods for measuring $V^2/R$, and then those for measuring time.

### Measurement of $V^2/R$

Potential differences may be determined by comparison, by means of a potentiometer, with a standard cell. The resistance of the calorimeter heater may be determined by a Wheatstone bridge such as those used in resistance thermometry. A more usual procedure is to compare the potential drop across the heater with that across a standard resistor connected in series with it. If the resistance of the standard resistor is $R_s$ and the potential drop across it is $V_s$, it is evident that the electrical power input is $V^2/R = VV_s/R_s$. Both the standard resistor and the calorimeter heater are equipped with four leads, two of which are for measuring the potential drops. The potential leads to the calorimeter heater are attached to the current-carrying leads between the calorimeter and the jacket surrounding it at such a point that the measured potential will include that part of the potential drop in the current leads which supplies energy to the calorimeter. Ordinarily, the proper point of attachment will be approximately halfway between the calorimeter and the jacket. The potential leads can be of relatively fine wire; in deciding on the size of the current leads, consideration must be given to the opposing requirements of minimizing both the energy dissipation in the leads and heat conduction by the leads.

Potential measurements are most conveniently made by means of digital voltmeters, of which there are many now available with accuracies of 0.05% or better and sufficiently high input impedances so that no significant loading of the heater circuit is introduced.

The accurate evaluation of $V^2/R$ by ordinary means requires that $V$ be reasonably constant. A frequently used source of constant low voltage is the lead storage cell. A sufficient number of these are used in a series-parallel arrangement in order to obtain the desired voltage with a current drain on each cell not exceeding about 0.1 A. The batteries should be discharged into a "dummy" resistor of approximately the same resistance as the calorimeter heater for a period of an hour or more before use, so that they will come to a constant voltage. A more convenient source of constant voltage is a regulated power supply of which a large selection is available commercially, with regulation to as little as $\pm 0.005\%$ for $10\%$ line voltage changes.

## Measurement of Time

An ordinary stop watch may be employed for time measurements in many cases. It should be calibrated by comparison over periods of a few hours with an accurate (e.g., electric) clock. It should be established that the calibration of the stop watch is independent of the amount the watch is run down. While a stop watch may be read to 0.2 sec, it is very doubtful that the manual closing and opening of the switch controlling the calorimeter heater can by synchronized with the starting and stopping of the watch, with this degree of precision. The physiological error involved can be removed by mechanical coupling between the switch and the watch. Where the frequency of the ac power supply is sufficiently defined, an electric clock can be employed. An electric stop clock can be conveniently synchronized with the switch which controls the calorimeter heater by using a multipole switch. With a well-constructed switch, the clock circuit will be synchronized with the heater circuit within 0.05 sec or less. If the precision of timing thus obtained is insufficient, recourse can be had to crystal controlled interval timers available from various manufacturers.

Useful procedures have been devised which obviate the need for direct timing of the heating period. For example, Foss and Pitt [8] employed an analog-to-frequency converter and counter in a calibration system having an accuracy of 0.01%. With the counter gated for a precise number of seconds and the converter connected across the calorimeter heater during a non-calibration period, the value of the heater voltage is obtained; during a calibration period the converter is connected across a standard resistor and the counter gating is controlled by the heater switch, so that a value for the number of coulombs passing through the heater is obtained. The product of these two figures then gives the energy dissipated in the heater during the calibration period. It is evident that this system requires a highly constant source of power.

Schemes which involve dividing the heater energy into pulses of constant amplitude which are then counted are outlined in the next section.

## Integration of a Varying Electrical Power

In some situations it is necessary to determine the electrical energy introduced into a calorimeter when the heater voltage and current vary. This requires the use of some form of integrating watt-second meter or joulemeter. Several systems can be thought of in this connection. For example, the voltages across the heater and a resistor in series with it might be amplified, after conversion to ac by a reed converter if dc is used in the heater, and applied, respectively, to the voltage and current coils of a modified commercial watt-hour meter [7].

A scheme [9] developed in the writer's laboratory for use in an automatic energy feedback system gives, under somewhat restricted conditions as to the rate of variation of the heater input power, a precision of the order of a few tenths %. This scheme is somewhat cumbersome, but is inherently reliable in its operation; it is applicable only in the case of a heater of constant resistance with dc applied to it. The varying voltage across the heater is squared by an electromechanical servo computer [10]. The result is integrated by a device which is essentially a rate servo [11], such that the speed of a small motor is made accurately proportional to the output of the squaring computer. A revolution-counter attached to the shaft of the motor gives a reading which is proportional to $\int V^2/R\,dt$. As actually developed, the registration of the revolution-counter is recorded on a strip-chart recorder along with temperature information of importance. A modernized version of this scheme, utilizing solid state electronics, has been described by Clem et al. [12].

Christensen et al. [13] employed pulsed heating in an energy feedback system in which the calorimeter heater received a varying heat input. The output of a voltage-controlled oscillator was frequency divided by a factor of two and used to operate a relay which alternately connected a high-quality thermostated capacitor to a regulated dc supply and discharged it through the heater, the number of pulses being counted to evaluate the total heat effect. A very different type of pulsed system, designed to perform the same function, has been described by Albert and Gill [14].

## 3   GENERAL REMARKS ON CALORIMETRY

Heat energy is difficult to control and therefore to measure. No heat insulator corresponds in perfection to the insulating materials for electrical energy, nor are there conductors of heat energy which do not absorb significant quantities of heat. These and other difficulties have brought about extensive elaboration of calorimetric equipment to meet more exacting requirements of precision, and the development of corrections for the remaining errors. In this section, some of the more important types of calorimeters and calorimetric methods are briefly characterized. Further discussions are given in later sections. For a general treatment, particularly of the theory, of calorimetric methods, the reader is referred to White [15], and McCullough and Scott [16]. The reader is also referred to the list of general references at the end of this chapter.

## Uniformity of Temperature

In general, the temperature of a body can be observed only in a single region, or at best in a few selected regions. Therefore, not only the calorimeter, but also its surroundings, should be at uniform and measurable temperatures, so that heat losses can be accurately evaluated and corrected for. Temperature uniformity is obtained in two ways: by thorough stirring in a liquid; and by taking advantage of the high thermal conductivity of metals, particularly copper. Efficient stirring of a liquid is accomplished in most cases by a propeller, preferably mounted in a tube through which the liquid streams. A reciprocating stirrer, that is, a ring or series of rings which is moved up and down in the liquid, is inefficient if there is a relatively large central obstruction, such as a bomb, in the liquid. In order to avoid the direct connection between a calorimeter and its surroundings by a stirrer shaft, or to permit tight sealing of the calorimeter, stirring can be accomplished by rotating the calorimeter [17], or by a magnetically operated reciprocating stirrer [18]. In a long, thin calorimeter containing a liquid, rapid temperature equilization is obtained even in the absence of stirring [19], but this form of calorimeter is unfavorable from the point of view of thermal leakage to the surroundings because of its large surface–volume ratio. It is important that temperature uniformity be rapidly established after the system has undergone a change. This requires that all material beyond the immediate range of the stirring, or other temperature-equalizing mechanism, should have as low a heat capacity and lag as possible. In particular, metal parts should be used in the construction of the calorimeter wherever possible, and insulating supports should be as light as possible. A strong support for a calorimeter, which is practically free of lag, is furnished by properly spaced loops of nylon thread.

## Liquid and Aneroid Calorimeters

A quantity of heat is in most cases measured by the temperature rise which it produces in a suitable calorimeter. The most common form of calorimeter is a metal shell filled with a stirred liquid. For some purposes, particularly at very high and very low temperatures, an aneroid calorimeter replaces the liquid-filled calorimeter. In the aneroid type, the quantity of heat to be measured raises the temperature of a metal block, usually constructed of copper because of the high thermal conductivity of this metal.

The lower the "dead" heat capacity of the calorimeter, the higher is the temperature change produced by a given amount of heat. Thus, in the determination of heat capacities by observing the temperature change produced by a known amount of electrical energy, the substance under investigation itself usually constitutes the calorimeter filling; in the case of a solid, thin metal vanes are placed inside the calorimeter to aid in heat distribution. In a bomb calorimeter the amount of water surrounding the bomb should be no more than is needed for efficient stirring. In the case of an aneroid calorimeter, the optimum dimensions of the metal block are such as to give sufficiently rapid temperature equalization without undue dead heat capacity.

## Nonisothermal Calorimeters

In nonisothermal calorimeters, heat quantities are estimated by the temperature changes they produce.

### Constant Temperature Environment

The calorimeter is completely enclosed by a jacket of uniform temperature. The necessary degree of uniformity and accuracy of measurement of the jacket temperature depend on the leakage modulus of the calorimeter and on the desired accuracy in the over-all calorimetric measurement [20]. It is convenient, though not necessary, to have the jacket temperature held constant by automatic regulation.

JACKET DESIGN

A constant temperature environment may be supplied by a jacket containing a stirred liquid, or by a relatively massive metal shield. The former type of jacket is employed in most work at ordinary temperatures, the latter type at low or high temperatures. In both cases, temperature uniformity is more readily obtained when the jacket heater is well-distributed throughout the jacket; with aneroid jackets, this is particularly important. Complete distribution of the heating throughout the jacket may be accomplished by Daniels' procedure [21]. The inner and outer (metal) walls of the jacket are insulated from each other. The jacket liquid is an electrolyte of low conductivity. Heating results from an alternating potential applied to the jacket walls. The jacket itself should be reasonable well insulated thermally from its surroundings.

HEAT EXCHANGE BETWEEN CALORIMETER AND JACKET [22]

The heat exchange between the calorimeter and its surrounding jacket has several causes. Those which depend chiefly on the temperature difference between the calorimeter and the jacket (the thermal head) are conduction by solid connections between them, convection and conduction by the intervening gas, and radiation. In addition, thermal leakage from the calorimeter may in part be due to evaporation from the calorimeter, a factor which is not solely dependent on the thermal head. Since the thermal leakage is always determined by direct observation, it is convenient to include in the *apparent* thermal leakage the heat of stirring, which is independent of the thermal head.

Thermal leakage through solid connections, such as supports and electrical connections, should be very small compared with that due to other causes, provided the connections coming in from outside the jacket are in good thermal contact with the jacket before they go to the calorimeter. Evaporation should be eliminated as completely as possible; whenever possible, the calorimeter should be tightly closed. If this is impossible, evaporation is lessened by having the jacket always warmer than the calorimeter.

If the thermal head does not exceed a few degrees, heat exchange due to radiation and gas conduction follows Newton's law. If $T$ is the calorimeter temperature at time $t$ and $T_j$ is the jacket temperature.

$$\frac{dT}{dt} = K(T_j - T), \tag{7.22}$$

where $K$ is the leakage modulus of the system. Heat exchange due to convection does not follow this law. Since it is very difficult to evaluate the thermal leakage unless (7.22) with a constant value of $K$ is followed during an entire experiment, it is important to keep heat transfer due to convection to a minimum. According to White [15], this requires that the air gap (at atmospheric pressure) between calorimeter and jacket should not exceed 1–1.5 cm.

The heat loss during an experiment need be evaluated with lower relative accuracy if it is kept small. There are two general lines of attack in accomplishing this. Either $K$ may be made small, or the thermal head may be kept small. We shall consider here some of the methods of reducing the leakage modulus.

The radiation contribution to the leakage modulus is minimized by having the outside surface of the calorimeter and the inside surface of the jacket highly polished. The conduction contribution [23] is decreased by evacuating the space between the calorimeter and jacket. This is almost universally done in low temperature calorimeters, but seldom at ordinary temperatures, particularly with liquid calorimeters and jackets, because of constructional difficulties. Conduction may also be reduced by using a vacuum-jacketed glass container (Dewar flask) as the calorimeter itself. The chief objection to this type of calorimeter is its relatively high lag, although, in spite of this, it has been employed in very precise work [24]. The leakage modulus can be decreased by increasing the width of the air gap between the calorimeter and the jacket, provided an increase in convection is prevented by placing in the gap one or more very thin, metal convection shields. The heat capacity of these shields must be as small as possible in order to avoid increasing the lag of the calorimeter.

CORRECTION FOR HEAT EXCHANGE

As pointed out by White [15], the term "correction" here is somewhat misleading. It is preferable to consider the unknown quantity of heat as measured in two parts, one of which produces a change in the temperature of the calorimeter. The other part flows either from or to the calorimeter, and its magnitude is deduced from the physical laws governing its flow. This point of view emphasizes the fact that it is possible to carry out precise experiments in a calorimeter which undergoes considerable heat exchange with its surroundings, provided that due care is exercised in controlling and evaluating this heat leakage. Detailed discussions of various procedures for applying corrections for heat exchange have been given by numerous authors [25]. One point which is frequently overlooked may be mentioned: the "rating" periods preceding and following the measuring period, during which the heat-exchange behavior of the calorimeter is observed, should in general be as long in duration as the measuring period itself.

### Adiabatic Method

Heat exchange between the calorimeter and its environment is eliminated if the thermal head is zero at all times. This fact is the basis of the adiabatic method carried to a high degree of development by Richards [26] and his collaborators.

The maintenance of a zero thermal head is an ideal situation which cannot be realized in actual practice. This simple fact has been often neglected in considerations of the adiabatic method. Actually, the observation of the smooth calorimeter-temperature–time curve, which is needed for the evaluation of the heat-leakage correction in the ordinary method is replaced in the adiabatic method by a manual or automatic adjustment of the jacket temperature. It may happen, particularly in experiments of short duration, that the evaluation of the heat-loss correction can be carried out in the ordinary method with greater accuracy than can be attributed to the "elimination" of the heat loss in the adiabatic procedure. In other words, it is usually not the size of the heat-loss term which is important, but the accuracy with which it can be determined.

A very instructive discussion of the relative merits of the ordinary and adiabatic methods has been given by White [27]. We shall mention here four of the advantages of the adiabatic procedure which have led to its very widespread adoption:

1. In the evaluation of the heat-leakage correction in nonadiabatic calorimetry, the assumption is made that the leakage modulus, $K$, remains constant during an experiment. Errors due to variation in $K$, and to deviation of

heat exchange from Newton's law, are reduced in the adiabatic method, since the multiplier of $K$, the thermal head, is small.

2. Maintenance of a small thermal head reduces convection in the air gap. It is thus permissible to use considerably wider air gaps, with correspondingly smaller leakage moduli, in the adiabatic method. This advantage can, however, be realized in the ordinary method, as mentioned above, by employing convection shields.

3. Troubles due to evaporation from an incompletely sealed calorimeter are less serious in the adiabatic method. However, it should again be emphasized that complete elimination of evaporation by tight sealing is always desirable if it can be accomplished.

4. The greatest advantage of the adiabatic procedure comes in its application to prolonged [28] calorimetric experiments (see Section 8). In such cases, the heat-loss correction in the ordinary method might be as large as, or even larger than, the total quantity of heat to be measured, and it could only be evaluated on the basis of very long rating periods. On the other hand, in the adiabatic method, the heat-loss term can be held to a small fraction of the total and correction for any deviation from strict adiabaticity can be based on relatively short rating periods. In fact, with constant stirring heat, the deviation from adiabaticity will, in some cases, be sufficiently constant so that it need be determined only occasionally. This is because, as mentioned above, variations in the magnitude of $K$ are of less importance when the thermal head is always small.

In protracted experiments the change of the temperature of the calorimeter is slow enough that the temperature of the jacket can be held equal to that of the calorimeter much more accurately than in experiments on rapid processes. It is particularly advantageous in such cases to employ automatic regulation of the jacket temperature [29].

It should be noted here that heat-leak calorimeters (Section 3, p. 367), which are purposely constructed with a relatively rapid heat transfer through a many-junction thermel from calorimeter to surroundings, have been successfully applied to the study of very slow processes extending over periods of many hours.

## Twin Calorimeters

For some purposes, the twin calorimeters first employed by Joule [30] and later by Pfaundler [31] are very valuable. The apparatus consists of two calorimeters as nearly identical in construction as possible, supported in as nearly identical surroundings as possible. The calorimeters can be electrically heated by two heaters (usually of equal resistances) either separately or connected in series. If during an experiment in which the heaters are connected in series, the temperatures of both calorimeters change at very nearly the same rate, the thermal leakages of the two calorimeters are very nearly the same. While this does not necessarily mean that the error due to thermal leakage vanishes [32], it does follow that any small residual error can be accurately evaluated by suitable rating periods. When the adiabatic method is applied to twin calorimeters, virtually complete elimination of the need for any heat-leakage correction can be accomplished. Of course, in this case, it is still necessary to make suitable corrections for any differences between the two calorimeters with respect to stirring heats, evaporation effects, and so on.

The chief advantage of twin calorimeters is that they permit relative measurements to be made in such a way that the observations of temperature upon which the results are based are essentially differential observations. It thus becomes possible to utilize most advantageously the properties of thermo-couples as indicators of temperature differences. For example, in the deter-mination of the heat capacity of a solution, one calorimeter contains pure solvent while the other contains solution. When both calorimeters are heated with precisely the same electrical energy input, through a certain temperature interval, a small temperature difference between the calorimeters will result. This difference can be measured with great precision and, after correction for slight differences in heat losses, stirring heats, and so on, leads to a very accurate evaluation of the heat capacity of the solution relative to that of the solvent. This application of twin calorimeters illustrates the fact that they essentially constitute a thermal balance, in which the thermal properties of two different substances may be directly compared, or the heat produced or absorbed in some process may be directly compared with an electrical heat-ing effect.

Twin calorimeters have been very successfully applied to measurements of rapid processes, particularly to heats of dilution in the range of low concen-trations within which the heat effects are very small. Illustrations of these applications will be found in Section 7. Twin calorimeters are especially valuable in the observation of very slow heat effects [33]. In the case of an exothermic process, the tare calorimeter is heated sufficiently to maintain its temperature equal to that of the reaction calorimeter, while, in the case of an endothermic process, the working calorimeter is heated to hold its tempera-ture equal to that of the tare. At the end of the experiment, the two calorim-eters are heated with their heaters connected in series to determine the differ-ence in the heat capacities of the two calorimetric systems. The successful application of this procedure to cases where the heat effect varies with time requires the development [34] of an accurate method for evaluating the in-tegral, $\int_{t_1}^{t_2} (V^2/R)dt$, where $V$, the potential impressed on the heater of resist-ance $R$, is a function of time $t$. The advantage of differential temperature measurements characteristic of twin calorimeters is obtained to some extent with a single calorimeter when the temperature measurements are made with a thermel the reference junctions of which are immersed in a "cold calorimeter" such as that described by White [35].

### Heat-Flow Calorimeters

The adiabatic method, which attempts to make the heat exchange of the calorimeter with its surroundings as small as possible, was discussed above. In contrast to this method, a number of calorimetric equipments have been described in which direct use is made of the heat exchange in evaluating the heat evolution or absorption in the calorimeter or in studying the thermal properties of the material inside the calorimeter. Some of the instruments employing heat flow are properly classed as isothermal, and are discussed in the following section. An early example of a calorimeter of this type is the "radiation" calorimeter of Thomas and Parks [36] which was used for determining specific heats at temperatures up to 350°. A small calorimeter containing the material under study is suspended in a massive copper block the temperature of which is automatically held at a value differing by a constant amount from that of the calorimeter. The rate of change of the calorimeter temperature under such conditions is inversely proportional to the heat capacity of the calorimeter and its contents, provided that errors due to the thermal gradients within the calorimeter can be neglected. The chief advantage to be expected in a system of this type is its relative simplicity, in that no calorimeter heater is required. On the other hand, greater attention must be paid to the control of thermal gradients within the calorimeter and to the reproducibility of the control of the actual temperature difference producing the heat exchange than is needed in the conventional procedures employing electrical heating of the calorimeter.

## Isothermal Calorimeters

### Phase-Change Calorimeters

In one form of isothermal calorimeter, a quantity of heat is measured by the amount of isothermal phase change which it produces in the calorimetric material. The most common type of isothermal calorimeter is the Bunsen [37] ice calorimeter. In its simplest form, the calorimeter consists of a glass tube, $A$ (see Fig. 7.1), sealed into another tube, $B$, which is filled with air-free water and mercury. The bottom of the outer tube connects by means of a small glass tube with a mercury reservoir, $C$, and a capillary tube leading to a removable weighed vessel containing mercury. Tube $B$ is completely immersed in a mixture of ice and water in a Dewar flask. A portion of the water in $B$ is frozen, after it has been cooled to 0°, by temporarily placing a freezing mixture in calorimeter tube $A$. An exothermic reaction occurring in $A$ causes some of the ice in $B$ to melt, giving rise to a contraction in volume and a decrease in the weight of the mercury in $D$. The very slight heat exchange of the

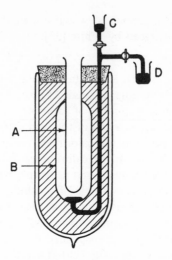

**Fig. 7.1** Ice calorimeter.

calorimeter with its surroundings is determined in a rating period. The apparatus is calibrated by an electrical heater, or by dropping into it a body of known heat capacity previously brought to a known temperature. Because of the isothermal characteristic of the ice calorimeter, it is possible to keep the heat exchange with the surroundings very low. The apparatus is therefore particularly useful in the determination of small and very slow heat effects. Its chief disadvantage is that it can be employed only at 0°. This disadvantage is to a large extent removed by employing other substances in place of ice. Cruikshank et al. [37] list, with references to the literature, various substances which have successfully replaced water in phase-change calorimeters. It may be noted that a calorimeter employing diphenyl ether (mp, 26.9°), for example, is inherently three times as sensitive as one employing water because of the smaller heat of fusion of the organic liquid.

The vaporization of a liquid or the condensation of a vapor may also be employed for the measurement of heat in an isothermal calorimeter. For example, Simon and Ruhemann [38] have described a calorimeter for determining heat capacities at low temperatures by the method of mixtures, the heat given up to the calorimeter by a body of slightly higher temperature dropped into it being measured by the increase in vapor pressure of the liquefied gas, such as hydrogen, contained in the calorimeter. Kraus [39] and his collaborators have determined the heat effects of rapid processes in liquid ammonia at its boiling point by measuring the amount of gas evolved by the heat of the reaction. Stout and Jones [40] determined the heat evolution due to the $\alpha$-decay of plutonium by measuring the evolution of nitrogen gas from a calorimeter containing liquid nitrogen.

Lacher et al. [41] have described a calorimeter in which the heat evolved by a vapor-phase reaction is absorbed in a volatile liquid surrounding the reaction chamber and through which an inert gas is bubbled to cause vaporization. The stream of gas is regulated to maintain a steady state, and the heat of reaction is determined by measurement of the electrical heating needed to duplicate this steady state in the absence of chemical reaction.

## Labyrinth-Flow Calorimeter

Junkers [42] used a flow calorimeter in the determination of the heat of combustion of gases (see Section 5, p. 395). Swietoslawski [43] and his collaborators have further developed this type of calorimeter. The calorimeter proper is located within a labyrinth consisting of several concentric passageways through which water (or other liquid) from a thermostat flows. The design is such as to ensure that a precise measure of the heat effect in a time interval in the body under investigation is obtainable from a knowledge of the rate of liquid flow, the specific heat of the liquid, and the temperature difference between the ingoing and outflowing liquid. The liquid flow is usually made rapid enough to maintain essentially isothermal conditions in the calorimeter, even in cases where the heat effect to be measured varies with time. Under such conditions the heat capacity of the calorimeter does not enter into the computations.

## Quasi-Isothermal Heat-Flow Calorimeter

In cases where the heat absorbed or liberated by a process taking place in a calorimeter is readily exchanged between the calorimeter and a jacket of relatively very large heat capacity, the calorimeter and jacket can be considered to constitute an essentially isothermal system. If the heat exchange can be accurately measured, the heat evolved or absorbed by the process taking place within the calorimeter can be evaluated by measuring the total heat which the isothermal jacket receives from or gives to the calorimeter during the process. For example, if the heat exchange follows Newton's law, the rate of gain of heat by the jacket from the calorimeter is

$$\frac{dq}{dt} = k(T - T_j), \tag{7.23}$$

where $T$ and $T_j$ are, respectively, the temperatures of the jacket and the calorimeter and $k$ is a constant. It is evident that the total heat absorbed by the jacket between times $t_1$ and $t_2$ is

$$\Delta q = k \int_{t_1}^{t_2} (T - T_j)dt, \tag{7.24}$$

where the integration will usually be performed graphically. If $T = T_j = 0$ at $t_1$, and at $t_2$, the heat of the process occurring between these two times is equal to $\Delta q$. If $T - T_j = 0$ at $t_1$, but is different from zero at $t_2$, account must be taken of the heat capacity of the calorimeter and its contents. If this quantity is represented by $C$, the heat of the process is given by $\Delta q - C(T_2 - T_j)$.

It was pointed out by Tian [44] that, if the temperature difference $T - T_j$ is measured by a multijunction thermocouple so designed that essentially all the heat transfer takes place through the thermocouple, (7.23) will be accurately followed. The calorimeters described by Tian and by Calvet have made use of this fact in the calorimetry of very slow processes. It is evident that the principle may also be applied to systems [45] in which only a fraction of the heat exchange takes place through the thermocouple, provided this fraction remains constant. In such a system the efficiency of utilization of the thermocouple output is decreasesd.

The calorimeter originally developed by Tian [44] and extensively improved by Calvet and his co-workers [46] has been employed, in more or less modified form by many workers, especially in recent years [47]. It has been found to be adaptable to a wide range of applications, in both batch and flow (see Sections 7, p. 407 and 8, p. 413) operation, and is now available commercially from Societé D.A.M., Lyon, France.

An interesting calorimeter which employs the heat-flow principle has been described by Kitzinger and Benzinger [48]. They employ a thermel of several thousand junctions, and the geometry of their calorimeter is such that the solution in which reaction takes place is present in a thin layer so that heat transfer to the thermel is relatively rapid. They thus achieve a very efficient utilization of the thermocouple output in the case of processes of short duration, for which the apparatus is particularly well suited. This calorimeter has been further developed by Beckman Instruments, Inc., of Palo Alto, Calif., and marketed by them under the designation Model 190 B Microcalorimeter. A flow modification of this heat-leak calorimeter has been developed by Sturtevant [49].

The heat-flow principle has been employed by Wadsö [50] in the design of a microcalorimeter of the batch type. A similar instrument of the flow type has been described by Monk and Wadsö [51]. Both of these instruments are commercially available from LKB Produkter AB, Stockholm, Sweden.

## Importance of Experimental Calibration

In the earlier days of calorimetry, the heat capacity of a calorimeter was frequently estimated from a knowledge of the masses and specific heats of the materials used in its construction. It cannot be too strongly emphasized, however, that calorimetric measurements should be based on electrical calibration wherever this is possible. Even in cases in which the use of an electrical heater for calibration purposes is impossible, calibration should be accomplished by making observations on a process with a known heat effect. Electrical, or, more generally, experimental, calibration is particularly important because a calorimeter cannot be a perfectly isolated body. The material connections between it and its surroundings have a finite heat capacity, and the only way to make proper allowance for the fraction of this heat capacity which is to be added to that of the calorimeter is through experimental calibration under conditions as nearly as possible like those in the subsequent measurements.

## Calorimeters of Special Types

The variety of designs in calorimetric work is evidence of the fact that it is impossible to state a few simple principles of design which apply to all cases. We shall mention briefly a few calorimeters of special design.

Chall and Doepke [52] have measured heat effects in liquid ammonia at room temperature. Calorimeters for use at more or less elevated temperatures have been described by several authors [53]. Daniels et al. [54], devised a calorimeter having quartz windows with which the heats of photochemical processes can be determined.

Beckett and Cezairliyan [55] have reviewed the methods available for high-speed calorimetric measurements. A number of calorimeters designed for application to rapid processes have been described by Berger and his colleagues [56].

Savage [57] has developed a calorimeter in which one can safely handle alpha-active and pyrophoric materials. Nowicki [58] has published the design of a calorimeter for the accurate measurement of laser output power.

Boyd [59] has described a calorimeter especially designed for the determination of heats of the processes going on in ion exchange columns.

## Automatic Calorimetry

As is true in other fields of scientific activity, there is an increasing tendency to adopt methods of automatic control and measurement in the design of calorimeters, and to utilize forms of data collection and analysis which are directly compatible with digital computers. Experience makes it clear that the advantages to be gained from automatic operation may extend beyond the saving of manpower or relief from tedious activities. Thus the amount of useful information which can be obtained from a given piece of equipment may be significantly increased if a degree of automatic operation is adopted. More important, the reliability of the information may be considerably improved by decreasing the opportunities for the introduction of human errors.

Numberous experimenters have used automatic devices for controlling the temperature of an adiabatic jacket. The twin calorimetric procedure has been made semiautomatic (see Section 4, p. 384) in its operation by having the electrical energy required for balancing the energy of a chemical process continuously and automatically adjusted and recorded.

Stull [60] has developed an automatic adiabatic calorimeter for the determination of heat capacities and the heats of phase transitions in the range 5–330°K. The general calorimetric design of this instrument is similar to that described in Section 4, p. 376. Automatic control of the temperature of the adiabatic shield and the other elements with which the calorimeter proper exchanges heat, automatic programming of a complete run, and automatic recording of pertinent data make it possible to carry through a run extending as long as 100 hr with only occasional operator attention.

## 4    HEAT CAPACITIES

The heat capacity of a system at temperature $T$ is the limit, as $\delta T$ approaches zero, of the ratio $\delta Q/\delta T$, where $\delta Q$ is the amount of heat which must be introduced into the system to increase its temperature from $T$ to $T + \delta T$. In accordance with the discussion in Section I, we must distinguish the cases in which the system is heated at constant volume and at constant pressure. Thus the heat capacity at constant volume is

$$C_V = \left(\frac{\partial U}{\partial T}\right)_V, \tag{7.25}$$

and the heat capacity at constant pressure is

$$C_P = \left[\frac{\partial(U + PV)}{\partial T}\right]_P = \left(\frac{\partial H}{\partial T}\right)_P. \tag{7.26}$$

The unit of mass in these expressions is the mole. If the unit of mass is the gram, $c_V$ and $c_P$ are the specific heats at constant volume and at constant pressure. For an ideal gas $C_P - C_V = R$, the gas constant; in general,

$$C_P - C_V = -T\frac{(\partial V/\partial T)_P{}^2}{(\partial V/\partial P)_T} \tag{7.27}$$

a quantity which is small in many condensed systems.

### Importance

The absorption of energy by a substance as its temperature is increased is the result of increasing the populations in the higher-energy states available for the molecules. At the relatively low temperatures available to heat-capacity measurements, these higher states are states of higher translational and rotational energy of the molecules, and higher energies of internal motions such as vibrations and rotations of atoms and groups relative to each other. In the case of solids, lattice vibrations and rotations are also of importance. It is thus evident that the study of heat capacities is of the greatest importance in connection with theories of molecular structure and of the various states of aggregation of matter.

Several applications of heat-capacity data may be cited to show their great importance, such as the calculation of the temperature coefficients of heats of reaction, the evaluation of entropies and free energies by means of the third law of thermodynamics, the study of a wide variety of transitions ranging from phase transitions in crystalline solids to conformational transitions in macromolecules, and the establishment of purity by determination of the heat capacity of a solid substance just below its melting point.

### Change of Heat of Reaction with Temperature

Since the value of $\Delta H$ for a reaction is the difference between the sums of the heat contents of the reactants and of the products, it follows from (7.26) that

$$\left(\frac{\partial(\Delta H)}{\partial T}\right)_P = \Delta C_P, \tag{7.28}$$

where $\Delta C_P$ is the sum of the molal heat capacities of the products less that of the reactants. This relation is known as the Kirchhoff, or Person-Kirchhoff law. Methods of integrating this equation are discussed in treatises on thermodynamics.

The integrated forms of (7.28) enable one to calculate $\Delta H$ at any temperature from a measurement of the heat of reaction at a single temperature and a knowledge of the heat capacities of the reactants and products.

### Evaluation of Entropies by Means of the Third Law

A restricted statement of the Third Law of thermodynamics is as follows: The entropy of all "perfect" crystals is zero at the absolute zero, and their heat capacities approach zero asymptotically near the absolute zero. A perfect crystal is one in which a geometrical arrangement of atoms is repeated *without modification* throughout the crystal.

In an infinitesimal process carried out reversibly at constant pressure at temperature $T$, the entropy change $dS$ is

$$dS = \frac{dH}{T}. \tag{7.29}$$

It follows from (7.26) that

$$dS = \frac{C_P}{T} \, dT. \tag{7.30}$$

Integration gives for the entropy of the system at temperature $T$

$$S - S_0 = \int_0^T \frac{C_P}{T} \, dT = \int_0^T C_P d \ln T. \tag{7.31}$$

According to the Third Law, if the system under consideration is composed of a pure substance which forms perfect crystals at $T = 0$, then $S_0 = 0$. If the substance undergoes any phase transitions in the interval $T = 0$ to $T = T$, (7.31) must be modified:

$$S = \int_0^{T_1} C_P d \ln T + \frac{\Delta H_1}{T_1} + \int_{T_1}^{T_2} C_P d \ln T + \frac{\Delta H_2}{T_2} + \ldots \quad (7.32)$$

Here $T_1$ is the temperature of the first phase transition, $\Delta H_1$ is the isothermal heat of that transition, and so on. In many cases, several phase transitions will be involved, including transitions between different crystalline modifications, melting, and vaporization.

It is obviously impossible to extend measurements of $C_P$ to $T = 0$, so that an extrapolation has to be performed from some low temperature. In the most reliable determinations of entropy, the heat-capacity measurements are extended down to temperatures obtainable by evaporating liquid helium under reduced pressure. At these low temperatures, the heat capacity is in general very small, and in most cases can be assumed to be proportional to $T^3$. According to theoretical considerations, it is $C_V$ which should be proportional to $T^3$ at very low temperatures. However, in the region in which this proportionality holds, the difference between $C_V$ and $C_P$ is usually negligible. With this assumption it readily follows that

$$\int_0^T C_P d \ln T \approx \tfrac{1}{3} C_P(T), \quad (7.33)$$

where $C_P(T)$ is the value of the heat capacity at some low temperature $T$ near $10°K$. The integrals in (7.32) above the lowest temperature at which measurements are made are evaluated graphically.

If the approximation referred to above is not sufficiently accurate, considerably more elaborate extrapolation procedures must be adopted. These are based on the theoretical treatments of Einstein, Debye, and Born and von Kármán. The reader is referred to discussions such as those by Aston [61] and Stull et al. [62], for details concerning these extrapolation methods.

### Free Energies

From a practical point of view, one of the most important applications of entropies evaluated from heat-capacity data, or by statistical methods, is in the calculation of free energies of formation. The standard entropy of formation of a substance, $\Delta S^0$, is the entropy increase accompanying the formation of one mole of the substance in its standard state from its elements, taken in their standard states. Entropies of formation usually refer to the standard temperature of 25°C. The entropy increase accompanying a chemical reaction with reactants and products in their standard states is readily obtained as the sum of the absolute entropies, or of the entropies of formation, of the products minus the corresponding sum for the reactants.

For any isothermal process, the important relation

$$\Delta G = \Delta H - T\Delta S \qquad (7.34)$$

applies. Here $\Delta G$ and $\Delta H$ are the increases in free energy and enthalpy accompanying the reaction. If the process is a reaction in which all the substances involved are in their standard states, (7.34) becomes

$$\Delta G^0 = \Delta H^0 - T\Delta S^0. \qquad (7.35)$$

This relation is the basis of one of the most important methods of evaluating free energies of formation. Values of $\Delta H$ of formation are usually evaluated from calorimetric data, chiefly from heats of combustion (Section 5). The standard free energy change for any reaction can be computed from free energies of formation in exactly the same way as entropies of reaction.

The equilibrium constant, $K$, for a reversible reaction can be calculated from the standard free energy change for the reaction, and thus also from the changes in enthalpy and entropy accompanying the reaction with the reactants and products in their standard states, by means of the relation

$$-\Delta G^0 = RT \ln K. \qquad (7.36)$$

The choice of standard states used in these considerations must be clearly specified. In connection with heats of formation (Section 5), the standard state usually employed is that state which is stable at 25° and a pressure of 1 atm. Thus for hydrogen the standard state is represented by $H_{2(g,\ 1\ atm)}$. In entropy and free energy calculations, the standard state of a gaseous substance is usually taken as the ideal gas state at unit fugacity. This standard state would more properly be employed also in connection with the heat contents of gases. However, for most gases, this change in standard state corresponds to a difference in heat content which is beyond the limits of present experimental accuracy. The heat content of a gas in the ideal gas state at unit fugacity is the same as the heat content of the real gas at zero pressure.

### Conformational Transitions of Macromolecules

We may illustrate the application of heat capacity measurements to the study of phase transitions and similar phenomena by the estimation of the enthalpies of transition of various macromolecules by means of sensitive differential heat capacity determinations. If a solution of any one of a variety of biopolymers, such as the protein ribonuclease [63] or the polynucleotide poly(dA-dT) [64] the alternating copolymer of deoxyadenylic acid and deoxythymidylic acid), is heated in aqueous medium, the biopolymer undergoes a change in three dimensional structure, in many cases fully reversible, with absorption of heat. This heat absorption manifests itself as a heat capacity anomaly, in these cases extending over temperature ranges from a few degrees to a few tens of degrees. In the case of ribonuclease [63], heat absorption amounting to 6.5 cal/g of protein takes place at pH 2.8 over a temperature range of roughly 35°; actual measurements have to extend from 10° or more below the transition range to 10° above it. If a protein solution of 0.1 % concentration is employed [63], the heat absorption due to the transition will amount to only 0.01 % of the total heat input during the experiment. It is thus evident that extraordinarily sensitive calorimetry is required for this purpose (p. 384). The significance of calorimetric data on conformational transitions in biopolymers has been briefly reviewed by Tsong et al. [63].

### Heat Capacities of Solids Just Below their Melting Points

An important criterion of purity is obtained from measurements of the heat capacity of a substance just below its melting point, since the apparent heat capacity of an impure substance shows a rapid increase below the melting point [65]. This phenomenon is called premelting, since it can be attributed to absorption of heat as a result of the melting of a portion of the material. Figure 7.2 shows a typical heat-capacity curve for such a case. The dashed portion of the lower curve (for the solid state) represents the behavior to be expected for a pure substance. If the heat of fusion is determined by heating the impure substance from $T_1$ to $T_2$, it is of course necessary to deduct from the observed heat input the quantities, $\int_{T_1}^{T_m} C_{P(\text{solid})} dT$ and $\int_{T_m}^{T_2} C_{P(\text{liquid})} dT$. In the first integral, the value of $C_{P(\text{solid})}$ employed must be corrected for premelting.

It can be shown [66] on the basis of the laws of dilute solutions, in cases in which no solid solutions are formed, that the observed specific heat of the solid mixture at $T°K$ is given by

$$C_{P(\text{obs})} \approx C_P + \frac{n_2 R T_0^2}{(T_0 - T)^2} \qquad (7.37)$$

$$\approx C_P + \frac{\Delta H_f(T_0 - T_m)}{M_1(T_0 - T)^2} . \qquad (7.38)$$

In these equations, $n_2$ is the number of moles of solute (impurity)/g of solvent, $T_0$ is the melting point of the solvent (°K), $T_m$ is the temperature at which the solid solvent is in equilibrium with liquid having the composition of the mixture, $\Delta H_f$ is the molar heat of fusion of the solvent, and $M_1$ is the molecular weight of the solvent. If $\Delta H_f$ is measured in calories, $R$ has the value 1.986. These equations are derived on the assumption that equilibrium between solid and liquid phases is maintained during the heat-capacity determination. Experimental [66] results show that it is possible to approach this ideal situa-

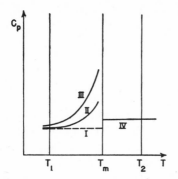

**Fig. 7.2**   Typical premelting behavior: I and IV, pure solid and liquid, respectively; II and III, impure samples.

tion quite closely. In most cases, $C_P$ can be expressed with sufficient accuracy in the form $A + BT$ in the short region below the melting point which is of interest in the present connection. Thus the mole fraction of impurity can be estimated by fitting the observed specific heats to an equation of the form

$$C_{P(\text{obs})} = \frac{A + BT + C}{(T_m - T)^2} \tag{7.39}$$

(since $T_m \approx T_0$). Obviously $n_2 \approx C/RT_m^2$.

It should be noted that (7.38) shows that premelting heat capacities give a much more sensitive criterion of purity than does the lowering of the melting point. Thus, in the case of water, the observed specific heat at $-1°C$ will exceed the true heat capacity by $80 (T_0 - T_m)$ cal/(deg)(g); an impurity concentration which causes a melting-point lowering of only $0.001°$ will result in an apparent specific heat at this temperature which is approximately $15\%$ larger than the true specific heat. Pure organic substances generally show smaller differences between the specific heats of the solid and liquid forms than does water, so that a sample containing a quite small amount of impurity ($T_0 - T_m$ of the order of $0.1°$) will usually have a specific heat, at temperatures slightly below the melting temperature, greater than that of the liquid form, the specific heat of which is essentially unaffected by small amounts of impurities. This behavior is illustrated in Fig. 7.2. On this basis, Skau [67] has proposed as a "noncomparative criterion of purity" that the specific heat of the solid form just below the melting temperature should be less than that of the liquid form.

### Determination of Heat Capacities

It is evident from (7.25) and (7.26) that the true heat capacity is not usually determined directly by calorimetric means because this quantity is the slope of the energy- or heat-content curve. The quantity determined experimentally is a *mean* heat capacity between two temperatures:

$$\tilde{C}_P = \frac{\Delta Q}{T_2 - T_1} = \frac{H_2 - H_1}{T_2 - T_1} \tag{7.40}$$

in the case of constant pressure. If the curve of $C_P$ versus $t$ is nearly linear between $T_1$ and $T_2$ and $T_2 - T_1$ is sufficiently small, the value of $\tilde{C}_P$ can be identified with $C_P$ at $(T_1 + T_2)/2$. This approximation can be safely made in most cases if $T_2 - T_1$ does not exceed 5°. Due allowance must be made for any change in the distribution of the material between the phases present. Thus, in determining the heat capacity of a liquid, some of the energy introduced is used in vaporizing a small fraction of the liquid [68].

Most of the important procedures for determining heat capacities have recently been very thoroughly discussed in *Experimental Thermodynamics*, Vol. 1, edited by J. P. McCullough and D. W. Scott, and published in 1968. We shall present here only a brief outline of these methods so that a general idea of procedures can be obtained. Anyone proposing experimentation in this area should study carefully the references given here.

### Determination of Heat Capacities at Low Temperatures

Much of the work at low temperatures has been carried out by the Nernst [69] method. The material under investigation is contained in a metal calorimeter which is equipped with a heater and a thermometric device (either resistance thermometer or thermocouple). The calorimeter is supported inside a jacket of relatively large heat capacity and accurately measurable temperature. This jacket is contained within another vessel (frequently a Dewar flash) which is cooled by liquid air or liquid hydrogen. Provision is made for evacuating the space between the jacket and the calorimeter to a very low pressure to improve the isolation of the calorimeter. In the Nernst type of apparatus, the jacket is held at a constant temperature during a heat-capacity determination, and correction is made for the small heat leakage from the calorimeter. A detailed discussion of this type of calorimeter, with several designs taken from the literature has been presented by Stout [70].

Many experimenters prefer an adiabatic to an isothermal shield. Low temperature adiabatic calorimetry is covered thoroughly by Westrum et al. [71]. A simplified schematic diagram of an adiabatic calorimeter, adapted from the paper by Westrum et al. [71], is shown in Fig. 7.3. The sample under study is contained in the sample vessel or calorimeter, 7, which is equipped with a heater and a thermometer contained in a reentrant well, 8. The sample vessel is suspended within an evacuated submarine, 5, immersed in an appropriate refrigerant, 4, such as liquid nitrogen, contained in a Dewar vessel, 3. The adiabatic shield, 6, reduces the heat leak from the sample vessel and per-

mits operation at temperatures considerably higher than that of the refrigerant. The submarine is evacuated through connection 2, and electrical leads
are brought out at a sealed plate, 1.

The special problems which arise in calorimetry at extremely low temperatures are discussed by Hill et al. [72].

### Determination of Heat Capacities at Moderate Temperatures

Determination of the heat capacities of gases, liquids, and solids will be
considered separately. The classification of experimental methods according
to temperature is quite arbitrary, and most of the methods considered in these
paragraphs can be extended to low or high temperatures by suitable modifications in design and procedure.

GASES AND VAPORS

*Heat Capacity at Constant Pressure.* The continuous-flow method used by
Callendar and Barnes [73] for liquids has been extensively employed in the
determination of the heat capacities, at constant pressure, of gases and vapors.
Its adaptation to this purpose was first undertaken by Scheel and Heuse [74].
The method is very simple in principle, though accurate results can be obtained only if great care with regard to experimental details is taken. Gas or
vapor is passed at a known constant rate over an electrical heater, and the
temperature of the gas stream is measured just before and just after passing
over the heater, after a steady state has been reached. If the rate of flow of the

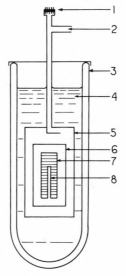

**Fig. 7.3** Simplified representation of a cryogenic calorimeter with electrical heating (from
[72], p. 136).

gas is $F$ mole/sec, the power input to the heater is $W$ watts, and the (corrected) temperature rise is $\Delta T$, the mean heat capacity of the substance is

$$\tilde{C}_P = \frac{W}{4.1840 \, \Delta TF} \tag{7.41}$$

in cal/mole. Since the value of $\Delta T$ is usually small, the heat capacity obtained can in most cases be identified with the true heat capacity at the mean temperature of the experiment. The heat capacity of the calorimeter does not appear in this expression because measurements are made only after a steady state has been reached. If the heat capacity of the calorimeter is unduly large, excessive lengths of time are required for the establishment of the steady state.

It is impossible, in this method, to work under strictly adiabatic conditions because the calorimeter itself is not at a uniform temperature. Consideration must be given to minimizing both the heat exchange of the calorimeter with its surroundings and the heat exchange between the various parts of the calorimeter resulting from causes other than the flow of the gas. It is not feasible to eliminate completely these sources of error. The relative effect of heat leakage from the calorimeter and of heat exchange between the heater and the thermometers resulting from conduction by the material of the calorimeter and from radiation should decrease with increasing rate of flow. The procedure is therefore usually followed of measuring the apparent heat capacity with different rates of flow and extrapolating to infinite rate of flow. The apparent heat capacity in most cases varies linearly with $1/F^n$, where the power, $n$, depends on the type of apparatus, so that the extrapolation is easily carried out. In general, the sources of internal heat exchange are minimized by providing poorly conducting material for the gas path between the first thermometer and the heater, and by interposing suitable radiation shields between the thermometers and the heater. The heater should have sufficient surface in good thermal contact with the gas so that its temperature does not exceed that of the gas by more than a few degrees. If the elimination of errors due to heat exchange has been satisfactorily accomplished, experiments in which the rate is varied and $\Delta T$ is held constant should give values for the apparent heat capacity falling on the same straight line when plotted against $1/F^n$ as that given by holding $W$ constant and allowing $\Delta T$ to vary.

Further discussion of vapor-flow calorimetry, with description of highly developed apparatus, is given in a review by McCullough and Waddington [75].

The determination of $C_P$ of a gas can be accomplished by a modification of the method of mixtures. The gas, flowing at a constant rate, is warmed or cooled by passage through a heat exchanger held at constant temperature and is then led through a calorimeter. The calorimeter temperature is usually considerably different from that to which the gas is initially brought, so that a mean heat capacity is obtained. This method has been applied by several investigators [76].

The most troublesome source of error in the method of mixtures as applied to gases is the conduction of heat from (or to) the calorimeter along the tube connecting the precooler (or preheater) with the calorimeter. The assumption frequently made, that this heat exchange can be evaluated from observations made when the gas is not flowing through the calorimeter, is incorrect [76].

A method for the determination of $C_P$ of gases which has received considerable attention in recent years is that first applied by Lummer and Pringsheim [77]. The temperature change resulting from an isentropic (reversible adiabatic) expansion of a gas is related to the heat capacity by the expression

$$C_P = \frac{T(\partial V/\partial T)_P}{(\partial T/\partial P)_S}. \tag{7.42}$$

In observing the temperature–pressure coefficient, a sample of the gas contained in a large flask (capacity 10–60 liters) is allowed to expand suddenly to a pressure slightly below the original pressure; the temperature change, measured at the center of the flask by a thermometer having the lowest possible heat capacity and time lag, must be observed within a very short interval of time after the expansion, before appreciable heat loss to the walls can take place. The expansion must be carried out in such a way that convection currents are not set up in the gas. The temperature-measuring device usually consists of an extremely thin platinum wire employed as a resistance thermometer in one arm of a Wheatstone bridge, the bridge off-balance being indicated by a galvanometer of very short period. The mean value of $C_P$ obtained in this way is practically equal to the true heat capacity at the mean temperature and pressure of the experiment.

*Heat Capacity at Constant Volume.* The heat capacity of a gas at constant volume can be evaluated from its thermal conductivity at very low pressures. The thermal conductivity of a gas is approximately independent of its pressure until the pressure is reduced to the region at which the mean free path of the gas molecules is of the same order of magnitude as the distance between the surfaces involved in the heat exchange. In this region, thermal conductivity becomes proportional to the pressure. If the pressure is reduced to such a low value that the possibility of molecular collisions in the space between the surfaces is essentially excluded (i.e., the mean free path is large compared with the separation of the surfaces), the theoretical relation [78] between heat capacity and thermal conductivity becomes relatively simple. The experimental procedure [79–81] consists essentially in measuring, by means of its resistance, the temperature of an electrically-heated platinum wire which is supported in a cylindrical space of small radius. The temperature of the cylindrical wall is also observed. Measurements are made with the space between the wire and the wall evacuated, and then filled with gas at very low pressure, the former measurement allowing correction for radiation to be made. One of the chief difficulties with this method is that a gas molecule does not acquire the temperature of the wire or the wall after a single collision with either, so that the conduction of heat is not as large as it would be in the ideal case. An empirical evaluation of the *accommodation coefficient* [81], which allows for this deviation from ideality, depends on measurements made under conditions such that the results can be compared with $C_V$ values determined by other methods. This method is of importance in that it is the only procedure by which one can obtain heat-capacity data directly at very low pressures where the gas behaves very nearly in accordance with the perfect gas law.

The heat capacity at constant volume and the ratio, $C_P/C_V(=\gamma)$, of many gases have been evaluated from measurements of the velocity of sound [82]. Cornish and Eastman [83] have discussed this method and have described an apparatus which they employed with hydrogen from 80 to 370°K.

LIQUIDS

The application of heat-capacity data of pure liquids to the evaluation of entropies is discussed above. The heat capacities of liquid mixtures and of solutions, both of nonelectrolytes and of electrolytes, are of great significance in connection with the physics and thermodynamics of such systems. For example, the heat capacities of very dilute aqueous solutions of electrolytes have been frequently studied in connection with the Debye-Hückel theory of such solutions. The interest which attaches to the heat capacity anomalies observed with solutions of certain macromolecules has been mentioned earlier.

*Direct Method.* It has been frequently demonstrated [84] that values of $C_P$ of liquid systems accurate to 1% or better may be obtained with simple apparatus. One type of equipment is shown in Fig. 7.4. A silvered Dewar flask serves as the calorimeter. It is closed by a rubber stopper and submerged in a water bath which may be kept either at constant temperature or at the same temperature as the calorimeter. The calorimeter heater is a coil of constantan or manganin wire enclosed in a glass tube. It is supplied with both potential and current leads. The liquid in the calorimeter is stirred at a constant rate by a propeller stirrer. Heat leakage along the stirrer shaft is reduced by a section of poorly conducting material such as Bakelite or Lucite. The temperature of the liquid in the calorimeter is measured by a Beckman thermometer, or, better, by a resistance thermometer or a thermel.

The reader is referred to chapters by Ginnings and Stimson [85], and Cruickshank et al. [86] for further discussions of methods and apparatus for the determination of the heat capacities of liquids and solutions.

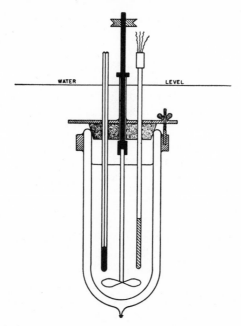

WATER                              LEVEL

**Fig. 7.4**  Simple calorimeter for determining heat capacities of liquids.

*Method of Mixtures.*   The method of mixtures has been employed only rarely in determining heat capacities of liquids at ordinary temperatures. An example is furnished by the work of Nelson and Newton [87] on glucose glass. A sample of material was brought to a temperature in the neighborhood of 25° and was then quickly transferred to a calorimeter at about 60°. The calorimeter consisted of a pair of Dewar flasks in a container jacketed with chloroform vapor. Mercury was used as the calorimetric liquid.

*"Piezothermometric" Method.*   This method is of interest because measurements can be made on small samples (5 ml) of material over practically the entire liquid temperature range. It is based on the same thermodynamic relation as the Lummer-Pringsheim method for determining the heat capacities of gases:

$$C_P = \frac{T(\partial V/\partial T)_P}{(\partial T/\partial P)_S}. \tag{7.42}$$

The method has been investigated very carefully by Burlew [88], who obtained results, to which he assigned a probable error of a few tenths of a per cent, for benzene and toluene over a wide temperature range. In the last paper listed in [88], Burlew includes a consideration of all previous values for the heat capacities of benzene and toluene. It is very instructive to note the discordance of these data, particularly some of the older data. This illustrates the great need for the redetermination of many of the thermal quantities already reported in the literature.

SOLIDS

*Direct Method.*   As in the case of liquids, the Nernst type of calorimeter can be applied to the determination of the heat capacities of solids at ordinary temperatures. In the case of a compact solid having a high thermal conductivity, such as a metal, it is possible to dispense with the calorimeter shell and temperature-equalizing vanes usually employed. Accurate measurements of the specific heat of copper have been made by Harper [89] in which 50 m of rather heavy copper wire coiled in a helix served as calorimeter, heater, and resistance thermometer.

*Method of Mixtures.*   The work of Andrews, Lynn, and Johnston [90] on various organic substances may be cited as an early example of the application of this method at moderate temperatures, using very simple apparatus. The sample, contained in a glass container, was heated to the desired temperature, and then rapidly dropped into a calorimeter at about 25°. The calorimeter consisted of a Dewar flask filled with kerosene which was stirred by a reciprocating stirrer. The temperature of the calorimeter was measured relative to that of a similar Dewar flask by means of a thermal. It was found that, if a pair of Dewar flasks having very similar thermal leakage rates was selected, results of moderate accuracy could be obtained without jacketing the flasks. The heat capacities were obtained relative to that of a silver rod used as reference material.

## Determination of Heat Capacities at Elevated Temperatures

The thermodynamic properties of some organic compounds at elevated temperatures are of considerable interest; thus, a knowledge of the free energies of paraffin hydrocarbons up to high temperatures will permit predictions regarding the composition of equilibrium mixtures resulting from isomerization reactions involving these compounds. Such reactions are of great technical importance. However, it is usually not possible to determine by direct observations the data from which such high-termperature thermodynamic properties can be evaluated, largely because nearly all organic substances undergo decomposition (or at least isomerization) when heated to elevated temperatures. Considerable progress has been made in calculating the thermodynamic properties of relatively simple molecules, such as the paraffin hydrocarbons, at high temperatures from molecular and spectroscopic data combined with values for the thermodynamic functions at lower temperatures.

The thermodynamic properties of materials at extremely high temperatures have recently become of great importance in connection with nuclear and space technology. For a consideration of the techniques used in high-temperature calorimetry the reader is referred to articles by Douglas and King [91], West and Westrum [92], and Beckett and Cezairliyan [93].

### THE METHOD OF MIXTURES FOR SOLIDS AND LIQUIDS

If $T_1$ is the temperature to which the sample is initially heated, and $T_0$ and $T_0'$ are the initial and final temperatures of the calorimetric system having an energy equivalent $E$, it is readily seen that the *mean* heat capacity of the sample, in the temperature interval $T_0$ to $T_1$, is given by

$$C_s = E \frac{T_0' - T_0}{T_1 - T_0'} - \tilde{C}_c, \qquad (7.43)$$

where $\tilde{C}_c$ is the mean heat capacity of the container in the same temperature interval. If the pressure on the sample is the same at both the initial and final temperatures,

$$H_{T_1} - H_{T_0'} = C_s(T_1 - T_0') \frac{M}{m_s}, \qquad (7.44)$$

where $m_s$ is the weight of the sample and $M$ is the molecular weight of the substance. The heat function may then be differentiated either graphically or analytically to give the true heat capacity:

$$\left[ \frac{\partial (H_{T_1} - H_{T_0'})}{\partial T} \right]_P = C_P. \qquad (7.45)$$

The furnace for preheating the sample must obviously contain a region of sufficiently uniform temperature. This can be accomplished by using a furnace with a large ratio of length to bore. In some cases, the sample is preheated in a vessel surrounded by a vapor or liquid bath of known temperature. It cannot be assumed that the heat loss during the fall from the preheater into the calorimeter is negligible. If the apparent heat capacity of the empty sample container is evaluated by observations made under the same conditions as those existing during the determination of the heat capacity of the sample, it may be assumed that the heat losses during the drop cancel out, provided the sample container itself has a sufficiently large heat capacity. From this point of view, it is advantageous for the container to be constructed of material having not too high a thermal conductivity. Heat loss during the drop can be materially reduced by evacuating the furnace, calorimeter, and connecting tube to a pressure of about 1 mm; at this pressure, the static thermal conductivity of a gas is approximately the same as at atmospheric pressure, so the attainment of thermal equilibrium in the furnace and calorimeter is not slowed down, while at the same time the sample during its fall will lose much less heat to the surrounding gas.

For the receiving calorimeter, one can employ any sufficiently precise apparatus of convenient design, such as those described elsewhere in this chapter. The only modification of ordinary design which is needed is the provision either of openings in the calorimeter and jacket covers which are large enough for the sample to fall through, or of means for temporarily removing these covers. Receiving calorimeters are of two types, either a stirred-liquid or an aneroid (metal-block) calorimeter. With a liquid receiver, precautions must be taken to avoid splashing of the liquid (the amount of liquid in the calorimeter, or the energy equivalent of the calorimeter, should in any case be determined after, rather than before, the drop) when the sample drops into it. This difficulty, as well as errors caused by possible vaporization of the calorimeter liquid by a hot sample, is avoided by the use of an aneroid receiver. It is evidently advantageous to have the "dead" heat capacity of the calorimeter as small as is consistent with good design so far as other factors, such as temperature equalization and heat loss to the calorimeter surroundings, are concerned. In the case of an aneroid calorimeter, the consideration of temperature equalization is particularly important.

An apparatus (Fig. 7.5) described by Southard [94] illustrates several of the points mentioned above. The copper block calorimeter, $C$, is enclosed in a brass case inmersed to level $H$ in a thermostat at $25°$. The receiving well is closed by a circular copper gate except at the instant of the drop, the gate being manipulated by eccentric shaft $E$. The sample, contained in a platinum-rhodium capsule, $A$, is heated in furnace $B$, which is surrounded by a water-cooled jacket. The capsule is suspended in the furnace by a fine platinum-rhodium wire, so that it is easily replaced in the furnace for another determination without dismounting the apparatus in any way. The temperature of the sample is determined by a noble-metal thermocouple situated just above it in the furnace tube. A notable feature of Southard's design is the water-cooled gate, $D$, mounted on shaft $E$, which prevents heat leakage to the calorimeter even when the furnace is at $1500°$.

## Differential Heat Capacity Calorimetry

It was mentioned above (Section 4, p. 374) that the determination of the enthalpy changes accompanying conformational transitions in macromolecules in dilute solution requires highly sensitive calorimetry. Several instruments have been developed for this purpose [95], some of which are in effect extremely sensitive differential scanning calorimeters (Chapter VIII). An apparatus developed in the author's laboratory [96] is shown in schematic cross-section in Fig. 7.6. Two platinum cells are supported by means of their platinum filling tubes within an aluminum adiabatic jacket, which is in turn

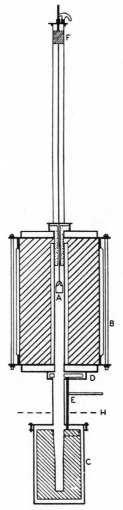

**Fig. 7.5** Calorimeter for determination of heat capacities by the method of mixtures (Southard).

contained in a submarine submerged in a water bath. The cells are heated by heaters contained in reentrant tubes, and are completely filled, one with solution and the other with solvent, by means of Teflon tubes connecting to syringe fittings above the level of the water in the bath. The temperature difference between the cells is measured by a 10-junction thermel the junctions of which are distributed over the surfaces of the cells. A thermel with one junction on each cell and two immersed in the adiabatic jacket actuates

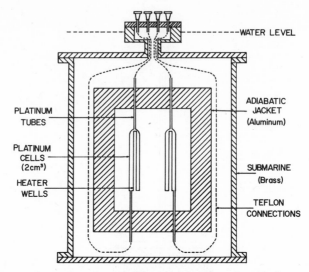

**Fig. 7.6** Schematic cross-section of differential heat-capacity calorimeter, showing the disposition of platinum calorimeter cells within an adiabatic jacket [97].

a control system which holds the jacket temperature close to that of the cell. A resistance-thermometer controller [97] holds the water bath temperature closely equal to that of the adiabatic jacket. The cells are usually heated at a rate of 18°C/hr. When a transition takes place in one cell, the resulting heat capacity anomaly requires additional heat input into that cell to maintain zero temperature difference between the cells. This extra heat input is automatically controlled and recorded [98], and constitutes the principal data output of the apparatus. This equipment, and others designed for this application [95], afford excellent examples of the power of the differential method in calorimetry. A transition heat amounting to only 0.01% of the total heat supplied to either calorimeter during the measurement can be estimated with an accuracy of 2–3%.

## 5   HEATS OF COMBUSTION

The determination of heats of combustion, particularly of organic compounds, has long occupied an important place in thermochemistry. Interest in the heats of combustion of substances used as fuels is of obvious origin; of much greater scientific interest are the applications of combustion data to the evaluation of other quantities such as heats and free energies of formation, and heats of reaction. It is important to note that, with respect to the experimental accuracy required, there is an essential difference between these two types of application of combustion data. In the comparison of the calorific value of fuels, data having an accuracy of the order of 1 % will usually suffice, while the calculation of important thermodynamic quantities from heats of combustion requires data of the highest quality. Since apparatus for the determination of heats of combustion of a relatively low order of precision may be readily purchased and is easily used, attention in this section will be concentrated on modern methods by which data having an accuracy of a few hundredths of a percent may be obtained. We shall first discuss some of the more important applications of heats of combustion, and then describe the experimental procedures employed in their determination.

## Applications

Modern heats of combustion refer, in most cases, to a reaction taking place isothermally, with all the substances involved being in their standard states. The standard state of each substance is taken as the form which is stable at 25° under a pressure of 1 atm. (In most of the older work on heats of combustion the standard temperature was 18°.) Since the halogens and sulfur become aqueous halogen acid or aqueous sulfuric acid, respectively, in the combustion process, it is customary to report heats of combustion of substances containing these elements with the halogen or sulfur in the products in the form of the corresponding aqueous acid. Of course, in each case the concentration of the acid should be clearly specified.

## Heats of Formation

The results of combustion calorimetry are conveniently summarized in the form of heats of formation. The standard heat of formation of a substance may be defined as the heat content of one mole of the substance less the sum of the heat contents of the appropriate amounts of the elements from which the substance is formed, the substance and the elements being in their standard states. It is customary to take the following standard states for the elements: graphite; gaseous hydrogen, oxygen, nitrogen, and halogen at 1 atm; and rhombic sulfur. The calculation of a heat of formation requires a knowledge of the heats of formation of the products of the combustion of the substance, namely, carbon dioxide, water, and aqueous sulfuric and halogen acid. Thus in the case of the substance $C_aH_bO_c$, it readily follows from Hess's law that

$$a C_{graphite} + \frac{b}{2} H_{2(g,\ 1\ atm)} + \frac{c}{2} O_{2(g,\ 1\ atm)} = C_aH_bO_{c(sec\ 1,\ org.\ 1\ atm)}$$

$$\Delta H_f = -\Delta H_c + a\Delta H_{f(CO_2)} + \frac{b}{2} \Delta H_{f(H_2O)} \qquad (7.46)$$

where $\Delta H_c$ is the standard heat of combustion of the substance, and $\Delta H_f$, $\Delta H_{f(CO_2)}$, and $\Delta H_{f(H_2O)}$ are, respectively, the standard heats of formation of the substance, of carbon dioxide, and of water.

Very careful work, largely at the National Bureau of Standards, has furnished reliable data for the heats of formation [99] of water and carbon dioxide at 25°. (The enthalpy of a solid or a liquid is practically independent of pressure at ordinary pressures. For this reason it is permissible to omit the specification of pressures after the formulas of liquids and solids.):

$$H_{2(g,\ 1\ atm)} + \tfrac{1}{2}O_{2(g,\ 1\ atm)} = H_2O_{(l)}$$

$$\Delta H_{f(H_2O)} = -285,790 \text{ J/mole} = -68,317 \text{ cal/mole} \qquad (7.47)$$

$$C_{graphite} + O_{2(g,\ 1\ atm)} = CO_{2(g,\ 1\ atm)}$$

$$\Delta H_{f(CO_2)} = -393,448 \text{ J/mole} = -94,052 \text{ cal/mole} \qquad (7.48)$$

As an example of the corresponding values needed when other elements are present, we may cite the case of sulfur [100]. The combustion data given by Mansson and Sunner lead to the result

$$S_{rhombic} + \tfrac{3}{2}O_{2(g,\ 1\ atm)} + 116\ H_2O_{(l)} = H_2SO_4 \cdot 115\ H_2O_{(l)}$$

$$\Delta H_{f(H_2SO_4)} = -143,880 \text{ cal/mole.} \qquad (7.49)$$

### Compilations of Heats of Combustion and Formation

The heats of formation of many organic compounds containing one or two carbon atoms are given by Bichowsky and Rossini [101] in their important thermochemical tables. Kharasch [102] made a rather complete critical compilation of the heats of combustion of organic compounds, which was also published in *International Critical Tables* [103]. Additional tabulations have been made [104] which cover the more recent literature. A number of compilations of data have been indexed in convenient form [105]. The most recent tabulation for organic compounds, by Stull et al. [106], is a very useful updating of the pioneering monograph of Parks and Huffman [107].

### Free Energies

The standard free energy of formation is the free energy increase accompanying the formation of a substance in its standard state from its elements in their standard states. This quantity is frequently spoken of simply as the free energy of the substance. The important role played by heats of formation in the evaluation of free energies of formation has already been indicated in Section 4 in connection with the use of the fundamental relation,

$$\Delta G_f^0 = \Delta H_f^0 - T\Delta S_f^0, \qquad (7.50)$$

for this purpose. The standard entropy of formation, $\Delta S_f^0$ is the entropy of the substance less the sum of the entropies of its elements, the individual entropies being evaluated either from heat-capacity measurements extending to low temperatures or from molecular data. This application of heats of formation illustrates most clearly the great need for attaining the highest possible precision in the determination of the heats of combustion from which they are derived. According to Rossini [108], ". . . in most cases the relative magnitudes and absolute accuracies of the values of $\Delta S_f{}^0$ and $\Delta H_f{}^0$ are such that the resulting uncertainty in $\Delta G_f{}^0$ is practically equal to the error in $\Delta H_f{}^0$." A large fraction of the combustion data appearing in the literature before 1930 is of insufficient accuracy to be of much value in the calculation of thermodynamic quantities. For example, in the case of the isomerization of cyclohexane to methylcyclopentane, the equilibrium measurements of Glasebrook and Lovell [109] gave a value for the entropy change in fairly good agreement with the value derived from heat-capacity data. However, the changes in enthalpy and free energy differed even with respect to sign from the values deduced by Parks and Huffman [107] from older combustion heats. A redetermination [110] of these heats with modern precision has eliminated the discrepancies.

Free energies may be added and subtracted just as may $\Delta H$ values. Thus, if the free energies of formation of all the substances except one involved in a reaction are known, the free energy of that one can be evaluated from a knowledge of the free energy change accompanying the reaction, with all the substances in their standard states. This standard free energy for the reaction is given by

$$\Delta G^0 = -RT \ln K, \tag{7.51}$$

where $K$ is the equilibrium constant. In cases in which it is difficult to obtain a sufficiently precise equilibrium constant at a single temperature, considerable improvement may be obtained by correlation of equilibrium constants determined over a range of temperature with appropriate thermal data. The heat-capacity change accompanying the reaction may be expressed by the empirical equation,

$$C_P = a + bT + cT^2, \tag{7.52}$$

which on integration gives

$$\Delta H = \Delta H_0 + aT + \tfrac{1}{2}bT^2 + \tfrac{1}{3}cT^3. \tag{7.53}$$

Substituting this equation in

$$\frac{d(\Delta G/T)}{dT} = -\frac{\Delta H}{T^2} \tag{7.54}$$

and integrating, we obtain

$$\frac{\Delta G^0}{T} = -R \ln K = \frac{\Delta H_0}{T} - a \ln T - \tfrac{1}{2}bT - \tfrac{1}{6}cT^2 + I, \tag{7.55}$$

where $\Delta H_0$ and $I$ are integration constants. If the function,

$$\Sigma \equiv -R \ln K + a \ln T + \tfrac{1}{2}bT + \tfrac{1}{6}cT^2, \qquad (7.56)$$

is plotted against $1/T$, a straight line should be obtained having the slope $\Delta H_0$ and intercept $I$. In general, the precision of drawing the $\Sigma$ plot is considerably increased if a reliable value of $\Delta H$ at some temperature within the range under consideration is available, for then the slope of the plot can be determined by (7.53). Such a value of $\Delta H$ can be obtained from combustion data.

### Heats of Reaction

The most general method for evaluating the heats of organic reactions is by means of combustion heats. This method, however, leads to rather inaccurate results in many cases. The sum of the heats of combustion of the products is usually a large number which differs only by a small amount from the corresponding sum for the reactants so the difference between these two sums, the heat of the reaction, may be very seriously in error. For example, to obtain with an accuracy of only 10% the heat of mutarotation of glucose [111] in the solid state from the heats of combustion of the $\alpha$ and $\beta$ forms, one would have to determine each heat of combustion with an accuracy of 0.01%.

Heats of reaction calculated from heats of combustion refer only to reactions with all the substances in their combustion standard states. Additional data, such as heats of solution and vaporization, are necessary to convert such heats of reaction to conditions which may be of more practical interest.

Because of these limitations, much attention has been given to the direct calorimetric determination of reaction enthalpies. This subject is discussed in Section 8. There are, however, many processes investigation of which will remain beyond the scope of direct calorimetry because of limitations imposed by side reactions or calorimetric difficulties. In such cases, the use of combustion heats will continue to be of importance.

Prosen and Rossini [112] have developed a method for the evaluation of heats of isomerization from combustion experiments which considerably reduces the inaccuracies inherent in the usual procedure. In principle, the method consists in determining the ratio of the masses of two isomers (in the work cited, involving the hexanes, the masses of carbon dioxide formed in the combustions were actually employed), the combustion of which produces equal temperature rises in the calorimetric system. Since all the experiments are carried out under nearly identical conditions, calorimetric uncertainties are minimized. It may be noted that the reactions studied by Prosen and Rossini, namely, the isomerization of hydrocarbons, cannot possibly be investigated by the direct calorimetric method.

### Bond Energies

Various attempts have been made to discover additive relations between heats of combustion, or values derived from them, and molecular structure. The consideration of bond energies affords a fruitful, if not very quantitative, basis for correlating such data. The total heat of dissociation of a polyatomic substance into atoms, a value which can be calculated from combustion data, is equal to the sum of the energies of the bonds in the substance. It is not, however, possible to deduce individual bond energies from these sums as it is in the case of diatomic molecules; only mean values can be obtained. Table 7.1 gives a selection of the bond energy values published by Pauling. These values are so chosen that their sums represent, with an accuracy of about $\pm 2$ kcal/mole, the $-\Delta H$ values for the formation of gaseous substances from gaseous atoms at 1 atm and 18°; most of the thermochemical data used in deducing these bond energies were taken from Bichowsky and Rossini [101]. The values used for the heats of formation $(-\Delta H)$ of the elements in their usual standard states from monatomic gases are given in Table 7.2.

**Table 7.1**    Bond Energies [a]

| Bond | Energy (kcal/mole) | Bond | Energy (kcal/mole) | | Bond | Energy (kcal/mole) |
|------|------|------|------|------|------|------|
| H—H | 103.4 | O—H | 110.2 | | C=N | 94 |
| C—C | 58.6 | S—H | 87.5 | | C≡N | 144 (HCN) |
| N—N | 20.0 | H—F | 147.5 | | C≡N | 150 (cyanides) |
| O—O | 34.9 | H—Cl | 102.7 | | C—S | 54.5 |
| S—S | 63.8 | H—Br | 87.3 | | C=S | 103 |
| F—F | 63.5 | H—I | 71.4 | | C=C | 100 |
| Cl—Cl | 57.8 | C—O | 70.0 | | C≡C | 123 |
| Br—Br | 46.1 | C=O | 142 (formaldehyde) | | C—F | 107.0 |
| I—I | 36.2 | C=O | 149 (aldehydes) | | C—Cl | 66.5 |
| C—H | 87.3 | C=O | 152 (ketones) | | C—Br | 54.0 |
| N—H | 83.7 | C—N | 48.6 | | C—I | 45.5 |

[a] L. Pauling, *Nature of the Chemical Bond and the Structure of Molecules and Crystals*, Cornell University Press, Ithaca, N. Y., 1940.

Numerous more elaborate schemes [113], some based on the concept of bond energies alone and others on group contributions or a combination of both types of contributions, have been developed which enable prediction of a wide range of enthalpies of combustion or of formation with accuracies of the order of 0.5 kcal/mole.

**Table 7.2**    Heats of Formation of Elements in Their Standards States From Monatomic Gases

| Element | Heat of Formation (kcal/mole) | Element | Heat of Formation (kcal/mole) |
|---------|-------------------------------|---------|-------------------------------|
| $H_2$   | 103.4                         | $F_2$   | 63.6                          |
| C       | 124.3 [a]                     | $Cl_2$  | 57.8                          |
| $N_2$   | 170.2                         | $Br_2$  | 53.8                          |
| $O_2$   | 118.2                         | $I_2$   | 51.2                          |
| S       | 66.3                          |         |                               |

[a] This is the heat of sublimation of diamond.

Despite the approximate nature of the heats of formation estimated from these empirical schemes, they are of considerable interest in connection with problems of molecular structure. For example, a molecule whose structure cannot be represented satisfactorily by a single classical valence bond formula can, in one type of approximation method, be more accurately considered as a resonance hybrid of several such structures. In such a case, the difference between the observed heat of formation and that calculated for one of the contributing structures with the help of the table of bond energies gives an empirical value for the resonance energy relative to the structure considered. For example, the heat of combustion of benzene leads to the value $-\Delta H_f = 1039$ kcal/mole; one would predict for a single Kekulé structure the value $-\Delta H_f = 6(C-H) + 3(C-C) + 3(C=C) = 1000$ kcal/mole. The difference, 39 kcal/mole, is the resonance energy of benzene relative to the Kekulé structure. Extensive discussions of this type of application of heats of combustion will be found in various treatises on physical organic chemistry.

## Calorific Values of Explosives

The "calorific value" of an explosive is defined as the heat evolution, in cal/g, when the substance is exploded in the absence of oxygen except for what it contains itself. This quantity [114] differs but little from the heat evolved when the substance is exploded under practical conditions. Experimental techniques differ somewhat from those employed in ordinary combustion calorimetry. For example, an especially strong bomb is used, and various methods of detonation are employed.

## Bomb Calorimetry

### Description of Apparatus

The heats of combustion of solids and liquids are usually determined by combustion in a bomb in an excess of oxygen. The bomb is immersed in a calorimeter containing suitable liquid, usually water. The calorimeter is surrounded by a jacket; in the ordinary method the temperature of the jacket is held constant, while in the adiabatic method the temperature of the jacket is maintained equal to that of the calorimeter. Combustion experiments are almost without exception of relatively short duration, so that there is little to be gained from the added complications of the adiabatic method (Section 3). Bomb calorimeters are, fundamentally, devices for comparing the heats of combustion processes with known amounts of electrical energy. However, since very careful measurements of the heat of combustion of benzoic acid have been made, it is possible, without significant loss of accuracy, to use the combustion calorimeter to compare the heat of combustion of the substance under investigation with that of benzoic acid, thus eliminating the need for a calorimeter heater (a crude heater will still be desirable to facilitate the adjustment of the temperature of the calorimeter) and the accessory apparatus for the measurement of electrical energy. In order for the comparison between combustion heat and electrical energy, or between two different combustion heats, to be made as accurately as possible, it is important that the time–temperature curves for the two processes be as nearly alike as possible. In this way inaccuracies in the calorimetric observations, particularly those resulting from heat leakage to the surrounding, will be minimized.

Bomb calorimetry has been developed to the point where *calorimetric* accuracies of the order of 0.01 % can be obtained. At this level of calorimetric accuracy, limitation of the overall accuracy is likely to be of chemical origin, primarily arising from inadequate definition of the final state of the combustion products. For example, with organic chloro compounds, 20–30% of the chlorine may appear as free $Cl_2$, the rest being converted to HCl. A reducing agent such as arsenic trioxide may be added to reduce the $Cl_2$, but the reduction may not be complete and the HCl formed may be at different concentrations in different parts of the bomb. Difficulties of this sort have been largely eliminated by the development of moving-bomb calorimetry, in which an appropriate motion of the bomb after combustion ensures complete equilibration of the combustion products between the various phases present in the bomb.

The applicability of bomb calorimetry in the determination of enthalpies of formation has been very significantly extended by the substitution of fluorine for oxygen as the oxidizing agent. This change naturally involves extensive changes in the instrumentation for bomb calorimetry.

The classical and modern aspects of bomb calorimetry have been exhaustively considered in several volumes published in the last decade and a half [115]. Oxygen-bomb calorimeters of modern design which are capable of very high precision are exemplified by those described by Keith and Mackle [116], Good et al. [117], and Bjellerup [118].

## Reduction to Standard Conditions

Washburn [119] pointed out that the actual bomb process had not been completely defined by previous workers, and that the precision obtainable by modern calorimetric methods is such as to make this lack of definition significant. Furthermore, the initial and final states in an actual combustion process, even when completely defined, are thermodynamically uninteresting or trivial, and suitable corrections should be employed to refer the calorimetric data to a convenient set of standard states. Washburn discussed in detail the corrections which must be included, in the case of compounds containing C, H, and O, in order to obtain the quantity of primary interest, $\Delta U_R$, from a bomb combustion experiment. $\Delta U_R$ is the heat absorbed when one mole of substance in its standard state, at the standard temperature (usually 298.16°K), reacts isothermally with an equivalent amount of oxygen gas under a pressure of 1 atm, to form pure carbon dioxide gas and pure liquid water, both under a pressure of 1 atm, the reaction taking place without the production of any external work. It is of course of no significance that this process is not experimentally realizable. The so-called Washburn corrections have been extended by later workers [115] to include combustions of compounds containing elements in addition to C, H, and O.

## Calculation of $\Delta H_c$

For most thermodynamic calculations, the quantity of greatest interest to be derived from combustion experiments is the difference in enthalpy between the products and reactants. This quantity, which is denoted by $\Delta H_c$ if all the substances are in their standard states, is obtained from $\Delta U_R$ by means of the equation (see Section 1, p. 351).

$$\Delta H_c = \Delta U_R + \Delta(PV). \tag{7.57}$$

Since oxygen and carbon dioxide under a pressure of 1 atm behave approximately as perfect gases, this expression can, with sufficient accuracy, be written:

$$\Delta H_c = \Delta U_R + \Delta nRT, \tag{7.58}$$

where $\Delta n$ is the difference between the number of moles of carbon dioxide formed and oxygen consumed when one mole of substance is burned (if nitrogen is present, allowance has to be made for the fact that this element also appears in the gaseous state in the products of the combustion). In (7.58), the volume changes in condensed phases are neglected. At 298.16°K, $RT$ has the value of 592.3 cal/mole.

## Flame Calorimetry

The heats of combustion of substances which are gaseous or have high vapor pressures at room temperature are most conveniently determined in a flame calorimeter. The substance is burned in oxygen or in fluorine at constant pressure. Rossini [120] has described an oxygen-flame calorimeter which gives results of about the same accuracy as realizable in oxygen-bomb calorimetry. The subject of fluorine-flame calorimetry has been reviewed by Armstrong [121]. The data obtained from flame calorimetry have the same general significance as the data emanating from bomb-combustion calorimetry.

# 6  HEATS OF FUSION, TRANSITION, AND VAPORIZATION

## Heats of Fusion

The most accurate data on heats of fusion have been determined in the Nernst type of calorimeter (Section 4, p. 376) which is employed in determining heat capacities. The heat necessary to raise the temperature of a known mass of substance from slightly below its melting point to slightly above is measured; this information, together with a knowledge of the heat capacities of the substance in the solid and liquid states, is sufficient to determine the heat absorbed in the melting process, provided the sample under investigation is pure. It is evident that the heat absorbed/mole at constant pressure in going from temperature $T_1$ to temperature $T_2$ is

$$\Delta Q = \int_{T_1}^{T_m} C_{P(\text{solid})} dT + \Delta H_f + \int_{T_m}^{T_2} C^{(\text{liquid})} dT, \qquad (7.59)$$

where $\Delta H_f$ is the heat of fusion/mole. During the actual melting of the sample, the temperature will remain practically constant at the melting temperature, $T_m$, of the substance.

In the method of mixtures (p. 383), observations of the heat evolved when the material is cooled to the temperature of the calorimeter are made, with initial temperatures above and below the melting point. The calculation of the heat of fusion from such data is carried out by means of (7.59), where the integrals are evaluated using true heat capacities.

It is necessary to be certain that the crystalline modification of the material used in these measurements is the modification which exists in equilibrium with the liquid substance at the melting point. If the liquid is rapidly cooled to a temperature below its melting point, a metastable form may be obtained which may change to the stable form only very slowly. In the case of certain organic compounds, glasses are obtained rather than crystalline solids, and transition to the crystalline state may be extremely slow. It is obvious that the method of mixtures cannot be applied in such cases. In the ordinary calorimetric method, indication of these abnormalities, in the case of pure substances, is obtainable from the behavior of the temperature of the sample during melting; if equilibrium between solid and liquid forms is not established, there will be no period of approximately constant temperature during melting.

Equation 7.59 will lead to erroneous results with substances which are not pure, since in such cases the melting process is not strictly isothermal. In particular, the phenomenon of "premelting" (p. 374) may be of considerable importance even with relatively pure substances. The first integral in (7.59) will be too large if observed heat capacities are utilized without regard for premelting; in many cases, the observed heat capacities below the melting point can be corrected for premelting as outlined on p. 374.

Useful information about the purity of a very nearly pure substance can be obtained during the determination of its heat of fusion in a calorimeter of the Nernst type. In an ideal solution, or in a sufficiently dilute solution, van't Hoff's law of the lowering of the freezing point holds:

$$\frac{dT}{dN_2} = \frac{RT^2}{\Delta H_f}\left(\frac{k}{k'} - 1\right).$$ (7.60)

In this expression $k/k'$ is the distribution coefficient of the solute between solid and liquid phases, $N_2$ is the mole fraction of the solute, and $T$ is the melting temperature. If $\Delta H_f$, the molal heat of fusion of the solvent, is measured in calories, $R$ has the value 1.986. In cases in which solid solutions are not formed, $k/k' = 0$; in such cases, integration of (7.60) ($\Delta H_f$ can be considered to be constant over a short range of temperature) gives

$$T_0 - T = \frac{RN_2T_0^2}{\Delta H_f},$$ (7.61)

where $T_0$ is the melting point of the pure solvent and $T$ is the temperature at which the liquid mixture is in equilibrium with pure crystalline solvent. If $N_s$, the mole fraction of the solute (impurity) is small, and $r$ is the fraction of the sample melted at $T$, then

$$N_2 \approx \frac{N_s}{r}.$$ (7.62)

Substitution of this expression in (7.61) gives

$$T = T_0 - \left(\frac{RT_0^2N_s}{\Delta H_f}\right)\left(\frac{1}{r}\right).$$ (7.63)

Thus a plot of $T$ versus $1/r$ should be a straight line, with intercept at $(1/r) = 0$ equal to $T_0$ and slope $-RT_0^2N_s/\Delta H_f$. From the observed value of $\Delta H_f$ and the extrapolated value of $T_0$, it is thus possible to calculate $N_s$.

This method has been discussed, and a calorimeter designed specifically for its application has been described, by Tunnicliff and Stone [122].

Heats of fusion can be deduced from measurements of solubility and of freezing-point lowering. In a dilute solution,

$$N_2 \approx \frac{n_2 M_1}{w_1} = \frac{m_2}{1000} M_1, \qquad (7.64)$$

where $w_1$ is the weight of solvent present, $M_1$ the molecular weight of the solvent, $n_2$ is the number of moles of solute, and $m_2$ the molality of the solute (mole/1000 g of solvent). Substitution of this relation in (7.61) gives

$$\Delta H_f \approx \frac{m_2 M_1 R T_0^2}{1000 (T_0 - T)} . \qquad (7.65)$$

This equation cannot be applied to the determination of $\Delta H_f$ unless the molecular weight of the solute *in the solution* is known; it is thus necessary that the simple solute molecules do not undergo association or dissociation under the conditions of the experiment.

### Time–Temperature Curves

The behavior of a substance on being gradually cooled or heated through its transition between the liquid and solid states gives valuable information concerning its purity. It is probable that this behavior serves as one of the most critical and most widely applicable criteria of purity [123]. The equipment involved, such as that described by Skau [11], is much simpler and easier to use than the apparatus involved in heat capacity determinations.

### Heats of Transition

Heats of transition between different solid phases are in general determined by methods similar to those employed in the determination of heats of fusion. It should be noted, however, that, whereas the general pattern of behavior of all pure substances at their melting points is relatively constant, a wide variety of patterns is observed in the case of transitions. For example, the heat capacity of solid ethylene dichloride [124] rises to a rather sharp maximum at 177°K, but no sharp isothermal transition takes place, while the heat capacity of solid ethylene dibromide rises rapidly below 249.5°K, at which temperature there is a sharp transition to a form having a much lower heat capacity.

The heat capacity anomalies due to conformational transitions in macromolecules were briefly discussed earlier (p. 373). The highly specialized calorimetric equipment required for quantitative observations of such transitions was illustrated in Section 4, p. 384.

## Heats of Vaporization

Heats of vaporization [125] are usually evaluated either by direct calorimetric observation of the amount of heat necessary to vaporize a known amount of material, or by calculation from the change of vapor pressure with temperature. These two methods are discussed briefly below. Generally less satisfactory results can be obtained from measurement of the heat evolved on condensing a known amount of vapor [126], and by various modifications of the method of mixtures.

### Calorimetric Determination

Heats of vaporization may be determined in calorimeters similar to those used for determining heat capacities, the only important addition to the apparatus resulting from the necessity of measuring the amount of material vaporized. Flow calorimeters for the precise determination of both the enthalpy of vaporization and the heat capacity of the vapor have been described by McCullough and Waddington [127]. The calorimetry of saturated fluids, including the determination of enthalpies of vaporization, has been authoritatively treated by Ginnings and Stimson [128]. Very effective equipments, for use with small quantities of liquids, have been developed by Wadsö and by Morowetz [129].

### Evaluation from Vapor Pressures

The change with temperature of the vapor pressure, $P$, is related to the molar heat of vaporization $\Delta H_{vap}$ by the Clapeyron expression.

$$\frac{dP}{dT} = \frac{\Delta H_{vap}}{T\Delta V}, \tag{7.66}$$

where $\Delta V$ is the increase in volume when one mole of substance is vaporized. At sufficiently high temperatures (low vapor pressures), the vapor may be assumed to obey the perfect gas law, and the volume of the liquid may be neglected in comparison with that of the vapor. Under these conditions (7.66) becomes

$$-\frac{d \ln P}{d(1/T)} = \frac{\Delta H_{vap}}{R}. \tag{7.67}$$

At temperatures high enough so that the vapor pressure exceeds $1/10$ to $1/4$ atm, (7.66) must be applied in conjunction with state data from which $\Delta V$ may be evaluated.

It is beyond the scope of this section to consider in detail the methods used for determining vapor pressures. Suffice it to state that two general methods are available: (a) the static method consists of direct observation of the pressure resulting from equilibration of liquid and vapor; as an example of a very careful application of this method we may cite the work of Osborne, Stimson, Fiock, and Ginnings [130], and (b) in the dynamic method, the vapor pressure is deduced from the amount of material removed by a known volume of an inert carrier gas at known temperature and total pressure.

### Trouton's and Hildebrand's Rules

A rough approximation to the molar heat of vaporization of *nonpolar* liquids may be obtained from Trouton's rule [131], which states that, for such compounds, the entropy of vaporization at the normal boiling point on the absolute scale ($\Delta H_{vap}/T_B$) is approximately 21 cal/(mole) (°C). The actual Trouton "constants" show a definite trend toward higher values for liquids of higher boiling point. Hildebrand [132] has shown that, if the comparison is made at temperatures at which the vapor concentrations are equal (he chose the arbitrary value of 0.005 mole/liter), this trend is largely eliminated. At this vapor concentration the entropy of vaporization is approximately 27 cal/(mole) (°C). Polar substances such as ammonia and water have larger values, about 32 or 33.

### Heats of Adsorption

Two distinct types of adsorption of gases on solids are recognizable. *Physical* or *van der Waals* adsorption is the more common, and is due to the general type of forces which hold molecules together in the solid and liquid states. *Activated* adsorption, or *chemisorption*, is the result of relatively much stronger forces of a chemical nature between the adsorbent and the adsorbed molecules. Physical adsorption is characterized by low activation energies, probably of the order of 1 kcal/mole, while activated adsorption, like other chemical reactions, has activation energies of the order 10–20 kcal/mole. It would thus be expected that activated adsorption would be more important at higher temperatures.

A widely accepted theory [133] of physical adsorption postulates that the heat evolution accompanying the adsorption is close to the bulk heat of liquefaction of the gas being adsorbed, except for the heat evolved during the formation of the first layer of adsorbed molecules, which is also assumed to be constant and to have a value somewhat larger than the heat of liquefaction. Direct calorimetric measurements [134] have given important confirmation to the general validity of these assumptions, though it appears that in some cases, at least, the heat of adsorption of the first layer is not constant but decreases as the fraction of the surface covered by the monolayer increases. This indicates that the surface is not uniform in adsorption activity, the heat of adsorption being larger on the more active areas.

Activated adsorption is accompanied by heat evolution of the order of 20–100 kcal/mole, depending on both the adsorbent and adsorbate. Since heats of physical adsorption are always below about 10 kcal/mole, calorimetric observations afford a means for distinguishing between the two types.

Calorimeters for the accurate determination of heats of adsorption have been described by Beebe et al. [135]. The most recent discussion of this type of calorimetry was published by Chihara and Morrison [136] in 1968.

### Heats of Radioactive Transitions [137]

If a radioactive material is immersed in a calorimeter having walls in which the emitted particles or rays are completely absorbed without producing any further chemical effect, the rate of heat evolution gives an accurate measure of the radioactivity of the material. This measure of radioactivity has been found to be especially useful in cases where millicuries to curies are involved.

The rate of heat evolution also leads to an estimate of the half-life provided that the energy per emission and the amount of radioactive isotope present are known. This procedure is useful in the case of very long-lived isotopes where the rate of decrease of counting rate observed with a given sample is too slow for accurate measurement.

Since very heavy walls are needed to absorb $\gamma$-rays, the calorimetric method is less useful with $\gamma$-emitters than with $\alpha$- or $\beta$-emitters.

## 7   HEATS OF SOLUTION, DILUTION, AND MIXING

There is no fundamental distinction to be drawn between enthalpies of solution, dilution, and mixing. According to common usage, one speaks of the heat effect accompanying the solution of a solid or a gas in a liquid as a heat of solution, while that accompanying the solution of one pure liquid in another is called a heat of mixing. A heat of dilution refers to the mixing of a solution with the corresponding pure solvent, or to the mixing of two solutions containing the same components in different concentrations. A heat of solution, dilution, or mixing is said to be positive if heat is absorbed in the process.

It is obviously essential that complete mixing of the contents of a calorimeter designed for the determination of the enthalpy change resulting from that mixing must be achieved, either by diffusion or by mechanical stirring. It is primarily this requirement that makes it appropriate to give special attention to the calorimetry of processes initiated by the mixing of two initially separated phases.

No useful distinction can be made between calorimeters appropriate for the determination of heats of solution, dilution, and mixing and those applied to the estimation of the enthalpy changes accompanying chemical reactions (Section 8), and we shall therefore consider in this section all types of calorimeters which incorporate provision for the mixing of two different phases to initiate whatever process is to be studied. Three major types of mixing calorimeters can be distinguished: (a) batch calorimeters in which samples of each of the two phases are held apart until thermal equilibration is reached and are then rapidly mixed; (b) titration calorimeters in which one phase is present in the calorimeter and the other phase is introduced from an external supply through an inlet tube in various amounts; and (c) flow calorimeters, in which both (usually liquid) phases are supplied from external supplies; the calorimeter itself may be a stirred vessel or a flow tube.

## Applications

### Heats of Solution

The heat of a reaction can be evaluated from the heats of combustion of the reactants and products, by the application of Hess's law. It can likewise be evaluated from the heats of solution of the reactants and of the products in some solvent in which the reactants give precisely the same solution as the products. Where it can be applied, this procedure is likely to be more satisfactory than that based on heats of combustion since the heat of reaction is obtained as the difference between relatively small heats of solution. Because of a change of the solvent by each individual solute, the heat effect accompanying simultaneous solution of several reactants is not always the sum of the various individual heats of solution, though in most cases of dilute solutions additivity will be nearly fulfilled. The same restriction applies also, of course, if more than one product is formed. The method under consideration is of particular value when direct measurements of the desired heat effect are ruled out by calorimetric limitations. It has been applied to several inorganic problems [138]. Heiber and his collaborators [139] have used it to evaluate the heats of formation from their components of numerous metal complexes with amines, and Campbell and Campbell [140] estimated the heats of formation of various organic molecular compounds. Evans and Richards [141] determined the heat of transformation of $\alpha$- to $\beta$-quinol, and the heat of formation [142] of some clathrate compounds of the $\beta$-form. Solution calorimetry, where the solvent is molten tin or other liquid metal, has been employed in determinations of the enthalpies of formation of alloys [143].

Heats of solution are of further importance in converting the heat of a reaction in solution, where the heat may be more conveniently measured, to the heat which would be observed if the reaction took place in the absence of solvent. Thus Sturtevant [144] measured the heat of mutarotation of $\alpha$- and $\beta$-D-glucose in aqueous solution, and employed the heats of solution of the two isomers to evaluate the heat of isomerization of solid $\alpha$-glucose to solid $\beta$-glucose. Williams [145] determined the heats of hydrogenation of some substances in acetic acid and evaluated the hydrogenation heats of the undissolved compounds by means of the appropriate heats of solution.

Occasionally, heats of solution have been used to evaluate the relative heat contents of substances in solution, though heats of dilution are usually to be preferred for this purpose.

### Enthalpies of Dilution

Since there is no change in enthalpy associated with the dilution of an ideal solute, it is evident that a nonvanishing heat of dilution indicates a significant deviation from ideal behavior, and may lead to useful insight into the solvent–solute system under study.

If the total enthalpy of a solution containing $n_1$ moles of solvent and $n_2$ moles of solute is represented by $H$, then the *partial molal enthalpies* of the solvent and solute are defined as $\bar{H}_1 = \partial H/\partial n_1$ and $\bar{H}_2 = \partial H/\partial n_2$. Since it is impossible to state the absolute value of enthalpies, it is necessary to consider enthalpies relative to a specified reference state. Thus the partial molal enthalpies of the solvent and solute at infinite dilution are represented by $\bar{H}_1{}^0$ and $\bar{H}_2{}^0$ and *relative* partial molal enthalpies are defined by the expressions $\bar{L}_1 = \bar{H}_1 - \bar{H}_1{}^0$ and $\bar{L}_2 = \bar{H}_2 - \bar{H}_2{}^0$. The relative enthalpy of the solution is given by

$$L = n_1\bar{L}_1 + n_2\bar{L}_2. \tag{7.68}$$

Evidently, at infinite dilution $\bar{L}_1{}^0 = \bar{L}_2{}^0 = \bar{L}^0 = 0$. It is convenient to define the *apparent* relative molal enthalpy of the solute by the equation

$$\varphi L_2 \equiv \varphi H_2 - \varphi H^0 \equiv \frac{L - n_1\bar{L}_1{}^0}{n_2} = \frac{L}{n_2}. \tag{7.69}$$

In the case of aqueous solutions, concentrations are usually expressed in molalities (moles solute/1000 g water), represented by $m$. If $L$ is the relative enthalpy of an amount of solution containing 1000 g (55.51 moles) of water, then

$$L = 55.51\,\bar{L}_1 + m\bar{L}_2 \tag{7.70}$$

and

$$\varphi L_2 = \frac{L}{m}. \tag{7.71}$$

Suppose a solution of molality $m_1$ is diluted with solvent to molality $m_2$. If $\Delta H$ is the heat absorbed in this process per mole of solute,

$$\Delta H = (\varphi H_2)_{m_2} - (\varphi H_2)_{m_1}. \tag{7.72}$$

Successive dilutions can be carried out with decreasing values of $m_2$, and finally an extrapolation to $m_2 = 0$ performed to give

$$\Delta H_{m_2=0} = \varphi H_2{}^0 - (\varphi H_2)_{m_1} = -(\varphi L_2)_{m_1}. \tag{7.73}$$

If a theoretical relation for the slope of the $\Delta H$ curve at $m_2 = 0$ is available, such as that provided by the Debye-Hückel limiting slope for strong electrolytes of sufficiently simple valence type, this extrapolation can frequently be performed with considerable precision. The dilutions to intermediate concentrations can then be employed to establish the variation of $\varphi L_2$ with concentration up to $m_1$. The relative partial molal enthalpies can be evaluated by graphical or analytical differentiation:

$$\overline{L}_2 = \frac{\partial L}{\partial m} = \varphi L_2 + m\left(\frac{\partial \varphi L_2}{\partial m}\right)$$

$$\overline{L}_1 = \frac{m(\varphi L_2 - \overline{L}_2)}{55.51} . \tag{7.75}$$

Similar expressions for $\overline{L}_1$ and $\overline{L}_2$ in other concentration units are readily obtained.

The method outlined here for evaluating the relative heat contents is based on so-called "long chord" dilutions—dilution of a small amount of solution with a relatively large amount of water. Young and his collaborators [146] have shown that partial enthalpies, in particular, can be more accurately evaluated by means of "short chord" dilutions, in which the concentration of the solute is changed only slightly.

It should be noted that, in dilution experiments extending to very low concentrations, complications may arise because of heat effects resulting from the disturbance of the dissociation equilibrium of the solvent [147].

The *apparent molal heat capacity* of the solute is defined as

$$\varphi C_{P2} = \frac{C_P - 55.51 \,\overline{C}_1^0}{m} .$$

*Partial molal heat capacities* can be evaluated from $\varphi C_{P2}$ in just the same way as the corresponding enthalpies:

$$\overline{C}_{P2} = \frac{\partial C_P}{\partial m} = \varphi C_{P2} + m\left(\frac{\partial \varphi C_{P2}}{\partial m}\right) \tag{7.76}$$

$$\overline{C}_{P1} - \overline{C}_{P1}^0 = \frac{m(\varphi C_{P2} - \overline{C}_{P2})}{55.51} . \tag{7.77}$$

It can be shown that

$$\varphi C_{P2} = \frac{\partial}{\partial T} (\varphi L_2) \tag{7.78}$$

and

$$\frac{\partial}{\partial m} (\varphi C_{P2}) = \left(\frac{\partial}{\partial T}\right)\left(\frac{\partial}{\partial m} (\varphi L_2)\right) . \tag{7.79}$$

These equations allow the evaluation of apparent and partial molal heat capacities from heats of dilution determined at two temperatures. Heat capacities determined in this way will usually be very precise.

The formal treatment outlined above obviously provides no interpretation of the actual causes of a nonvanishing dilution heat. In the case of very dilute solutions of strong electrolytes, important substantiation of the Debye-Hückel theory has been obtained from heats of dilution. The enthalpy change on diluting a weak electrolyte will in general include a contribution from the change of the extent of dissociation as well as from changes in the activity coefficients of the various solute species. A similar situation obtains in the dilution of any dissociating solute, although in cases of nonelectrolytes it may be possible to obtain a satisfactorily quantitative interpretation of experimental data on the assumption of activity coefficients equal to unity for all solute (and solvent) species. For example, Stoesser and Gill [148] found that heats of dilution of 6-methylpurine, measured in a flow calorimeter (Section 7) could be adequately accounted for either by unlimited association of the solute, with each successive step characterized by the same intrinsic association constant and enthalpy change, or by an association equilibrium limited to dimerization. Shiao and Sturtevant [149] found a similar treatment to apply to the enthalpies observed on dilution of $\alpha$-chymotrypsin.

### Heats of Mixing

The enthalpy of mixing two liquids is of considerable interest in connection with theories of liquid mixtures. According to the Scatchard-Hildebrand solubility parameter theory [150], the molar excess energy of mixing at constant volume, $\Delta U_v$, is given by the expression

$$\Delta U_v = (N_1 V_1 + N_2 V_2)(\delta_1 - \delta_2)^2 \phi_1 \phi_2, \tag{7.80}$$

where $N_1$, $V_1$, $\delta_1$, and $\phi_1$ are the mole fraction, partial molar volume, solubility parameter, and volume fraction for component 1, respectively. The solubility parameter is a property of the pure liquid, not at all dependent on the other component of the mixture. Since the quantity $(N_1 V_1 + N_2 V_2)$ is the molar volume of the mixture, (7.80) can be put in the form

$$\Delta Q_v / \phi_1 \phi_2 = (\delta_1 - \delta_2)^2 = \text{constant}, \tag{7.81}$$

where $\Delta Q_v$ is the excess energy of mixing at constant volume per unit volume of mixture. It is expected that volume changes on mixing for mixtures approximating the so-called regular behavior predicted by (7.81) will be small, so that values of the energy change observed at constant pressure, $\Delta Q$, will be a good measure of $\Delta Q_v$. Although it has not been possible to set up a table of values of solubility parameters which permits accurate predictions of enthalpies of mixing, an interesting indication of the validity of the Scatchard-Hildebrand approach is furnished by the fact that plots of experimental values of $\Delta Q$ against $\phi_1$ are generally more nearly symmetrical about $\phi_1 = 0.5$ than are plots of $\Delta H_m = (N_1 V_1 + N_2 V_2)\Delta Q$ against $N_1$ about $N_1 = 0.5$ [151]. Equation 7.81 predicts a perfectly symmetrical plot of $\Delta Q$ against $\phi_1$ for cases of zero volume change on mixing.

In liquid systems in which molecules of different types undergo specific interactions, the simplifying assumptions on which (7.80) is based are not valid. In such cases heats of mixing may yield interesting indications as to the sources of nonideality. In cases where hydrogen bonding between the two components occurs, heats of mixing will in general be many times larger than those observed with regular mixtures, and of opposite sign. Zellhoefer and Copley [152] found that the maximum heat in the system $CHCl_3 : CH_3OCH_2\text{-}CH_2CH_2OCH_3$ occurs at a mole ratio of one, and in the system $CHCl_3 : CH_3\text{-}O(CH_2CH_2O)_4CH_3$ at a mole ratio of three. These results indicate that only alternate oxygens in the ether molecules are available for hydrogen bonding, presumably because of steric interferences.

## Mixing Devices

A calorimeter designed for use with liquids can be applied to the determination of enthalpies of solution, dilution, and mixing, as well as of chemical reactions (Section 8), after incorporation of a mixing device. The problem is to hold separately two phases until the calorimetrically appropriate time to mix them, for example, after thermal equilibration, and then to secure complete mixing of the phases without the generation of an unduly large extraneous thermal effect. As might be expected, a great deal of ingenuity has gone into the search for solutions to this problem.

The most commonly used mixing device consists of a thin, sealed glass bulb which contains one of the phases and is submerged in the other. At the appropriate moment the bulb is fractured by some suitable means. This method is discussed by Skinner et al. [153].

Permanent "dilution cups" have been employed by many workers. White and Roberts [154] have described a device particularly suited for use with solids. A "boat" of Bakelite is supported from the calorimeter cover in a vertical position. It is covered with a thin sheet of nitrocellulose which is waxed on. The cover is pulled up to open the boat by means of a platinum wire going out through the cover of the calorimeter. Pitzer [155] used a vertical glass tube with waxed-on closures of thin silver. A glass rod extending down through the hollow stirrer shaft was used to rupture the silver sheets. Partington and Soper [156] employed glass pipettes closed at the bottom with ground-glass stoppers which could be opened by a rod extending up through the elongated neck of the pipette. Lange [157] employed submerged metal pipettes which could be opened at both ends so that good circulation of the liquid in the calorimeter through the pipette would be ensured. The design of the pipettes was such that they could be emptied, cleaned, and refilled without dismantling the calorimeter. In the design of Young and Machin [158] the pipettes were vertical silver tubes closed at each end with waxed-on silver covers which were pulled off by wires controlled from outside the calorimeter. Gucker, Pickard, and Planck [159] used tantalum dilution cups with ground-in closures at each end. The cups were opened in a very uniform way by a rather complicated lifting device so that the heat effect accompanying this process would be sufficiently constant. Lipsett, Johnson, and Maas [160] placed one of the substances to be mixed in a separate metal container within their calorimeter, which calorimeter was stirred by being rotated about its axis; the inside container was so arranged that its cover fell off during the first revolution of the calorimeter. Provision was made for catching the cover in clips so that it would be held from further motion inside the calorimeter. A somewhat similar mixing arrangement was employed by Benzinger and Kitzinger [161] in their "heat burst" calorimeter, and has been retained in the Beckman Instruments Co., Model 190B Microcalorimeter, based on the Benzinger-Kitzinger design. Another commercially-available calorimeter, the Model 10700-2 manufactured by LKB Produkter AB, Stockholm, Sweden, also uses calorimeter rotation to secure mixing of two liquids previously in communication only by way of the vapor phase.

Several mixing calorimeters have been developed in which the two liquids to be mixed are kept apart by a mercury barrier. Representative of such instruments are those described by Scatchard et al. [162], and Brown and Horowitz [163]; in the former case mixing is achieved by rotating the calorimeter and thus disrupting the barrier, while in the latter case the open lower end of a cylindrical vessel is lifted out of a mercury pool at the bottom of the calorimeter vessel to bring about mixing.

A calorimeter incorporating a ball mixer was developed by Berger et al. [164]. With this mixer 1 ml of one liquid is mixed with 3 ml of another, with a mixing artifact of about 5 mcal.

Several mixing schemes, primarily for use in the determination of enthalpies of mixing of liquids, have been summarized by McGlashan [165].

It is appropriate to include mention of titration calorimetry and flow calorimetry in this discussion of mixing devices. In a number of titration calorimeters [166], one reagent is contained in the calorimeter and the other one, carefully brought to the correct temperature, is introduced with stirring from an external reservoir. The calorimeter described by Christensen et al. [166], has been further developed for commercial production by Tronac, Inc., of Provo, Utah.

An interesting system for *continuous* isothermal titration has been developed by Becker and Kiefer [167]. This consists essentially in a stirred-flow reactor of the heat-conduction type, of 25-ml volume, containing one of the liquids, with no vapor phase. The other liquid is added continuously at a rate of 1 ml/min. The stirring is sufficiently effective that the contents of the calorimeter can be considered to be of uniform composition at all times. The chief advantages of this method are that the entire composition range is covered continuously rather than discontinuously and that this operation requires a relatively short time.

Flow calorimeters are of rapidly increasing importance, and are considered in detail below. In current forms of these instruments, the two liquids to be mixed, after being equilibrated at the calorimeter temperature, are led through a simple mixing junction into the flow tube proper, where the small inside diameter of one millimeter or less ensures a short diffusion path for completion of the mixing.

### Batch Calorimeters

A very large number of mixing calorimeters of the batch type have been developed over the years. Many examples of these are included in various articles in *Experimental Thermochemistry*, Vol. 2 [168], and in *Biochemical Microcalorimetry* [169]. A versatile calorimeter of the batch type was described by Benjamin [170]. Instruments of the highest sensitivity have been constructed primarily for the determination of heats of dilution of very dilute solutions [171]. Several commercially available batch calorimeters were mentioned earlier.

### Titration Calorimeters

The subject of titration calorimeters and their use has been fully reviewed by Tyrrell and Beezer [172]. A very precise instrument of this type has been developed by Christensen et al. [166], and has been made commercially available by Tronac, Inc., Provo, Utah.

### Flow Calorimeters

An important class of flow calorimeter has been designed for the measurement of enthalpies of combustion (p. 395). The flow method has also been employed for other gas-phase reactions. For example, Kistiakowsky et al. [173], determined the enthalpies of hydrogenation of several unsaturated hydrocarbons by flowing a mixture of hydrogen and hydrocarbon over a bed of catalyst contained in a calorimeter.

The first precise flow microcalorimeter for use with liquids was built by Stoesser and Gill [174]. This instrument demonstrated the extreme convenience and versatility of the flow method, and has stimulated the further development of the method. The problem of bringing the two liquid streams exactly to the temperature of the calorimeter is readily solved in heat flow calorimeters (Section 3) having a massive heat sink with which the entering liquids can be equilibrated before being mixed. Two recent designs have made use of this principle. One of these, by Monk and Wadsö [175], utilizes solid state thermoelectric modules for measuring the rate of heat transfer between a flow tube and the heat sink; this instrument is commercially available as Model 10700-1 from LKB Produkter AB, Stockholm, Sweden. The other [176], which is a modification of the Beckman Model 190 microcalorimeter, employs a 10,000-junction electroplated thermopile for monitoring the heat transfer. In recent work in the author's laboratory it has been found to be possible, by means of very careful thermostating of the calorimeter and some of the associated equipment, to achieve a useful sensitivity of $10^{-8}$ cal/sec with the modified Beckman calorimeter. This sensitivity exceeds by a factor of approximately 10 what has previously been accomplished with flow calorimeters. With a typical flow rate of 0.002 ml/sec, the sensitivity is $5 \times 10^{-6}$ cal/ml. This is considerably below the sensitivity achieved with batch calorimeters such as those devised by Lange and Gucker and their colleagues [166], but it is realized with far smaller volumes of solutions.

A feature of recent flow calorimeters which makes them especially suitable for a variety of purposes is the fact that there is no gas phase in contact with the liquids either before or after mixing. This is a very desirable situation in the study of the mixing of volatile organic liquids, and it affords the possibility of strictly anaerobic operation with readily oxidizable systems. Additional important aspects of flow calorimeters are their extreme simplicity of operation and the rapidity with which successive experiments can be performed.

The enthalpy changes on mixing gases or vapors are accurately determined in a flow calorimeter described by Wormald [177].

## 8   HEATS OF CHEMICAL REACTIONS

The direct measurement of the heats of chemical reactions, other than combustion (Section 5), is a field of calorimetry which has received relatively little attention until recent years. One of the chief reasons for this is the difficulty of making calorimetric observations on processes which are not complete within a few minutes. Another reason is that direct measurements cannot lead to significant results in cases which are complicated by side reactions. Furthermore, many reactions take place only under conditions which are outside the scope of direct calorimetry. There are, however, many reactions which are susceptible to direct calorimetric observation, and the data obtained from such measurements are of considerable importance. We shall first point out some of the applications which have been made of directly determined heats of reaction, and then discuss the experimental methods available for measuring the heats of fast and of slow processes. Methods for the indirect estimation of heats of reaction will be summarized in the last part of this section.

It was pointed out in Section 1 that of the three thermodynamic parameters, $\Delta G^0$, $\Delta S^0$, and $\Delta H^0$, which are frequently determined for chemical reactions, only the standard enthalpy change is in general independent of the activity units employed. This generalization holds in all cases where the usual selection of standard states, based on taking the activities of all solutes equal to their concentrations at infinite dilution and the activity of the solvent equal to unity at infinite dilution, is employed.

## Applications

### Thermodynamics of Chemical Reactions

The three most important thermodynamic parameters for a chemical reaction are the standard changes in free energy, $\Delta G^0$, entropy, $\Delta S^0$, and enthalpy, $\Delta H^0$, related by the equation

$$\Delta G^0 = \Delta H^0 - T\Delta S^0. \qquad (7.82)$$

It frequently happens that some insight into the nature of a reaction may be derived from a knowledge of the enthalpic and entropic contributions to the standard free energy change in the reaction. For example, the extremely strong binding of Zn(II) ions in aqueous solution to cyclohexylenediamine tetraacetate, $C_6H_4[N(CH_2COO^-)_4]_2$, $\Delta G^0 = -25.9$ kcal/mole at 20°C, is due only in minor degree to a favorable energy change, $\Delta H = -1.9$ kcal/mole. The complex is highly stable primarily because of the very large favorable entropy change, $\Delta S^0 = 81.8$ cal/(deg)(mole), in its formation [178], and this entropy increase must arise in large part from "liberation" of water molecules which solvate the reactant ions before the reaction takes place.

An essential step in evaluating the thermodynamic parameters for a reaction is the determination of the enthalpy change. Although this can be accomplished by indirect means (p. 411), a much more accurate value can in general be obtained by direct calorimetric measurement when this method is applicable.

There are many examples in the literature of the calorimetric determination of equilibrium constants [179]. In view of the relation between the equilibrium constant and the standard free energy change for a reaction,

$$\Delta G^0 = -RT \ln K, \qquad (7.83)$$

it is obvious that in such cases calorimetry suffices for the evaluation of both the standard entropy and the enthalpy changes.

It may be noted here that a closely related application of calorimetric data leads to a calorimetric estimate of molecular weight. Application of the van't Hoff equation, see (7.8), gives an enthalpy change in calories/mole;

since the calorimetric heat of reaction is in calories/g, the ratio of the two heats is equal to the g/mole. The first application of this procedure was by Barcroft and Hill [180], and Brown and Hill [181] to the hemoglobin molecule and the stoichiometry of its binding of oxygen.

The calorimetrically determined enthalpy change in a reaction gives in general the most accurate estimate available for the dependence of the free-energy change on temperature, see (7.18).

### Heats of Formation

If, in a given reaction, the heats of formation of all the substances involved but one are known, the heat of formation of that substance can be calculated from the heat of reaction. In view of the great importance of heats of formation for the "master table of chemical thermodynamics" [182] (Section 4, p. 373), this constitutes a most important general application of reaction heats.

### Heats of Hydrogenation

Kistiakowsky and co-workers [183] pioneered in the precise determination of the heats of hydrogenation of unsaturated compounds. The following examples illustrate the conclusions that can be drawn from such measurements:

1. An evaluation of the heats of hydrogenation of benzene, cyclohexadiene, and cyclohexane, together with estimates of the appropriate entropy changes, give the following free energies (at 298°K, 1 atm);

$$benzene_{(g)} + H_{2(g)} = cyclohexadiene-1,3_{(g)}$$

$$\Delta G^0 = 13.6 \text{ kcal/mole}$$

$$cyclohexadiene-1,3_{(g)} + H_{2(g)} = cyclohexene_{(g)}$$

$$\Delta G^0 = -17.7 \text{ kcal/mole}.$$

Since $\Delta G^0 = -RT \ln K$, where $K$ is the equilibrium constant for a reaction, it is evident that cyclohexadiene is not obtainable from benzene by hydrogenation because the free energy change is such that only a negligible quantity of the product is present at equilibrium. It is, therefore, unnecessary to make assumptions concerning the relative rates of the successive steps in the hydrogenation of benzene to explain the absence of cyclohexadiene in the reaction product.

2. Support was found for the conclusion that rather large repulsive interactions exist between nonbonded atoms in a molecule. For example, the heat of hydrogenation of cyclopentene would seem to indicate that this molecule is less strained than cyclopentane, a conclusion strikingly at variance with the prediction based on the tetrahedral structure of the carbon atom. This apparently anomalous result can be understood when account is taken of the

fact that, in the five-membered rings, the hydrogen atoms are held in the "eclipsed" position, where their mutual repulsions are greater than when they are farther apart as in the more usual "staggered" position.

3. Heats of hydrogenation have been employed in the computation of resonance energies [184], the results obtained being in general more reliable than those obtained from heats of formation. It is applications of this type which are primarily responsible for the continuing interest shown by various workers in the determination of heats of hydrogenation [185].

### Heats of Reactions of Biochemical Interest

The calorimetric determination of the heats of biochemical reactions has in recent years become the most active area in the field of reaction calorimetry. Reviews of this activity have been published by Sturtevant [186] and by Brown [187]. From the instrumental point of view, work in this area is particularly challenging since the combination of small heat effects per mole and large molecular weights frequently overtaxes the capabilities of currently available equipment.

### Calorimetric Determination of Reaction Rates

Under usual reaction conditions, both in the gas phase and in solution, the heat evolved or absorbed during a reaction is proportional to the extent of reaction. Since most reactions are accompanied by a appreciable enthalpy change, this proportionality serves as the basis for a very general method [188] for the determination of reaction rates. Recent contributions to this application of calorimetry have been made by Berger et al. [189], and by Meites et al. [190].

Bell and Clunie [191] have developed an interesting method for determining the rates of reactions in solution having half-times down to 0.2 sec. The method depends on observing the temperature maximum (or minimum) reached when an exothermic (or endothermic) reaction takes place in a chamber which loses heat to its surroundings at a controlled rate.

Temperature rises have also been employed in the study of the kinetics of rapid gas-phase reactions [192].

## Indirect Determination of Heats of Reaction

### From Calorimetric Data

Heats of reaction are frequently calculated from heats of combustion (see Section 5). It was pointed out previously that the results of such calculations are in most cases relatively inaccurate because of the large magnitude of heats of combustion compared with other heats of reaction, and that this situation is somewhat improved when heats of reaction are inferred from heats of solution because of the smaller magnitude of heats of solution.

Heats of reaction are sometimes evaluated from other heats of reaction. For example, the hydrogenation data of Kistiakowsky et al. [183], for butene-1 and *cis*-butene-2 give a value for the heat of isomerization of butene-1 to *cis*-butene-2 of $-1.77$ ckal/mole. It is interesting to note that this difference is about 6% of either heat of hydrogenation, while it is only 0.3% of either heat of combustion.

### From Other Thermodynamic Data

The heat of a reaction can be evaluated from other thermodynamic data in several ways. If the free energy and entropy changes accompanying an isothermal process are known, the change in heat content is given by

$$\Delta H = \Delta G + T\Delta S. \tag{7.84}$$

If this expression is differentiated with respect to temperature, and use is made of the relations

$$\left(\frac{\partial H}{\partial T}\right)_P = C_P = T\left(\frac{\partial S}{\partial T}\right)_P, \tag{7.85}$$

it follows that

$$\left(\frac{\partial \Delta G}{\partial T}\right)_P = \frac{\Delta G - \Delta H}{T}. \tag{7.86}$$

Here, the subscript $P$ means that the pressure on each substance is to be the same whether we are working at one temperature or another; the process under consideration is not necessarily one which takes place at constant pressure. Thus, if the process $A = B$, and the pressure on substance $A$ is 1 atm, and that on substance $B$ is 10 atm at one temperature, these same pressures must be used at other temperatures. From (7.86) it is readily found that

$$\frac{\partial(\Delta G/T)}{\partial(1/T)} = \Delta H. \tag{7.87}$$

This equation (see (7.18)) provides a method for evaluating heats of reaction from free energy data which is equivalent to that based on the van't Hoff equation (7.8).

An important method for determining heats of reaction is based on the observation of the emf of a reversible galvanic cell. According to the Gibbs-Helmholtz equation,

$$\frac{\Delta H}{nF} = T\frac{\partial E}{\partial T} - E. \tag{7.88}$$

In this equation, $F$ is the Faraday equivalent, $n$ is the number of Faraday equivalents passing through the cell when the cell reaction takes place, $\Delta H$ is the heat of this reaction, and $E$ is the reversible electromotive force of the cell. Numerous illustrations of the application of this method will be found in the literature.

## Determination of the Heats of Fast Reactions

The heat effect accompanying a reaction can be determined by the methods of calorimetry discussed in previous sections, provided the process is complete in a reasonably short time. The allowable magnitude of this period depends on several factors, such as the size of the heat effect, the desired accuracy, and the design of the calorimeter, particularly with respect to heat leakage. In the brief discussion which follows, attention will be limited to calorimeters designed for the study of reactions in the liquid phase. Calorimeters suitable for gas-phase reactions have been referred to above.

### Batch Calorimeters

The design and operation of batch-type reaction calorimeters has been considered in detail by Skinner et al. [193]. Successful calorimeters of this type are now commercially available (Section 7). Reaction times as long as one hour can be handled without excessive loss of accuracy.

### Flow Calorimeters

Flow calorimetry is of increasing importance in the determination of heats of reaction. Although the sensitivity in terms of cal/unit mass attainable with batch calorimeters is in general higher than that characteristic of flow calorimeters, the latter have impressive advantages in ease and rapidity of operation. When continuous flow is employed, it is obvious that the reaction must be completed within the residence time of the solution in the calorimeter, which is usually of the order of a few minutes, unless one is interested only in the *rate* of heat evolution or absorption. Slower processes can be handled by the stopped-flow procedure, with a considerable loss in accuracy.

An ingenious modification of the flow method permits the estimation of both the velocities and the heats of reactions having half-periods as short as a few milliseconds. Two liquid streams moving at high velocity are very rapidly mixed and then flow through an observation tube which is fitted with thermocouples at appropriate intervals. This method was developed largely by Hartridge and Roughton and their collaborators [194], and by LaMer and Read [195]. Particular attention must be given to ensuring very rapid mixing of the solutions. The results obtained have demonstrated that, at the very high rates of flow used in these experiments, the heat loss from the flowing solution to the surroundings is negligible.

## Determination of the Heats of Slow Processes

The calorimetric observation of any process which requires a reaction period greater than one hour poses difficult technical problems not encountered in the great majority of reaction calorimetric studies [196].

## Adiabatic Calorimeters

### EXTRAPOLATION METHODS

In cases where the heat evolution or absorption follows a discoverable kinetic law, and the apparent calorimeter temperature does not differ significantly from the mean temperature of the calorimeter and its contents, extrapolation procedures can be employed [197]. The condition regarding calorimetric lags will usually be met if an all-metal calorimeter of low heat capacity is employed, and temperature measurements are made by thermocouples or resistance thermometers of low lag. In some cases it is not necessary to stir the calorimeter after the initial mixing process is complete. Extrapolation procedures should be employed whenever possible since they offer several advantages: the heat of mixing the reactants may be distinguished from the heat of reaction and may be separately evaluated; the rate constant of the reaction is obtained; and it is frequently possible to shorten materially the duration of the calorimetric experiment and thus to decrease calorimetric errors.

### OTHER METHODS

If the heat change does not follow a simple kinetic law, it becomes necessary either to continue calorimetric observations long enough to ensure complete reaction, or to determine the amount of reaction in a given time interval by analysis of samples withdrawn from the calorimeter or from a reaction run under identical conditions outside the calorimeter. The work of Tong and Kenyon [198] on polymerization reactions illustrates the problems met with when processes following complicated kinetics are studied. As mentioned above, these methods do not give as much information as do the extrapolation methods. Several of the batch calorimeters referred to in Section 7, especially those of the twin type, can with suitable care be adapted to the study of relatively slow processes.

## Heat-Flow Calorimeters

The heat-flow calorimeter first constructed by Tian [44] and later modified and improved by Calvet and his colleagues [46] and by others [47], is without question the most suitable instrument for the observation of extremely slow processes, extending even to months in duration. Although preeminently adapted to the calorimetry of extremely slow processes, this instrument has been successfully applied to reactions of intermediate duration as well as to fast processes. Other heat-flow batch calorimeters mentioned earlier can also be utilized in the study of reactions of intermediate duration.

# References

1. G. H. Hess, *Ann. Phys.* **50**, 385 (1840); W. Ostwald's *Klassiker der exakten Wissenschaften, No.* 9, Englemann, Leipzig, 1890.
2. O. B. Ptitsyn and T. M. Birshtein, *Biopolymers* **7**, 435 (1969).
3. R. A. Alberty, *J. Amer. Chem. Soc.* **91**, 3899 (1969).
4. F. D. Rossini, *J. Res. Natl. Bur. Std.* **6**, 1 (1931); **12**, 735 (1934); E. F. Mueller and F. D. Rossini, *Amer. J. Phys.* **12**, 1 (1944); H. F. Stimson, *ibid.* **23**, 614 (1955).
5. H. F. Stimson, *Amer. J. Phys.* **2**, 617 (1955).
6. Various aspects of the measurement of electrical energy in calorimetry are discussed by several authors in *Experimental Thermodynamics*, Vol. I, J. P. McCullough and D. W. Scott, Eds., Butterworths, London, 1968. See especially Chap. 3 by L. Hartshorn and A. G. McNish, and Chap. 5, by E. F. Westrum, Jr., G. T. Furukawa, and J. P. McCullough.
7. M. Dole et al., *Rev. Sci. Instr.* **22**, 818 (1951).
8. G. D. Foss and D. A. Pitt, *Rev. Sci. Instr.* **39**, 1375 (1968).
9. A. Buzzell and J. M. Sturtevant, *J. Amer. Chem. Soc.* **73**, 2454 (1951).
10. E. C. Pollard and J. M. Sturtevant, *Microwaves and Radar Electronics*, Wiley, New York, 1948, Chap. 9.
11. A. Buzzell and J. M. Sturtevant, *Rev. Sci. Instr.* **19**, 688 (1948).
12. T. R. Clem, R. L. Berger, and P. D. Ross, *Rev. Sci. Instr.* **40**, 1273 (1969).
13. J. J. Christensen, H. D. Johnston, and R. M. Izatt, *Rev. Sci. Instr.* **39**, 1356 (1968).
14. H. Albert, Ph.D. Thesis, University of Colorado, 1969.
15. W. P. White, *The Modern Calorimeter*, Chem. Catalog. Co., New York, 1928.
16. J. P. McCullough and D. W. Scott, Eds., *Experimental Thermodynamics*, Vol. 1, Butterworths, London, 1968. See especially Chap. 1, by D. C. Ginnings, and Chap. 4, by D. C. Ginnings and E. D. West.
17. S. G. Lipsett, F. M. G. Johnson, and O. Maas, *J. Amer. Chem. Soc.* **49**, 925, 1940 (1927); K. L. Wolf and H. Frahm, *Z. phys. Chem.* **A178**, 411 (1937); J. M. Sturtevant, *J. Phys. Chem.* **45**, 127 (1941); C. Kitzinger and T. Benzinger, *Z. Naturforsch.* **10b**, 365 (1955); I. Wadsö, *Acta Chem. Scand.* **22**, 927 (1968).
18. B. H. Carroll and J. H. Mathews, *J. Amer. Chem. Soc.* **46**, 30 (1924); J. H. Awbery and E. Griffiths, *Proc. Phys. Soc. (London)* **52**, 770 (1940).
19. J. M. Sturtevant, *Physics* **7**, 232 (1936); *J. Amer. Chem. Soc.* **59**, 1528 (1937).
20. W. P. White, *op. cit.*, p. 50.
21. F. Daniels, *J. Amer. Chem. Soc.* **38**, 1473 (1916); J. W. Williams and F. Daniels, *ibid.*, **46**, 903 (1924).
22. D. C. Ginnings and E. D. West, in *Experimental Thermodynamics*, Vol. 1, J. P. McCullough and D. W. Scott, Eds., Butterworths, London, 1968, p. 85.
23. D. R. Harper, *Natl. Bur. Std. (U.S.), Bull.* **11**, 319 (1915).
24. J. J. Christensen, R. M. Izatt, and L. D. Hansen, *Rev. Sci. Instr.* **36**, 779 (1965); J. J. Christensen et al., Ref. 14.
25. W. P. White, *op. cit.*; H. C. Dickinson, *Natl. Bur. Standards (U.S.) Bull.*

**11,** 189 (1915); T. J. Coops and K. van Nes, *Rec. Trav. Chim.* **66,** 161 (1947); J. W. Stout, in *Experimental Thermodynamics,* Vol. 1, J. P. McCullough and D. W. Scott, Eds., Butterworths, London, 1968, p. 215; H. J. V. Tyrrell and A. E. Beezer, *Thermometric Titrimetry,* Chapman and Hall, London, 1968, Chap. 2.

26. T. W. Richards, *J. Amer. Chem. Soc.* **31,** 1275 (1909).

27. W. P. White, *op. cit.,* p. 116. See also D. C. Ginnings, in *Experimental Thermodynamics,* Vol. 1, J. P. McCullough and D. W. Scott, Eds., Butterworths, London, 1968, p. 1.

28. F. Barry, *J. Amer. Chem. Soc.* **44,** 899 (1922).

29. M. Dole et al., *Rev. Sci. Instr.* **22,** 812 (1951); A. Buzzell and J. M. Sturtevant, *J. Amer. Chem. Soc.* **73,** 2454 (1951); D. R. Stull, *Anal. Chim. Acta* **17,** 133 (1957); M. G. Zabetakis, R. S. Craig, and K. F. Sterrett, *Rev. Sci. Instr.* **28,** 497 (1957); E. D. West and D. C. Ginnings, *ibid.* **28,** 1070 (1957); T. M. Gayle and W. T. Berg, *ibid.* **37,** 1740 (1966); E. D. West and E. F. Westrum, Jr., in *Experimental Thermodynamics,* Vol. 1, J. P. McCullough and D. W. Scott, Eds., Butterworths, London, 1968, p. 333

30. J. P. Joule, *Mem. Proc. Manchester Lit. Phil. Soc.* **2,** 559 (1845).

31. L. Pfaundler, *Sitzber. Akad. Wiss. Wien, Math. Naturw. Kl.* **59,** 145 (1869); **100,** 351 (1891).

32. W. P. White, *op. cit.,* p. 126; D. C. Ginnings, in *Experimental Thermodynamics,* Vol. 1, J. P. McCullough and D. W. Scott, Eds., Butterworths, London, 1968, p. 1.

33. A. V. Hill, *J. Physiol.* **43,** 261 (1911); H. H. Dixon and N. G. Ball, *Sci. Proc. Roy. Dublin Soc.* **16,** 153 (1920); W. Swietoslawski, *Microcalorimetry,* Reinhold, New York, 1946, Chap. 4; E. Calvet, *The Microcalorimetry of Slow Phenomena,* Report to the Commission on Thermochemistry of the International Union of Pure and Applied Chemistry, May 1952.

34. M. Dole et al. *Rev. Sci. Instr.* **22,** 818 (1951); A. Buzzell and J. M. Sturtevant, *J. Amer. Chem. Soc.* **73,** 2454 (1951); T. R. Clem, R. L. Berger, and P. D. Ross, *Rev. Sci. Instr.* **40,** 1273 (1969); J. J. Christensen, H. D. Johnston, and R. M. Izatt, *ibid.* **39,** 1356 (1968); H. Albert, Ph. D. Thesis, University of Colorado, 1969.

35. W. P. White, *op. cit.* p. 131.

36. S. B. Thomas and G. S. Parks, *J. Phys. Chem.* **35,** 2091 (1931). Calorimeters based on the same principle have been described by D. H. Andrews, *J. Amer. Chem. Soc.* **48,** 1287 (1926); E. Haworth and D. H. Andrews, *ibid.,* **50,** 2998 (1928); R. H. Smith and D. H. Andrews, *ibid.,* **53,** 3644 (1931); D. R. Stull, *ibid.,* **59,** 2726 (1937); W. T. Ziegler and C. E. Messer, *ibid.,* **63,** 2694 (1941).

37. R. Bunsen, *Ann. Physik* **141,** 1 (1870). For more modern descriptions of ice calorimeters, see W. Ostwald-R. Luther, *Physikochemische Messungen,* 4th ed., 1925, p. 393; W. Swietoslawski, A. Zmaczynski, I. Zlotowski, J. Salcewicz, and J. Usakiewicz, *Compt. Rend.* **196,** 1970 (1933); *Roczniki Chem.* **14,** 250 (1934); W. Swietoslawski, *Microcalorimetry,* Chap. 5; H. A. Skinner, J. M. Sturtevant and S. Sunner, in *Experimental Thermochemistry,* Vol. 2, H. A. Skinner, Ed., Interscience, New York, 1962, Chap. 9; O. Kubaschewski and

R. Hultgren, *ibid.* Chap. 16; J. Opdycke, C. Gay, and H. H. Schmidt, *Rev. Sci. Instr.* **37,** 1010 (1966); T. B. Douglas and E. G. King, in *Experimental Thermodynamics*, Vol. 1, J. P. McCullough and H. A. Skinner, Eds., Butterworths, London, 1968, Chap. 8; A. J. B. Cruickshank, T. Ackermann, and P. A. Giguiere, *ibid.* Chap. 12.

38. F. Simon and M. Ruhemann, *Z. physik. Chem.* **129,** 321 (1927). See also C. A. Taylor and W. H. Rinkenbach, *J. Amer. Chem. Soc.* **46,** 1505 (1924).

39. C. A. Kraus and J. A. Ridderhof, *J. Amer. Chem. Soc.* **56,** 79 (1934); C. A. Kraus and R. F. Prescott, *ibid.*, **56,** 86 (1934); C. A. Kraus and F. C. Schmidt, *ibid.*, **56,** 2297 (1934); F. C. Schmidt, F. J. Studer, and J. Sottysiak, *ibid.*, **60,** 2780 (1938).

40. J. W. Stout and W. M. Jones, *Phys. Rev.* **71,** 582 (1947).

41. J. R. Lacher et al., *J. Amer. Chem. Soc.* **71,** 1330 (1949); *J. Phys. Chem.* **60,** 492 (1956).

42. H. Junkers, *J. Gasbeleucht.* **50,** 520 (1907).

43. W. Swietoslawski, *op. cit.*, Reinhold, New York, 1946, Chapter 10.

44. A. Tian, *Bull. Soc. Chim. France* **33,** 427 (1923); *J. Chim. Phys.* **30,** 665 (1933); E. Calvet, *op. cit.*, Ref. 33. See also the recent discussions by H. A. Skinner, J. M. Sturtevant, and S. Sunner, in *Experimental Thermochemistry*, Vol. 2, H. A. Skinner, Ed., Interscience, New York, 1962, p. 198; and H. A. Skinner, in *Biochemical Microcalorimetry*, H. D. Brown, Ed., Academic, New York, 1969, p. 7.

45. W. Swietoslawski and J. Salcewicz, *Compt. Rend.* **199,** 935 (1934); R. Sandri, *Monatsh.* **68,** 415 (1936). J. L. Magee and F. Daniels, *J. Amer. Chem. Soc.* **62,** 2825 (1941), have employed this method with their calorimeter designed for photochemical reactions.

46. E. Calvet, *op. cit.*; E. Calvet and H. Prat, *Recent Progress in Microcalorimetry*, translation by H. A. Skinner, Pergamon, London, 1963.

47. R. W. Attree, R. L. Cushing, J. A. Ladd, and J. J. Pieroni, *Rev. Sci. Instr.* **29,** 491 (1958); J. B. Darby, R. Kleb, and O. J. Kleppa, *ibid.*, **37,** 164 (1966); G. Rialdi and P. Profumo, *Bipolymers* **6,** 899 (1968); W. J. Evans in *Biochemical Microcalorimetry*, H. D. Brown, Ed., Academic, New York, 1969, p. 257.

48. C. Kitzinger and T. H. Benzinger, *Z. Naturforsch.* **10b,** 365 (1955); T. H. Benzinger, *Fractions*, No. 2 (1965) (published by Beckman Instruments, Inc.); T. H. Benzinger and C. Kitzinger in *Temperature: Its Measurement and Control in Science and Industry*, Vol. 3, C. M. Herzfield, Ed., Reinhold, New York, 1963, Part 3, Chap. 5.

49. J. M. Sturtevant and P. A. Lyons, *J. Chem. Thermodynam.* **1,** 201 (1969); *Fractions*, No. 1 (1969) (published by Beckman Instruments, Inc.).

50. I. Wadsö, *Acta Chem. Scand.* **22,** 927 (1968).

51. P. Monk and I. Wadsö, *ibid.*, **22,** 1842 (1968).

52. P. Chall and O. Doepke, *Z. Elektrochem.* **37,** 357 (1931).

53. O. J. Kleppa, *J. Phys. Chem.* **59,** 175, 354 (1955); F. E. Wittig and F. Huber, *Z. Elektrochem.* **60,** 1181 (1956); *ibid.* **62,** 529 (1958); J. B. Darby, R. Kleb, and O. J. Kleppa, *Rev. Sci. Instr.* **37,** 164 (1966); E. D. West and E. F. Westrum,

Jr., in *Experimental Thermodynamics*, Vol. 1, J. P. McCullough and D. W. Scott, Eds., Butterworths, London, 1968, Chap. 9; D. L. Hildenbrand, *Adv. High Temp. Chem.* **1**, 193 (1967); E. F. Westrum, Jr., *ibid.* **1**, 239 (1967).

54. J. L. Magee, T. W. DeWitt, C. E. Smith, and F. Daniels, *J. Amer. Chem. Soc.* **61**, 3529 (1939); J. Tonnelat, *Ann. Phys.* **20**, 601 (1945).

55. C. W. Beckett and A. Cezairliyan, in *Experimental Thermodynamics*, Vol. 1, J. P. McCullough and D. W. Scott, Eds., Butterworths, London, 1968, Chap. 14.

56. R. L. Berger, in *Temperature: Its Measurement and Control in Science and Industry*, Vol. 3, C. M. Herzfeld, Ed., Reinhold, New York, 1963, Part 3, Chap. 6; R. L. Berger and L. C. Stoddart, *Rev. Sci. Instr.* **36**, 78 (1965); R. L. Berger et al., *ibid.* **39**, 486, 493, 498 (1968).

57. H. Savage, *Rev. Sci. Instr.* **37**, 1062 (1966).

58. R. Nowicki, *Electron. Lett.* **4**, 404 (1968).

59. G. E. Boyd, in *Analytical Calorimetry*, R. S. Porter and J. F. Johnson, Eds., Plenum, New York, 1968, p. 141.

60. D. R. Stull, *Anal. Chim. Acta* **17**, 133 (1957).

61. J. G. Aston, in *Treatise on Physical Chemistry*, 3rd ed., Vol. 1, H. S. Taylor and S. M. Glasstone, Eds., Van Nostrand, Princeton, N. J., 1942, p. 620.

62. D. R. Stull, E. F. Westrum, Jr., and G. C. Sinke, *The Chemical Thermodynamics of Organic Compounds*, Wiley, New York, 1969, pp. 24 ff.

63. T. Y. Tsong, R. P. Hearn, D. P. Wrathall, and J. M. Sturtevant, *Biochemistry* **9**, 2666 (1970).

64. I. E. Scheffler and J. M. Sturtevant, *J. Mol. Biol.* **42**, 577 (1969).

65. E. F. Westrum, Jr., in *Analytical Calorimetry*, R. S. Porter and J. F. Johnson, Eds., Plenum, New York, 1968, p. 231; P. Clechet and J.-C. Merlin, *Bull. Soc. Chim. (France)* **1964**, 2644.

66. H. C. Dickinson and N. S. Osborne, *Natl. Bur. Standards (U.S.) Bull.* **12**, 49 (1915); M. Le Blanc and E. Möbius, *Ber. Verhandl. Sächs. Akad. Wiss. Leipzig, Math. Phys. Kl.* **85**, 75 (1933). E. L. Skau, *J. Amer. Chem. Soc.* **57**, 243 (1935).

67. E. L. Skau, *Bull. Soc. Chim. Belg.* **43**, 287 (1934); *J. Phys. Chem.* **39**, 541 (1935).

68. H. J. Hoge, *J. Res. Natl. Bur. Std.* **36**, 111 (1946).

69. A. Eucken, *Physik. Z.* **10**, 586 (1910), W. Nernst, *Sitzber. Preuss. Akad. Wiss. Physik.-Math. Kl.* **1910**, 262; *Ann. Phys.* **36**, 395 (1911).

70. J. S. Stout, in *Experimental Thermodynamics*, Vol. 1, J. P. McCoullough and D. W. Scott, Eds., Butterworths, London, 1968, Chap. 6.

71. E. F. Westrum, Jr., G. T. Furukawa, and J. P. McCullough, *ibid.*, Chap. 5.

72. R. W. Hill, D. L. Martin, and D. W. Osborne, *ibid.*, Chap. 7.

73. H. L. Callendar, *Trans. Roy. Soc. (London)* **A199**, 55 (1902).

74. K. Scheel and W. Heuse, *Ann. Physik* **37**, 79 (1912); **40**, 473 (1913); W. Heuse, *ibid.* **59**, 86 (1919).

75. J. P. McCullough and G. Waddington, in *Experimental Thermodynamics*, Vol. 1, J. P. McCullough and D. W. Scott, Eds., Butterworths, London, 1968, Chap. 10.

76. A. Eucken, *Energie- und Wärmeinhalt*, Vol. 3, Part 1, of Wien-Harms, *Handbuch der Experimentalphysik*, pp. 391 *et seq.*

77. O. Lummer and E. Pringsheim, *Ann. Physik* **64**, 555 (1898).

78. M. Knudsen, *Ann. Physik* **34**, 593 (1911).

79. A. Eucken and K. Weigert, *Z. physik. Chem.* **B23**, 265 (1933).

80. G. B. Kistiakowsky and F. Nazmi, *J. Chem. Phys.* **6**, 18 (1938); G. B. Kistiakowsky, J. R. Lacher, and F. Stitt, *ibid.* **7**, 289 (1939).

81. E. R. Grilly, W. J. Taylor, and H. L. Johnston, *J. Chem. Phys.* **14**, 435 (1946).

82. J. R. Partington and W. G. Shilling, *The Specific Heat of Gases*, Van Nostrand, Princeton, N. J., 1924.

83. R. E. Cornish and E. D. Eastman, *J. Amer. Chem. Soc.* **50**, 627 (1928).

84. F. R. Bichowsky, *J. Amer. Chem. Soc.* **45**, 2225 (1923); B. C. Hendricks, J. H. Dorsey, R. LeRoy, and A. G. Moseley, *J. Phys. Chem.* **34**, 418 (1930); F. S. Stow and J. H. Elliott, *Anal. Chem.* **20**, 250 (1948); F. E. Blacet, P. A. Leighton, and E. P. Bartlett, *J. Phys. Chem.* **35**, 1933 (1931).

85. D. C. Ginnings and H. F. Stimson, in *Experimental Thermodynamics*, Vol. 3, J. M. McCullough and D. W. Scott, Eds., Butterworths, London, 1968, Chap. 11.

86. A. J. B. Cruickshank, T. Ackermann, and P. A. Giguere, *ibid.*, Chap. 12.

87. E. W. Nelson and R. F. Newton, *J. Amer. Chem. Soc.* **63**, 2178 (1941).

88. J. S. Burlew, *J. Amer. Chem. Soc.* **62**, 681, 690, 696 (1940).

89. D. R. Harper, *Natl. Bur. Standards (U.S.) Bull.* **11**, 259 (1914).

90. D. H. Andrews, G. Lynn, and J. Johnston, *J. Amer. Chem. Soc.* **48**, 1274 (1926).

91. T. B. Douglas and E. G. King, in *Experimental Thermodynamics*, Vol. 1, J. P. McCullough and D. W. Scott, Eds., Butterworths, London, 1968, Chap. 8.

92. E. D. West and E. F. Westrum, Jr., *ibid.*, Chap. 9.

93. C. W. Beckett and A. Cezairliyan, *ibid.*, Chap. 14.

94. J. C. Southard, *J. Amer. Chem. Soc.* **63**, 3142 (1941). See also K. K. Kelley, B. F. Naylor, and C. H. Shomate, *U.S. Bur. Mines, Tech. Papers, no. 686*, Washington, D. C., 1946; T. B. Douglas and E. G. King, in *Experimental Thermodynamics*, Vol. 1, J. P. McCullough and D. W. Scott, Eds., Butterworths, London, 1968, Chap. 8; A. J. B. Cruickshank, T. Ackermann, and P. A. Giguere, *ibid.*, Chap. 12.

95. T. Ackermann, *Ber. Bunsenges. physik. Chem.* **62**, 411 (1958); P. L. Privalov, *Biofizika* **8**, 308 (1963); T. Ackermann and H. Rüterjans, *Ber. Bunsenges. phys. Chem.* **68**, 850 (1964); P. L. Privalov, D. R. Monselidze, G. M. Mrevlishvili, and V. A. Magaldadze, *J. Exptl. Theoret. Phys. (U.S.S.R.)* **47**, 2073 (1964); S. J. Gill and K. Beck, *Rev. Sci. Instr.* **36**, 274 (1965); F. E. Karasz and J. M. O'Reilly, **37**, 255 (1966); R. Danforth, H. Krakauer, and J. M. Sturtevant, *ibid.*, **38**, 484 (1967); A. J. B. Cruickshank, T. Ackermann, and P. A. Giguere, in *Experimental Thermodynamics*, Vol. 1, J. P. McCullough and D. W. Scott, Eds., Butterworths, London, 1968, Chap. 12; T. R. Clem, R. L. Berger, and P. D. Ross, *Rev. Sci. Instr.* **40**, 1273 (1969).

96. T. Y. Tsong, R. P. Hearn, D. P. Wrathall, and J. M. Sturtevant, *Biochemistry* **9**, 2666 (1970); Danforth et al., Ref. 95.

97. Hallikainen Instruments, Richmond, Calif.
98. A. Buzzell and J. M. Sturtevant, *J. Amer. Chem. Soc.* **73**, 2454 (1951).
99. F. D. Rossini, D. D. Wagman, W. H. Evans, S. Levine, and I. Jaffe,"Selected Values of Chemical Thermodynamic Properties," *Natl. Bur. Std. Circ. No. 500*, Washington, D. C., 1952.
100. M. Mansson and S. Sunner, *Acta Chem. Scand.* **17**, 723 (1963).
101. F. R. Biochowsky and F. D. Rossini, *The Thermochemistry of Chemical Substances*, Reinhold, New York, 1936.
102. M. S. Kharasch, *J. Res. Natl. Bur. Std.* **2**, 359 (1929).
103. *International Critical Tables*, Vol. 5, McGraw-Hill, New York, 1929, p. 162.
104. *Tables Annuelles de Constantes et Donnees Numeriques*, Vol. 7, p. 150 (1925–1926); Vol. 8, p. 177 (1927–1928); Vol. 9, p. 122 (1929); Vol. 10, p. 119 (1930); Vol. 11, Section 12 (1931–1934). Landolt-Börnstein, *Physikalischchemische Tabellen*, 5th ed., Vol. 2, p. 1586; Suppl. Vol. 1, p. 866; Suppl. Vol. 2b, p. 1633; Suppl. Vol. 3c, p. 2893 (up to March, 1936). "Selected Values of Chemical Thermodynamic Properties," *Natl. Bur. Standards Circ. No. 500*, Washington, D. C., 1952, and Supplements.
105. "Consolidated Index of Selected Property Values," *National Academy of Science—National Research Council, Publication 976*, Washington, D. C., 1962.
106. D. R. Stull, E. F. Westrum, Jr., and G. C. Sinke, *The Chemical Thermodynamics of Organic Compounds*, Wiley, New York, 1969.
107. G. S. Parks and H. M. Huffman, *The Free Energies of Some Organic Compounds*, Chemical Catalog Co., New York, 1932.
108. F. D. Rossini, *Chem. Rev.* **18**, 233 (1936).
109. A. L. Glasebrook and W. G. Lovell, *J. Amer. Chem. Soc.* **61**, 1717 (1939).
110. G. E. Moore and G. S. Parks, *J. Amer. Chem. Soc.* **61**, 2561 (1939).
111. J. M. Sturtevant, *J. Amer. Chem. Soc.* **59**, 1528 (1937); *J. Phys. Chem.* **45**, 127 (1941).
112. E. J. Prosen and F. D. Rossini, *J. Res. Natl. Bur. Std.* **27**, 289 (1941).
113. D. R. Stull, E. F. Westrum, Jr., and G. C. Sinke, *The Chemical Thermodynamics of Organic Compounds*, Wiley, New York, 1969, Chap. 6.
114. J. Taylor, C. R. L. Hall, and H. Thomas, *J. Phys. Colloid Chem.* **51**, 580, 593 (1947).
115. F. D. Rossini, Ed., *Experimental Thermochemistry*, Vol. 1, Interscience, New York, 1956; H. A. Skinner, Ed., Vol. 2, 1962. *The Chemical Thermodynamics of Organic Compounds*, D. R. Stull, E. F. Westrum, Jr., and G. C. Sinke, Eds., Wiley, New York, 1969. V. P. Kolesov, *Proc. First Internatl. Conf. on Calorim. and Thermodynamics*, Warasw, 1969.
116. W. A. Keith and H. Mackle, *Trans. Faraday Soc.* **54**, 353 (1958).
117. W. D. Good, D. R. Donslin, D. W. Scott, A. George, J. L. Lacina, J. P. Dawson, and G. Waddington, *J. Phys. Chem.* **63**, 1133 (1959); W. D. Good, D. W. Scott, J. L. Lacina, and J. P. McCullough, *ibid.* **63**, 1139 (1959).
118. L. Bjellerup, *Acta Chem. Scand.* **13**, 1511 (1959).
119. E. W. Washburn, *J. Res. Natl. Bur. Std.* **10**, 525 (1933).
120. F. D. Rossini, *J. Res. Natl. Bur. Std.* **8**, 119 (1932); **12**, 735 (1934); *Experi-*

*mental Thermochemistry*, Vol. 1, F. D. Rossini, Ed., Interscience, New York, 1956, Chap. 4. C. B. Miles and H. Hunt, *J. Phys. Chem.* **45**, 1346 (1941).

121. G. T. Armstrong, in *Experimental Thermochemistry*, Vol. 2, H. A. Skinner, Ed., Interscience, New York, 1962, Chap. 7.

122. D. D. Tunnicliff and H. Stone, *Anal. Chem.* **27**, 73 (1955). See also, R. F. Westrum, Jr., in *Analytical Calorimetry*, R. S. Porter and J. F. Johnson, Eds., Plenum, New York, 1968, p. 231; G. L. Driscoll, I. N. Daling, and F. Magnotta, *ibid.*, p. 271; C. Plato and A. R. Glasgow, Jr., *Anal. Chem.* **41**, 330 (1969); S. V. R. Mastrangelo and R. W. Dornte, *J. Amer. Chem. Soc.* **77**, 6200 (1955).

123. W. P. White, *J. Phys. Chem.* **24**, 393 (1920); E. W. Washburn, *Ind. Eng. Chem., Ind. Ed.* **22**, 985 (1930); and E. L. Skau, *Proc. Amer. Acad. Arts Sci.* **67**, 551 (1932); *J. Phys. Chem.* **37**, 609 (1933).

124. K. S. Pitzer, *J. Amer. Chem. Soc.* **62**, 331 (1940).

125. A. Eucken, *Energie-und Wärmeinhalt*, Vol. 8, Part 1, of Wien-Harms *Handbuch der Experimentalphysik*, p. 527.

126. A. Eucken, *op. cit.*, p. 543. See also A. S. Coolidge, *J. Amer. Chem. Soc.* **52**, 1874 (1930).

127. J. P. McCullough and G. Waddington, in *Experimental Thermodynamics*, Vol. 1, J. P. McCullough and D. W. Scott, Eds., Butterworths, London, 1968, Chap. 10.

128. D. C. Ginnings and H. F. Stimson, in *Experimental Thermodynamics*, Vol. 1, J. P. McCullough and D. W. Scott, Eds., Butterworths, London, 1968, Chap. 11.

129. I. Wadsö, *Acta Chem. Scand.* **20**, 536 (1966); E. Morowetz, *ibid.* **22**, 1509 (1968).

130. N. S. Osborne, H. F. Stimson, E. F. Fiock, and D. C. Ginnings, *J. Res. Natl. Bur. Std.* **10**, 155 (1933).

131. R. Pictet, *Ann. Chim. Phys.* **9**, 180 (1876); F. Trouton, *Phil. Mag.* **18**, 54 (1884).

132. J. H. Hildebrand, *J. Amer. Chem. Soc.* **37**, 970 (1915).

133. S. Brunauer, P. H. Emmett, and E. Teller, *J. Amer. Chem. Soc.* **60**, 309 (1938); A. B. D. Cassie, *Trans. Faraday Soc.* **41**, 450 (1945); R. B. Anderson, *J. Amer. Chem. Soc.* **68**, 686 (1946).

134. R. A. Beebe, J. Biscoe, W. R. Smith, and C. B. Wendell, *J. Amer. Chem. Soc.* **69**, 95 (1947). H. Chihara and J. A. Morrison, in *Experimental Thermodynamics*, Vol. 1, J. P. McCullough and D. W. Scott, Eds., Butterworths, London, 1968, Chap. 13.

135. R. A. Beebe, J. Biscoe, W. R. Smith, and C. B. Wendell, *J. Amer. Chem. Soc.* **69**, 95 (1947).

136. H. Chihara and J. A. Morrison, in *Experimental Thermodynamics*, Vol. 1, J. P. McCullough and D. W. Scott, Eds., Butterworths, London, 1968, Chap. 13.

137. The application of calorimetry to radioactivity has been reviewed by O. E. Myers, *Nucleonics* **5** (5), 37 (1949). See also, S. R. Gunn, *Univ. of Calif. Rad. Lab. Report 5375*, Nov., 1958.

138. M. M. Popov, A. Bundel, and W. Choller, *Z. physik. Chem.* **A147**, 302 (1930);

J. C. Southard, *Ind. Eng. Chem., Ind. Ed.* **32**, 442 (1940); S. G. Lipsett, F. M. G. Johnson, and O. Maas, *J. Amer. Chem. Soc.* **50**, 1030 (1928); D. H. Torgeson and T. H. Sahama, *ibid.* **70**, 2156 (1948); W. A. Roth and H. Troitzsch, *Z. anorg. Chem.* **260**, 337 (1949); E. F. Westrum, Jr., and L. Eyring, *J. Amer. Chem. Soc.* **74**, 2045 (1952); J. P. Coughlin, *Experimental Thermodynamics*, Vol. 2, H. A. Skinner, Ed., Interscience, New York, 1962, Chap. 14.

139. W. Hieber et al., *Z. anorg. Chem.* **186**, 97 (1930); *Z. Elektrochem.* **40**, 256 (1934); **44**, 881 (1938); **46**, 556 (1940).

140. A. N. Campbell and A. J. R. Campbell, *J. Amer. Chem. Soc.* **62**, 291 (1940).

141. D. F. Evans and R. E. Richards, *J. Chem. Soc.* **1952**, 3932.

142. D. F. Evans and R. E. Richards, *Proc. Roy. Soc. (London)* **A223**, 238 (1954).

143. O. Kubaschewski and R. Hultgren, *Experimental Thermochemistry*, Vol. 2, H. A. Skinner, Ed., Interscience, New York, 1962, Chap. 16; J. B. Darby, Jr., R. Kleb, and O. J. Kleppa, *Rev. Sci. Instr.* **37**, 164 (1966).

144. J. M. Sturtevant, *J. Phys. Chem.* **45**, 127 (1941).

145. R. B. Williams, *J. Amer. Chem. Soc.* **64**, 1395 (1942).

146. T. F. Young and O. G. Vogel, *J. Amer. Chem. Soc.* **54**, 3030 (1932); T. F. Young and W. L. Groenier, *ibid.* **58**, 187 (1936); T. F. Young and P. Seligman, *ibid.* **60**, 2379 (1938).

147. E. Doehlemann and E. Lange, *Z. physik. Chem.* **A170**, 391 (1934); J. M. Sturtevant, *J. Amer. Chem. Soc.* **62**, 3519 (1940).

148. P. R. Stoesser and S. J. Gill, *J. Phys. Chem.* **71**, 564 (1967); S. J. Gill and E. L. Farquhar, *J. Amer. Chem. Soc.* **90**, 3039 (1968).

149. D. F. Shiao and J. M. Sturtevant, *Biochemistry*, **8**, 4910 (1969).

150. J. H. Hildebrand and R. L. Scott, *Regular Solutions*, Prentice-Hall, Englewood Cliffs, N. J., 1962.

151. J. M. Sturtevant and P. A. Lyons, *J. Chem. Thermodynam.* **1**, 201 (1969).

152. G. F. Zellhoefer and M. J. Copley, *J. Amer. Chem. Soc.* **60**, 1343 (1938).

153. H. A. Skinner, J. M. Sturtevant, and S. Sunner, in *Experimental Thermochemistry*, Vol. 2, H. A. Skinner, Ed., Interscience, New York, 1962, Chap. 9.

154. W. P. White and H. S. Roberts, *J. Amer. Chem. Soc.* **59**, 1254 (1937).

155. K. S. Pitzer, *J. Amer. Chem. Soc.* **59**, 2365 (1937).

156. J. R. Partington and W. E. Soper, *Phil. Mag.* **7**, 209 (1929).

157. E. Lange and A. L. Robinson, *Chem. Rev.* **9**, 89 (1931). See also J. B. Conn, G. B. Kistiakowsky, and R. M. Roberts, *J. Amer. Chem. Soc.* **62**, 1895 (1940); T. Davis, S. S. Singer and L. A. K. Staveley, *J. Chem. Soc.* **1954**, 2304.

158. T. F. Young and J. S. Machin, *J. Amer. Chem. Soc.* **58**, 2254 (1936).

159. F. T. Gucker, H. B. Pickard, and T. W. Planck, *J. Amer. Chem. Soc.* **61**, 459 (1939).

160. S. G. Lipsett, F. M. G. Johnson, and O. Maas, *J. Amer. Chem. Soc.* **49**, 935, 1940 (1927).

161. T. H. Benzinger and C. Kitzinger, in *Temperature: Its Measurement and Control in Science and Industry*, Vol. 3, C. M. Herzfeld, Ed., Reinhold, New York, 1962, Part 3, Chap. 5; H. A. Skinner, in *Biochemical Microcalorimetry*, H. D. Brown, Ed., Academic, New York, Chap. 1.

162. G. Scatchard, L. B. Ticknor, J. R. Goates, and E. R. McCartney, *J. Amer. Chem. Soc.* **74**, 3721 (1952).

163. H. C. Brown and R. H. Horowitz, *J. Amer. Chem. Soc.* **77**, 1730 (1955).

164. R. L. Berger, Y.-B. F. Shick, and N. Davids, *Rev. Sci. Instr.* **39**, 362 (1968).

165. M. L. McGlashan, in *Experimental Thermochemistry*, Vol. 2, H. A. Skinner, Ed., Interscience, New York, 1962, Chap. 15.

166. I. Danielsson, B. Nelander, S. Sunner, and I. Wadsö, *Acta Chem. Scand.* **18**, 995 (1964); C. G. Savini, D. R. Winterhalter, L. H. Kovach, and H. C. van Ness, *J. Chem. Eng. Data* **11**, 40 (1966); J. J. Christensen, H. D. Johnston, and R. M. Izatt, *Rev. Sci. Instr.* **39**, 1356 (1968); R. H. Stokes, K. N. Marsh, and R. P. Tomlins, *J. Chem. Thermodynam.* **1**, 211 (1969).

167. F. Becker and M. Kiefer, *Proc. First Internat. Conf. on Calorim. and Thermodynamics*, Warsaw, 1969.

168. H. A. Skinner, Ed., *Experimental Thermochemistry*, Vol. 2, Interscience, New York, 1962. See especially, H. A. Skinner, Chap. 8; H. A. Skinner, J. M. Sturtevant, and S. Sunner, Chap. 9; F. S. Dainton and K. J. Ivin, Chap. 12; M. L. McGlashan, Chap. 15; O. Kubaschewski and R. Hultgren, Chap. 16; E. Calvet, Chap. 17.

169. H. D. Brown, Ed., *Biochemical Microcalorimetry*, Academic, New York, 1969. See especially, H. A. Skinner, Chap. 1; W. J. Evans, Chap. 14; R. L. Berger, Chap. 15; H. D. Brown, Chap. 16.

170. L. Benjamin, *Can. J. Chem.* **41**, 2210 (1963).

171. E. Lange and A. L. Robinson, *Chem. Rev.* **9**, 89 (1931); F. T. Gucker, H. B. Pickard and T. W. Planck, *J. Amer. Chem. Soc.* **61**, 459 (1939).

172. H. J. V. Tyrrell and A. E. Beezer, *Thermometric Titrimetry*, Chapman and Hall, London, 1968. See also J. Jordon, in *Treatise on Analytical Chemistry*, Vol. 8, Part 1, I. M. Kolthoff and P. J. Elving, Eds., Interscience, New York, 1968, p. 5206.

173. G. B. Kistiakowsky, H. Romeyn, J. R. Ruhoff, H. A. Smith, and W. E. Vaughan, *J. Amer. Chem. Soc.* **57**, 65 (1935).

174. P. R. Stoesser and S. J. Gill, *Rev. Sci. Instr.* **38**, 422 (1967).

175. P. Monk and I. Wadsö, *Acta Chem. Scand.* **22**, 1842 (1968).

176. J. M. Sturtevant and P. A. Lyons, *J. Chem. Thermodynam.* **1**, 201 (1969); J. M. Sturtevant, *Fractions*, No. 1 (1965) (Published by Beckman Instruments, Inc.).

177. C. J. Wormald, *Proc. First Internat. Conf. on Calorim. and Thermodynamics*, Warsaw, 1969.

178. G. Anderegg, *Helv. Chim. Acta* **46**, 1833 (1963).

179. J. M. Sturtevant, *J. Amer. Chem. Soc.* **59**, 1528 (1937); *J. Phys. Chem.* **45**, 127 (1941); P. Ohlmeyer, *Z. physiolog. Chem.* **282**, 37 (1945); T. H. Benzinger and R. Hems, *Proc. Natl. Acad. Sci., U.S.* **42**, 896 (1956); I. E. Scheffler and J. M. Sturtevant, *J. Mol. Biol.* **42**, 577 (1969).

180. J. Barcroft and A. V. Hill, *J. Physiol.* **39**, 411 (1910).

181. W. E. L. Brown and A. V. Hill, *Proc. Roy. Soc.* (*London*) **94B**, 297 (1923).

182. F. D. Rossini, *Chem. Rev.* **18**, 233 (1936).

183. J. B. Conant and G. B. Kistiakowsky, *Chem. Rev.* **20**, 181 (1937); G. B.

Kistaikowsky et al., *J. Amer. Chem. Soc.* **60,** 440, 2764 (1938); **61,** 1868 (1939). See also R. B. Williams, *ibid.* **64,** 1395 (1942).

184. G. W. Wheland, *Resonance in Organic Chemistry,* Wiley, New York, 1955, p. 78.

185. J. R. Lacher et al., numerous papers; see especially, J. R. Lacher, in *Experimental Thermochemistry,* Vol. 2, H. A. Skinner, Ed., Interscience, New York, 1962, Chap. 10. R. B. Turner et al., numerous papers of which a recent one is *J. Amer. Chem. Soc.* **90,** 4315 (1968). H. A. Skinner et al., *Trans. Faraday Soc.* **53,** 784 (1957); *ibid.* **54,** 47 (1958); *ibid.* **55,** 404 (1959).

186. J. M. Sturtevant, in *Experimental Thermochemistry,* Vol. 2, H. A. Skinner, Ed., Interscience, New York, 1962, Chap. 19.

187. H. D. Brown, Ed., *Biochemical Microcalorimetry,* Academic, New York, 1969.

188. J. Duclaux, *Compt. Rend.* **146,** 120 (1908); V. Chelintzev, *J. Russ. Phys.-Chem. Soc.* **44,** 865 (1912); A. Tian, *Bull. Soc. Chim.* **33** (4), 427 (1923); H. Hartridge and F. J. W. Roughton, *Proc. Cambridge Phil. Soc.* **22,** 426 (1925); E. Calvet, *The Microcalorimetry of Slow Phenomena,* report to the Commission on Thermochemistry of the International Union of Pure and Applied Chemistry, May, 1952; G. Akerlöf, *J. Amer. Chem. Soc.* **49,** 2955 (1927); J. M. Sturtevant, *ibid.* **59,** 1528 (1937); L. K. J. Tong and W. O. Kenyon, *ibid.* **67,** 1278 (1945); F. H. Westheimer and M. S. Kharasch, *ibid.* **68,** 1871 (1946); M. J. Rand and L. P. Hammett, *ibid.* **72,** 287 (1950); R. Livingston, in *Technique of Organic Chemistry,* A. Weissberger, Ed., Vol. 8, Interscience, New York, 1953, p. 58; T. L. Smith, *J. Phys. Chem.* **59,** 385 (1955).

189. R. L. Berger et al., *Rev. Sci. Instr.* **39,** 486, 493, 498 (1968). See also R. L. Berger, in *Temperature, Its Measurement and Control in Science and Industry,* Vol. 3, C. M. Herzfeld, Ed., Reinhold, New York, 1962, Part 3, p. 61.

190. T. Meites and L. Meites, *J. Amer. Chem. Soc.* **92,** 37 (1970).

191. R. P. Bell and J. C. Clunie, *Proc. Roy. Soc. (London)* **A212,** 16 (1952); R. P. Bell, V. Gold, J. Hilton, and M. H. Rand, *Discussions Faraday Soc.* **17,** 151 (1954).

192. A. B. Callear and J. C. Robb, *Discussions Faraday Soc.* **17,** 21 (1954); D. Garvin, V. P. Guinn, and G. B. Kistiakowsky, *ibid.* **17,** 32 (1954).

193. H. A. Skinner, in *Experimental Thermochemistry,* Vol. 2, H. A. Skinner, Ed., Interscience, New York, 1962, Chap. 8; H. A. Skinner, J. M. Sturtevant and S. Sunner, *ibid.,* Chap. 9.

194. H. Hartridge and F. J. W. Roughton, *Proc. Cambridge Phil. Soc.* **22,** 426 (1925); **23,** 450 (1926); F. J. W. Roughton, *Proc. Roy. Soc. (London)* **A126,** 439, 470 (1930); *J. Amer. Chem. Soc.* **63,** 2930 (1941); J. B. Bateman and F. J. W. Roughton, *Biochem. J.* **29,** 2622 (1935). See also R. L. Berger et al., *Rev. Sci. Instr.* **39,** 486, 493, 498 (1968).

195. V. K. LaMer and C. L. Read, *J. Amer. Chem. Soc.* **52,** 3098 (1930).

196. H. A. Skinner, J. M. Sturtevant and S. Sunner, in *Experimental Thermochemistry,* Vol. 2, H. A. Skinner, Ed., Interscience, New York, 1962, pp. 196 ff.

197. J. M. Sturtevant, *J. Amer. Chem. Soc.* **59,** 1528 (1937); *Physical Methods of Organic Chemistry,* 3rd ed., Vol. 1, Part 1, A. Weissberger, Ed., Interscience, New York, 1959, pp. 639 ff.

198. L. K. J. Tong and W. O. Kenyon, *J. Amer. Chem. Soc.* **67**, 1278 (1945); **68**, 1355 (1946).

## General

American Institure of Physics, *Temperature, Its Measurement and Control in Science and Industry*, Vol. 3, Parts 1–3, C. M. Herzfeld, Ed., Reinhold, New York, 1962.

Bichowsky, F. R., and F. D. Rossini, *The Thermochemistry of Chemical Substances*, Reinhold, New York, 1936.

Brown, H. D., Ed., *Biochemical Microcalorimetry*, Academic, New York, 1969.

Burton, E. F., H. Grayson Smith, and J. O. Wilhelm, *Phenomena at the Temperature of Liquid Helium*, Reinhold, New York, 1940.

Calvet, E., and H. Prat, *Recent Progress in Microcalorimetry*, translated by H. A. Skiner, Pergamon, New York, 1963.

Eucken, A., *Energie-und Wärmeinhalt*, Vol. 8, Part 1, of Wien-Harms *Handbuch der Experimentalphysik*, Akadem. Verlagsgesellschaft, Leipzig, 1929.

McCullough, J. P., and D. W. Scott, Eds., *Experimental Thermodynamics*, Vol. 1, Buterworths, London, 1968.

Ostwald, W.-R. Luther, *Physikochemische Messungen*, 4th ed., Akadem. Verlagsgesellshaft, Leipzig, 1925.

Parks, G. S., and H. M. Huffman, *The Free Energies of Some Organic Compounds*, Reinhold, New York, 1932.

Partington, J. R., and W. G. Shilling, *The Specific Heats of Gases*, Van Nostrand, Princeton, N. J., 1924.

Rossini, F. D., Ed., *Experimental Thermochemistry*, Vol. 1, Interscience, New York, 1956.

Roth, W., and F. Becker, *Kalorimetrische Methoden zur Bestimmung chemischer Reaktionswärmen*, Vieweg, Brunswick, Germany, 1956.

Skinner, H. A., Ed., *Experimental Thermochemistry*, Vol. 2, Interscience, New York, 1962.

Stull, D. R., E. F. Westrum, Jr., and G. C. Sinke, *The Chemical Thermodynamics of Organic Compounds*, Wiley, New York, 1969.

Swietoslawski, W., *Microcalorimetry*, Reinhold, New York, 1946.

Taylor, H. S., and S. M. Glasstone, Eds., *Treatise on Physical Chemistry*, 2nd and 3rd eds., Van Nostrand, Princeton, N. J., 1931 and 1942.

Tyrrell, H. J. V., and A. E. Beezer, *Thermometric Titrimetry*, Chapman and Hall, London, 1968.

Wadsö, I., "Microcalorimetry," *Quart. Rev. Biophys.*, **3**, 383 (1970).

Weber, R. L., *Temperature Measurement and Control*, Blakiston, Philadelphia, 1941.

Wenner, R. R., *Thermochemical Calculations*, McGraw-Hill, New York, 1941.

White, W. P., *The Modern Calorimeter*, Chem. Catalog Co., New York, 1928.

# Chapter VIII

# DIFFERENTIAL THERMAL ANALYSIS

Bernhard Wunderlich

# 1    INTRODUCTION

## The General Principle of Differential Thermal Analysis

The difficulty of exact measurement of heat lies at the heart of the differential thermal analysis method. As mentioned in the chapter on calorimetry [1], there is no perfect insulator for heat and every conductor for heat absorbs a large amount of it, making the measurement of heat a technique beset by "loss" determinations and empirical calibrations. Differential thermal analysis, or DTA, is a method which masters these difficulties by taking a loss of accuracy, but gains speed and convenience of measurement. In principle, the unknown sample is brought into an identical environment as a known reference sample which is calorimetrically fully analyzed. On heating or cooling through the temperature region of interest, differential thermal effects are measured by detecting temperature differentials. Temperature measurements can be made practically instantaneously with high precision [2]. After correction for asymmetry, the thermal properties of the unknown are evaluated from the known thermal properties of the reference.

Although the above description makes DTA look like a poor or lazy man's calorimeter, it has its unique place among the techniques of chemistry. It is used widely as an instrument for qualitative analysis. Also, there are many substances which one can not characterize well enough to warrant high-precision quantitative calorimetric analysis. In these cases quantitative DTA is useful. The accuracy of DTA has increased in the last 10 years so much that heat capacities, for example, can be determined to about 2% accuracy which compares to an accuracy of 0.1% routinely possible in precision adiabatic calorimetry. Even more unique is the application of DTA to problems involving thermal measurements as a function of time. Because of its easy adaption to sample sizes of milligrams or less, heating rates as fast as 100°C/min are easily achieved. Without major change in standard equipment, the heating rate may be raised to 1000°C/min. These fast analysis rates enable the thermal investigation of metastable and transient states.

DTA has developed from the tool which allows a quick, approximate thermal analysis to an instrument which enables reasonably accurate measurements, impossible by any other technique.

### Historical Development of DTA

Cooling curves as a means of thermal analysis of materials were used as early as the first half of the 19th century. Differential temperature measurement between the sample and an equally placed inert reference presented a major advance of thermal analysis, since it eliminated nonuniformity due to temperature fluctuations in the environment. Differential measurements are even more important for heating curves in which uniform furnace control is inherently more difficult. The stage for differential thermal analysis as we know it today was set when it became possible to record automatically thermocouple outputs continuously as a function of time. The early development of instrumentation has been reviewed in detail by Burgess [3].

The early work of LeChatelier [4] on the action of heat on clays set the pattern of development and application of DTA. LeChatelier used the change in rate of temperature rise of the sample, recorded photographically, to determine exothermic and endothermic transitions. The first sensitive, thermocouple-based differential thermal analyses were carried out by Roberts-

Austen [5] in 1899. Saladin [6] finally, in 1903, combined the differential thermocouple measuring mode with the simultaneous photographic recording of sample temperature and differential temperature. Data was used in the following 50 years mainly for investigation of metals, oxides, salts, ceramics, glasses, minerals, and soils. The first 1000 research papers on the general area of DTA were published by 1952. They dealt mainly with the determination of phase diagrams and transition temperatures, the detection of chemical reactions, and the analysis of unknown samples. Although it was recognized early that DTA could, by comparison with known samples, be used to measure heat capacities [7] the only paper describing such application seems to have been the one by Sykes [8] published in 1935. Sykes measured heat capacities on cooling by comparison with a reference material. He found, however, that an adiabatic heating curve gave better results.

A big advance in DTA was made possible by the use of electronics during the last 20 years. Sensitivity has been enhanced and more quantitative measurements have become possible. This new development has led to the invention of the name differential scanning calorimetry, DSC, for the quantitative use of DTA.

Classical differential calorimetry had its beginning also in the 19th century. The first differential calorimetry with twin calorimeters was described by Joule [9] and Pfaundler [10]. For a long time these twin calorimeters were closely related to adiabatic calorimeters and as such are properly described in the chapter on calorimetry [1]. With the development of constant heating-rate calorimeters [11] differences between DTA and calorimetry decreased. Heating rates of 1–2°C/min were used in earlier calorimeters which still used rather large amounts of sample to achieve an accuracy in heat capacity of 1–5%. Temperature lags in these instruments were estimated to be about 1°. Calorimeters more adapted to a continuous-heating mode have been developed mainly for heat-capacity work on linear high polymers [12]. The modern development of DSC began with the work by Müller and Martin [13] in 1960 who reduced sample sizes to 1g without a decrease in accuracy. The next step was the invention of a differential scanning calorimeter in which heat flow from a linearly heated block was replaced by electronically governed heat transfer [14].

The present trends in DTA are in the direction of increased specialization and sensitivity. A large effort is directed to couple DTA to other determinations such as, for example, thermogravimetry, dielectric measurements, X-ray analysis, optical analysis, and effluent-gas analysis by gas-phase chromatography or mass spectrometry. The development of DTA over the years has been a continuously accelerating one. While during the 1930's approximately 14 papers were published each year in the field, about 300 papers were published annually between 1960 and 1970.

### Scope of the Chapter on DTA

The organization of this chapter on DTA calls in Section 2 for the description of typical instrumentation for the different fields of application. Qualitative operational parameters and standard calibration procedures necessary for their use in DTA and DSC will be discussed in Section 3 of this chapter. The section on theory of DTA is an attempt to bring together selected developments which have led or may lead to meaningful application. For much of the thermodynamic theory and the description of measurement of temperature and heat, the reader is directed to Chapter I and VII of this volume. A more detailed discussion of electronic components used in DTA and DSC can be found elsewhere in this treatise. Section 5 of this chapter on DTA and DSC finally deals with a series of typical applications as they have developed over the years.

### Literature on DTA

The literature on differential thermal analysis is found in a wide variety of journals. Fortunately extensive bibliographies are available. The *Handbook of Differential Thermal Analysis* by Smothers and Chiang [15] contains a listing of 4248 publications on DTA which had appeared by the middle of 1965. To update this list one can make use of the excellent biannual review articles by Murphy [16]. A special effort is being made by the newly formed International Confederation for Thermal Analysis to evaluate the needs for publication in the thermal analysis field. Up to the present, proceedings of the conferences of this organization have been published every 3 years [17, 18]. Recent new publications of general interest include the start of two new journals [19, 20], *Thermochimica Acta* and *Journal of Thermal Analysis*, and several books and reviews [21–33].

## 2 INSTRUMENTATION

### Basic Design

#### Classical DTA

The classical design of differential thermal analysis equipment is schematically represented in Fig. 8.1. Reference and sample are placed in an identical environment and their temperatures determined by thermocouples. By proper combination of thermocouples, an electrical potential proportional to $\Delta T$, the temperature difference between sample and reference, is recorded as a function of time, or is used as the $y$-axis signal of an $x$–$y$–recorder. The potential proportional to the sample temperature (or in some cases the reference temperature) is recorded either simultaneously with the $\Delta T$ signal as a function of time, or is used as the $x$-axis signal of an $x$–$y$–recorder. Heating

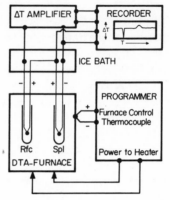

**Fig. 8.1**  Schematic diagram of a classical DTA apparatus. All wires from the ice-bath to the $\Delta T$ amplifier and the recorder are copper wires.

or cooling of the sample and reference environment (furnace) is regulated by a programmer actuated by, for example, a third thermocouple.

The complete set-up with the exception of the DTA-furnace is easily assembled from commercial components. The DTA-furnace and sample placement have to be adapted to individual needs, and some special designs will be discussed below (p. 438).

### Thermocouples

The choice of thermocouples is dictated by the useful temperature range, output, linearity, and stability of particular thermocouples. A detailed description of thermocouple properties can be found in the *National Bureau of Standards Monograph 40* [34]. Table 8.1 lists the emf output/deg temperature rise for five common thermocouples. The letter designation refers to the nomenclature of the Instrument Society of America [35].

**Table 8.1**  Thermocouple EMF at Various Temperatures ($\mu V/°C$)

| Thermocouple | Temperature | | | | |
|---|---|---|---|---|---|
| | −190°C | 0°C | 500°C | 1000°C | 1500°C |
| Copper–constantan (T) | 17 | 38.4 | — | — | — |
| Iron–constantan (J) | 26 | 50.1 | 56 | — | — |
| Chromel–alumel (K) | 23 | 40.0 | 43 | 39 | — |
| Pt+10%Rh–Pt (S) | — | 5.6 | 10 | 12 | 12 |
| Pt+13%Rh–Pt (R) | — | 5.5 | 11 | 13 | 14 |

The copper-constantan thermocouple is noted for its well-established and reproducible temperature–emf relation [36]. The standard limits of error are 0.8°C in the range −59 to 93°C. Its usual temperature range is given as −190 to 300°C, although applications from −260 to 600°C can be found.

Iron-constantan couples were the first widely accepted thermocouples. The usual range is −190 to 760°C with a maximum temperature of 980°C [34]. The certified accuracy of calibrations of such thermocouples at the National Bureau of Standards is ±1°C. Outside of lower price and the slightly higher emf, both of which are of little importance for DTA-application, there is little advantage of iron-constantan thermocouples over the chromel-alumel couple.

The chromel-alumel thermocouples may be used up to 1350°C. The certified accuracy of calibration at the National Bureau of Standards is ±1°C in the range from 0 to 1100°C. It is perhaps the most widely used thermocouple for DTA.

The thermocouples using Pt–Rh alloys and Pt have a distinctively lower change in emf/°C. They have extreme stability and freedom from internal changes. Their useful temperature range begins at about 0°C and is limited by the melting point of platinum at 1769°C. The Pt + 10%Rh–Pt thermocouple is used to maintain the International Practical Temperature Scale (IPTS-68) in the temperature range 3 (630.74–1064.43C) [37]. The limits of error of both thermocouples is usually given by suppliers to be 3°C (or $\frac{1}{2}$% in the upper temperature range). Accuracies based on calibration at the National Bureau of Standards are 0.3°C from 0 to 1100°C and up to 2°C at 1450°C.

Tables of thermocouple emf versus temperature for all discussed thermocouples are given in Ref. 36 and can be found in many handbooks. The change from the IPTS-48 to IPTS-68 would affect only high-precision thermometry and should be insignificant for practically all DTA. The changes up to 550°C are less than 0.1°C, at higher temperatures they increase almost linearly, reaching 3°C at 2000°C.

The size of thermocouple wires is arbitrary as long as the sample size is large. For small samples where heat conduction through the thermocouple wires becomes important, the wire should be as small as possible; the practical limit being set by the mechanical rigidity of the thermocouple. The actual junction of the couple must also be fitted to the case at hand. A butt-welded couple adds the lowest heat capacity to the sample; a larger bead or disk might, however, be preferred because of its ability to average temperature over a larger sample. The best arrangement from the point of view of temperature measurement is direct contact between thermocouple and sample. It must be known in this case that there is no chemical attack by the sample on the thermocouple material.

Since the potentials to be measured are quite small, care must be taken to avoid extraneous thermal emf's. The schematic diagram of Fig. 8.1 indicates continuous wires into the ice bath where all leads are joined to copper wires. If it is necessary to have interchangeable thermocouples, switches, binding posts, and so on, in the measuring circuit, it is important to make sure that these do not introduce spurious potentials.

Thermistors have not been used extensively in DTA despite their much greater sensitivity. The main problem is the difficulty in matching the non-linear response of the thermistors with temperature. Resistance thermometers can be made small enough to be usable for DTA; large-scale application seems to be restricted, however, to the scanning calorimetry to be described below. Thermocouples made of refractory metals or carbon and carbides capable of higher temperatures may be used for special applications. Metal-sheathed, ceramic-packed thermocouples are used advantageously where rigidity and reproducibility of placement of the thermocouples are of importance. The characteristics of this group of thermocouples are identical to the nonsheathed couples. The thermal lag can be avoided if the thermo-couple tip is made to protrude out of the sheath and ceramic packing. The total diameter of thermocouple wires, packing, and metal sheath can be made to fit the particular apparatus, with diameters as small as 0.02 cm commercially available.

### Other Components

The reference temperature of the thermocouples is traditionally set by an ice bath. Its temperature is reproducible and it is easily set up. A Dewar vessel filled with a slurry of distilled water and crushed, distilled-water ice is sufficient for this purpose. The thermocouples must be positioned within sufficient ice to truly be at 0°C. Problems arise when, after longer use, the warmer layer of water at the bottom of the Dewar vessel reaches the thermo-couple and changes the reference temperature. An automatic ice-point reference system based on Peltier cooling needs less attention; it maintains temperature to better than ±0.05°C. In these systems automatic control is achieved by sensing the volume of the water–ice mixture containing jacket. As soon as a decrease in volume indicates too low an ice-level, cooling is called for. Since the temperature of the sealed water–ice mixture stays constant over a wide range of degree of freezing, very little accuracy of control is needed.

The two electrical signals, proportional to the temperature differential and temperature are widely different in magnitude. It may be desirable to detect temperature differentials of 0.01°C or less for small samples, while it is rare even for large sample sizes to have temperature differentials above ±10°C. According to Table 8.1 these conditions ask for a noise level of less than 0.5 $\mu V$ and a possible recording range of $\pm 500$ $\mu V$. The temperature recording range, in contrast, may be from $-200$ to 2000°C with an equivalent voltage range of almost 100 m$V$. Accurate temperature recording of $\pm 1$°C requires a noise level of less than $\pm 50$ $\mu V$. Present-day practice brings the two signals to about equal levels by preamplification of the $\Delta T$ signal (see Fig. 8.1). A low-noise, solid-state preamplifier of a gain of 1000 with sufficient linearity is suitable for this application.

The recording of the $T$ signal and the preamplified $\Delta T$ signal can be done in two ways. Either both signals are fed into a two-point strip-chart recorder for a time-based DTA trace, or the $T$ signal is recorded as the $x$-axis signal and the $\Delta T$ signal is recorded as the $y$-axis signal of an $x$–$y$-recorder. Because of the wide range of both signals, it is necessary to provide multiple steps of attenuation and a wide range of zero shifts. Besides requirements of noise level and linearity, which can be determined from the temperature-range and sample-size limits to be used, it is necessary to employ a recorder of sufficiently fast response and high input impedance. Fast response is necessary for use of faster heating rates and the proper recording of sharp transitions. Full-scale deflections should be achieved in one second or less. In the case of time-based recording, a sufficiently variable recording speed must be available to accommodate different heating rates. Typical recording speeds are 1–120 in./hr in suitable steps. High input impedance for the recorder as well as the preamplifier is necessary to eliminate emf changes in the thermocouple system due to current flowing. Any emf drop in the sample thermocouple produces a significant drift in the $\Delta T$ recording (change of $\Delta T$ with $T$). For recorders with insufficient input impedance special thermocouple arrangements have been suggested [38]. A less desirable solution, at least for small samples sizes, is to separate the $\Delta T$ and $T$ thermocouple system by using two thermocouples in the sample. A simple calculation of the $\Delta T$ drift shows that in a 10-$\Omega$ sample thermocouple at 5.0 $mV$ output ($\approx 120$°C) the recorder impedance must be $10^5$ $\Omega$ to keep the effect on the $\Delta T$ thermocouple at the noise level 0.5 $\mu V$. Although the limits can in most cases be relaxed somewhat, this calculation indicates that a potentiometric recorder with practically infinite input resistance is most suitable for the $T$-signal.

The final item of Fig. 8.1 is the programmer. It serves to heat and, if possible, cool the furnace at a controlled rate. The furnace-control thermocouple emf is compared to a linearly-changing reference voltage generated either mechanically, by driving a potentiometer, or electrically, by applying a constant voltage to an integrator circuit. Any imbalance is used to increase or decrease the electrical power input into the DTA furnace. Since the response of the furnace mass to a change in heater voltage is noninstantaneous, it is necessary that the dynamics of the programmer be matched to the time constant of the furnace. To fit a controller to any particular furnace, it must have an adjustable proportional band to achieve optimum response to the changing temperature command. A typical specification for the adjustable range of the proportional band would be between 1 and 10°C. In addition to the proportional action, the rate of power output must also be adjustable to counter overshooting the control point. Older methods of achieving linear temperature changes can be found in the literature [15, 21, 22]. Additional sophistication is necessary to avoid the small nonlinearity introduced by the nature of the furnace-control thermocouple. For cooling control and operation below 40°C one usually immerses the furnace in a constant-temperature cold bath and corrects the cooling rate by supplying controlled heat. Furnaces with controlled Peltier cooling and heating have not been designed as yet.

The functioning of the automatic ice-bath, preamplifier, recorder, and programmer has been discussed only in outline form. These instruments are being constantly improved and developed. For the latest advances in precision and convenience several reputable instrument manufacturers should be contacted to get the latest technical descriptions. The given outline specifications can serve as a guide for final selection.

### Differential Scanning Calorimeter

Since about 1964 differential thermal analysis mainly concerned with the quantitative measurement of heat has become known as differential scanning calorimetry, DSC. This label was introduced mainly by commercial makers of such instruments. There is no sharp dividing line between DSC and classical DTA on the one hand, and DSC and twin calorimetry on the other. One expects the DSC to be more quantitative than classical DTA, although with care most modern DTA equipment can be made to perform similar functions.

A new principle of differential thermal analysis was described in 1964 [14, 39]. A block-diagram of this differential scanning calorimeter presently marketed worldwide by the Perkin-Elmer Corp. is shown in Fig. 8.2 [40]. Sample and reference, each of 1–100 mg, are contained in small, disposable aluminum cups which in turn fit into the sample and reference holders. The latter are small, steel cups each containing a heater and a temperature-sensing, Pt-resistance winding. Sample and reference are heated by the output of the average-temperature amplifier such that the average temperature of the two always stays at the level set by the programmer. The differential-temperature amplifier senses any temperature difference between sample and reference and transfers energy from the hotter steel cup to the cooler one to keep the temperature difference small. The differential power transfer, proportional to the measured temperature difference, is recorded as a function of time alongside temperature markings generated by the programmer.

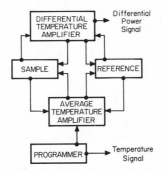

**Fig. 8.2**  Schematic circuit diagram of the Perkin-Elmer differential scanning calorimeter.

The DSC curve presents in this way a recording of the power differential necessary to keep reference and sample heating or cooling at the same rate.

A DSC arrangement based on classical DTA, marketed at present by the duPont Company [41], is illustrated in Fig. 8.3a. Only the sample arrangement is illustrated. The heating block, made of silver, takes the place of the standard DTA furnace, and the thermocouple wires are connected to the three ice-bath junctions as illustrated in Fig. 8.1. The $\Delta T$ junction is moved from the ice bath to the thermoelectric disk which provides a direct short circuit. All other details are identical to Fig. 8.1. The function of the average-temperature amplifier is taken over by the thermoelectric disk which carries most of the heat from the controlled-temperature furnace. The temperature differential between sample and reference is recorded and heat-capacity differences are calculated from this data as will be discussed in Section 4. For many applications the temperature differential can be kept below 1°C.

The International Confederation for Thermal Analysis (ICTA) definition of DSC restricts this label (Appendix 2 of Ref. 18) to energy-recording techniques. This definition should certainly be changed since no present instrument measures energy directly. Both instruments described above rely in the end on the detection of temperature differences between sample and reference.

## Special Details

### Standard Furnaces and Sample Holders

A DTA cell which covers a temperature range from about liquid nitrogen temperature or room temperature to perhaps 500–600°C, and which is able to be used in vacuum or up to several atmospheres gas pressure, may be called a standard unit. The placement of sample and the design of furnace can be adjusted to the particular need without special difficulty. Figures 8.3b to 8.3h illustrate a selection of recently-published design diagrams.

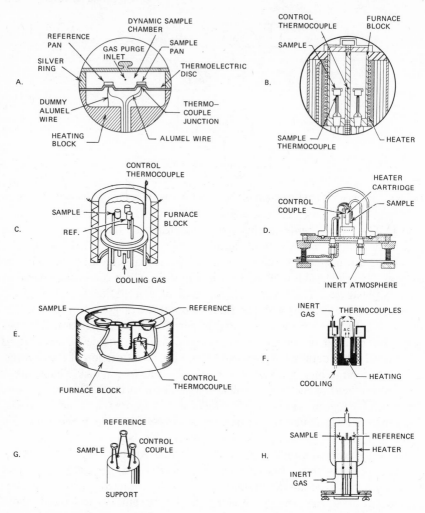

**Fig. 8.3** DTA furnace and sample arrangements. (*a*) DSC cell as marketed by the duPont Co. (1968). (*b*) DTA cell suggested for calorimetric work by Sarasohn (1964). (*c*) DTA furnace and sample arrangement as suggested by E. M. Barrall, II, R. S. Porter, and J. F. Johnson, *Anal. Chem.* **36,** 2172 (1964). (*d*) DTA Arrangement described by J. Chiu, *Anal. Chem.* **34,** 1841 (1962). The same arrangement is used in the standard duPont 900 Differential Thermal Analyzer. The sample is contained in a glass capillary into which the thermocouple dips. The furnace block is made of silver. The reference, not shown in the figure, is arranged next to the sample. (*e*) DTA by D. J. David, *Anal. Chem.* **36,** 2162 (1964). (*f*) DTA of D. A. Vasallo and J. C. Harden, *Anal. Chem.* **34,** 132, (1962). (*g*) DTA arrangement for microsamples by J. Mazieres, *Anal. Chem.* **36,** 602 (1964). Note that the ceramic tube supporting the sample, reference and furnace control thermocouples is only about 0.2 mm in diameter. The thermocouple is an integral part of the cup. (*h*) DTA by W. W. Wendlandt, *Anal. Chim., Acta.,* **27,** 309 (1962).

438

Their functioning can easily be deduced from the figure and legend. More about the aims of the design of a DTA cell and furnace will become obvious from the discussions in Sections 3 and 4. Special designs to be discussed below are high-temperature DTA, high-pressure DTA, and DTA combined with other measurements.

### High-Temperature DTA

High-temperature DTA is mainly a question of temperature-measuring technique and furnace design. Up to 1600°C it is possible to use the Pt-based thermocouples described above. The samples may be contained in Pt or alumina crucibles in an arrangement similar to Fig. 8.3$h$. The furnace can be wire-wound, using Pt–Rh alloys. Commercial equipment claims a $\pm2$°C temperature precision with standard $\Delta T$ sensitivity [41].

At higher temperatures other thermocouples have to be developed. For example W–Mo thermocouples have been used up to 2200°C [42]. Graphite-borated graphite [43] and tantalum carbide graphite thermocouples [44] have been suggested for temperatures up to 3000°C. Pyrometry, which is used to maintain the IPTS-68 above the melting point of Au (1064.43°C), has also been applied for temperature measurement [45, 46] and differential temperature measurement [31]. The changes from the scheme of Fig. 8.1 are described in detail in Ref. 45.

Furnace materials and sample holder are restricted to high-meltings substances as, for example, $Al_2O_3$ (mp 2015°C), BeO (mp 2530°C), $ThO_2$ (mp 3050°C), W (mp 3410°C), Mo (mp 2610°C), Ta (mp 2996°C), and graphite (subl 3562°C). Heating may be supplied by a W-wire-wound furnace or by induced radio-frequency current. Errors in temperature measurement in the upper temperature range have been quoted to be 0.2–0.8%.

### High-Pressure DTA

Differential thermal analysis at elevated pressures has been mainly applied to phase-diagram work. Up to about 10kbar* and 500°C the design of the pressure DTA furnace and sample holders can follow along the lines indicated in Fig. 8.3. Figure 8.4$a$ illustrates such a DTA cell [47]. The DTA cell is a cylinder of René 41 high-temperature alloy of 55% Ni, 19% Cr. and 10% Mo. The thermocouples are of the ceramic-insulated, metal-sheathed type. They are symmetrically placed and pressure-sealed by soldering into the cone-in-cone high-pressure closures. Separate thermocouple systems for $\Delta T$ and $T$ recordings are used. Heating was accomplished by standard strip heaters clamped to the outside of the cell. The normal heating rate was 4°C/min. The pressure transmitting medium can be either gas or hydraulic oil depending on the nature of the sample. The hydrostatic pressure is

* 1 kbar $= 10^9$ dyne/cm$^2$ $= 986.923$ atm $= 1019.716$ kg/cm$^2$ $= 14,503.8$ lb/in.$^2$.

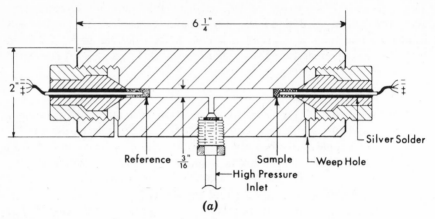

Fig. 8.4(a)   High-pressure DTA. (a) High-pressure DTA cell usable up to 4.5 kbar and 500°C. [Davidson and Wunderlich, *J. Polymer Sci.*, *Part A–2* 7, 377 (1969)].

measured by a Bourdon tube, manganin gage, or dead-weight tester [48, 49]. The generation of pressure is done by standard intensifier techniques [48]. By keeping the volume in the DTA cell small and providing a reservoir of pressurized fluid, the variation in pressure due to temperature change can be made negligible.

The thermal emf of the thermocouple generated is, as one would expect, dependent on pressure [50, 51]. Fortunately chromel-alumel is one of the least pressure-sensitive thermocouples. At 10 kbar the effects amount to less than 0.2°C at 100°C [50], negligible for most applications. At higher pressures and temperatures, and for other thermocouples, the corrections can be appreciable.

To go to higher pressures and temperatures the design must be changed. Instead of external heating, it becomes necessary to heat inside the pressurized cavity in order to preserve the mechanical strength of the pressure cylinder. Pressure is applied by a piston and cylinder apparatus using high-crushing-strength, cobalt-bonded, tungsten-carbide pistons capable of withstanding 45–50 kbar. Even higher pressures are possible by double staging, in other words, immersing the pressure-generating piston in a suitable prepressurized medium.

Above 10 kbar there are no usable pressure-transmitting liquids. The whole DTA set-up (furnace, and sample and reference capsules), is made of solid material of low shear strength so that the pressure is transmitted almost hydrostatically. Insulator material can be talc, pyrophyllite, boron nitride, or silver chloride. Conductors for heaters can be graphite, stainless steel, or platinum. The heater is usually shaped in the form of a tube into which the sample, reference, and thermocouple assembly are inserted.

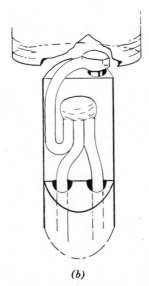

*(b)*

**Fig. 8.4(***b***)**   DTA thermocouple arrangement for high pressure piston and cylinder apparatus usable up to 50 kbar. [Cohen, Clement, and Kennedy, *J. Phys. Chem. Solids* **27,**
179 (1966)].

Figure 8.4*b* shows the arrangement of thermocouples of a high-pressure
piston and cylinder set-up [52]. The upper thermocouple junction touches
the sample container, while the three-wire reference junction is solidly
embedded in alundum cement used for wire insulation. The three thermocouple wires are connected to a measuring circuit similar to the one shown
in Fig. 8.1. The recorded temperature in this design is the reference temperature and needs to be corrected by $\Delta T$ to obtain the sample temperature.

The pressure generated in the sample can be calculated from geometrical
consideration, but must be corrected for friction which decreases the actual
pressure on the pressurizing stroke and increases the actual pressure on the
depressurizing mode. Typical friction values [53] are (measured with a $\frac{1}{2}$-in.
piston): 6.1 kbar friction at 450°C and 15.2 kbar, 6.8 kbar friction at 600°C
and 34.2 kbar. Besides knowing the friction values, it is important to know at
all times whether the sample is under compression or decompression conditions. During heating, cooling, or phase transitions, volume changes may
occur which are unrelated to the geometrically calculated pressure which
may change the sample condition from compression to decompression or
vice versa. Overall precision of careful DTA in the 8–45 kbar region has been
estimated to ±0.5 kbar and ±2°C using chromel-alumel thermocouples.

## DTA Combined with Other Techniques

Because of the simplicity of DTA, its small requirement of space in the vicinity of the sample, and its highly informative value, it has recently been coupled with a large number of other measuring techniques.

Perhaps the most obvious combinations arise out of pyrolysis applications. The loss of weight needs to be monitored to determine whether exothermic or endothermic reactions indicated by DTA correspond to evolution of gases. Also, the evolved gases need to be identified. The first objective is solved by simultaneous thermogravimetry, TG, and DTA [54, 55], and the second, by simultaneous DTA and rapid-scan mass spectrometry. It is even possible to carry out all three techniques simultaneously [55].

A typical DTA head for a thermobalance [56] is shown in Fig. 8.5. The DTA head is placed in a wire-wound furnace and is supported on an alumina capillary by the balance arm. The sample and reference are placed in thin alumina or platinum cups which slide into the holders which also serve as a differential thermocouple system. Using Pt/Rh–Pt thermocouples, the temperature range is 25–1600°C. The temperature recording is obtained from the outside enclosure which is also made of Pt/Rh–Pt with a junction at the same level as the differential system. The system has been used for samples between 5 and 50 mg. Heating and cooling rates as small as 0.5°C necessary for following slow weight losses, still gave sufficient sensitivity to separate overlapping processes. The thermobalance is sensitive enough to indicate changes in weight as small as $\pm0.01$ mg.

Effluent-gas analysis has been coupled with the DTA cell shown in Fig. 8.3h. The cell is continuously swept by a helium stream whose conductivity is monitored by a thermistor thermal-conductivity cell [57]. Similar effluent-gas analysers are available for the DSC illustrated in Fig. 8.2 [40]. The set-up of Fig. 8.3d has been coupled to a mass spectrometer for monitoring water evolution [58].

Electrical conductivity is another property which has been measured simultaneously with DTA. An apparatus recording the DTA curve together with conductivity plots of salts has been described by Berg and Burmistrova [59]. Electrical conductivity is measured by placing the sample between two platinum electrodes between 1 and 10 mm apart. Fusion of salts leads to sharp changes in conductivity, while polymorphic transitions give only small changes.

Visual inspection during the heating of a sample is a useful technique which is possible for many DTA set-ups with minor changes, often with only a small loss in uniformity of sample and reference placement. Even more useful is the application of microscopy to establish, for example, the crystallographic parameters of crystalline materials. By using thin-film thermocouples desposited on microscope slides, hot-stage DTA is possible [60, 61].

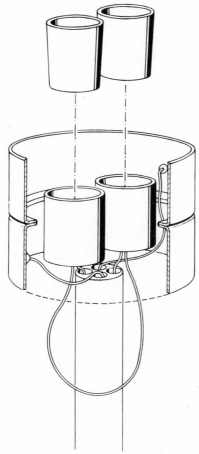

**Fig. 8.5** DTA head to be mounted on thermogravimetric balance, designed by Wiedemann, *Z. Anal. Chem.*, **233**, 161 (1968).

The use of thin-film thermocouples for DTA has been described by King, Findeis, and Camilli [61] who deposited 100–200 $g/cm^2$ nickel–gold thermocouples on quartz and glass slides. Heating curves of up to 3000°C/min were achieved with electrically conducting slides under an interference microscope [62]. In both cases, measurements should be adaptable to extremely small quantities.

As the few descriptions above have shown, there is no limitation to the possible combination of DTA with other measurements. For example, the DTA thermocouple arrangement shown in Fig. 8.4*b* for high-pressure analysis may have a diameter as small as ⅛ in., and need only touch the sample with the top thermocouple junction. Any measurement which allows the sample to be touched with such a thermocouple and involves a heating rate of the proper magnitude can be extended to simultaneous DTA. Reports of simultaneous DTA and resistivity [63], X-ray analysis [63, 64], thermal diffusivity [65], and dilatometry [66], have been presented in the literature.

## Commercial Equipment

In the past, most workers in the field have constructed their own DTA equipment. As equipment has become more sophisticated this trend has decreased. More and more only highly specialized DTA apparatus or variations on commercial equipment such as different furnace and sample holder designs are made individually.

A brief comparison of several DTA instruments is given in Chapter 8 of Ref. 15. This equipment is continually changed and improved, so that information is best obtained from the manufacturer at the time of interest. The *Science* "Guide to Scientific Instruments" (1968), lists 17 manufacturers with offices in the U.S. [67]. Prices for the standard differential thermal analyzers range from below $1000 to about $10,000 with even higher prices for simultaneous- or parallel-measurement capability of other variables. Accuracy reaches and occasionally surpasses custom-made equipment. Convenience of measurement and quality of workmanship usually is better than in custom-built equipment.

The next stage of development is the design of equipment with digital output to be linked to computers for instantaneous evaluation.

## 3   OPERATIONAL PARAMETERS AND STANDARDS

### Instrument Parameters

Differential thermal analysis investigations may range from qualitative scanning to quantitative measurement. For all of them it is of importance to separate any instrument effects from the obtained DTA curve. To a large degree, instrument factors have to be evaluated empirically by analysis of standard materials described later in this section (p. 455). The following instrument parameters will be described: furnace parameters, sample and reference holder parameters, and thermocouple parameters. A more detailed analysis of these problems can be found in Chapter 3 of the book by Garn [24] and Section Vc of the book by Wendlandt [25] which have served together with experience developed in our laboratory as a guide for the following discussion.

### Furnace

The furnace should be a heat source with smoothly increasing temperature which produces identical environments for sample and reference. A linear temperature rise is most convenient and usually preferred. Small deviations from linearity remain, however, from the nonlinearity of the furnace-control elements. This nonlinearity causes a temperature dependence of the calibration constant for heat-capacity and heat-of-transition or reaction measurements. For work directed toward the detection of temperatures of transitions only, this change in calibration constant is insignificant. Instead of controlled linear temperature increase, it is often sufficient to apply a constant voltage to a furnace of relatively large thermal mass and observe the endothermic and exothermic peaks superimposed on the slowly curving baseline. The curvature is caused by the constantly changing heating rate. Different heating rates can be achieved in this case by applying different voltages. Only a relatively sophisticated programmer-controller can duplicate the smoothness of such a simple set-up.

Fluctuations of the furnace temperature will give rise to spurious peaks or an irregular base line. This effect precludes the use of on–off controllers for DTA and indicates the need of sufficient insulation from the temperature fluctuations of the environment. The most desirable situation is the location of the thermal equipment in a draft-free, closely temperature-controlled room.

The shape of the furnace can be varied as is shown in Fig. 8.3. Most frequently a cylindrical elongated shape is preferred in order to have a zone of relatively constant temperature in the middle of the furnace for sample placement. Supports of the furnace block are made of insulating materials for minimum imbalance at the furnace bottom. Little investigation of the most desirable geometry of the furnace has been made.

To achieve constant temperature over the furnace surface facing the sample holder and quick response for faster heating rates, the furnace material is chosen to be highly conductive. Metal furnaces for lower temperatures are best made of gold, silver, or pure aluminum. For higher temperatures, high-thermal-conductivity ceramics may be used. Furnace blocks with separate holes for sample and reference should be checked for uniformity of temperature gradients. Garn [24] suggests that thin-walled blocks with separate symmetrically placed holes for sample and reference should have a metal wall surrounding the holes of at least the well diameter at their closest point. This should avoid influence of sample reaction on reference behavior. It should also not be forgotten that heat conductivity of block material changes with temperature.

The three types of heat transfer; conduction, radiation, and convection, may contribute varying amounts in differently designed furnaces and in different temperature regions. The arrangements shown in Figs. 8.3a, d, and f have the largest amount of heating by conduction through solids. High reproducibility and precision relies in these DTA cells upon good and constant contact between sample and reference container and furnace block or thermoelectric disk. All other arrangements shown in Fig. 8.3 rely on gases and, at higher temperatures, increasing amounts of radiation as means of heat transfer. Heat transfer by gases is usually unstable because of the un-reproducibility of convection currents. A solution to this problem is the use of a dynamic inert atmosphere which sets up a steady-state flow thus removing the unreproducible convection currents in a static atmosphere. Gas flow must be controlled and strategically directed to achieve fluctuation-free heat transfer. Heat transfer by radiation becomes increasingly important at temperatures above about 100°C. The factors governing heating by radiation are the absolute temperature and emissivity of the furnace and sample- and reference-container walls. Uniformity (cleanliness) of these surfaces is an often overlooked prerequisite for symmetrical and fluctuation-free heat transfer by radiation.

Most furnaces are primarily designed for heating. The cooling necessary for low-temperature operation is usually introduced by immersion into a cold bath or by conducting cold gases or liquids through cooling coils as indicated in Fig. 8.3f. For control, the cooling is partially compensated by electrical heating. Obviously the arrangement of Fig. 8.3f is not able to achieve the controlled cooling necessary for DTA with decreasing temperature because the temperature gradients set up by the cooling coils and heater add, rather than modulate. Similarly poor for DTA with decreasing temperature is the arrangement shown in Fig. 8.3d. The cooling medium should be placed such that the temperature gradient can be modified by the heater. The best location for the cooling coils is behind the heater as seen from the sample. Using a cold inert atmosphere for cooling is also suitable only for uncontrolled cooling after a heating run and cannot give a precise DTA with decreasing temperature.

### Sample and Reference Holder

The instrumental parameters deriving from the sample and reference holder are not too well understood and are the subject of some controversy. Two situations can be separated: the apparatus is built either to have close to constant temperature throughout the sample, or to leave an appreciable temperature gradient within the sample. DTA with almost-constant-temperature samples lends itself more to quantitative interpretation as will be shown in Section 4, while DTA with a larger temperature gradient within the sample may have greater sensitivity for determination of transition temperatures.

A small temperature variation within the sample can be achieved by going to small sample sizes or by enclosing the sample in good conducting cups. In order to have a reproducible heat flux which is relatively unaffected by changes of the sample properties, thermal resistance (steep temperature gradient) must be set up between furnace wall and sample. In the apparatus shown in Fig. 8.3*d* and *f*, small sample masses are chosen and the thermal resistance is furnished by a glass capillary which serves as disposable sample and reference holder. The apparatus of Fig. 8.3*a* relies on metal cups for temperature equilibration and provides the thermal resistance in the form of the thermoelectric disk. All other set-ups shown in Fig. 8.3 rely on a gas gap as thermal resistance. The questions of sensitivity and resolution of thermal effects is in all cases a matter of sample mass, specific heat, thermal conductivity, and heat effects in transition regions balanced against holder-material heat capacity and conductivity in the temperature field set up by the furnace.

For high resolution a sample with small temperature gradient and a high flux of heat is necessary such that the thermal gradient which existed before the thermal reaction is recovered fast. For qualitative work recovery can be speeded up if sample and reference are placed closer together so that heat exchange between sample and reference becomes possible (see set-up shown in Fig. 8.4*b*). Sample and reference are then included in the same deformed temperature field caused by a sample reaction and a quicker return of the temperature difference recording to the baseline results, allowing a second process to be better separated from the first. It is clear that the larger the overlap of the sample heat flow to the reference heat flow, the more removed is the DTA from differential calorimetry which is based on separate heat flux to reference and sample. The gain in resolution is accompanied with a loss in sensitivity.

### Response Time

In discussing response time parameters it is assumed that the time lag between sensing the temperature of the thermocouple and recording the emf has been reduced to an insignificant amount by proper instrument design. The remaining time factor is the thermal response to different heating rates. More quantitative aspects will be discussed in Section 4. Basically a faster heating rate requires a steeper temperature gradient, which sets up larger temperature differentials, which increase the sensitivity. The larger temperature differentials, however, require a wider temperature region in the DTA curve for their decay after endothermic or exothermic reactions in the sample which results in loss of resolution of DTA peaks.

Figure 8.6 shows a plot of the ratio of baseline deflection to sample heat-capacity change at the glass transition temperature of polystyrene (about 100°C) for two decades of heating and cooling rates [68]. The plot shows that at higher heating rates the proportionality increases somewhat more than linearly. A similar effect can be found in the peak areas of changes involving latent heat. In discussing peak areas it is important to note that the tempera-

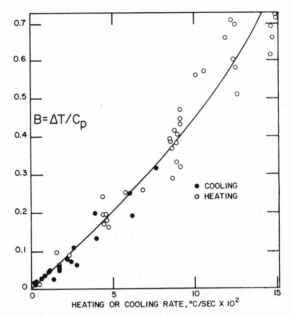

**Fig. 8.6**  Change of Sensitivity as expressed by the ratio of baseline deflection, $\Delta T$, to heat capacity, $C_p$, as well as a function of heating rate. The cylindrical DTA cell operated with temperature gradient inside the sample which was in direct contact with the metal block. Radius of the sample holder about 0.4 cm (see Section 4 for additional calculations on cells of this type). [Reproduced from B. Wunderlich and D. Bodily, *J. Polymer Sci., Part C* **6,** 137 (1964).]

ture difference change is transient in this case. Plotting the temperature difference as a function of reference temperature results in an increase of the peak area with rate of temperature change similar to Fig. 8.6; plotting the temperature difference as a function of time keeps the peak area almost constant; plotting the temperature difference as a function of sample temperature gives uninterpretable results since the sample temperature changes more slowly during an endothermic change than programmed, and during an exothermic change the sample temperature changes faster than programmed.

Peak temperatures should be measured versus the sample temperature, otherwise their true location must be calculated from the measured $\Delta T$ recording which in many cases remains uncalibrated. For most precise measurements small sample masses and direct thermocouple–sample contact are necessary. If these conditions are fulfilled, the peak temperature does not change with heating rate. In our laboratory constancy of melting peak temperatures was shown with DTA cells similar to Fig. 8.3*d*. Up to 100°C/

min heating-rate changes in peak temperatures on melting were less than 0.2°C for samples weighing about 1 mg. If the thermocouple does not touch the sample, however, as in the DSC cell shown in Fig. 8.3a a large lag develops. Table 8.2 shows the effect of heating rate on start of melting ($T_m$) and peak temperature ($T_p$) in such a cell containing in one case 0.38 mg benzoic acid and in another 13.01 mg of indium. The two sample weights have approximately the same heats of fusion and their temperature lag is also of the same order of magnitude. The equilibrium melting temperatures of benzoic acid and indium are 122.35 and 156.6°C, respectively.

**Table 8.2** Change in Start of Melting and Peak Temperature with Heating Rate

| Heating Rate (°C/min) | Benzoic Acid | | Indium | |
|---|---|---|---|---|
| | $T_m$(°C) | $T_p$(°C) | $T_m$(°C) | $T_p$(°C) |
| 5 | 123.5 | 124.1 | 158.6 | 159.6 |
| 10 | 123.6 | 124.7 | 158.9 | 160.5 |
| 50 | 124.4 | 126.4 | 160.0 | 162.6 |
| 100 | 126.0 | 128.6 | 161.6 | 165.2 |

The change in the temperature of the beginning of melting indicates the lag in temperature of the sample during normal heating. During the melting this lag increases considerably.

Figure 8.7 shows the loss of resolution on increase of heating rate. Closely-lying peaks may be completely masked at fast heating rates if the time for recovery (see Section 4) overlaps significantly into the second transition region.

### Thermocouples

The inherent properties of thermocouples have already been discussed in Section 2, at this point the effect of thermocouple size relative to sample size and the effect of thermocouple placement will be considered.

The size of the thermocouple has two effects, it contributes to the heat capacity of the sample and it may conduct heat into or out of the sample through its wires. Both effects do not seriously impair measurements as long as the placement of the sample is reproducible and the heat effect can be incorporated in the calibration. The maximum signal will be derived when the thermocouples, sample as well as reference, are inserted so deeply (relative to their size) that complete isolation from external effects is achieved at the position of the junction.

The position of the thermocouple is particularly critical if the instrument has a sizable temperature gradient within the sample. In the case of the

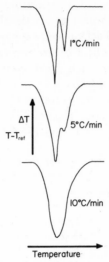

**Fig. 8.7** Loss of resolution of the two close-lying melting transitions with increasing heating rate. Sample masses were changed to achieve about equal peak sizes.

usual central positioning in a cylindrical sample holder, the thermocouple is at the position where the reaction occurs last. If the transition occurs over a narrow temperature range, as in melting of a crystal, the peak temperature is close to the transition temperature. The beginning of a noticeable deviation from the baseline signifies only the start of melting of the outside layer of the sample. The central location of the thermocouple has an additional advantage: since it is at the shallowest portion of the temperature gradient, it is least sensitive to physical movement during measurement. Positioning the thermocouples outside of the sample and reference makes for a considerable increase in convenience of sample handling and removes problems of chemical reaction with the thermocouple material. It increases, however, the lag as was shown above (Table 8.2). In partially filled sample containers with outside placement of thermocouples, lags can be enhanced if the sample comes to rest away from the thermocouple location.

Even more involved are problems connected with shifting of sample material or thermocouples during measurement. Quite striking artifacts can be produced in a DTA curve when a sample is allowed to draw away from the thermocouple or to contact it more intimately. The only way to recognize such artifacts (usually in the form of baseline shifts and occasionally in the form of peaks) is to carry out multiple analyses under different conditions.

### Sample Parameters

In this section parameters connected with sample size, particle size, reference nature, and interaction with inert atmosphere will be discussed. These parameters, similar to the instrument parameters discussed above are irrelevant to most answers sought by DTA and must be understood (and hopefully eliminated) for proper interpretation of DTA curves.

### Sample Size

As long as the analysis is run under calorimetric conditions, the sample size should be directly proportional to the obtained signal. For any one apparatus this is at best true only for a limited size range. The peak area seems to decrease [69] at high sample masses. Probably this decrease in sensitivity is due to an increased heat flow between reference and sample caused by the larger temperature differential.

The peak temperature shows increased lag with increased sample size if it is not measured in the center of the sample. Also, the recovery to the baseline after a transition will take longer, reducing the resolution for large sample masses. The detrimental effects of too-large sample sizes go parallel with the effects discussed for higher heating rates. In practical application, the optimum range of sample size is coupled with an optimum range of heating rates.

In a DTA cell with a larger temperature gradient within the sample, it is possible to modify the signal per mass unit by incorporation of an inert diluent into the sample. The temperature gradient within the sample is determined (see Section 4) by the heat capacity (mass dependent) and the thermal conductivity (and packing density if the sample is not a solid block). Table 8.3 reproduces data by Barrall and Rogers [70] on the effect of inert materials mixed into the sample on the melting peak area of salicylic acid. Larger peak areas seem to be caused by higher thermal conductivity of the diluent.

**Table 8.3**   Melting Peak Areas for Salicylic Acid–Diluent Mixtures

| Diluent | Salicylic Acid (%) | Peak Area/0.01 g Acid (mm²) |
|---|---|---|
| Carborundum | 6.87 | 306 |
| Iron | 8.82 | 710 |
| $Fe_2O_3$ | 3.40 | 280 |
| Glass beads, 0.029 mm | 4.57 | 322 |
| Glass beads, 0.29 mm | 5.58 | 289 |
| Alumina | 8.60 | 313 |
| Nujol | 20.00 | 92 |

### Particle Size

The influence of particle size in DTA with small temperature gradients should again be minimal until the sizes are reduced such that sample properties are impaired. With decreasing particle diameter the specific surface area (cm$^2$/g) is increased. Surface effects usually show up when particle sizes get to the order of magnitude of microns. The most widely studied samples of small particle size are crystalline, linear high polymers. Frequently these materials form crystals in the 50–500 Å size range. The surface effects are sufficient in these cases to cause melting-point lowerings by as much as 50°C. In addition, small particles are quite metastable and will frequently undergo reorganization to larger particle sizes while being analyzed. Measurements on reorganizing samples are quite meaningless if not controlled carefully. A more detailed description of the application of fast-heating-rate DTA to analyze particle sizes will be given in Section 5. Since grinding is a frequent method of preparation of samples for DTA, it is imperative to be aware of possible size effects. Besides size reduction, the mechanical grinding stress may induce transitions or strain which may be reversed with accompanying heat effect on DTA.

DTA cells with larger temperature gradient within the sample show, in addition, effects arising from packing of the sample. Tighter packing increases the heat conduction and generally increases the reproducibility. Loose packing may cause drawn-out peaks because of the larger temperature lag. Packing of samples is an even more important parameter in the analysis of samples decomposing to a gaseous product or reacting with the furnace atmosphere. More will be said about these effects below.

### Reference Material

The reference material should be chosen to give close to zero temperature difference between sample and reference under steady-state heating conditions with no transition in the sample. This can be achieved by choice of material as well as adjustment of amount of reference material. In DTA with sizable temperature gradient, the thermal conductivity has an effect on the measurement and, because of their influence on thermal conductivity, the particle size and tightness of packing also are variables.

Frequently used as reference materials are calcined $Al_2O_3$, MgO, powdered, fused quartz, or glass beads. It is self-evident that the reference material must be thermally inert. Since no two different materials have the same heat capacity and thermal conductivity over a large range of temperature, a matched sample and reference will still show a sloping baseline over a larger temperature range. Although commercial equipment often offers electronic compensation of baseline slopes, it is useful to match sample and reference initially and use electronic compensation only as a minor correction.

Special analysis problems might require a different choice of reference material. Small changes in samples may for example be shown up by using the unchanged sample as a reference. For example, glass transitions of materials which can also be obtained crystalline are best analyzed with the crystalline form as reference since heat capacity and, to a lesser degree, thermal conductivity are similar in glasses and crystals. At the glass transition, only the sample will show the change to the melt.

### Atmosphere

As long as the sample does not interact with the furnace atmosphere or does not evolve a gas itself, the effect of atmosphere has no influence on the DTA beyond the instrument effect described above.

In case a gas is involved in a sample change, the partial pressure enters into the equilibrium constant and changes the transition temperature. Reproducible effects can only be expected as long as equilibrium can be established. Dynamic atmospheres with a controlled partial pressure of the reacting gas out of or into the sample holder are essential for the establishment of equilibrium.

Long, narrow sample holders, beneficial for uniform, two-dimensional heat flow in the middle portion, show a large effect due to a gas evolved during heating. The local particle pressure increases considerably forcing the transition to uncontrolled higher temperature. On the other hand, reactions caused by the furnace atmosphere, like oxidations, can penetrate only slowly from the top and are not reproducible. In case of unwanted reactions such sample holders may be advantageous by keeping the reaction at the surface where the influence on the sample close to the much lower thermocouple is negligible.

Packing of different particle sizes becomes an important parameter when considering diffusion of gases into or out of the sample. The same diffusion problem must be taken into account in deciding whether a sample holder should be covered or not. The advantage of a covered sample is a more uniform temperature environment, particularly if the cover touches the sample as for example in a sample pan usable for the DSC cell of Fig. 8.3a; a possible disadvantage is the decrease of diffusion. When diffusion is of no concern, sample pans with hermetic seals can be used for analysis under reproducible conditions

Differential thermal analysis under conditions of reactions or transitions involving gases thus needs special care. A shallow layer of sample, well-granulated in an open sample holder, with a controlled dynamic atmosphere, and reasonably slow heating rate, are conditions for operation close to equilibrium.

### Standards

### Standardization of Technique

From the discussion of the operational parameters it becomes abundantly clear that the major problem in DTA is the separation of instrument and extraneous sample effects from the thermal properties of the sample. The larger the temperature gradient within the sample, the more difficult is this separation and with it, the interpretation of the DTA curve and the inter-comparison of curves from different sources. The International Confedera-tion for Thermal Analysis (ICTA) has established a Standardization Com-mittee. The following recommendations about information which should accompany any DTA record were made [71].

1. Identification of all substances (sample, reference, diluent) by a definitive name, an empirical formula, or equivalent compositional data.

2. A statement of the source of all substances, details of their histories, pretreatments, and chemical purities, so far as they are known.

3. Measurement of the average rate of linear temperature change over the temperature range involving the phenomena of interest.

4. Identification of the sample atmosphere by pressure, composition, and purity; whether the atmosphere is static, self-generated, or dynamic through or over the sample. Where applicable the ambient atmospheric pressure and humidity should be specified. If the pressure is other than atmospheric, full details of the methods of control should be given.

5. A statement of the dimensions, geometry, and materials of the sample holder; the method of loading the sample where applicable.

6. Identification of the abscissa scale in terms of time or of temperature at a specified location. Time or temperature should be plotted to increase from left to right.

7. A statement of the methods used to identify intermediates or final products.

8. Faithful reproduction of all original records.

9. Wherever possible, each thermal effect should be used identified and supplementary supporting evidence stated.

10. Sample weight and dilution of the sample.

11. Identification of the apparatus, including geometry and materials of the thermocouples and the locations of the differential and temperature-measuring thermocouples.

12. The ordinate scale should indicate deflection per degree Celsius at a specified temperature. Preferred plotting will indicate upward deflection as a positive temperature differential, and downward deflection as a negative temperature differential, with respect to the reference. Deviations from this practice should be clearly marked.

The ASTM recommends, in addition, for thermal analysis of metallic materials (E 14–63), recording the:

13. Maximum temperature attained, and time of temperature.

14. Temperature of minimum rate of change of specimen at recognized and identified inflection point.

15. Direction of change.

16. Temperatures at the beginning of heat effect, at the maximum or at the end of the effect, or both.

17. Maximum and minimum temperature of decalescence and recalescence if either is observed.

Both ICTA [54] and ASTM [72] are at present involved in further study of nomenclature, definitions, standard reference materials, test methods, and recommended practices.

Although the above recommendations are a step forward in allowing intercomparison of DTA curves, it still remains up to the operator to evaluate the particular instrument and sample parameters of his experiment. Hope of improvement is in sight due to the rapid increase in the use of commercial set-ups which, with the help of manufacturers research and publications [73, 74], are being better characterized.

The arbitrary choice of abscissa and ordinate of the DTA curves has been a constant source of confusion. There are, besides the possibilities of labeling in voltage, temperature, heat capacity, or arbitrary units, four combinations for the direction of plotting the measured quantities. ICTA recommendation 12 removes the arbitrariness of the temperature abscissa, but retains room for confusion for the ordinate. On heating, a positive temperature differential $(T_{sample} - T_{reference})$ will show an endothermic process as a minimum and an exothermic process as a maximum. This recommendation requires, as a large number of workers in the field of qualitative DTA have done in the past, DTA curves on heating to be plotted reverse from the practice in calorimetry. An endothermic process measured by adiabatic calorimetry is always plotted as a maximum [1]. There seems little hope for agreement on the labeling of the ordinate in units proportional to the heat capacity of the sample, the only logical choice [75]. In the meantime, it will be good practice to label in addition to any other unit the direction of endothermic and exothermic processes.

### Standard Materials

Standard materials are quite essential for the evaluation of a given DTA instrument. Depending on the mode of operation up to three calibrations are necessary. Most fundamental is the temperature calibration. Although calibration of thermocouples outside of the DTA is frequently possible, a better procedure is to calibrate by using melting point standards under conditions similar to later measurements of unknowns. For a DTA cell with a small temperature gradient within the sample, the beginning of melting extrapolated from the linear portion of the endotherm to the baseline is most reproducible. For a DTA cell with larger temperature gradient, it is better to calibrate the peak temperature. For later use, it must be remembered that in the first case the calibration includes the (small) temperature gradient within the sample while the latter calibrates the coolest spot in the DTA cell (if the thermocouples are placed properly). Although efforts are underway to generate recommended standards, most laboratories are presently using their own sets. From $-190$ to $+500°C$ melting points of 3213 organic compounds have been tabulated [76]. Similar tables can be found in compilations of melting points for microscopy [77] or in general data sources [78, 79]. During the last 10 years zone-refined chemicals have become commercially available in larger numbers. Their melting temperatures are usually quite reproducible and are recommended for calibration.

Table 8.4 lists a set of calibrations carried out in our laboratory with the DSC cell shown in Fig. 8.3a at a heating rate of 10°C/min. The measured temperature was taken in this small temperature gradient cell at the extrapolated beginning of melting. The standard thermocouple corrections [36] were applied before comparison with literature data. Each sample was run up to 7 times. The standard deviation of the error of a single run of all 15 reference compounds was $\pm0.2°C$. The average deviation of the melting temperature of all 15 reference compounds from the literature value was 1.06°C with a standard deviation of $\pm0.2°C$. As described in the section on

**Table 8.4**  Temperature Calibration

| Reference Compound | Literature $t_m(°C)$ | Measured $t_m(°C)$ | $\Delta t(°C)$ |
|---|---|---|---|
| Water (distilled) | 0.0 | 1.4 | 1.4 |
| p-Chlorotoluene (Litton) | 7.5 | 8.4 | 0.9 |
| p-Nitrotoluene (Fisher therMetric standard) | 51.5 | 52.4 | 0.9 |
| p-Dichlorobenzene (Litton) | 53.0 | 54.3 | 1.3 |
| Cyclododecane (Litton) | 60.8 | 61.9 | 1.1 |
| Durene (Litton) | 79.4 | 80.3 | 0.9 |
| Naphthalene (Fisher therMetric standard) | 80.3 | 81.2 | 0.9 |
| Vanillin (Fisher) | 80.5 | 81.7 | 1.2 |
| Benzoic acid (Fisher therMetric standard) | 122.4 | 123.6 | 1.2 |
| Salicylic acid (Fisher) | 159.5 | 161.0 | 1.5 |
| o-Iodobenzoic acid (Fisher) | 163.0 | 163.6 | 0.6 |
| Succinic acid (Fisher) | 187.5 | 188.4 | 0.9 |
| Anthracene (Fisher) | 216.2 | 217.3 | 0.9 |
| Bismuth (Fisher 99.99% pure) | 271.3 | 272.5 | 1.2 |
| Lead (Fisher pure) | 327.5 | 328.5 | 1.0 |

response time, the constant value of $\Delta t$ is an instrument effect. (See also Table 8.2.) Repeating calibrations with the same reference compounds using the DTA cell shown in Fig. 8.3d reduced the average value at $\Delta t$ to zero.

Table 8.5 lists the literature value and DTA data on solid I $\rightleftharpoons$ solid II-type transition temperatures derived as an initial set of standards by the Committee on Standardization of the ICTA, published as Appendix 3 of Ref. 18. The systems were selected from 200 solid $\rightleftharpoons$ solid transitions listed by the National Bureau of Standards. For each system a large master sample was obtained, and participants in the analysis were provided with smaller quantities from it. Maximum sample size per run was 300 mg. Two rates of heating were requested, close to 3°C/min (A), and close to 10°C/min (B). Twenty-five workers contributed to the effort, using 18 different instruments. Table 8.5 lists only the intersection temperature of the peak extrapolated back to the baseline and the peak temperatures measured on heating. Data were reported after all instrument corrections had been applied by the individual workers, the means and standard deviations are reproduced in Table 8.5

**Table 8.5** Mean Transition Temperature Data

| Compound | Heating Rate[a] | Data Points (#) | Intersection (°C) | Peak (°C) |
|---|---|---|---|---|
| $KNO_3$ (127.7°C) | A | 36 | 128 ± 4 | 134 ± 5 |
| | B | 62 | 129 ± 5 | 137 ± 6 |
| $KClO_4$ (299.5°C) | A | 41 | 299 ± 6 | 304 ± 6 |
| | B | 62 | 300 ± 5 | 309 ± 6 |
| $Ag_2SO_4$ (412°C) | A | 40 | 426 ± 7 | 433 ± 7 |
| | B | 67 | 429 ± 6 | 438 ± 6 |
| $SiO_2$ (573°C) | A | 39 | 572 ± 5 | 575 ± 5 |
| | B | 71 | 569 ± 6 | 575 ± 4 |
| $K_2SO_4$ (583°C) | A | 38 | 584 ± 4 | 588 ± 4 |
| | B | 66 | 583 ± 4 | 589 ± 5 |
| $K_2CrO_4$ (665°C) | A | 37 | 668 ± 5 | 671 ± 4 |
| | B | 59 | 668 ± 5 | 674 ± 5 |
| $BaCO_3$ (810°C) | A | 72 | 807 ± 6 | 815 ± 7 |
| | B | 41 | 809 ± 4 | 819 ± 8 |
| $SrCO_3$ (925°C) | A | 37 | 928 ± 6 | 934 ± 5 |
| | B | 57 | 928 ± 6 | 938 ± 7 |

[a] Heating rates A were below or equal to 4.9°C/min while heating rates B were above or equal to 5.0°C/min.

The present optimum consistency is obviously only $\pm 5$–$6°C$ irrespective of temperature. Workers using similar sample holders and measuring locations were consistent to within $\pm 2.5°C$, while individual workers achieved still higher precision. A comparison with Table 8.4 indicates that much better calibration can be achieved with melting points of zone-refined chemicals using DTA of smaller samples and accordingly smaller temperature gradient within the sample.

Calibration of the baseline deflection ($\Delta T$) due to the heat capacity of the sample is best done with one of the substances with well-established heat capacities [79]. For most ceramics and also organic materials $Al_2O_3$ in the form of corundum or synthetic sapphire is an ideal standard. Its heat capacity is known [80] to better than $\pm 0.5\%$ between $-260$ and $1000°C$, and it shows no thermal transitions over the whole range. We were able to achieve $\pm 2\%$ heat-capacity accuracy with DSC using $Al_2O_3$ as calibration material [81, 82].

Endothermic or exothermic phase transitions or chemical reactions are calibrated best with known processes in the same temperature region. Again, no standards have been established as yet and general reference works must be consulted for suitable calibration materials [78, 79]. For heat of fusion measurements in our laboratory [83] benzoic acid, urea, indium, and anthracene have proven useful. Table 8.6 gives data on their melting temperature and heat of fusion.

**Table 8.6**  Heat of Fusion Reference Materials

| Reference Compound | Melting Temperature (°C) | Heat of Fusion (cal/g) |
|---|---|---|
| Benzoic acid | 122.4 | 35.2 |
| Urea | 132.7 | 57.8 |
| Indium | 156.4 | 6.80 |
| Anthracene | 216.3 | 38.7 |

## 4  MATHEMATICAL TREATMENT OF DTA

### General Discussion

The mathematical treatment of DTA has been the subject of many frustrations and errors. As can be judged from the previous discussion, it is quite difficult to take the experimental conditions into account with the necessary precision for mathematical analysis. In addition, instrument as well as sample parameters change continuously as temperature changes. As a result, only simplified models have been treated. If one defines DTA as a technique in which heat is allowed to flow from the sample to the reference and vice versa (as in Fig. 8.4b), as was proposed by Garn [24], one must admit that no mathematical treatment has been proposed as yet. On the other hand, if one defines DTA broadly, as was done in this chapter, and includes DSC as a technique in which one can neglect the temperature gradient within the sample, a quite detailed mathematical description is possible as will be shown in a later section (p. 469). The intermediate case of negligible crossflow of heat between sample and reference, but sizable temperature gradient within sample and reference has been treated partially and will be discussed below.

Before applying any of the mathematical expressions derived below, it is important to judge whether the particular apparatus used for measurement fulfills the often stringent simplifications necessary for the mathematical description.

Additional discussions of this topic can be found in Chapter 2 and 5 of Ref. 24 and in Chapter 5 of Ref. 15. Assumptions made for all mathematical treatments described below are the following:

1. Heat conduction by and heat capacity of thermocouples are always neglected.

2. Contacts between sample and holder, holder and furnace, and so on, are always assumed perfect.

3. The sample and reference packing is always assumed to be homogeneous and its effect included in the heat conductivity.

4. Specific heat, density, and heat conductivity of sample and reference are assumed to be constant except for transitions.

5. Heat transfer by other than conduction mechanisms is not considered.

## DTA with Temperature Gradient within the Sample

### Heat-Flow Equation

In case a sizable temperature gradient exists within the sample, measured effects in $\Delta T$ depend not only on heat capacity and changes in enthalpy, but also on the heat conductivity. A detailed discussion of the problem has been given by Ozawa [84]. Detailed treatments of the heat conductivity problem can be found in Refs. 85–87.

If a temperature gradient exists in a material, heat will flow from the higher temperature according to the following equation:

$$\vec{u} = - \kappa \operatorname{grad} T. \tag{8.1}$$

Vector $\vec{u}$ represents the heat flow in cal/(cm²)(sec), $\kappa$ is the thermal conductivity in cal/(cm)(sec)(deg), and $T$ represents the temperature. The net rate of flow of heat out of any volume $V$ bounded by the closed surface $S$ is given by the integral of the normal component of $\vec{u}$ at the surface, $u_n$, over the surface $S$:

$$\int u_n dS = \int \text{div } \vec{u} \, dV. \tag{8.2}$$

The rate of heat flow can also be expressed in terms of temperature rise. The heat absorbed/$(\text{cm}^3)(\text{sec})$ is given by:

$$\frac{\partial Q'}{\partial t} = \frac{\rho c_p \partial T}{\partial t}. \tag{8.3}$$

The density $\rho$ is expressed in $\text{g/cm}^3$, $c_p$ is the specific heat in $\text{cal/(g)(deg)}$, and $t$ represents time. Equations 8.1–8.3 must hold for any arbitrary region so that by combination and integration one finds that:

$$\frac{\partial T}{\partial t} = \frac{\kappa}{\rho c_p} \text{ div grad } T = k\nabla^2 T, \tag{8.4}$$

where $k$ is the thermal diffusivity expressed in $\text{cm}^2/\text{sec}$, and $\nabla^2$ is the Laplacian operator. Equation 8.4 is known as the Fourier equation [85–87]. It expresses the conditions that govern the flow of heat in a body. The solution of the problem of heat conduction in DTA with temperature gradient within the sample must satisfy this equation.

## General Solution

For a general solution a cylindrical DTA cell is assumed. Its length is chosen such that mathematically it can be taken to be infinitely long, in other words, end effects are neglected. Discussion of the end effect has shown that this is already justified if the temperature is measured 1–2 radii away from the end [85]. Figure 8.8 illustrates the cross-section through the sample cell and block. The reference has identical geometry. The outer shaded area represents the metal block heated or cooled at constant rate $q$. No effects inside the cell are assumed to affect the block temperature $T_0 + qt$ at the radius $R_0$. The next region with a radius $r$ between $R_i$ and $R_0$ represents either a low-thermal-conductivity cell-holder or an air space. The sample finally lies between a value of $r$ of zero and $R_i$.

The temperature in the aggregate of furnace, holder, and sample is according to its origin divided into three parts [84]:

$$T(r) = T_1(r) + T_2(r) + T_3(r), \tag{8.5}$$

the steady-state temperature $T_1(r)$, the initial transient temperature $T_1(r) + T_2(r)$, and the additional effect due to a change in parameters or a heat of transformation, $T_3(r)$. First the steady state will be considered; then the

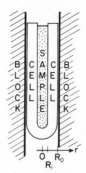

**Fig. 8.8** Cross-section through a typical DTA sample arrangement. The block is metal, the cell a relatively high heat resistance. The thermocouple for measuring is not drawn. The reference cell is identical and symmetrically placed to the sample cell.

transient temperature and the heat effects will be treated. This separation is permissible since (8.4) is a linear homogeneous differential equation.

At steady state, the heat per unit time $Q$ necessary to heat 1 cm of the inner cylinder of radius $r$ with a rate $q$ is given by (sample $= s$):

$$Q = \int_0^r 2\pi r c_s q \rho_s \, dr \qquad \text{for } 0 \le r \le R_i. \qquad (8.6)$$

The $Q$ necessary to achieve steady state in the cell holder can be calculated as (holder $= h$):

$$Q = \int_{R_i}^r 2\pi r c_h q \rho_h \, dr + R_i^2 c_s q \rho_s \pi \qquad \text{for } R_i \le r \le R_0. \qquad (8.7)$$

The temperature gradients in turn are

$$\frac{dT_1}{dr} = \frac{Q}{2\pi r \kappa_s} \qquad \text{for } 0 \le r \le R_i, \qquad (8.8)$$

and

$$\frac{dT_1}{dr} = \frac{Q}{2\pi r \kappa_h} \qquad \text{for } R_i \le r \le R_0, \qquad (8.9)$$

where $\kappa_s$ and $\kappa_h$ are the thermal conductivities of the sample, and holder, respectively. Equations 8.6–8.9 can be combined in pairs and integrated to result in the temperature differences at different locations:

$$T_1(r) - T_1(R_i) = q c_s \rho_s \frac{r^2 - R_i^2}{4\kappa_s} \qquad \text{for } 0 \le r \le R_i, \qquad (8.10)$$

and

$$T_1(r) - T_1(R_0) = qc_h\rho_h \frac{r^2 - R_0^2}{4\kappa_h} + \frac{qR_i^2}{2\kappa_h}(c_s\rho_s - c_h\rho_h) \log \frac{r}{R_0}$$

$$\text{for } R_i \leq r \leq R_0. \quad (8.11)$$

Taking the temperature at the metal block surface at any time equal to $qt$ by starting the experiment at $T_0 = 0$ allows the development of the final expression for the temperature distribution at the steady state [84] in the sample ($0 \leq r \leq R_1$):

$$T_1(r) = qt - \frac{q}{4\kappa_h}\left[ c_h\rho_h(R_0^2 - R_i^2) + 2R_i^2(c_s\rho_s - c_h\rho_h) \log \frac{R_0}{R_i}\right] -$$

$$\frac{qc_s\rho_s}{4\kappa_s}(R_i^2 - r^2). \quad (8.12)$$

For the temperature distribution at the steady state in the cell holder ($R_i \leq r \leq R_0$) one obtains

$$T_1(r) = qt - \frac{q}{4\kappa_h}\left[ c_h\rho_h(R_0^2 - r^2) + 2R_i^2(c_s\rho_s - c_h\rho_h) \log \frac{R_0}{r}\right]. \quad (8.13)$$

The second step in the calculation involves the initial transient period until steady state is reached which was designated in (8.5). as $T_1(r) + T_2(r)$. For this period $dT/dt$ is not equal to zero. Fourier's equation, (8.4), must hold for sample and holder regions. The boundary conditions are

$$T_2(r) = 0 \quad \text{at } r = R_0, \quad (8.14)$$

$$\frac{\kappa_s \partial T(r)}{\partial r} = \frac{\kappa_h \partial T(r)}{\partial r} \quad \text{at } r = R_i. \quad (8.15)$$

The initial conditions are set by the solutions of the steady state (8.12) and (8.13) such that at any $r$

$$T_1(r) + T_2(r) = 0 \quad \text{at } t = 0. \quad (8.16)$$

The initial value of $T_2(r)$ can be obtained by substitution of (8.12) or (8.13) into (8.16). The solutions of (8.4) with conditions (8.14) and (8.15) and initial condition (8.16) have been discussed in detail in general terms [85–87]. The solutions of the differential equations for the two regions are [84]

$$T_2(r) = \sum_{n=0}^{\infty} A_n(r) \exp \frac{-t}{\tau_n} \quad \text{for } 0 \leq r \leq R_i, \quad (8.17)$$

and

$$T_2(r) = \sum_{n=0}^{\infty} B_n(r) \exp \frac{-t}{\tau_n} \quad \text{for } R_i \leq r \leq R_0. \quad (8.18)$$

The $A_n(r)$'s and $B_n(r)$'s are functions of $r$ only:

$$A_n(r) = \alpha_n\, \mathbf{J}_0(ry)\, [\mathbf{J}_0(R_i x)\, \mathbf{Y}_0(R_0 x)] - \mathbf{J}_0(R_0 x)\, \mathbf{Y}_0(R_i x) \tag{8.19}$$

$$B_n(r) = \beta_n\, \mathbf{J}_0(R_i y)\, [\mathbf{J}_0(rx)\, \mathbf{Y}_0(R_0 x)] - \mathbf{J}_0(R_0 x)\, \mathbf{Y}_0(rx). \tag{8.20}$$

Constants $\alpha_n$ and $\beta_n$ must be chosen to express the initial temperature distribution. While $\mathbf{Y}_0$ and $\mathbf{Y}_1$ are Neumann functions (Bessel functions of the second kind) of zero and first order, $\mathbf{J}_0$ and $\mathbf{J}_1$ are Bessel functions of the first kind of zero and first order, respectively. The symbols $x$ and $y$ stand for the following expressions:

$$x = \sqrt{c_h \rho_h / \kappa_h \tau_n} \tag{8.21}$$

$$y = \sqrt{c_s \rho_s / \kappa_s \tau_n}. \tag{8.22}$$

In order to fulfill the initial and boundary conditions the $\tau_n$'s are the roots of the following equation:

$$\sqrt{\kappa_h c_h \rho_h}\, \mathbf{J}_0(R_i y)\, [\mathbf{J}_1(R_i x)\, \mathbf{Y}_0(R_0 x) - \mathbf{J}_0(R_0 x)\, \mathbf{Y}_1(R_i x)] =$$
$$\sqrt{\kappa_s c_s \rho_s}\, \mathbf{J}_1(R_i y)\, [\mathbf{J}_0(R_i x)\, \mathbf{Y}_0(R_0 x) - \mathbf{J}_0(R_0 x)\, \mathbf{Y}_0(R_i x)]. \tag{8.23}$$

The largest $\tau_n$, ($\tau_0$), determines the response time of the system; it gives the time necessary to attain steady state. A large $\tau_0$ results in poor resolution. It is of interest to note that $\tau_n$ does not depend on $r$ or $q$. Only the geometry ($R_i$ and $R_0$) and the thermal properties and densities enter into (8.23).

A set of equations analogous to (8.12), (8.13), and (8.17) and (8.18) can be derived for the reference and reference-holder system. By combining the proper equation for the point of temperature detection for reference and sample a complete generation of the DTA baseline is possible [84]. It becomes obvious that a difference in heat capacity between reference and sample will show a deviation of $\Delta T$ from zero at steady state as well as in the approach to steady state, while a difference in heat conductivity will enter into the approach to steady state only. Any effects of fluctuations in heating rate will thus only be cancelled by the reference if the match in heat capacity and heat conductivity is sufficiently good.

At this point of the development of mathematical description of the DTA curve, exact description has reached its present day end, although computer-generated solutions of the further steps can be envisioned. To calculate $T_3$, the temperature effect of changes in the sample, stringent simplifications must be made. The difficulty lies with the progression of an endothermic or exothermic reaction or a change in heat capacity from a radius $r = R_i$ to zero. The change occurs at different radii at different times. The problem of changes which are not infinitely sharp is even more complicated.

Ozawa [84] has discussed the decay of a transient effect with given initial temperature distribution. This discussion is relatively simple since the same

$\tau_n$'s as in (8.23) are arrived at. Again the $\tau_n$'s are independent of the initial temperature distribution. Thus, any transient temperature $T_3$ disappears in the same order of magnitude of time, about $5\tau_0$. If a second transformation occurs within the temperature range of about $5q\tau_0$, an overlap occurs with the first transformation.

Finally, peak areas were considered by Ozawa [84]. The equation which must be solved in this case is

$$\frac{\partial T_3}{\partial t} = \kappa \nabla^2 T_3 - \frac{\Delta H}{c_s} \frac{\partial m}{\partial t}, \tag{8.24}$$

where $dm/dt$ is the rate of transformation expressed in fraction of mass per unit time, and $\Delta H$ is the heat transformation per unit mass. Integrating (8.24) over the time interval in which the deviation from steady state was taking place due to a finite $T_3$, Ozawa derived for constant $\kappa_s \rho_s$, and $c_s$, an equation for the peak area $A(r)$ ($T_3$ integrated at position $r$ over the total time interval):

$$A(r) = \frac{M\Delta H}{2\pi l} \left( \frac{\log R_0/R_i}{\kappa_h} + \frac{R_i^2 - r^2}{2\kappa_s R_i^2} \right). \tag{8.25}$$

$M$ is the total mass in the sample undergoing transition and $l$ is the length of the cell. Of importance for this discussion is the result that if $\Delta T$ is measured at $r = 0$ or $r = R_i$, and also in the case of measurement at $r = 0$ for no holder ($R_i = R_0$), $M\Delta H$ is directly proportional to the area $A(r)$. If $\kappa_s$, $\rho_s$ and $c_s$ are not constant, the peak area must be corrected. In usual practice this correcton for first order transitions is often small and can be reduced empirically by proportioning the baseline between the initial and final steady state according to the % conversion measured by the area at any given time (see below). The rather complicated exact calculation is given by Ozawa [84]. It is of interest to note that in case temperature is measured at $r = R_i$, $A(r)$ is dependent on the geometry of the apparatus ($R_0$ and $R_i$) and the thermal conductivity of the holder $\kappa_h$ only, not on the thermal conductivity of the sample. The cell holder serves in this case as a heat-flow meter. The positioning of the measuring thermocouples at $r = R_i$ has been proposed by Ozawa [84] for quantitative DTA. Simplifications can also be noted for such thermocouple placement in (8.12). The sensitivity of Ozawa's apparatus is easily derived from (8.25). The optimum ratio of $R_i$ to $R_0$ is exp $(-\frac{1}{2})$. For fixed $R_i$, the sensitivity increases logarithmically with an increase in $R_0$.

## Simplifications

The most common simplification of the mathematical treatment of DTA with a temperature gradient within the sample involves the omission of the high-thermal-resistance cell holder, assuming mathematically $R_i = R_0$. In

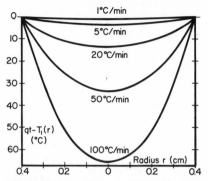

**Fig. 8.9** Temperature distribution in a DTA Sample of 0.4-cm radius in direct contact with the furnace block as a function of heating rate. For a sample of 0.04-cm radius (abscissa multiplied by 0.1) (8.26), multiply ordinate by 0.01.

this case, which is often duplicated by conventional DTA using a metal block with tightly fitting metal holders, the sample has to play two conflicting roles: it has to act as source or sink of heat to be measured, and it has to act as measuring resistance, setting up the temperature gradient necessary for detection. The temperature within the sample due to steady state heating simplifies from (8.12) to

$$T_1(r) = qt - \frac{q(R_i^2 - r^2)}{4k_s}, \tag{8.26}$$

where $k_s$ is the thermal diffusivity, $\kappa_s/c_s\rho_s$, of the sample. Typical calculations of steady state temperature profiles for varying heating rates are shown in Fig. 8.9. The $R_i$ is assumed to be 0.4 cm, as can be found in a cell using about 1 g sample; $k_s$ was given the value $10^{-3}$ cm²/sec, which is the order of magnitude of the thermal diffusivity of most organic polymers and also $Al_2O_3$; the latter is often taken as reference material. For a smaller cell, using perhaps 1–2 mg sample, a typical value for $R_i$ is 0.04 cm. As can be seen from (8.26), the abscissa of Fig. 8.9, $r$, must for this case be multiplied by 0.1, and the ordinate, $T_1(r) - qt$, must be multiplied by 0.01.

The initial transient temperature contribution $T_2(r)$ which is to be added to $T_1(r)$ to give the temperature at distance $r$ from the center (8.5), simplifies considerably over (8.17). The boundary condition and initial condition are again given by (8.14) and (8.16). $T_2(r)$ changes from $q(R_i^2 - r^2)/4k_s$ at $t = 0$ (8.26) to zero at sufficiently long time:

$$T_2(r) = \frac{2q}{R_i k_s} \sum_{n=1}^{\infty} \frac{J_0(r\zeta_n) \exp(-k_s\zeta_n^2 t)}{\zeta_n^3 J_1(R_i\zeta_n)}, \tag{8.27}$$

where $\zeta_n$ is the $n$th root of $\mathbf{J}_0(R_i\zeta) = 0$ tabulated for example in [86]. $\mathbf{J}_n$ is the Bessel function of the first kind of order $n$. At the center of the sample ($r = 0$) the initial transient contribution is

$$T_2(0) = \frac{2q}{R_i k_s} \sum_{n=1}^{\infty} \frac{\exp(-k_s\zeta_n^2 t)}{\zeta_n^3 \mathbf{J}_1(R_i\zeta_n)}. \tag{8.28}$$

The first four terms of (8.28) are

$$T_2(0) = \frac{qR_i^2}{4k_s}\left[1.108 \exp \frac{-5.783k_s t}{R_i^2} - 0.140 \exp \frac{-30.47k_s t}{R_i^2}\right.$$
$$\left. + 0.046 \exp \frac{-74.9k_s t}{R_i} - 0.021 \exp \frac{-139.0k_s t}{R_i}\right]. \tag{8.29}$$

The exponential factors decrease rapidly with increasing $n$ so that it is often sufficient to use the first term only. The initial temperature change is thus approximately:

$$T(0) = T_1(0) + T_2(0) = qt - \frac{qR_i^2}{4k_s}\left(1 - \exp\frac{-5.78k_s t}{R_i^2}\right). \tag{8.30}$$

Similarly, the decay of the originally parabolic steady-state temperature distribution established with heating rate $q$ in a medium of thermal diffusivity $k_s$ in an infinite cylinder can be approximated at the center of the sample by:

$$T(0) = T_1'(0) + T_2'(0) = q(t - t^*) - \frac{qR_i^2}{4k_s}\exp\frac{-5.78k_s t^*}{R_i^2}, \tag{8.31}$$

where $t^*$ is counted from the start of decay of temperature. The surface temperature stays constant during this decay at $[q(t - t^*)]$.

With the two approximate equations, (8.30) and (8.31), an estimate of the time-dependent changes in a DTA experiment is possible by making use of the Boltzmann superposition principle [88]. The simplest effect is an abrupt change in thermal diffusivity on heating at a certain time which resembles for example the change at a glass transition temperature. Note that the more accurate description of the glass transition would call for thermal diffusivity changing over a narrow temperature range. This change would occur at different radii $r$ at different times. This latter case has not been solved as yet, although an estimate was given by Haly and Dole [89] for the glass transition temperature of polypropylene. Haly and Dole divided the sample into eight concentric cylinders for approximate calculation. The time-dependent change of temperature can be calculated by adding the decay of the steady-state temperature gradient as expressed in (8.31) to a build-up of the new steady state in the liquid material, (8.30). The constant time $t'$

marks the moment of the change of thermal diffusivity throughout the sample. At times $t \geq t'$, $T(0)$ is given by:

$$T(0) = qt - \frac{qR_i^2}{4k_s'} \left[ 1 - \exp \frac{-5.78k_s'\{t - t'\}}{R_i^2} \right]$$
$$- \frac{qR_i^2}{4k_s} \exp \frac{-5.78k_s\{t - t'\}}{R_i^2}. \qquad (8.32)$$

Figure 8.10 shows the time dependent change of $\Delta T$, the temperature difference between the block temperature $T(R_i) = qt$ and the temperature at the center of the sample $T(0)$ as expressed by (8.23):

$$\Delta T = T(R_i) - T(0). \qquad (8.33)$$

The same cell parameter $R_i = 0.4$ cm was used as in Fig. 8.9. The thermal diffusivity at times less than $t'$ was taken to be $0.855 \times 10^{-3}$ cm$^2$/sec, similar to glassy poly (methyl methacrylate); the thermal diffusivity at times above $t'$ was taken to be $0.733 \times 10^{-3}$ cm$^2$/sec, similar to molten poly(methyl methacrylate). The heating rate was 20°C/min. Again it is possible to use the graph for smaller cells by changing the coordinates; the $\Delta T$ and time coordinates must be multiplied by 0.01 for a cell of 0.04-cm radius. Since the thermal conductivity and density of poly(methyl methacrylate) changes only little during the glass transition interval, Fig. 8.10 gives a good representation of the heat capacity jump at $T_g$ within the approximations used. The actual DTA curve would have a $\Delta T$ shift at a level close to zero because of the recording of the difference $\Delta T$ (sample) $-$ $\Delta T$ (reference). Only with the capillary-type sample cell of 0.04-cm radius is the broadening of the transition negligible against the natural glass transition region which is 10–20°C (30–60 sec in Fig. 8.10).

It is also possible to simulate a peak in a DTA curve by the same method. Up to $t'$ steady state with $k_s$ is assumed. This is followed between $t'$ and $t''$ by an approach to steady state with an altered $k_s'$. Final approach to the new

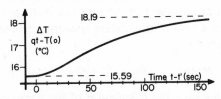

**Fig. 8.10** Change of the temperature at the center of the sample cell due to a jump of thermal diffusivity typical for the glass transition at 20°C/min heating rate. The change occurred uniformly throughout the sample at $t'$; sample cell of 0.4-cm radius. For a 0.04-cm cell, ordinate and abscissa must be multiplied by 0.01 (8.32).

steady state with a value of the thermal diffusivity $k_s''$ closer to $k_s$ is assumed after $t''$. The following set of equations need to be solved:

1. Equation 8.26 at $r = 0$, for $t < t'$.
2. Equation 8.32, for $t' < t < t''$    (8.34)
3. The following equation, for $t > t''$:

$$T(0) = qt - \frac{qR_i^2}{4k_s''} + \left[ \frac{qR_i^2}{4k_s''} - \frac{qR_i^2}{4k_s'} \right.$$

$$\left. + \left( \frac{qR_i^2}{4k_s'} - \frac{qR_i^2}{4k_s} \right) \exp \frac{-5.78k_s'\{t - t'\}}{R_i^2} \right] \exp \frac{-5.78k_s''\{t - t''\}}{R_i^2}. \quad (8.35)$$

Figure 8.11 illustrates the change of $\Delta T$, the temperature difference between the block temperature $T(R_i) = qt$ and the temperature of the sample $T(0)$ as a function of the temperature of the sample $T(0)$ as calculated using (8.35) [90]. As in Figs. 8.9 and 8.10, the sample holder radius is 0.4 cm. The thermal diffusivities $k_s$, $k_s'$, and $k_s''$, were chosen to be $0.9 \times 10^{-3}$, $0.4 \times 10^{-3}$, and $0.7 \times 10^{-3}$ cm²/sec, respectively. The values of $k_s$ and $k_s''$ correspond again approximately to the glassy and liquid thermal diffusivity of poly (methyl methacrylate). The width of the region of $k_s'$ was chosen to be 3°C, similar to the hysteresis effect in annealed glasses (see Section 5). To display the results for heating rates of 6 and 0.6°C/min in the same graph the ordinate was expressed in units of $\Delta T/q$ ($q$ in °C/sec). The abscissa represents $T(0)$ measured from an arbitrary zero and two arrows mark the temperature of the block $[T(R_i)]$ at $t'$ and $t''$. The limitations imposed by the double function of the sample, source of the effect and measuring device, become quite obvious. At fast rate the effect, a square function decrease in thermal diffusivity, is almost lost; at slow rate the sensitivity expressed as the magnitude of $\Delta T$ is low.

Finally, it would be of interest to simulate heat absorption as in melting or endothermic chemical reaction, or heat evolution as in crystallization or exothermic chemical reaction. Again, in order to get a single expression, it is necessary to have the change occurring simultaneously throughout the sample. The amounts of heat absorbed or evolved are in these cases so large that one quickly reaches a period of constant transition temperature for the fast transitions or reactions:

$$T(r) = T(0) \text{ constant} \quad \text{for } t' \leq t \leq t'' \text{ and } 0 \leq r < R_i. \quad (8.36)$$

Of interest is, in this case, the calculation of the decay of the original constant temperature and the reestablishment of the new steady-state temperature gradient of the transformed samples. The solution for the temperature at the center of an infinite cylinder of constant surface temperature ($qt''$)

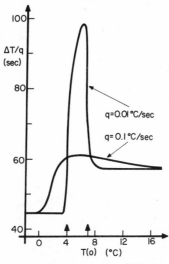

**Fig. 8.11** Change of $\Delta T/q$, the ratio of temperature difference to heating rate between the block temperature and the temperature of the sample at the center (8.33) as a function of the sample temperature at the center for changes of thermal diffusivity $k$ to $k'$ at $t'$ and $k'$ to $k''$ at $t''$ (8.35). The changes are chosen to simulate hysteresis peaks in glasses at the glass transition range. The two arrows on the abscissa indicate the temperature of the block at $t'$ and $t''$ (0.4-cm sample holder).

and at constant initial internal temperature difference $(qt' - qt'')$ is given in Refs. 85–87. Representing it, as before in (8.30) and (8.31), by one term only leads to:

$$T_3(0) = (qt' - qt'') \exp \frac{-5.78k''\{t - t''\}}{R_i^2} \quad \text{for } t \geq t''. \quad (8.37)$$

The new steady-state gradient is set up as expressed by (8.30), the time in the exponent replaced by $t - t''$. Figure 8.12 represents the time change of $\Delta T$, (8.33), after $t''$ as calculated from (8.5), (8.30), and (8.37). The parameters were chosen to simulate the melting of a polymer in a sample cell of 0.4-cm radius at 20°C/min heating rate as in the previous illustrations. At smaller radii the final steady-state gradient becomes smaller and the approach to steady state is more and more governed by (8.37).

## DTA without Temperature Gradient within the Sample

### General Discussion

The foregoing discussion of DTA with temperature gradient within the sample has shown that by reduction of sample-holder diameter the gradient within the sample can be reduced drastically (see for example Figs. 8.9–8.10 for $R_i = 0.04$ cm). The problems of measurement with a sample which

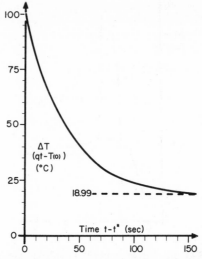

**Fig. 8.12** Approach of the temperature of the sample at the center to steady state after sharp melting (8.31) and (8.37). Heating rate 20°C/min, 0.4-cm radius sample holder. The dotted line at 18.99°C indicates final steady state.

contains a temperature gradient are based on the use of the unknown thermal conductivity of the sample as means for measurement. A more quantitative approximation removes the thermal resistance (which acts as measuring barrier for heat) from the sample, as was discussed in the general case of DTA (Fig. 8.8), but in addition reduces the sample temperature gradient to a negligible level. The mathematical description of this arrangement, which should be called a differential scanning calorimeter, has been developed by Müller and co-workers [91–93]. The heat flow into the sample holder can be approximated by:

$$\frac{dQ}{dt} = K(T_b - T), \tag{8.38}$$

where $K$ is the geometry-dependent thermal conductivity of the thermal resistance layer around the sample, assumed to be temperature independent. Changes of $K$ over larger temperature intervals are corrected by calibration. The temperature $T$ represents the sample, $T_b$ is the block temperature which changes with constant rate $q$:

$$T_b = T_0 + qt. \tag{8.39}$$

An equation analogous to (8.38) can be written for the reference:

$$\frac{dQ_{\text{ref}}}{dt} = K(T_b - T_{\text{ref}}). \tag{8.40}$$

The thermal conductivity of the reference is identical to that of the sample by design of the instrument.

To illustrate the property of such a DSC, it is useful to analyze the heat flow and temperature change of a sample with constant heat capacity, $C_p$. The heat absorbed on heating between $T_0$ and $T$ is in this case:

$$Q = C_p(T - T_0). \tag{8.41}$$

Combination of (8.38), (8.39), and (8.41) gives for the differential equation for heat flow:

$$\frac{dQ}{dt} = K\left[-\frac{Q}{C_p} + qt\right]. \tag{8.42}$$

With the initial conditions:

$$t = 0, \quad T_b = T = T_0 = 0, \quad Q = 0, \tag{8.43}$$

the solution of (8.42) is represented by:

$$Q = qC_pt - \frac{qC_p{}^2}{K}\left[1 - \exp\frac{-Kt}{C_p}\right]. \tag{8.44}$$

$$T = qt - \frac{qC_p}{K}\left[1 - \exp\frac{-Kt}{C_p}\right]. \tag{8.45}$$

The first part of (8.45) represents the linear increase of the block temperature, while the second represents the increasing lag of the sample temperature. Equation 8.45 can also be written:

$$T_b - T = \frac{dT}{dt}\frac{C_p}{K}. \tag{8.46}$$

For the reference system of heat capacity $C_p{}'$ a similar equation can be derived:

$$T_b - T_{\text{ref}} = \frac{dT_{\text{ref}}}{dt}\frac{C_p{}'}{K}. \tag{8.47}$$

At sufficiently long time steady state has been reached and the heat capacity is proportional to the temperature difference between sample and block. Recording $\Delta T = T_{\text{ref}} - T$ versus sample temperature allows then easy evaluation of the heat-capacity difference since both sample and reference have the same heating rate $q$:

$$\Delta T = \frac{q(C_p{}' - C_p)}{K} \tag{8.48}$$

Heat-flow equations derived specifically for the differential scanning calorimeter shown in Fig. 8.3a (duPont DSC cell) have been published by Baxter [94]. The differential scanning calorimeter based on recording heat flow to keep reference and sample temperatures equal (Perkin-Elmer DSC, see Fig. 8.2) can be treated similarly [95]. Although sensing of temperature differences between sample and reference holders and correction by heat transfer from the hotter to the cooler holder can be thought of as practically instantaneous, limited thermal conductivity, $K$, exists between sample pan and sample holder.

### Heat Capacity

For heat-capacity measurements it is generally assumed that changes occur sufficiently slowly that steady state is practically maintained throughout the measurement. If both sample and reference heat capacities stay constant, (8.48) is the relationship between $\Delta T$ and $C_p$. If not, the heating rates in (8.46) and (8.47) are not identical [92]. An equation correcting for this difference can be derived with the additional simplifications:

$$C_p = C_p' + mc_p, \tag{8.49}$$

$$\frac{dT_{\text{ref}}}{dt} = q, \tag{8.50}$$

$$\frac{d\Delta T}{dT_b} = \frac{d\Delta T}{dT_{\text{ref}}} = 1 - \frac{1}{q}\frac{dT}{dt}. \tag{8.51}$$

Equation 8.49 states that the sample heat capacity, $C_p$, is divided into the reference heat capacity $C_p'$ and an excess quantity $mc_p$. If the reference heat capacity is that of an empty identical holder as the sample holder, $m$ represents the mass of the sample and $c_p$ the specific heat. Equations 8.50 and 8.51 state that $T_{\text{ref}}$ and $T_b$, the reference and block temperatures change identically. Using (8.46), (8.47), (8.49), (8.50), and (8.51), a final equation for the sample specific heat can be derived:

$$c_p = \frac{K}{m}\left[\frac{\Delta T}{q} + \left(\frac{\Delta T}{q} + \frac{C_p'}{K}\right)\frac{d\Delta T/dT_{\text{ref}}}{1 - d\Delta T/dT_{\text{ref}}}\right]. \tag{8.52}$$

The correction term plays a significant role in the region of glass transitions and diffuse transitions. For polyethylene a correction of about 10% was necessary [92] in the vicinity of the melting point were $d\Delta T/dT_{\text{ref}}$ is about 0.065.

The heat-flow-measuring DSC (Perkin-Elmer DSC, see Fig. 8.2) does not have to be corrected in this fashion since the heat flux is measured directly. What remains unknown in this instrument is the temperature lag of the sample due to limited thermal conductivity. The special difficulties arising in this instrument at the foot of a broad melting peak have been discussed by Davis and Porter [96] and will be summarized in the section on purity analysis.

### Enthalpy Changes

Enthalpy calculations can be separated into two types. In the first type, the heat capacity remains finite in the whole temperature region and its change is slow enough to keep the DSC practically in steady-state condition, as is found in diffuse transitions. In the second type, a sharp transition occurs with an abrupt enthalpy change halting the change in temperature of the sample. Figure 8.13 shows schematically the change in $\Delta T$ versus time for a diffuse transition. The enthalpy change is given by:

$$\Delta H = \int_{T_i}^{T_f} mc_p dT. \tag{8.53}$$

A convenient expression for the enthalpy change can be derived from (8.46), (8.47), (8,49), and (8.53):

$$\Delta H = \int_{t_i}^{t_f} K\Delta T dt + C_p'(\Delta T_f - \Delta T_i). \tag{8.54}$$

The enthalpy change is represented by the shaded area of Fig. 8.13 *and* a correction term which arises from the different rates of heating of sample and reference holder. This correction term goes to zero if after the transition the same baseline level is reached.

The changes in $\Delta T$ as a function of time for a sharp transition are shown in Fig. 8.14. The transition temperature is reached at sample temperature $T_m$ and time $t_s$. The example is an endothermic transition as in a melting experiment. The sample temperature remains constant up to the completion of the transition at time $t_f$. The temperature difference $\Delta T$ increases linearly with the rate $q$ during this time (see Fig. 8.15). After the transition, $\Delta T$

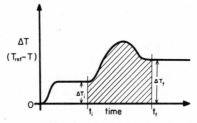

**Fig. 8.13**  Change in $\Delta T$ during a diffuse transition. The total enthalpy change $\Delta H$ between $t_i$ and $t_f$ is proportional to the shaded area corrected by $C_p'(\Delta T_f - \Delta T_i)$ (8.54). $C_p'$ = heat capacity of the holder.

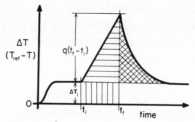

**Fig. 8.14**   Change in $\Delta T$ during a sharp transition. The enthalpy of transition is proportional to the area above the baseline (horizontally shaded and cross hatched) (8.55) and (8.57).

decays exponentially to the new steady state. It is assumed here that the new steady state is identical to the steady state before the transition (constant $c_p$). The treatment for different steady state levels can be found in Ref. 92. The heat of transition, absorbed between $t_i$ and $t_f$, is according to (8.54) and Fig. 8.14 simply:

$$\Delta H = K\Delta T_i(t_f - t_i) + \frac{Kq}{2}(t_f - t_i)^2 + C_p'q(t_f - t_i). \qquad (8.55)$$

The first term of (8.55) represents the vertically shaded area of Fig. 8.14 multiplied by $K$; the second term represents the horizontally shaded area. The third term represents the correction due to heating-rate differences in sample and reference. The cross-hatched area can be written according to (8.45) as:

$$A = \int_{t_f}^{\infty}\left[ q(t_f - t_i) \exp\frac{-K\{t - t_f\}}{C_p} \right] dt \qquad (8.56)$$

or after integration and substitution of (8.49),

$$A = \frac{q(t_f - t_i)(mc_p + C_p')}{K}. \qquad (8.57)$$

Since $mc_p q/K$ is equal to $\Delta T_i = \Delta T_f$ and $C_p'q(t_f - t_i)$ is the last term of (8.55), the enthalpy change $\Delta H$ is also given by $K$ times the peak area above the baseline. The usual evaluation of heats of transition or reaction makes use of this equality which has the added advantage that $C_p'$ need not be known.

An identical derivation for the heat-flow-measuring DSC has been given by Gray [95]. In case of constant $c_p$ throughout a sharp transition, the area above or below the baseline is proportional to the heat effect, but note the changes in peak due to the special electronically governed heat transfer described in the section on purity analysis below. It is of interest to analyze the effect of the thermal resistance for the two types of DSC described by

Fig's. 8.2 and 8.3a. Figure 8.15 compares curves of the two types of DSC for identical values of $K$. DSC relying on $\Delta T$ recording shows (Fig. 8.3a) a peak area proportional to the inverse of $K$ since the heat effect, $\Delta H$, is $K$ times the area above the baseline [8.54]. DSC relying on $dQ/dt$ recording on the other hand (Fig. 8.2) has a constant area for all values of $K$ since $K$ enters only into the lag between the sample temperature and sample holder temperature, $T_b$. The slope in the $dQ/dt$ recording increases with increasing $K$ as can be seen differentiating (8.38) at constant $T = T_m$:

$$\frac{d^2Q}{dt^2} = \frac{KdT_b}{dt}. \tag{8.58}$$

## Kinetics

The study of kinetics of a continuously heating or cooling system presents a special problem because the "rate constant" in most systems is strongly temperature dependent. Only first order kinetics has been worked out in detail [97, 98]. The rate of approach to equilibrium can be written:

$$\frac{dN}{dt} = \frac{1}{\tau}[N^*(T) - N], \tag{8.59}$$

where $N$ is the number of untransformed molecules at time $t$ while $N^*(T)$ is the temperature-dependent equilibrium number of untransformed molecules.

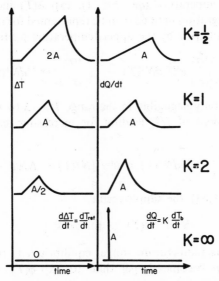

**Fig. 8.15** Response of heat-flow measuring (right) and temperature-difference (left) measuring DSC for different thermal conductivity $K$ between sample and block. (Equations refer to time of melting $T_{\text{sample}} = T_m = \text{constant}$).

The first order rate-constant is expressed as the inverse of the temperature-dependent relaxation time $\tau$. Introduction of the heating rate $q$ allows the calculation of the change in $N$ with temperature:

$$\frac{dN}{dT} = \frac{1}{q\tau} [N^*(T) - N].\tag{8.60}$$

If $q$, $\tau$, and $N^*(T)$ are known as functions of temperature, the solution of (8.60) is [98]:

$$N = N_a e^{-\phi(T)} + e^{-\phi(T)} \int_{T_a}^{T} \frac{N^*(T')}{q\tau} e^{\phi(T')}\, dT',\tag{8.61}$$

where

$$\phi(T) = \int_{T_a}^{T} \frac{dT'}{q\tau},\tag{8.62}$$

with $N_a$ and $T_a$ indicating the number of untransformed molecules and the temperature at time zero. By partial integration (8.61) becomes:

$$N = N^*(T) - e^{-\phi(T)} \left\{ [N^*(T_a) - N_a] + \int_{N^*(T_a)}^{N^*(T)} e^{\phi(T)} dN^*(T) \right\}.\tag{8.63}$$

Equation 8.63 can be simplified for certain ranges of temperature. At sufficiently high temperature ($q\tau << 1$), exp $\phi(T)$ increases so rapidly with $T$ that the integration in (8.63) can be represented for increasing temperature with only small error by the upper-temperature portion of the integral:

$$\int_{N^*(T_a)}^{N^*(T)} e^{\phi(T)} dN^*(T) \approx \int_{N^*(T-\Delta)}^{N^*(T)} e^{\phi(T)} dN^*(T).\tag{8.64}$$

$N^*(T)$ itself varies only negligibly in the range $T - \Delta$ to $T$, so that exp $\phi(T)$ is almost independent of $N^*$, leading for increasing temperatures to the result:

$$\int_{N^*(T-\Delta)}^{N^*(T)} e^{\phi(T)} dN^*(T) < e^{(\phi T)} [N^*(T) - N^*(T - \Delta)].\tag{8.65}$$

On insertion into (8.63), the simple result

$$N \approx \overset{*}{N}{}^*(T) \qquad \text{for } q\tau \ll 1\tag{8.66}$$

is obtained. In this temperature range, equilibrium is reached almost instantaneously. If $q\tau$ becomes larger than 10, exp $\phi(T) = 1.0$ and for increasing temperatures:

$$N \approx N^*(T) - e^{-\phi(T)} [N^*(T) - N_a] \qquad \text{for } q\tau > 10.\tag{8.67}$$

For temperatures even lower, exp $[-\phi(T)]$ is so close to 1.0 that

$$N \approx N_a \qquad \text{for } q_T \gg 10 \qquad (8.68)$$

for heating experiments.

The corresponding enthalpy effects connected with the reaction can be arrived at by calculating the total enthalpy change per degree temperature change:

$$\frac{d\Delta H}{dT} = \left(\frac{dN}{dT}\right) h, \qquad (8.69)$$

where $h$ is the enthalpy change per transformed molecule. Substitution of (8.59) and (8.63) leads to the final result:

$$\frac{d\Delta H}{dT} = \frac{h}{q_T} e^{-\phi(T)} \left[ N^*(T_a) - N_a + \int_{N^*(T_a)}^{N^*(T)} e^{\phi(T)} dN^*(T) \right]. \qquad (8.70)$$

Above the region of fast reaction (8.70) goes over into:

$$\frac{d\Delta H}{dT} = \left(\frac{dN^*(T)}{dT}\right) h \qquad (8.71)$$

as can be derived from (8.66). Below the region of fast reaction (8.70) has no contribution to $d\Delta H/dT$ as can be seen from (8.68). The beginning of reaction can be simplified by using (8.67):

$$\frac{d\Delta H}{dt} = \frac{h}{q_T} e^{-\phi(T)} [N^*(T) - N_a]. \qquad (8.72)$$

Equation 8.72 applies also in case the reaction goes to completion over the temperature range of interest.

Application of these equations assumes that $d\Delta H/dt$ is evaluated by the methods discussed before. Further analysis using absolute reaction rate theory allows the calculation of activation energies from peak temperatures. A detailed description of the application of this treatment to the glass transition range has been given in Ref. 90 and 98.

More information on nonisothermal kinetics has been derived in the field of thermogravimetry. With little effort these derivations can be adapted to DSC. The subject has been reviewed recently by Flynn and Wall [99, 100], and will not be treated here. The use of computers for evaluation of nonisothermal thermogravimetric data has also been discussed recently [101]. The formalism could again be easily adapted to DTA.

## 5  APPLICATIONS

### Introduction

The previous sections have laid the groundwork for the application of DTA to problems in chemistry. Checking recent reviews and collections of papers [17–20] on DTA indicates that there is a seemingly unending variety of applications. It is quite hopeless to mention all possible uses, much less is it possible to describe them. In these last pages an attempt is made to illustrate the use of DTA in the large areas of qualitative and quantitative analysis, in the determination of thermodynamic quantities, and in time-dependent measurements. Somewhat more detailed is the purity analysis using van't Hoff's law since much erroneous information has appeared recently in this area. More is said about heat-capacity measurements since this seems to be a somewhat neglected application of modern DSC. Finally, the measurement of time-dependent quantities by DTA is a unique application and considerable advance is expected from this field. The illustrations for time-dependent DTA are mainly chosen from polymer science where much of this application has been pioneered, however, many nonpolymeric substances behave analogously.

### Qualitative Analysis

#### Simple Heating Curves

Routine thermal analysis of crystalline compounds via their melting point is well established. DTA advances this traditional technique considerably. Not only can melting points be established with precision as was shown in the section on standards (Table 8.4), but it is also possible to deduce information on purity and crystal perfection of the unknown sample as will be discussed in the sections on quantitative analysis and time-dependent DTA. Qualitative analysis via DTA allows, in addition to melting points, use of additional thermal effects of the unknown. Figure 8.16 shows typical DTA traces for four compounds: amyl alcohol, poly(ethylene terephthalate), iron, and barium chloride. Each compound shows quite characteristic heat effects, frequently sufficient for complete identification in case of pure substances, often also in case of mixtures.

Each of the heat effects illustrated in Fig. 8.16; glass transition, crystallization, solid–solid transition, melting, dehydration, evaporation, and pyrolysis, can, in addition, be the subject of separate studies. Since in most instances these changes of the compounds are well understood theoretically much valuable information can be gained from the analysis of variations of these transition temperatures with chemical composition.

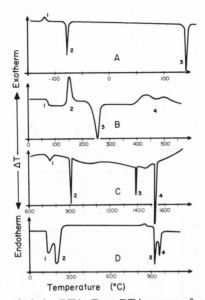

**Fig. 8.16** Qualitative analysis by DTA. Four DTA curves of pure substances redrawn after traces published in Ref. 74. (a) Amyl alcohol. DSC-cell as described in Fig. 8.3a, 2 microliter sample in air at atmospheric pressure, heating rate of 10°C/min, transitions: 1, some crystallization at −119°C; due to fast cooling before heating, incomplete crystallization was reached at the start of the experiment; 2, melting at −79°C; 3, boiling at 139°C. (b) Poly(ethylene terephthalate). DTA-cell as described in Fig. 8.3d, 10-mg sample in nitrogen at atmospheric pressure, heating rate 20°C/min, transitions: 1, glass transition at about 75°C; due to fast cooling of the sample before heating it was practically amorphous at the start of the experiment and shows thus a glass transition characteristic for noncrystalline materials; 2, crystallization at 145°C; 3, melting at 253°C; 4, exothermic decomposition above 360°C. (c) Iron. High temperature DTA cell, see Section 2, about 30 mg sample in helium at atmospheric pressure, heating rate 20°C/min; transitions: 1, $\alpha$ to $\beta$ iron transition at 757°C; 2, $\beta$ to $\gamma$ iron transition at 910°C; 3, $\gamma$ to $\delta$ iron transition at 1389°C: 4, melting at 1533°C, (d) Barium chloride, $BaCl_2 \cdot 2H_2O$. High temperature DTA cell, see Section 2, about 10-mg sample in air at atmospheric pressure, heating rate 20°C/min; transitions: 1, and 2, loss of water in two stages at 130 and 187°C; 3, rhombic to cubic transition at 323°C; 4, melting at 950°C.

## Chemical Reaction

Further analysis of unknowns and study of chemical reactions is possible by mixing the compound under investigation with a chemical reagent and observing the possible reactions by their heat effects and product characteristics during DTA. Figure 8.17 reproduces DTA curves on characterization of p-nitrophenylhydrazine. An illustration of pyrosynthesis [102] is given in Fig. 8.18. Into the same category falls the pyrolysis of samples in different

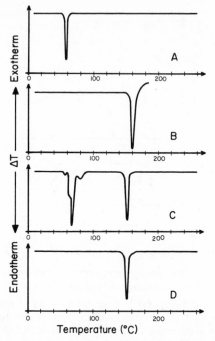

**Fig. 8.17** DTA curves showing formation of *p*-nitrophenylhydrazone from acetone, redrawn after traces published by J. Chiu, *Anal. Chem.* **34,** 1841 (1962). DTA cell described in Fig. 8.3*d*, static nitrogen at atmospheric pressure, 1–5 mg of sample, 15°C/min heating rate, mixing of reactants below 0°C inside the sample cell. (*a*) Pure acetone, boiling point 56°C. (*b*) Pure *p*-nitrophenylhydrazine; melting point 156°C followed by exothermic de-decomposition. (*c*) Reaction mixture of acetone (excess) and *p*-nitrophenylhydrazine, complex endotherms between 54 and 85°C; hydrazone formation and evaporation of excess acetone; melting of the formed hydrazone at 149°C. (*d*) Rerun of the sample formed during DTA curve *C* shows only the melting of the hydrazone.

atmospheres. Many special applications have been developed of which only the analysis of explosives and liquid crystals will be mentioned [17, 18].

Qualitative analysis by DTA can often be coupled with other techniques as was described earlier (p. 432); this possibility again widens the range of application of DTA. Many examples can be found in Refs. 17–20. Another advantage of qualitative analysis by DTA over other techniques is its relatively easy development into semiquantitative or quantitative analysis by calibration of the amplitude. The section on the mathematical treatment of DTA has shown that any DTA circuitry is basically able, with proper sample and reference placement, to do quantitative analysis, a fact still too often neglected.

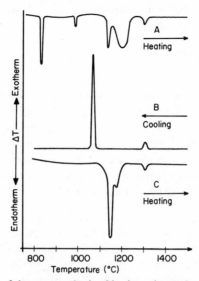

**Fig. 8.18** DTA curves of the pyrosynthesis of barium zincate from barium carbonate and zinc oxide, redrawn after traces published by McAdie [102]. Experiments were performed by simultaneous DTA and TG in an apparatus similar to the one described in Fig. 8.5, 0.1 cm³; sample, 10°C/min heating rate, oxygen at atmospheric pressure. (*a*) Initial heating of an equal molar mixture of $BaCO_3$ and ZnO, barium carbonate solid-solid phase transitions at 829 and 990°C ($\gamma$ to $\beta$ and $\beta$ to $\alpha$); loss of $CO_2$ (checked by TG) started at 916°C, the main loss is in the region of the double peak between 1100 and 1250°C. (*b*) Cooling after heating to 1500°C, a sharp exotherm occurred (without weight loss) at 1070°C, indicating crystallization of barium zincate, $BaZnO_2$. All barium carbonate transitions were missing. (*c*) Reheating after cooling to 700°C shows fusion of the zincate at 1150°C. No explanation was given for the small peaks at 1175°C and 1300°C.

## Applied Problems

The use of DTA in the applied sciences is even more widespread and steadily growing. A recent description of DTA [103] lists as some typical applications: the measurement of strain introduced in brass on cold working and examination of crystallization on heating; the quality control of polymers used in the automobile industry; the evaluation of time effectiveness of anti-oxidants compounded into experimental rubbers; the determination of binder content in asbestos gaskets to ensure sufficient adhesion; the control of the proper level of organic binder in diamond abrasive wheels; the study of resistance of automotive lacquers under heat; the detection of glass transitions in copolymer blends by nitrile rubber producers; the effect of monomer, accelerator, catalyst, and heating rate on foaming of plastisols; the evaluation of thermal history of extended polyethylene on heat sealability of packing film; the analysis of kaolinite–hectorite mixtures for clay mining;

and the measurement of solid-fat-content of hydrogenated soy bean oils, margarines, and shortening.

## Quantitative Analysis

### Peak Area

After establishing the nature of a sample by qualitative analysis (see previous Section), it is often relatively easy to derive quantitative information from DTA peak areas via the extensive thermodynamic parameters (see p. 469). Prerequisite for quantitative analysis of any component of the mixture is a phase transition or chemical reaction of known heat effect which is independent of all other components. The limits of precision are determined by the sensitivity of the DTA apparatus used.

### Phase Diagram

The use of the melting transition has received special attention in the quantitative application of DTA. As long as equilibrium is established quickly enough, DTA can be used advantageously to determine phase diagrams. Even metastable states can often be analyzed with faster heating rates as will be shown below (p. 489). The correlation of typical features in a binary system with the DTA curves is illustrated on a hypothetical phase diagram [104] in Fig. 8.19. The system includes incongruently and congruently

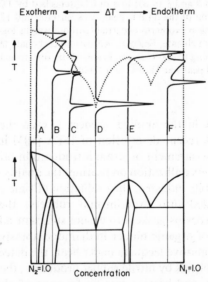

**Fig. 8.19** Schematic DTA curves taken on heating (upper half) as a function of composition for selected compositions of a binary system (with a phase diagram illustrated in the lower half).

melting compounds and solid solution, eutectic and liquidus reactions. The DTA traces of six selected compositions are depicted. All are heating traces because of frequent supercooling on cooling. It is assumed that equilibrium is approached throughout and that the eutectic crystals are large enough not to show excessive size effects on melting. Composition A shows the broadening of the melting range due to solid solution formation. Composition *B*, located at an incongruently melting compound has a sharp peak at the peritectic temperature where the sample undergoes an isothermal decomposition. Above this temperature melting occurs at an increasing rate until the liquidus is reached. Composition *C* is located on the eutectic side of the incongruent compound. Here the eutectic mixture given by the phase boundary melts at first, followed by increased dissolution of the incongruent phase into the melt until the second thermal arrest, where the remaining solid changes to the solid solution which continues melting at higher temperature as before. Composition *D* is chosen at the eutectic composition and shows a melting peak similar to that of a pure compound. *E* is chosen at the composition of the congruently melting compound, and again shows only one sharply defined endotherm. Finally, *F* represents the melting of a simple binary system with a eutectic, the first thermal arrest indicates the eutectic isotherm, the melt end determines the liquidus line. By proper choice of compositions it is thus possible to establish the complete phase diagram.

## Purity Analysis

The slope of the liquidus line of a phase diagram can often be expressed by van't Hoff's law [1]:

$$\frac{dT}{dN_2} = \frac{RT^2}{\Delta H_f} (K - 1),  \tag{8.73}$$

where $K$ is the distribution coefficient of the solute between solid and liquid, $N_2$ the mole fraction of solute, $R$ the gas constant, and $\Delta H_f$ the heat of fusion of the solvent. In case of no solid solution, $K$ is zero and the freezing point lowering from the value of the pure solvent $T_0$ is approximated by [1]:

$$T_0 - T = \frac{RT_0^2 N_2}{\Delta H_f}.  \tag{8.74}$$

If the mole fraction of solute is small and $r$ is the fraction of sample melted at temperature $T$, (8.74) can be used for purity determination if all impurities are insoluble in the solid, but soluble in the melt:

$$T = T_0 - \frac{RT_0^2 N_2/\Delta H_f}{r}.  \tag{8.75}$$

Although it looks like a simple task to calculate $r$ by integrating the melting peak up to a certain temperature and to divide by the total peak area, there are several sample and instrument effects which make purity analysis a frequently erroneous measurement. Because of the widespread use of purity analysis by DTA some of the more important points will be elaborated upon next.

Among the sample effects it must be remembered that (8.75) is an approximation which holds only for equal molecular size. In case of disparity of sizes, even as little as a factor of 2, but particularly for linear high polymer–solvent mixtures, the more elaborate equations derived by Flory and Huggins must be used [105].

Existence of equilibrium is another condition. If equilibrium is maintained throughout initial crystallization and final melting, the DTA curve should look like curve $F$ in Fig. 8.19. The sharper, low-temperature melting peak at the eutectic temperature contains all the impurity melting ($N_2$) and some of the solvent. The amounts can be estimated from the "lever rule:"

$$\% \text{ melt} = \frac{\text{impurity concentration}}{\text{eutectic impurity concentration}}. \qquad (8.76)$$

A convenient estimate is thus, that with a eutectic concentration of $N_1 = N_2 = 0.5$, the amount melting in the first sharp peak is all of the impurity plus an equal amount of solvent. A missing first peak, a broad initial peak, or a too small one is a good indication of nonattainment of equilibrium on crystallization or subsequent heating. Recrystallization under more favorable conditions should be tried before evaluation of the DTA trace.

To round out the sample effects, it should be remembered that (8.75) holds only for eutectic systems, in other words, the impurity must be soluble in the melt, but insoluble in the crystal. All impurities which do not fulfill this criterion are not determined by this method. Also, (8.75) is an approximation, in other words, a plot of $T - T_0$ versus $1/r$ is only a straight line for small concentrations, because $N_2$ is an approximation for $\ln N_1$ which does not permit the introduction of $r$ into (8.74), and $\Delta H_f$ is not constant over large ranges of temperature.

The instrument effects also introduce several possible errors. First, one must restrict analysis as outlined before to instruments with practically no temperature gradient within the sample. The total enthalpy change from the eutectic temperature to a temperature $T$ can be expressed by (8.54). The heat capacity contribution and the second term in (8.54) must be evaluated from the DTA trace. Under these circumstances, obtaining $r$ by integration of the peak area below the baseline up to $T$ and devision by the complete melting peak area is only an approximation. An estimation of the error is possible if one assumes little change in heat capacity during melting. In this case the area below the baseline represents the heat capacity contribution. The second term of (8.54) corresponds to a correction of $C'_p/\Delta H$ of the measured heat of fusion per degree difference in $\Delta T_f - \Delta T_i$. A typical value of $C'_p/\Delta H$ is perhaps 1% for the completely molten sample and correspondingly more for the integration over only part of the melting peak. Overall, it seems possible to keep the instrument effects within tolerable levels for useful purity analysis by DTA. Higher accuracy is possible if the impurity is known and quantitative relationships can be established by empirical calibration.

The DSC of the type shown in Fig. 8.2 presents special problems for purity analysis. A method which overcomes these has been proposed by Davis and Porter [96]. Figure 8.20 represents the heating curve of a cholesteryl propionate of 98.25% purity. A plot of the uncorrected values of

$$\frac{1}{r} = \frac{\text{area } AGC}{\text{area } AEH} \tag{8.77}$$

versus the sample temperature (lag corrected) for different positions of $EH$ is far from linear. At the smaller fraction molten (between $r = 0.02$ and $r = 0.10$), it has a slope of $-0.2$ instead of $-1.6$. As more sample melts, this discrepancy decreases, but has not reached an acceptable value at $r = 0.5$. The reason for this large discrepancy is mainly a supply of heat to the sample by the average-control loop which goes unrecorded initially. The functioning

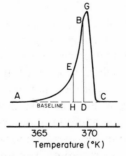

**Fig. 8.20** Copy of a DSC trace of cholesteryl propionate in the region of crystal-mesophase transition (endotherm). Perkin-Elmer DSC heating rate 2.5°C/min, sample weight 5.306 mg, $N_2$ atmosphere; $A$ is the start of visible melting; $EH$ is the amplitude at an arbitrarily chosen program temperature for analysis. The sample temperature at this program temperature is $T_s = T_p - R_0(EH)$, where $R_0$ is a constant evaluated by analysis of a substance with sharp melting point like indium, and $EH$ is the actual value of the amplitude in mcal/sec. $BD$ is the amplitude at the end of melting, its position on the temperature axis is evaluated by trial and error. At the proper choice (8.75) and (8.78) hold. $G$ is the melting peak, and $C$, the final attainment of the baseline.

of the DSC in this case becomes clear by inspection of Fig. 8.2. For a relatively pure sample the melting temperature varies only little and the fixed, programmed, average heating rate $dT_p/dt$ calls for a larger power input from the average-temperature amplifier since the sample temperature changes only little. The power transfer from the now faster heating reference to the sample by the differential-temperature amplifier cannot, however, reestablish equal distance of reference and sample temperature from the program temperature as long as melting is going on. As a consequence, power proportional to the increasing difference between reference and program temperature is added to reference and sample by the average-temperature amplifier. As soon as melting is completed, heat is now transferred by the differential temperature amplifier from the too hot reference to the sample in the amount identical to the previously unrecorded heat. Recognizing this fact and taking account of the difficulty of detecting the first melting as a deviation from the baseline, Davis and Porter [96] suggest for the corrected value of $r$ at point $E$ in Fig. 8.20, the following equation:

$$r = \frac{3/2\,(AEH) + X + Y}{AGC + Y}. \tag{8.78}$$

$Y$ is a correction term for a loss in heat of fusion which may arise because of missing the start of melting. It is to be evaluated by comparison of the heat of fusion of a pure sample with that of the heat of fusion due to the solvent of the impure sample. In Fig. 8.20 there is no evidence of the initial eutectic melting which should have included, besides the heat of fusion of the impurity, probably 1–3% [see (8.76)] of the cholesteryl propionate. Davis and Porter found a discrepancy of 1.5%, in line with the above estimate. The reason for not finding the eutectic melting portion in samples of about 95% purity may be the small amount of eutectic material crystallizing in small, defect crystals with a low-lying, broad-melting interval which is easily confused with background noise, or the small amount of eutectic composition possibly does not crystallize at all. The next change in (8.78) with respect to the uncorrected (8.77) is the addition of an extra area $1/2$ $AEH$. This is close to the amount of heat the average-temperature amplifier supplies initially to the sample without recording as outlined above. The factor $\frac{1}{2}$ derives from the assumption that the sample temperature stays almost constant. Since this is not the case in impure samples, and since lags of power input exist, the additional correction term, $X$, is necessary. It is arrived at by subtracting $\frac{1}{2}$ of the area $ABD$, the heat the average-temperature amplifier would have supplied in constant-temperature melting with no lags, from $BCD$. Point $B$ is found by trial and error and should correspond to the end of melting of the sample. With some experience it is possible to pick it at the first trial. A wrong value of $B$ would again destroy the relationship of (8.75). Equation 8.78 was tried on zone-refined benzoic acid with up to 5% of naphthalene and shown to be accurate to within a few per cent of the total impurity [96].

## Thermodynamic Quantities

### Heat Capacity

The basis for evaluation of thermodynamic quantities is given in Section 4. In general, one uses differential scanning calorimetry for such measurements [106, 107]. Despite the fact that many instruments capable of generating thermodynamic data are sold annually (in 1969—about 500), only very little data have been generated. It is hoped that the almost exclusive qualitative use of quantitative instruments will change in time.

The basic function from which enthalpy, entropy, and free energy of a substance can be derived is the heat capacity. Figure 8.21 presents a copy of a heat-capacity measurement using the DSC of Fig. 8.2. The basic principle is identical to that of all quantitative DTA. Initially a value for isothermal power input or temperature difference is established as an isothermal baseline on a time-based recorder (Fig. 8.21), or a point on a temperature $\Delta T$ recorder. Then the temperature is raised a preset amount at constant heating rate.

The interval of temperature measurable in one step is determined by the sample as well as by the instrument. It is necessary to establish that the baseline does not change significantly over the heating interval and that any change can be approximated by linear interpolation. The Perkin-Elmer DSC [40] seems to be run best in 15–20°C steps (Fig. 8.21), while the duPont

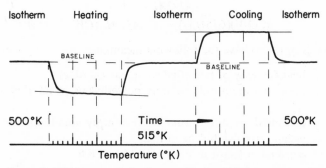

**Fig. 8.21** Copy of a DTA trace of aluminum oxide (sapphire) to measure heat capacity (calibrate). Sample weight 74.23 mg, reference empty Al cup, aluminum weight difference 0.25 mg, heating rate 5°C/min, plotted are uncorrected temperature signals as given by the instrument. Program mode: constant temperature heating 15°K, constant temperature, cooling 15°K.

DSC [41] (Fig. 8.3$a$) has a sufficiently linear baseline to be programmed over wider ranges. As Fig. 8.21 shows, steady state is reached, at 5° C/min heating-rate, after about 5 min. Instead of integrating the area between baseline and curve, it is permissible to read off amplitudes in 5°K steps [106]. Since it is almost impossible to repeat baselines from run to run, it is important to have a newly established baseline for each heat-capacity determination. In particular, truly isothermal conditions must be established before initial heatings; it may be necessary to wait as long as 30 min at the initial temperature to have all temperature gradients decay to a steady-state level. Also, the use of thermal-fluctuation-free locations for the instrument cannot be overemphasized. From experience, most loss of accuracy arises with base-line changes.

A determination of an unknown heat capacity involves the following steps:

1. Run of the empty pan (aluminum) versus the empty reference pan and evaluation of the amplitudes in steps of 5°C as illustrated in Fig. 8.21.

2. Run of a standard as $Al_2O_3$ versus the empty reference pan and again evaluation of the amplitudes at the same temperatures.

3. Run of the unknown as the two runs before.

4. Calculations of the heat capacities at the indicated program temperature (Perkin-Elmer DSC) or the recorded reference temperature (duPont DSC).

The heat-capacity calculation can be done by computer, using, for example, the following equations for polyethylene heat capacities:

$$C_p = \frac{PEAL - AL}{(K)\,(PEW)} - CPAL \left( \frac{ALPEW - ALW}{PEW} \right) \qquad (8.79)$$

$$K = \frac{(ALOAL - AL)}{(ALOW)(CPALO) + CPAL(ALAOW - ALW)}. \qquad (8.80)$$

The symbols of (8.79) have the following meanings:

$C_p$ = heat capacity of polyethylene in cal/(g) (deg)
$PEAL$ = measured amplitude running polyethylene in its aluminum cup (run 3).
$AL$ = measured amplitude running the empty aluminum cup (run 1).
$PEW$ = weight of polyethylene in mg.
$K$ = conversion factor as determined by (8.80).

The first portion of (8.79) calculates thus the heat capacity except for a small correction due to the difference in weight of the empty aluminum cup which contains the sample. The second term in (8.79) contains the following quantities necessary for the correction calculation:

$CPAL$ = heat capacity of aluminum from tables in cal/(g) (deg).
$ALPEW$ = aluminum cup weight which contains the polyethylene in mg.
$ALW$ = empty aluminum cup weight in mg.

The major factor of the conversion factor $K$ (8.80) is the first term of the denominator, while the second term is its aluminum weight difference correction. The additional symbols have the following meaning:

$ALOAL$ = measured amplitude running $Al_2O_3$ in its aluminum cup (run 2).
$ALOW$ = weight of $Al_2O_3$ in mg.
$CPALO$ = heat capacity of aluminum oxide from tables in cal/(g)(deg).
$ALAOW$ = aluminum cup weight which contains the $Al_2O_3$ in mg.

The final step of evaluation involves the calculation of the sample temperature from the reference or program temperature. The calculation involves the correction of the reference or program temperature from thermocouple emf or Pt-resistance reading to true temperature, and estimation of the lag. Since measurements are rather quickly done, it is an easy undertaking to make repeated runs and estimate internal precision. Standard deviations from the average of as little as 2% can be achieved on careful measurement.

### Enthalpy

The determinations of enthalpies of phase transitions or chemical reactions is straightforward as outlined in section 4; it involves establishment of baseline and curve, and comparison with a run of a reference substance in the same temperature region [108]. A comparison of Figs. 8.20 and 8.21 indicates that for phase transitions considerably less material (5–10 mg) is necessary than for heat capacities (30–100 mg). All quantitative heat measurements must have a time-based recording of the amplitude proportional to $T$. In case of a temperature–temperature difference recording the temperature recorded should be that of the reference which is unaffected by the thermal effects of the sample and thus proportional to time. The actual sample temperatures can be calculated using temperature difference and thermal lag information. The special corrections necessary for the Perkin-Elmer DSC have been outlined above in the discussion of purity analysis.

### Time-Dependent DTA

#### Glass Transition

Time-dependent measurements of thermal effects is an application most suited for DTA. No other instrument is available for this purpose, and again the full potential is not yet realized. It is possible with DTA to measure fast enough to often establish the thermodynamic parameters of unstable crystals and it is also possible to follow the kinetics of changes over a certain range of temperatures. The range of time-dependent changes measurable with any one instrument and sample placement must be evaluated by preliminary experiments. Of the almost unlimited applications, two examples from our laboratory will be described here, the investigation of the hysteresis phenomena of heat capacity on cooling and on heating through the glass transition temperature range [98], and the special effects found on polymer melting at different heating rates [109–111].

Measuring heat capacities on cooling, as in the right hand side of Fig. 8.21, one finds that amorphous substances experience a subtantial drop in heat capacities (10–50%) in the glass transition region. The effect is caused by a freezing of large-scale motion and as such, is cooling-rate dependent. On fast cooling the glass freezes at a higher temperature than on slow cooling. Changing the cooling rate of polystyrene for example from 0.1 to 0.01°C/sec lowers the glass transition from 103 to 98°C. Reheating such a sample will in turn unfreeze the large-scale motion at a temperature characteristic of the heating rate. If the heating rate is much faster than the cooling rate, unfreezing occurs at a temperature higher than the freezing, and the enthalpy lost on cooling at a higher heat capacity must be regained as an endotherm as soon as sufficient molecular motion is possible. Similarly, if the heating rate is much slower than the cooling rate, the system regains mobility at a temperature lower than the freezing temperature and the surplus enthalpy of the slowly unfreezing sample is lost in form of a broad exotherm. Figure 8.22 illustrates heat-capacity curves of polystyrene samples cooled at different rates, all heated at identical rates, covering the range from a shallow exotherm to a strong endotherm. Analysis of these peaks using equations derived in Section 4 has been described [98].

### Melting Transition

Kinetic data on processes involving crystalline solids can be divided into two groups. One set of data is obtained on cooling melts or solutions and following crystallization kinetics. This process will not be treated in detail since many accounts are available in the literature. The other group of data can be obtained on heating already crystallized polymers and observing melting, reorganization, and recrystallization [109-111].

Different polymer morphologies undergo different slow processes on heating. Equilibrium crystals have a fully extended polymer chain conformation within the crystal. On heating, the melting process is so slow, that even at slow heating rates more heat can be supplied to the crystals than can be used up in melting. As a result, the interior of the crystals superheats temporarily until the melt–crystal interface reaches the interior. DTA and DSC have been used to analyze this melting process [109]. Only little work

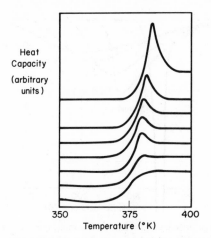

Heat
Capacity

(arbitrary
units)

350                375                400
Temperature (°K)

**Fig. 8.22**  Heat capacity of polystyrene in the glass transition region cooled at different rates measured by DTA on heating. Heating rate 0.09°C/sec, cooling rates from top to bottom: $1.4 \times 10^{-4}$, $3.2 \times 10^{-3}$, $1.8 \times 10^{-2}$, $4.1 \times 10^{-2}$, $8.7 \times 10^{-2}$, and 0.5°C/sec. Successive curves are displaced vertically.

had been done previously on the superheating of crystals in general and no calorimetry has been reported previously. The analysis of polymer crystals is hampered by the fact that the crystals are of different length in the chain direction as well as of different width. It has not yet been possible to do experiments on isolated crystals. Qualitatively it could be proven that length as well as width influences superheating [111].

Figure 8.23 illustrates the superheating of polyoxymethylene as a function of heating rate. The sample of only about 1 mg was placed around the thermocouple sandwiched between layers of $Al_2O_3$, the reference material. In this way lags could be reduced to a negligible level up to 100°C/min. The melting rates shown in Fig. 8.23 are slower than the response of the DTA so that at 100°C/min the last trace of crystal disappears only at about 215°C, more than 30°C above the equilibrium melting point. Peak temperatures of extended–chain polyethylene crystals showing superheating are plotted in Fig. 8.24, curve 1.

Extended-chain crystals of polymers are relatively difficult to produce. Most polymer samples contain metastable crystals which have their molecular chains folded every 50–500 Å to fit into the lamellar crystal shape. In accord with this, superheating is absent at heating rates presently possible; instead, reorganization is observed. The metastable crystals change during heating to a more stable crystal form with somewhat larger molecular-chain extension. Several effects have been observed by DTA at different heating rates and are illustrated in Figs. 8.24 and 8.25:

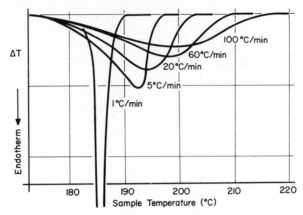

**Fig. 8.23** DTA traces of polyoxymethylene in the melting range at different heating rates. The peak areas have been adjusted to equal size. For all heating rates melting begins at about 175°C, but strong superheating exists at faster heating rates. DTA cell similar to Fig. 8.3*d*; sample size, about 1 mg.

1. The melting point decreases with increasing heating rate and stays constant at fast rates. This behavior is typical of reorganization on heating which diminishes with increasing heating rate, such that finally a path with a direct melting of the metastable crystals to a supercooled melt is reached. This limit has been accomplished in well-crystallized, folded-chain single crystals from solution at heating rates of 10–50°C/min (see Fig. 8.24, curve 4). The limit of constant-melting peak temperature can be used to get information on the surface free energy of these crystals.

2. The melting point stays constant with changing heating rates as would be found in a sample with melting peak temperatures intermediate to curves 2 and 3 of Fig. 8.24. More than one interpretation is possible for a completely unknown sample: (*a*) it could be an equilibrium crystal which shows no superheating; (*b*) it could be a metastable crystal with no rearrangement or recrystallization; or (*c*) the range of heating rates employed was too small and a compensation of superheating and reorganization occurred. Differentiation between (*a*), (*b*), and (*c*) can be achieved by systematically varying crystallizations of the polymer in question.

3. The melting temperature exhibits a maximum or minimum with changing heating rate. A combination of superheating and reorganization can give rise to such behavior (Fig. 8.24, curve 3). Since superheating changes mainly with crystal dimension in the molecular chain direction, it is often not difficult to recognize this effect if any other information on the crystals is available (such as low-angle X-ray data). A maximum in melting temperature can be found as a result of "cold crystallization" at different heating rates.

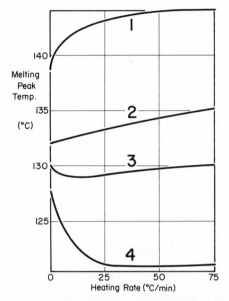

**Fig. 8.24** Plot of melting peak temperatures of polyethylene as a function of heating rate. All samples consist of chemically identical polyethylene. Curve 1, crystallized under high hydrostatic pressure to yield extended-chain, equilibrium-like crystals. Curve 2, slowly crystallized from the melt to yield metatable crystals which contain polymer chains folded every 250–300 Å. Curve 3, fast crystallized from the melt to yield metastable crystals which contain polymer chains folded about every 200 Å, reorganizing somewhat on slow heating to give a shallow minimum. Curve 4, crystallized from dilute solution to yield metastable crystals which contain polymer chains folded every 130Å, reorganizing strongly at the slower heating rates to higher melting crystals, above 25°C/min direct melting of the metastable crystals is reached.

Cold crystallization is observed whenever a quenched amorphous or only partially crystalline sample is heated slowly enough so that it can crystallize. The lower the temperature of crystallization, the less perfect will be the resultant crystals, and the lower will be the melting point of the produced crystals. Faster heating rates will shift the crystallization temperature to higher levels and thus increase the melting point of the resultant crystals. Increasing the heating rates to even higher levels, however, reduces the melting point again, because now there is not enough time available to complete even cold crystallization. The overall result is a maximum in melting temperature with heating rate. This seemingly variable melting point of polymers has caused much confusion in past years, but is now used routinely for analysis of crystal perfection. Reexamining the melting behavior of nonpolymeric materials, one frequently finds similar behavior for sufficiently small crystals.

4. A final observation on the melting of polymer crystals is the appearance of multiple peaks as is illustrated in Fig. 8.25. This situation can arise for several reasons. (*a*) There may be crystals of two or more degrees of perfection in the sample originally. In this case recrystallization of the molten sample or partial melting followed by recrystallization can prove the assumption. On heating at different rates each type of crystals behaves independently of the other. (*b*) There may be reorganization or rather metastable crystals after partial or complete melting before final melting. In case of complete melting before crystallization an exotherm can be observed by calorimetry before the final melting. Partial melting and recrystallization may, however, go on simultaneously. In this case a quantitative analysis of time-dependent calorimetry can give further information. Increasing the heating rate will always decrease the amount of higher melting polymers since the time left for recrystallization is shortened on increasing the heating rate. The variation of the separate melting points may be irregular because of different reorganization times and temperatures at different heating rates. (*c*) A portion of the polymer may be left amorphous or poorly crystallized. On heating slowly, this amorphous or poorly crystallized material will crystallize ("cold crystallization") or reorganize and melt at a different temperature from the crystals originally present. The amount of newly formed crystals during heating will decrease if heating rates are increased sufficiently, leaving,

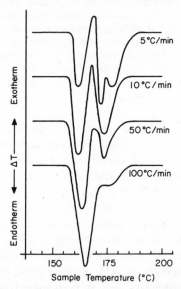

**Fig. 8.25** Schematic drawing of DTA traces of polyoxymethylene crystallized from solution (hedrites) at 5, 10, 50, and 100°C/min heating rate. Multiple peaks are developed due to reorganization and recrystallization at the slower heating rates.

at fast heating rates, only the melting of the crystals originally present. (*d*) Finally, there may be a true phase transition from one crystal form to another before melting as is found for example in polytetrafluoroethylene and polybutene-1 and is well known for low-molecular-weight substances.

Figure 8.25 shows schematically the melting endotherms of a poorly crystallized polyoxymethylene at different heating rates. In this case it is obvious from the disappearance of the higher melting endotherms at faster rates that the poorer crystals melted and reorganized at slow rates, but melted directly at faster rates (case *b*).

The examples of this final section on time-dependent measurements on DTA have shown that much further information on thermal properties can be developed. Characterization of materials in nonequilibrium states is possible. It is, in some cases, even feasible to derive thermodynamic data on heat capacity on unstable compounds.

### References

1. J. M. Sturtevant, "Calorimetry," Chap. 7, this book.
2. J. M. Sturtevant, "Temperature Measurement," Chap. 1, this book.
3. G. K. Burgess, *Electrochem. Metal. Ind.* **6**, 366, 403 (1908); also G. K. Burgess and H. LeChatelier, *The Measurement of High Temperatures*, 3rd ed., Wiley, New York, 1912.
4. H. LeChatelier, *Bull. soc. Franç. Minéral.* **10**, 204 (1887); *Compt. Rend.* **104**, 1443, 1517 (1887); *Z. Phys. Chem.* **1**, 396 (1887).
5. W. C. Roberts-Austen, *Proc. Inst. Mech. Eng.* (*London*) **1**, 35 (1889); *Metallographist* **2**, 186 (1899).
6. E. Saladin, *Iron Steel Metallurgy Metallography* **7**, 237 (1904), see also H. Le Chatelier, *Rev. Met.* **1**, 134 (1904).
7. A. Eucken, "Energie und Wärmeinhalt" in *Handbuch der Experimentalphysik*, Vol. 8, Part 1, W. Wien and F. Harms, Eds., Akadem. Verlagsges, Leipzig, 1929.
8. C. Sykes, *Proc. Roy. Soc.* (*London*) **148A**, 422 (1935).
9. J. P. Joule, *Mem. Proc. Manchester Lit. Phil. Soc.* **2**, 559 (1845).
10. L. Pfaundler, *Sitzber. Akad. Wiss. Wien, Math. Naturw. Kl.* **59**, 145 (1896).
11. For a series of constant heating rate calorimeters see: C. Sykes, *op. cit.*; H. Moser, *Phys. Z.* **37**, 737 (1936); J. D. Hoffman, *J. Amer. Chem. Soc.* **74**, 1696 (1952); B. Wunderlich and M. Dole, *J. Polymer Sci.* **24**, 201 (1957).
12. K. H. Hellwege, W. Knappe, and V. Semjonow, *Z. Angew. Phys.* **11**, 285 (1959); M. V. Volkenshtein and Yu. A. Sharanov, *Vysokomolekul. Soedin.* **3**, 1739 (1961); K. H. Hellwege, W. Knappe, and W. Wetzel, *Kolloid Z.* **180**, 126 (1962); H. Tautz, M. Glück, G. Hartmann, and R. Leuteritz, *Plaste Kautschuk* **11**, 657 (1964).

13. F. H. Müller, and H. Martin, *Kolloid Z.* **172,** 97 (1960).
14. M. J. O'Neill, *Anal. Chem.* **36,** 1238 (1964).
15. W. J. Smothers and Y. Chiang, *Handbook of Differential Thermal Analysis*, Chemical Publishing, New York, 1966.
16. C. B. Murphy, *Anal. Chem.* **36,** 374R (1964); **38,** 443R (1966); **40,** 391R (1968).
17. J. P. Redfern Ed., "Thermal Analysis '65" *Proc. of the First International Congress in Aberdeen*, Macmillan, London, 1965.
18. R. F. Schwenker, Jr., and P. D. Garn, Eds., "Thermal Analysis," Vols 1 and 2, *Proc. of the Second International Congress in Worcester* (1968), Academic, New York, 1969.
19. W. W. Wendlandt, Ed., *Thermochimica Acta*, American Elsevier, New York.
20. L. Erdey, Ed., *Journal of Thermal Analysis*, Akademiai Kiado, Budapest.
21. P. E. Slade, Jr., and L. T. Jenkens, Eds., *Techniques and Methods of Polymer Evaluation*, Vol. 1, Marcel Dekker, New York, 1966.
22. B. Ke, Ed., *Newer Methods of Polymer Characterization*, Interscience, New York, 1964.
23. L. G. Berg, N. P. Burmistrova, M. I. Ozerova, and G. G. Tsurinov, *Practical Handbook on Thermography*, Izd. Kazan University, Kazan, U.S.S.R., 1967.
24. P. D. Garn, *Thermoanalytical Methods of Investigation*, Academic, New York, 1965.
25. W. W. Wendlandt, *Thermal Methods of Analysis*, Interscience, New York, 1964.
26. K. W. Kohl, "Bibliography on Pyrolysis of Polymers, 321 Publications from 1862 to 1963," *U.S. At. Energy Comm. Rot. MLM-1271*, Monsanto Res. Corp., Miamisburg, Ohio, June 1965. A survey on thermal degradation, oxidation and thermal analysis of polymers was prepared by D. W. Levi (1963), and D. A. Teetsel and D. W. Levi (1966), *U.S. Clearinghouse Fed. Sci. Tech. Inform. AD 423546*, 1963, and *AD 631655*, 1966.
27. B. Ke, in *Encyclopedia of Polymer Science and Technology*, Vol. 5, Interscience, New York, 1967, p. 37.
28. H. Kambe, *Kogyo Kagaku Zasshi* **69,** 1603 (1966).
29. C. B. Murphy in *Encyclopedia of Industrial Chemical Analysis*, Vol. 1, Interscience, New York, 1966, p. 574.
30. J. Mitchell, Jr. and J. Chiu, *Anal. Chem.* **41,** 248R (1969).
31. A reference collection of thermograms has been published by Sadler Research Laboratories Inc., Philadelphia, Pa.
32. *Scifax DTA Index*, compiled by R. C. Mackenzie, contains several thousand DTA references on punched cards. Macmillan, London, England.
33. J. P. Redfern, Ed., *Thermal Analysis Review*, Battersea College of Technology, London (a tri-annual collection of abstracts since 1962).
34. F. R. Caldwell, "Thermocouple Materials," *U.S. Natl. Bur. Std. Monogr.* **40,** Washington, D. C., 1940.
35. "Recommended Practice for Thermocouples and Extension Wires," *ISA-RP 1.1–1.7*, July, 1959.
36. H. Shenker, J. I. Lauritzen, Jr., R. J. Corruccini, and S. T. Lonberger, *U.S. Natl. Bur. Stand. Circ.* **561,** Washington, D. C., 1955.

37. R. P. Benedict, *Fundamentals of Temperature, Pressure, and Flow Measurement*, Wiley, New York, 1969.

38. E. M. Barrall, II, J. F. Gernert, R. S. Porter, and J. F. Johnson, *Anal. Chem.* **35**, 1837 (1963).

39. E. S. Watson, M. J. O'Neill, J. Justin, and N. Brenner, *Anal. Chem.* **36**, 1233 (1964).

40. Perkin-Elmer Corporation, Instrument Division, Main Avenue, Norwalk, Conn. 06852.

41. E. I. duPont de Nemours and Co., Inc., Instruments Division, Wilmington, Del. 19898.

42. P. P. Budnikov and S. G. Tresvyatskii, *Ogneupory* **20**, 166 (1955).

43. R. P. Goton, U.S. Pat. 3,084,534, April 19, 1963.

44. L. Brewer and P. Zavitsanos, *J. Phys. Chem. Solids* **2**, 284 (1957).

45. G. N. Rupert, *Rev. Sci. Instr.* **36**, 1629 (1965).

46. N. A. Nedumov, *Zh. Fiz. Khim.* **34**, 184 (1960); *Russ. J. Phys. Chem.* **34**, 84 (1960).

47. T. Davidson and B. Wunderlich, *J. Polymer Sci.*, *Part A2* **7**, 377 (1969).

48. E. W. Comings, *High-Pressure Technology*, McGraw Hill, New York, 1956.

49. D. M. Newitt, *High Pressure Plant and Fluids at High Pressures*, Oxford University Press, Oxford, England, 1940.

50. R. E. Hanneman and H. M. Strong, *J. Appl. Phys.* **36**, 523 (1965); F. P. Bundy, *J. Appl. Phys.* **32**, 483 (1961).

51. D. Block and F. Chaisse, *J. Appl. Phys.* **38**, 409 (1967).

52. L. H. Cohen, W. Klement, Jr., and G. C. Kennedy, *J. Phys. Chem. Solids* **27**, 179 (1966).

53. L. H. Cohen, W. Klement, Jr., and G. C. Kennedy, *J. Phys. Chem. Solids* **27**, 171 (1966).

54. F. Paulik, J. Paulik, and L. Erdey, U.S. Pat. 3,045,472, July 24, 1962; L. Erdey, *Microchim. Acta* **1966**, 699.

55. The Mettler Recording Vacuum Thermoanalyzer, [H. G. Wiedemann, *Chem. Ing. Tech.* **36**, 1105 (1964)] can be used as TGA, DTA, and be coupled with Balzers Quadrupole High Frequency Mass Spectrometer QMG 101 for simultaneous operation. Mettler Instrument Corporation, Princeton, N. J. (or Stäfa, Switzerland); Balzers, A. G., Santa Ana, Calif. (or Principality of Liechtenstein).

56. H. G. Wiedemann, *Z. Anal. Chem.* **220**, 18 (1966).

57. W. W. Wendlandt, *Anal. Chim. Acta* **27**, 309 (1962).

58. R. Ryhage, S. Wickstrom, and G. R. Waller, *Anal. Chem.* **37**, 433 (1965).

59. L. G. Berg and N. P. Burmistrova, *Zh. Neorg. Khim.* **5**, 676 (1960); *Russ. J. Inorg. Chem.* **1960**, 326.

60. A. V. Tets and H. G. Wiedemann in *Thermal Analysis*, Vol. 1, R. F. Schwenker and P. D. Garn, Eds., Academic Press, New York, 1969, p. 121.

61. W. H. King, Jr., A. F. Findeis, and C. T. Camilli in *Analytical Calorimetry*, R. S. Porter and J. F. Johnson, Eds., Plenum, New York, 1968.

62. E. Hellmuth and B. Wunderlich, *J. Appl. Phys.* **36**, 3039 (1965).

63. A. F. Bessonov, V. M. Ustyantsev, and G. A. Takis, *Porosh. Met.* **7**, 92 (1967).

64. G. B. Ravich, V. Z. Kolodyazhnyi, V. G. Brodov, A. M. Kuchumov, and A. I. Zhemarkin, *Zh. Neorg. Khim.* **12,** 2256 (1967); *Russ. J. Inorg. Chem.* **1967,** 1190.

65. M. S. Yagfarov, *Zh. Neorg. Khim.* **6,** 2440 (1961); *Russ. J. Inorg. Chem.* **1961,** 1236.

66. F. Paulik, J. Paulik, and L. Erdey, *Microchim. Acta* **1966,** 894.

67. E. J. Scherago, *Science* **162A,** 45 (1968) lists the following DTA equipment sources:

American Instruments, 8030 Georgia Ave., Silver Spring, Md., 20910
Apparatus, Box 184, Kent, Ohio, 44240
Astro Industries, Box 938, Santa Barbara, Calif., 93102
E. I. duPont Instrument, Wilmington, Del., 19898
Dynatech, 17 Tudor St., Cambridge, Me., 02139
Eberback, 505 S. Maple Rd., Ann Arbor, Mich., 48106
Fisher Scientific, 711 Forbes Ave., Pittsburgh, Pa., 15219
Mettler Instruments, 20 Nassau St., Princeton, N. J., 08540
Perkin-Elmer, 723 G. Main Ave., Norwalk, Conn., 06852
Premco Instruments, 2006 E. First St., Austin, Tex., 78702
Schuco Scientific, 110 5th Ave., New York, N. Y., 10011
Technical Equipment, 917 Acoma St., Denver, Colo., 80204
Tem-Pres Research, 1401 S. Atherton, State College, Pa., 16801
Theta Industries, Box 79, Manhasset, N. Y., 11030
Tracor, 6500 Tracor Lane, Austin, Tex., 78721
Vari-Light, 9770 Conclin Rd., Blue Ash, Ohio, 45242
Voland, 27 Centre Ave., New Rochelle, N. Y., 10802

68. B. Wunderlich and D. Bodily, *J. Polymer Sci., Part C* **6,** 137 (1964).

69. M. Wittels, *Amer. Mineralogist* **36,** 615, 760 (1951).

70. E. M. Barrall and L. B. Rogers, *Anal. Chem.* **34,** 1106 (1962).

71. H. G. McAdie, *Anal. Chem.* **39,** 43 (1967); Appendix 1 in *Thermal Analysis,* Vols. 1 and 2, R. F. Schwenker, Jr., and P. D. Garn, Eds., Academic, New York, 1969.

72. New Provisional Subcommittee on Thermoanalytic Test Methods in Committee E 1, Jan. 28, 1968, *ASTM Proc.* **68,** 433 (1968).

73. *Thermal Analysis Newsletter,* Analytic Division Perkin-Elmer Corp., Norwalk, Conn.

74. *DuPont Thermogram,* E. I. duPont de Nemours and Co., Inc., Instrument Products Division, Wilmington, Del., 19898.

75. This choice is the recommended publication format for the Perkin-Elmer DSC, *Thermal Analysis Newsletter* **2** (1965).

76. W. Utermark and W. Schicke, *Melting Point Tables of Organic Compounds,* Interscience, New York, 1963.

77. W. C. McCrone, Jr., *Fusion Methods in Chemical Microscopt,* Interscience, New York, 1957.

78. *Handbook of Chemistry and Physics,* The Chemical Rubber Co., Cleveland, Ohio, Annual editions.

79. Landolt-Börnstein, *Zahlenwerte und Funktionen*, Vol. 2, Part 4, K. Schäfer and E. Lax, Eds., Springer, Verlag, Berlin, 1961.

80. D. C. Ginnings and G. T. Furukawa, *J. Amer. Chem. Soc.* **75**, 522 (1953).

81. B. Wunderlich, unpublished comparison of Perkin-Elmer and duPont instruments.

82. B. Wunderlich, *J. Phys. Chem.* **69**, 2078 (1965).

83. B. Wunderlich and C. M. Cormier, *J. Polymer Sci.*, *Part A2*, **5**, 987 (1967).

84. T. Ozawa, *Bull. Chem. Soc. Japan* **39**, 2071 (1966).

85. H. S. Carslaw and J. C. Jaeger, *Conduction of Heat in Solids*, 2nd ed., Claredon, Oxford, 1959.

86. P. J. Schneider, *Conduction Heat Transfer*, Addison-Wesley, Cambridge, Mass., 1955.

87. L. R. Ingersoll, O. J. Zobel, and A. C. Ingersoll, *Heat Conduction*, Univ. of Wisconsin Press, Madison, 1954.

88. S. Strella, *J. Appl. Polymer Sci.* **7**, 569, 1281 (1963).

89. A. R. Haly and M. Dole, *Polymer Letters* **2**, 285 (1964).

90. S. M. Wolpert, *Dynamic Differential Thermal Analysis of the Glass Transition Region*, Ph. D. thesis, Dept of Chemistry, Rensselaer Polytechnic Inst., presented 1970.

91. F. H. Müller and H. Martin, *Kolloid-Z.* **172**, 97 (1960).

92. G. Adam and F. H. Müller, *Kolloid-Z. Z. Polymere* **192**, 29 (1963).

93. H. Martin and F. H. Müller, *Kolloid-Z. Z. Polymere* **192**, 1 (1963).

94. R. A. Baxter, in *Thermal Analysis*, Vol. 1, R. F. Schwenker and P. D. Garn, Eds., Academic, New York, 1969, p. 69.

95. A. P. Gray, in *Analytical Chemistry*, R. S. Porter and J. F. Johnson, Eds., Plenum, New York, 1968.

96. G. J. Davis and R. S. Porter, *J. Thermal Anal.* **1**, 449 (1969).

97. M. V. Vol'kenshtein and O. B. Ptitsyn, *Zh. Tekh. Fiz.* **26**, 2204 (1956); *Soviet Phys. Tech. Phys.* **1**, 2138 (1957);.

98. B. Wunderlich, D. Bodily, and M. H. Kaplan, *J. Appl. Phys.* **35**, 95 (1964).

99. J. H. Flynn and L. A. Wall, *J. Res. Natl. Bur. Std.* **70A**, 487 (1966).

100. J. H. Flynn, in *Thermal Analysis*, Vol. 2, R. F. Schwenker, Jr., and P. D. Garn, Eds., Academic, New York, 1969, p. 1111.

101. For example, J. Šesták, A. Brown, V. R̆ihák, and G. Berggren in *Thermal Analysis*, Vol. 2, R. F. Schwenker, Jr. and P. D. Garn, Eds., Academic, New York, 1969, p. 1035.

102. H. G. McAdie in *Thermal Analysis*, Vol. 2, R. F. Schwenker, Jr. and P. D. Garn, Eds., Academic, New York, 1969, p. 717.

103. *Chem. Eng. News* **47**, 46 (1969).

104. Drawn after D. E. Etter, P. A. Tucker, and L. J. Wittenberg in *Thermal Analysis*, Vol. 2, R. F. Schwenker, Jr. and P. D. Garn, Eds., Academic, New York, 1969, p. 829.

105. For an application of the Flory-Huggins equation for multicomponent polymer mixtures see for example: B. Prime and B. Wunderlich, *J. Polymer Sci.*, *Part A2*, **7**, 2073 (1969). The derivation of the two component equation

can best be found in M. L. Huggins, *Physical Chemistry of High Polymers,* Wiley, New York, 1958.

106. B. Wunderlich, *J. Phys. Chem.* **69,** 2078 (1965).
107. M. J. O'Neill, *Anal. Chem.* **38,** 1331 (1966).
108. B. Wunderlich and C. M. Cormier, *J. Polymer Sci., Part A2,* **5,** 987 (1967).
109. E. Hellmuth and B. Wunderlich, *J. Appl. Phys.* **36,** 3039 (1965).
110. M. Jaffe and B. Wunderlich, *Kolloid-Z. Z. Polymere,* **216–217,** 203 (1967).
111. M. Jaffe and B. Wunderlich, in *Thermal Analysis,* Vol. 1,    R. F. Schwenker and P. D. Garn, Eds., Academic, New York, 1969, p. 387.

Chapter **IX**

# DETERMINATION OF SURFACE
# AND INTERFACIAL TENSION

A. E. Alexander*
and
John B. Hayter

* Deceased.

---

## 1  INTRODUCTION

The area of contact between any two phases is referred to as a "surface" or "interface," the former usually being restricted to systems in which one phase is gaseous. Various investigations have shown that the transition region between two bulk phases is normally extremely thin, often being no more than a few Ångstroms in thickness; nevertheless, the interface may play a dominant role in determining the behavior of many systems, particularly those of a colloidal nature (e.g., foams, emulsions, suspensions, detergent solutions, smokes, heterogeneous catalysts, and biological systems).

Certain macroscopic phenomena connected with liquids, such as the tendency of drops to become spherical, are readily explained if liquid surfaces are in a state of tension, or if the molecules in the surface are in a state of higher energy than those in the bulk. "Surface tension" and "surface Gibbs energy per unit area" are synonymous, and are usually expressed in dyne/cm or erg/cm². Although for the general case we should refer to "interfacial" or "boundary" tension, the term *surface tension* will be used for all fluid systems.

There are many reasons why surface tension may be studied:

1. It is an indirect but nonetheless extremely powerful method of estimating the concentration of adsorbed molecules at fluid interfaces. From measurements of surface tension and the use of the Gibbs adsorption isotherm (or other means, such as a suitable equation of state), surface concentrations

can be estimated, yielding particularly valuable information in the study of surface-active agents (surfactants) in aqueous systems (see Section 2).

2. As a physical property, it is useful for characterization purposes with pure liquids and solutions, and for giving an insight into intermolecular forces.

3. It is one method of testing theories of molecular interaction, particularly in surfaces.

## 2  GENERAL CONSIDERATIONS

### Reality of Surface Tension and Its Relation to Adsorption

That there is a true mechanical tension in the surface of a liquid is shown by many phenomena, and is demonstrated very simply by the Dupré frame (Fig. 9.1), in which a film of soap solution is stretched on a vertical rectangular framework of wire. The lower horizontal wire is free to move, and, provided the friction is sufficiently small, will move upward if the total tension $t$ of soap film is greater than the weight $W$ (which includes that of the wire), and downward if $W$ is greater. If $\gamma$ represents the surface tension of the film in dyne/cm, and $l$ the length of its attachment to the movable wire in centimeters, then, since the film has two surfaces, the length of the boundary between the two surfaces and the wire is $2l$, and

$$\gamma = \frac{t}{2l}. \tag{9.1}$$

If $l$ is 5 cm and $t$ is 250 dyne, then $\gamma = 25$ dyne/cm. Now, if the wire is pulled downwards 10 cm ($= d$), while $\gamma$ remains constant (which may occur if the

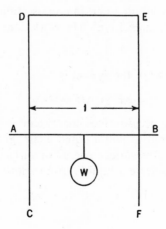

**Fig. 9.1**  Dupré frame.

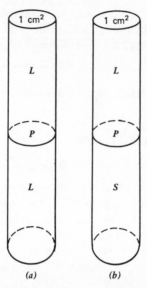

**Fig. 9.2** (a) If a cylinder of liquid 1 cm² in cross-section is pulled apart at the plane P, the work done ($W_c$) is equal to twice the free surface energy of the liquid (which is $2\gamma$). (b) If the liquid is pulled away to give a clean surface of the solid, the amount of work done is ($W_A = \gamma_a + \gamma_b - \gamma_{ab}$), since the surface of the solid and that of the liquid are formed and the interface between them disappears.

total area of the surface of the soap film is sufficiently large, and the concentration of the soap is sufficiently high), then the area of the surface increases by 100 cm², and the work done (i.e., the increase in free surface energy) is $t \times d = 250 \times 10 = 2500$ erg, or 25 erg/cm².

Consider a cylinder of liquid 1 cm² in cross-section (Fig. 9.2). Let this be pulled apart at a plane of area 1 cm². The work done on the system is

$$W_c = 2\gamma, \tag{9.2}$$

and the increase of energy of the system is

$$\Delta E = \Delta H = 2\epsilon, \tag{9.3}$$

where $\epsilon$ is the *internal* or *total* surface energy/cm².

While the work required to pull a liquid apart, given by the work of cohesion, $W_c$, is twice the free surface energy or surface tension of the liquid, the work required to separate a liquid *a* from another liquid (or solid) *b* is

$$W_A = \gamma_a + \gamma_b - \gamma_{ab}, \tag{9.4}$$

and

$$\epsilon_A = h_a + h_b - h_{ab}, \tag{9.5}$$

where $\gamma_{ab}$ is the interfacial tension between $a$ and $b$, and $h$ is the enthalpy change per unit increase in area.

Cohesion and adhesion, which are measures of intermolecular action, are among the most important physical properties of any condensed system, whether it consists of liquids, or solids, or of both.

Apart from the use of the Gibbs adsorption isotherm to calculate the surface concentration of adsorbed molecules, the thermodynamics of adsorption will not be considered further here. Several books which deal extensively with the thermodynamics of surfaces have been published recently [1-3].

The Gibbs adsorption isotherm, for a two-component system of solute $i$ and solvent $s$ is usually expressed in the form

$$-d\gamma = \Gamma_i \, d\mu_i + \Gamma_s \, d\mu_s, \tag{9.6}$$

where $\mu$ is the chemical potential and $\Gamma$ the "surface excess."

If we take a region in the solution bounded by the surface and a geometrical plane $t$ cm below the surface, where $t$ is chosen so that the solution at this plane has bulk-solution properties, then $\Gamma_i$ is defined as the difference between the number of moles of $i$ in unit cross-section of this region and the number of moles of $i$ in the same volume ($t$ cm$^3$) of the bulk solution. Since, provided $t$ is large enough to place the geometrical plane in the bulk, this definition is independent of $t$, $\Gamma_i$ is quoted in moles per unit area. We must, of course, define where the surface of the solution is; that is, we must choose a geometrical surface that divides the solution phase from the other (vapor) phase. This "Gibbs surface" is somewhat arbitrary, but we may choose it in such a position that the surface excess of solvent ($\Gamma_s$) is zero, in which case we will denote the *relative* surface excess of $i$ by $\Gamma_i^{(s)}$. Then

$$-d\gamma = \Gamma_i^{(s)} \, d\mu_i \tag{9.7}$$

$$= \Gamma_i^{(s)} \, RT \, d \ln N_i f_i, \tag{9.8}$$

since $\mu_i = \mu_i^0 + RT \ln N_i f_i$, where $N_i$ and $f_i$ are the molal concentration and activity coefficient of the solute $i$.

We are frequently concerned with relatively dilute solutions of capillary-active materials, in which case (9.8) reduces to

$$\Gamma_i^{(s)} = -\frac{1}{RT}\left[\frac{d\gamma}{d \ln C_i f_i}\right] \tag{9.9}$$

where $C_i$ is now the molar concentration of the solute.

Hence by plotting $\gamma$ against the logarithm of the bulk activity we can determine the *surface excess* $\Gamma_i^{(s)}$. (If $R$ is in erg/(mole)(deg), i.e., $8.3 \times 10^7$, then $\Gamma_i^{(s)}$ will be in mole/cm$^2$). For dilute solutions of capillary-active substances the surface concentration is so much greater than the bulk concentration that we can identify $\Gamma_i^{(s)}$ as the amount of solute/cm$^2$ in the adsorbed monolayer. Since $\Gamma_i^{(s)} NA \times 10^{-16} = 1$, the area per molecule ($A$, in Å$^2$ per molecule) of the adsorbed monolayer can be found, and hence the II-$A$ curve. (II is the surface pressure, $= \gamma_0 - \gamma$). From $\Gamma_i^{(s)}$ and an assumed thickness of the monolayer (usually of the order 10 Å), the true surface concentration in mole/cm$^3$ can also be calculated.

## Static and Dynamic Surface Tension

When a liquid surface (or liquid–liquid interface) is first formed, a finite time is required to establish equilibrium in the surface phase, and during this period the surface tension is time-dependent. In the case of fresh formation of pure liquid surfaces, it has not yet been found possible to measure the surface tension before the final value has been attained, since the times involved are so minute. With solutions, however, the position is quite different, and the equilibrium value is frequently reached only after a readily measurable time. Values of the surface tension taken during this period are termed "dynamic" to distinguish them from the final or "static" value. Dynamic surface tensions are of particular interest in connection with the general theory of diffusion of molecules into surfaces, which has practical importance in such problems as the speed of wetting by detergent solutions and the adsorption of stabilizer onto emulsion particles during preparation.

## Choice of Method

Before embarking on any experimental measurement of surface tension, it is advisable to consider the pros and cons of the various methods available. Methods vary greatly in accuracy, ease of setting up and operation, the amount of liquid needed, and ease of temperature control. The nature of the interface (e.g., air–water, or oil–water), whether the system studied is a solution or a pure liquid, the time required for equilibrium to be established, and the possibility of problems with contact angles are all relevant.

For "rough" measurements, say to get a quick estimate good to 5%, the capillary rise or ring methods are easily set up in simplified form. If a number of routine measurements of this type are going to be performed, it is probably worth while setting up the apparatus for measurements based on drop-shape. These have the advantage of being independent of contact-angle, and may be used over an extremely wide range of substances and surface tensions. They also may be modified to give higher accuracy without too much trouble, should this be needed at a later stage.

Any of the methods to be discussed in the following pages is capable of good accuracy, provided scrupulous care is taken to keep the apparatus clean and free from contamination. Since surfactant solutions are probably the most commonly studied systems, it is pertinent to comment on them here. With surfactants, and particularly cationic surfactants, contact-angle problems can be troublesome, and when studying solutions of this type, any method which entails the maintenance of zero contact angle (e.g., capillary rise) should be avoided. The authors have found the drop-weight (or drop-volume) method to be particularly useful, especially on small volumes of liquids, although care has to be taken to see that the drops fall from either the inner or the outer diameter of the tip (particularly when studying oil–water interfaces). The maximum bubble pressure and ring methods can be used as a rule, provided that the ring is kept hydrophilic when studying air–water interfaces. For oil–water interfaces, it may be made hydrophobic *or* hydrophilic, as discussed later (p. 527).

With highly viscous or non-Newtonian fluids, the above methods give rise to obvious difficulties, and static drop-shape measurements (avoiding shear on the material) are more applicable. For extremely low interfacial tensions (i.e., less than about 0.5 dyne/cm), the sessile-drop method is probably the *only* usable technique [4].

# 3 METHODS FOR DETERMINATION OF STATIC SURFACE TENSION

## Surface Tension by Calculation

Although the areas of application are still limited, a number of methods have been developed for the *estimation* of interfacial tensions, and it is worth considering the use of these before experimental work is contemplated, especially if high accuracy is not required. Paraffins, in particular, are amenable to this treatment [5].

Fowkes [6] has obtained good agreement between theory and experiment by assuming additivity of molecular interactions in surfaces. Dann [7] has shown that early attempts to derive rectilinear empirical relationships between contact angle and surface tension for homologous series of organic liquids are not, however, justified.

A method for calculating interfacial tensions indirectly from a knowledge of the surface tensions of the pure phases has also been proposed by Fowkes [8]. Good [9] has discussed this work for situations when mercury is one component of the interface.

Shereshefsky [10], and Hough and Warren [11] have extended the method to binary solutions. Surface tensions of liquids near the critical point [12], and of solid phases have also been estimated [13–15].

## Capillary-Height Method

This is the oldest known method and still finds a very important place for a number of both practical and theoretical reasons. It is one of the most accurate of absolute methods, capable of a precision better than 0.1%. The advent of drawn capillaries with their wide range of uniform diameters removed the main handicap to its general use. For practical purposes it is essential that the liquid wet the surface of the capillary, giving a zero contact angle. The theory given below is therefore based on the assumption of zero contact angle.

THEORY

The development of the classical theory of capillarity, and the basis of most methods of measuring surface tension, is the observation that if a liquid surface be curved, the pressure is greater on the concave side than on the convex, by an amount which depends on the curvature and the surface tension.

For a surface with principal radii of curvature $R_1$ and $R_2$, this pressure difference is given by the equation

$$\Delta P = p_1 - p_2 = \gamma\left(\frac{1}{R_1} + \frac{1}{R_2}\right), \qquad (9.10)$$

where $\gamma$ is the surface tension.

For the case where $R_1 = R_2 = r$ (e.g., a cylindrical capillary tube), then

$$\Delta P = p_1 - p_2 = \frac{2\gamma}{r} \qquad (9.11)$$

These equations were known to Young and Laplace over one and a half centuries ago. (Further historical details will be found in Refs. 16 and 17.)

One quantity which will find frequent mention is the "capillary constant," $a$, defined by the relation

$$a^2 = rh, \qquad (9.12)$$

where $h$ is the capillary height in a tube of radius $r$. Since $r$ and $h$ are linear dimensions, $a$ is also, and the quantity $r/a$, which finds extensive use in the theory of shapes of surfaces, is dimensionless.

The capillary of radius $r$ is assumed to be dipping vertically into the liquid in a vessel so wide that the large meniscus is not raised appreciably by capillarity (Fig. 9.3). If $h_0$ is the height of the liquid in the capillary (measured to the base of the meniscus), then

$$\gamma = \frac{1}{2} rh_0\rho g + \frac{mg}{2\pi r} \qquad (9.13)$$

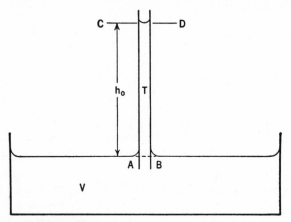

**Fig. 9.3**   Capillary-height method.

where $m$ is the mass (very small) of the liquid in the meniscus, $\rho$ the density of the liquid, and $g$ the acceleration due to gravity. If the capillary diameter is sufficiently small for the second term to be neglected then

$$\gamma = \frac{1}{2} r h_0 \rho g \quad \text{or} \quad \frac{2\gamma}{r} = \rho g h_0 \qquad (9.14)$$

If the vapor around the capillary tube has an appreciable density $\rho_0$, then part of the weight of the liquid is balanced hydrostatically by the vapor and (9.13) becomes

$$\gamma = \frac{1}{2} r h_0 g(\rho - \rho_0) + \frac{g v(\rho - \rho_0)}{2\pi r} \qquad (9.15)$$

where $v$ is the volume of liquid in the meniscus.

As a rule it is simpler to consider a total height $h = h_0 + h'$, where $h' \pi r^2 = v$. For the usual size of capillary tubes used in this technique $h' = r/3$ (valid for water if $r < 0.2$ mm), so that

$$h = h_0 + \frac{r}{3}, \qquad (9.16)$$

in other words, $\gamma = \frac{1}{2} r h g \, (\rho - \rho_0)$. [For interfacial tensions the term $(\rho - \rho_0)$ will be replaced by the difference in densities between the two liquids.]

For tubes of larger diameter, but less than 1 mm,

$$h = h_0 + \frac{r}{3} - 0.1288 \frac{r^2}{h_0} + 0.1312 \frac{r^3}{h_0^2} \ldots, \qquad (9.17)$$

or approximately,

$$h = h_0 + \frac{a^2 r}{3a^2 + r^2},\qquad(9.18)$$

where $a^2$, the "capillary constant," is equal to $rh$.

For tubes of considerably larger diameter, $r/a > 4.3$, an expression due to Rayleigh can be used, namely,

$$1.4142\,\frac{r}{a} - \ln\frac{a}{h_0} = 0.6648 + 0.19785\,\frac{a}{r} + \frac{1}{2}\ln\frac{r}{a}\ldots.\qquad(9.19)$$

For tubes of intermediate diameter the tables of Bashforth and Adams (F. Bashforth and J. Z. Adams, *An Attempt to Test the Theories of Capillary Action*, Cambridge University Press, London, 1883) should be used to obtain $h'$.

Recently [18] it has been shown that the surface area of a meniscus in a circular tube approximates (to about 2%) the area represented by an oblate spheroid which has the tube radius $r'$ and the meniscus height $h'$ as its major and minor half-axes, respectively. The meniscus area, $A_s$, is then related to the tube cross section $A_t$ approximately by $A_s = A_t(1 + r')$. The normalized radius $r'$ has been tabulated [18] as a function of the capillary radius and constant, and computer calculations of the area of such a nonspherical meniscus have shown that the error (*of up to 50%*) involved in the usual hemispherical approximation can be reduced to less than 1%.

EXPERIMENTAL

Provided sufficient liquid is available the apparatus shown in Fig. 9.4, with capillary and large-tube diameters of the order 0.5 mm and 8 cm, respectively, gives excellent results. The diameter of the capillary is determined from measurements of weight and length of a mercury column with the capillary in a horizontal position. The height of the capillary rise is measured with a suitable cathetometer.

The amount of liquid required can be reduced by modification of the above apparatus in various ways (e.g., Harkins and Jordan [17]).

Glass is commonly employed because of its transparency and because it is wet by most liquids. Cleanliness is essential, particularly when using aqueous systems. Use of a receding meniscus is advisable to ensure that the contact angle is strictly zero.

CAPILLARY HEIGHT METHOD AT INTERFACES

As pointed out above, the principles involved in measuring an interfacial tension are the same as for a surface tension. From the experimental standpoint the sole difference is that the upper liquid must have its upper surface above the upper opening of the capillary tube. Since the capillary rise is normally a great deal more at an interface than at a surface it is essential to

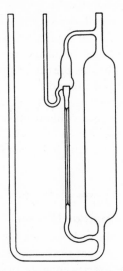

**Fig. 9.4** Capillary-height apparatus of Richards and Carver as modified by Young, Gross, and Harkins.

ensure that the diameter of the lower meniscus is correspondingly increased. Ensuring and maintaining a zero contact angle is usually also more of a problem at an interface, and great care must be taken to ensure it if this method is to be used. When studying an oil–water interface the glass should always be first wetted by the aqueous phase, since water wets glass better than an oil. (For further details see the papers by Harkins and Jordan [17], and Harkins and Humphery [19].)

VARIATIONS OF THE CAPILLARY-HEIGHT METHOD

Several variations of the capillary-height method, which are of value in certain circumstances, have been devised.

The method due to Ferguson and Dawson [20] measures the pressure necessary to force the capillary meniscus level with the outside liquid. This method largely overcomes the problem of uniformity of capillary bore and has potentialities as a differential method (e.g., for comparing water and dilute aqueous solutions), since the pressure difference can be measured with great accuracy.

A differential instrument which has achieved very high accuracy with dilute aqueous solutions was devised by Jones and Ray [21]. It consists of the usual wide tube and capillary, its novel feature being that the difference in level is found by weighing after the liquid in the capillary has been brought to a fixed mark.

Should only very small amounts of liquid be available, there are several micromethods based on the capillary rise principle. In one of these [22], the liquid is contained in a horizontal capillary, and the pressure necessary to force the meniscus, located at one end of the capillary tube, to a plane form, is measured. For a narrow tube this pressure equals $2\gamma/r$, so that knowing $r$, the surface tension can readily be found.

A ring method, which has been used to measure the temperature coefficient of surface tension [23] employs a column of liquid in a uniform capillary tube. If the two ends are at different and measured temperatures, the differential surface tension tends to cause flow which can just be prevented by suitable tilting of the tube.

## Drop-Weight and Drop-Volume Methods

### Drop-Weight Method of Harkins and Brown

If the drop which hangs at the end of a tip were cylindrical and of the same diameter as the tip, it is evident that the maximum weight of drop ($W$) which could be supported would be exactly equal to the weight of the liquid upheld in a capillary tube of the same diameter (i.e., $2r$), because, in both cases, the force of surface tension acts on a line $2\pi r$ long, so the force is $2\pi r\gamma$, and from Tate's Law,

$$W = mg = 2\pi r\gamma. \tag{9.20}$$

Both observation and theory indicate that, on tips of ordinary size, only a fraction of the drop falls, so the weight of the drop which falls must be less than that given by (9.20). (NOTE: for values of $r/a > 0$, even the weight of the hanging drop is less than $2\pi r\gamma$, since the surface of the liquid is not vertical where it meets the edge of the tip.) However, as the tip is made smaller, the *fraction* which falls becomes larger and larger, and extrapolation of the curve to zero diameter indicates that here all of the drop falls, or the hanging and the falling drop have the same weight. Both theory and the extrapolation of experimental values show that (9.20) gives the correct weight of the drop when $r/a$ or $r/V^{1/3}$ is zero, where $V$ is the volume of the drop.

THEORY

A drop of the weight ($2\pi r\gamma$) given by (9.20) has been designated by Harkins and Brown [24] as the *ideal drop*. The fraction of the ideal drop which falls was determined by them in an extensive series of experiments. The fraction of the ideal drop which falls is given by the ordinate of the curve in Fig. 9.5. While any value of $r/V^{1/3}$ may be used in a determination of surface tension, those between about 0.7 and 0.9 give the best results. Now the form of the maximum stable hanging drop is a function either of $r/a$ or $r/V^{1/3}$, and the

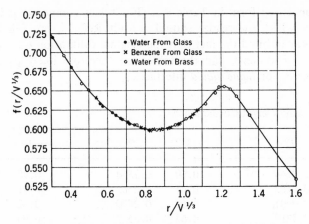

**Fig. 9.5** Drop-weight corrections. Fraction of ideal drop which falls versus ratio of tip radius to cube root of volume of drop.

form of the maximum hanging drop determines the fraction, $f(r/V^{1/3})$ or $f_1(r/a)$, which falls. Thus the weight, $W$, of the drop which falls is

$$W = 2\pi r \gamma \, f\left(\frac{r}{V^{1/3}}\right) = 2\pi r \gamma \, f_1\left(\frac{r}{a}\right), \qquad (9.21)$$

so, since $W = mg$

$$\gamma = \frac{mg}{2\pi r[f(r/V^{1/3})]} = \frac{mg}{2\pi r[f_1(r/a)]} = \frac{mg}{2\pi r\phi} = \left(\frac{mg}{r}\right)F. \qquad (9.22)$$

If the vapor is dilute, it is simplest to let $m$ represent the weight in grams of one drop as weighed in air. Now, since the drop, as it hangs, is also in air, no additional correction for buoyancy need be made. To determine the surface tension of a liquid, it is only necessary to determine the mass of one drop, calculate its volume, $V$, by dividing by a roughly determined density, also in air, multiply by $1/r^3$, look up the value of $F$ in the table and multiply this by $mg/r$. (The density of the liquid is involved only in determining the value of the correction, which is not very sensitive to density differences.)

Although the "sanctity" of the Harkins and Brown results has recently been questioned in a stimulating article by Drost-Hansen [25], it is unlikely that the monumental task of redetermining their data will be undertaken by contemporary workers. The analysis of their figures to obtain the correction factors, however, was only carried out by the original authors on their benzene and water data. Lando and Oakley [26] have recently submitted *all* of the original figures to multiple-regression analysis, and their substantially more accurate correction factors, which eliminate the need for interpolation, are given in condensed form in Table 9.1. (For tables computed at 0.001 unit intervals in $r/V^{1/3}$, the reader should consult Ref. 26, pp. 528–529.) One must still bear in mind the fact that the original calibration was carried out on *liquid–vapor* systems, and the tables are therefore likely to be several percent in error for *liquid–liquid* interfaces.

**Table 9.1**  Drop-Weight Corrections [a]

| $r/V^{1/3}$ | $F$ | $r/V^{1/3}$ | $F$ | $r/V^{1/3}$ | $F$ |
|---|---|---|---|---|---|
| 0.30 | 0.2166 | 0.60 | 0.2554 | 0.90 | 0.2644 |
| 0.31 | 0.2183 | 0.61 | 0.2562 | 0.91 | 0.2642 |
| 0.32 | 0.2201 | 0.62 | 0.2570 | 0.92 | 0.2640 |
| 0.33 | 0.2218 | 0.63 | 0.2577 | 0.93 | 0.2637 |
| 0.34 | 0.2235 | 0.64 | 0.2584 | 0.94 | 0.2634 |
| 0.35 | 0.2251 | 0.65 | 0.2590 | 0.95 | 0.2630 |
| 0.36 | 0.2267 | 0.66 | 0.2596 | 0.96 | 0.2626 |
| 0.37 | 0.2283 | 0.67 | 0.2602 | 0.97 | 0.2622 |
| 0.38 | 0.2299 | 0.68 | 0.2608 | 0.98 | 0.2618 |
| 0.39 | 0.2314 | 0.69 | 0.2613 | 0.99 | 0.2613 |
| 0.40 | 0.2328 | 0.70 | 0.2618 | 1.00 | 0.2608 |
| 0.41 | 0.2343 | 0.71 | 0.2622 | 1.01 | 0.2602 |
| 0.42 | 0.2357 | 0.72 | 0.2626 | 1.02 | 0.2597 |
| 0.43 | 0.2371 | 0.73 | 0.2630 | 1.03 | 0.2590 |
| 0.44 | 0.2384 | 0.74 | 0.2634 | 1.04 | 0.2584 |
| 0.45 | 0.2397 | 0.75 | 0.2637 | 1.05 | 0.2577 |
| 0.46 | 0.2410 | 0.76 | 0.2640 | 1.06 | 0.2570 |
| 0.47 | 0.2423 | 0.77 | 0.2642 | 1.07 | 0.2563 |
| 0.48 | 0.2435 | 0.78 | 0.2644 | 1.08 | 0.2555 |
| 0.49 | 0.2447 | 0.79 | 0.2646 | 1.09 | 0.2547 |
| 0.50 | 0.2458 | 0.80 | 0.2648 | 1.10 | 0.2538 |
| 0.51 | 0.2469 | 0.81 | 0.2649 | 1.11 | 0.2529 |
| 0.52 | 0.2480 | 0.82 | 0.2650 | 1.12 | 0.2520 |
| 0.53 | 0.2490 | 0.83 | 0.2650 | 1.13 | 0.2511 |
| 0.54 | 0.2501 | 0.84 | 0.2650 | 1.14 | 0.2501 |
| 0.55 | 0.2510 | 0.85 | 0.2650 | 1.15 | 0.2491 |
| 0.56 | 0.2520 | 0.86 | 0.2650 | 1.16 | 0.2480 |
| 0.57 | 0.2529 | 0.87 | 0.2649 | 1.17 | 0.2470 |
| 0.58 | 0.2538 | 0.88 | 0.2648 | 1.18 | 0.2459 |
| 0.59 | 0.2546 | 0.89 | 0.2646 | 1.19 | 0.2447 |
|  |  |  |  | 1.20 | 0.2435 |

[a] From [26], pp. 528–529.

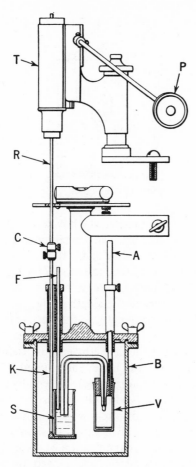

**Fig. 9.6** Drop-weight apparatus.

EXPERIMENTAL

The forms of the drop-weight apparatus described first are those designed to give a precision of 0.02–0.03% with pure liquids which are not too volatile. The apparatus may be greatly simplified, and very greatly reduced in cost, if a precision of 0.3%, which is adequate for a great many purposes, is considered satisfactory. Such an apparatus consists merely of an inverted U of capillary tubing, three weighing bottles, and two glass tubes.

The cross-section of an accurate drop-weight apparatus is shown in Fig. 9.6. For work of high precision, the apparatus is supported by a heavy iron rod, 1 in. in diameter, held upright by a heavy iron tripod which rests upon a

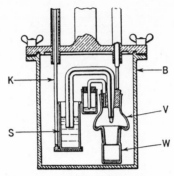

**Fig. 9.7**  Drop-weight apparatus for volatile liquids.

concrete pier. During a determination, the box is suspended in the water of a thermostat from its outside support. The most important part of the apparatus is the tip, shown in Figs. 9.6 and 9.7, which in this case is made from a straight, heavy-walled capillary of Pyrex glass, although other materials may be used (see below). The end of the tube is ground off, by the use of carborundum, until it is a plane perpendicular to the length of the tube. A piece of the same tube, only a few millimeters in length, is then cemented to this square-cut end for use as a support. (An epoxy resin such as Araldite, which may later be removed by dissolving in dichloroethylene, is useful for this purpose.) The tube is clamped in a precision lathe, and rotated in one direction, while being cut round to the proper diameter near the end by a wheel of fine carborundum which rotates at high speed in the opposite direction. It is then polished by rouge, on sealing wax, or pitch, while still rotating in the lathe. The end is not polished. A tip ground in this way should have an edge which appears perfectly sharp under a magnification of 40 diameters. The diameter of the tip is measured by a microscope moved by a micrometer screw (comparator).

Lang and Wilke [27] have developed an efficient method for making several dropping tips at once from hypodermic needles or stainless-steel tubing. The tips to be ground are inserted into a 7.5-cm brass block (Fig. 9.8) which is stood on a matching smaller block and the holes filled with molten Wood's metal. (A syringe is used to draw the metal into the tips.) After the Wood's metal has solidified, the small block is removed and the Wood's metal and embedded rough tips ground and then polished together. The dropping tips are removed from the block by melting the remaining Wood's metal with boiling water.

The tip should be illuminated properly if measurements of high accuracy are to be attained. The tip may be inverted, black paper used as a collar

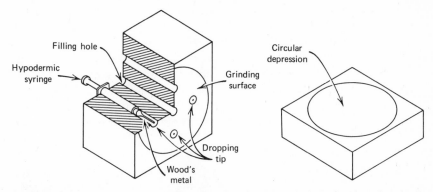

**Fig. 9.8** Brass blocks used in polishing dropping tips. Note the hypodermic syringe in the larger block, and the circular groove in the smaller block [27]. (Courtesy of Lang and Wilke.)

around the tube just below the tip, and the light adjusted until a sharp edge is clearly visible in the microscope.

The method is very accurate if the following conditions are met: (*a*) the drop is made to detach itself as slowly as possible; (*b*) if a solution is involved, the drop must hang at as nearly full extension as possible until the process of adsorption is completed, in other words, equilibrium is attained; and (*c*) with very volatile liquids additional precautions should be taken as described later. Hommelen [28] has discussed the last two factors in some detail.

It is important that the drop not be pulled over by suction, since this may disturb the equilibrium with the vapor. It is also important that the drop hang in the saturated vapor at almost full size until equilibrium is attained. For this purpose, a method of control of the drop devised by Harkins and Harkins [29] may be used. The supply bottle, *S*, in Figs. 9.6 and 9.7, which contains the phase under investigation, is made adjustable in height. To secure good adjustment, a stand with a ratchet and pinion from an old microscope stand is used. The supply bottle is held by a metal support, *K*, which is fastened to a metal rod, *R*, by means of connection piece *C*. Rod *R* is fastened to movable bar *T*. By turning *P*, a pinion wheel, *T* is raised or lowered and the height of *S* thus regulated. *F* is a tube used to regulate the height of the liquid in the supply bottle, *S*, which is raised to start the formation of the drop and is then lowered. Next, it is adjusted to such a height as to give the largest possible stable drop. The bottom of the largest stable drop is located by the position of the cross hair of a short-focus telescope which has been carefully adjusted by an earlier trial. The period during which the drops are suspended at full extension may usually be varied from 3–6 min/drop without any perceptible variation in the weight of the drop which falls. For some very dilute solutions, this suspension time must be greatly lengthened, and Sharma [30] has described a method for accurately controlling drop rate to $\pm0.006$ drop/min. In such cases, however, it is recommended that another method, such as the pendent or sessile drop, be used.

In some cases, depending on the vapor pressure and other factors, probably chiefly on the difference between the temperature of the room and that of the thermostat, it was found necessary to cool $V$, which contains the liquid for saturating the vapor phase, before adjusting the apparatus in the thermostat, in order to prevent distillation into $W$ while the desired temperature was being reached.

For determination of the drop weight of a very volatile liquid such as carbon disulfide, it was found necessary to devise a special type of vent tube, which is shown in Fig. 9.7. Instead of having a vent hole through the brass tube and stopper (to which $V$ is attached), a brass tube was soldered to the stopper, connecting with a small vent hole. The other end of the brass tube passed through another stopper provided with a vent hole. This second stopper supported a small weighing bottle containing just enough carbon disulfide so that its surface stood slightly above the outlet of the brass tube. In this way, the loss by vaporization from the bottle in which the drops are being collected is almost entirely eliminated, and more constant drop-weight results can be obtained than otherwise. (Any remaining correction due to vaporization loss is taken care of by determining the difference in weight between 30 drops and 5 drops of the liquid obtained under exactly similar conditions, according to the procedure of Morgan.) The box $B$, and the stand which carries the ratchet and pinion, are clamped to separate iron stands, in order that the drops may not be shaken by the adjustment of the height of bottle $S$.

After the dropping tip has been cleaned, $V$ is fitted onto the glass or brass stopper and $W$ is fitted onto $V$. The supply bottle which rests on $K$ is put into place. Support $K$ is temporarily prevented from slipping through the tube attached to the roof of the box by collar $C$. By fastening $C$ at the proper distance along $K$, the level of the liquid in $S$ may be adjusted to stand a few millimeters above the level of the tip surface from which the drops fall. By the application of suction at the end of the vent tube, enough liquid is forced into the capillary tube to fill it completely up to the tip surface where the drops are formed. Owing to capillary forces, the liquid will remain in the capillary tube as long as desired. If the level of the liquid in $S$ is not properly adjusted as described above, the liquid in the capillary will siphon back into $S$, or else drops will siphon over from $S$ into $W$, neither of which is desired until the apparatus has been in the thermostat long enough to attain the proper temperature. The vent-tube weighing bottle is next adjusted on its stopper. After the lower part of the box $B$ is attached by means of the wing nuts shown in the diagram, the apparatus is immersed in the thermostat. The apparatus is then leveled, the microscope stand is lowered into place, and rod $R$ is connected to $K$ by means of $C$. In moving $C$ to the end of $K$, care should be taken not to permit the liquid in the capillary tube to siphon out. After the apparatus has been in the thermostat the proper length of time, the drop formation is started by turning pinion wheel $P$, thus raising $S$. Details of a more complex apparatus for highly volatile or reactive liquids are given by Shits [31].

In almost all work the simple weighing bottle $W$ (Fig. 9.6) is attached directly to the stopper which supports the tip. The auxiliary upper section, $F$ (Fig. 9.7), is used in the determination of the surface tension of solutions such that the concentration may change by evaporation. Before the determination is begun, the annular depression in $V$, which is actually larger and deeper than shown, is filled with the solution so that the weighing bottle may be filled by the vapor of the solution before the first drop is formed.

### Drop-Weight-Volume Method of Harkins at Interfaces

The drop-weight method is directly applicable to the determination of the interfacial tension of oils or mercury against an aqueous phase, but for general

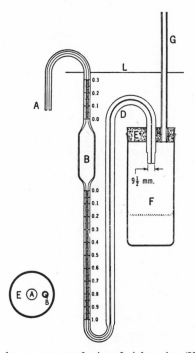

**Fig. 9.9**  Drop-weight-volume apparatus for interfacial tension (Harkins and Humphery).

use the drop-weight-volume method as developed by Harkins and Humphery [19] may be preferred. Their apparatus is shown in Fig.9.9. It is supported in the thermostat by a metal or hard-rubber back. After equilibrium has been established, the aqueous phase is put into pipette *ABD*. The tip is then placed under the surface of the other phase, which is in a vessel (not shown) supported in *F*. The edge of the aqueous phase is drawn exactly to the end of the capillary at the tip, and the reading of the meniscus of this phase above the bulb of the pipette is taken. Suction is applied carefully at *G* and drops are pulled off until the meniscus falls to the scale below the bulb. From the volume thus obtained, the surface tension may be calculated:

$$\gamma = \frac{V(\rho_1 - \rho_2)g}{2\pi r[f(r/V^{1/3})]} = \frac{V(\rho_1 - \rho_2)g}{2\pi r\phi} = \left[\frac{V(\rho_1 - \rho_2)g}{r}\right]F. \qquad (9.23)$$

A useful sophistication of the apparatus is a motor-driven device which gives accurate control of drop rate. This was developed by Roffia and Vianello [32] for use with mercury, but may be used for other liquids.

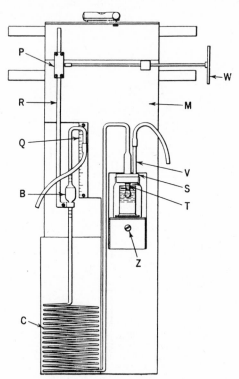

**Fig. 9.10**  Apparatus for interfacial tension (Harkins).

It may be noted that the phase which best wets the tip should be the one to be dropped either downward or upward into the other phase—if upward, the capillary must be bent so that the face of the tip is upward instead of downward. Thus an aqueous phase is dropped downward into an organic phase of lesser density, or upward into one of higher density. If sufficient care is taken, this apparatus gives accurate results, since the detachment of the drop is a slower process in dropping into a liquid than into a vapor. The pressure height can be adjusted much better by the use of either of the devices shown in Figs. 9.10 and 9.11. In the former, the height may be adjusted with great sensitivity because of the presence of a glass spring $C$.

The coiled glass tube, $C$, is made from a single length of ordinary 4-mm glass tubing by heating, drawing, and winding it around a hot asbestos cylinder. The coil has 16 turns, a diameter of 9.5 cm and an approximate internal volume of 5 ml. It may be pressed together until only a few centimeters high or else stretched to a height of 30 cm or more, without any appreciable change of internal volume. The upper end of the coil is sealed to pipette $B$, while the lower end is sealed to a capillary tube ending with a dropping tip $T$, the best diameter of which depends on the value of $r/V^{1/3}$.

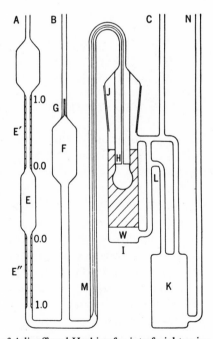

**Fig. 9.11**  Apparatus of Adinoff and Harkins for interfacial tension.

The mechanism for raising or lowering the pipette consists of a wheel and shaft, $W$, a rack, $R$, and pinion, $P$ (encased). Scale $Q$ acts as a guide for adjusting the level of the liquid in the pipette. Thus, for any desired speed of drop formation, the liquid level should stand at a certain height as indicated by the scale. These simple manipulations give complete control of the drops and eliminate the necessity of using the more difficult method of control by suction. Other parts are: $M$, metal plate; $S$, glass stopper; $V$, vent tube; and $Z$, screw.

The apparatus represented by Fig. 9.11, as devised by Adinoff [33], accomplishes the same result in another way. The pipette $E$ and the graduated tubes, $E'$ and $E''$, are connected by glass tubing to capillary $M$, which leads to dropping tip $H$, as in the apparatus of Harkins and Humphery. Parallel to this, a wider tube, $F$, is attached. At the top of tube $F$ is attached a very fine capillary tip $G$.

At the beginning of the experiment, water is drawn from $F$ into $G$ by suction. Because of capillary action it remains there after the suction is removed, even though the level of the water in $E'$ is somewhat below it. The small oil–water interface at the dropping tip is then adjusted to the reference position (described below) and the level of the water in $E'$ read. Then, by slight air pressure at $B$, the water is forced out of capillary $G$ into $F$. During the formation and detachment of the required number of drops, $B$ is kept open to the atmosphere and the level of the water in $F$ is kept constant by applying either suction or pressure at $A$. When the proper number of drops has fallen, the water is again drawn up to the initial level, in $G$, the interface is adjusted to the reference position, and the level in $E''$ read. Since the position of the interface and the amount of water in $F$ are the same at the beginning and end of the experiment, the volume of the drops is the difference between the two readings.

The receiving vessel, $I$, is constructed in such a way as to minimise any difficulties. As each drop of water falls, the water phase, $W$, overflows through side arm $L$ into the larger vessel, $K$, from which it is removed through $N$ from time to time by air pressure through $C$ or suction through $N$. Thus the level of the liquid in $I$ remains constant. The edge of ground-glass joint $J$ is coated with high-melting paraffin to prevent the entrance of any water from the thermostat, since it is considered inadvisable to have any grease present in the system. The paraffin used does not soften appreciably below 70°C.

In regard to the reference position of the interface, it was found that the most reproducible position of the interface was obtained by the reflection of a light placed in front of and below the dropping tip. It is also advisable to place the telescope slightly below the tip, so that the surface of the tip is visible. Under these conditions, the reflection pattern varies considerably with very small changes in volume when the interface is forced into the capillary of the dropping tip. As long as the light, the telescope, and the tip are kept in the same position, the position of the interface can be reproduced so that the level of the water in E changes by an amount which represents less than 0.001 ml. It was also found that forcing the water out of capillary G and pulling it back in again did not change the reading to any observable extent. Thus the difference in the volume of water in the apparatus could be determined to 0.001 ml, which was as close as the graduations could be read.

As in the drop-weight determination of surface tension, the drops are extended to nearly full size by suction, and then allowed to detach themselves under the influence of gravity only. The point of extension is chosen so that the remainder of the drop forms in about 20 sec.

## Other Modifications of the Drop-Weight-Volume Method

Other authors have published modifications of the drop-weight-volume method, for example the precision method of Ward and Tordai [34], and the less precise but very convenient method of Gaddum as modified by Adam [35]. The latter can be used for either surface or interfacial tension, and in view of its obvious advantages in construction and operation, together with the small volumes required (less than 0.5 ml) and fair accuracy (better than 0.5%), it will be described in some detail.

The experimental setup is shown in Fig. 9.12. The volume of the single drops is measured by means of a micrometer syringe (e.g., the "Agla") which can be read to ±0.00005 ml. Once the approximate size of the drop is known, subsequent drops can be expelled rapidly to over 90% of their final volume, the final stages being carried out with requisite slowness without extending the total time much beyond 1 min. The drops are formed from the glass tip, A, ground as described earlier, attached through the ground joint, B, to the syringe barrel. The syringe is first partially filled with the liquid, any air bubbles expelled, and then placed in position. By adjusting the liquid levels in C and D, the level in the U-tube, E and F, can be raised so that the syringe can be completely filled. (This is greatly facilitated by a rubber band between the end of the plunger and the thumb screw on the micrometer, as indicated in Fig. 9.12.)

For interfacial tensions it is only necessary to place a few milliliters of the less-dense liquid on top of the denser one in E, the tip dipping into the lighter liquid when forming the drops, and into the lower liquid when filling.

Another convenient form of thermostated syringe is described by Parreira [36].

For surface-active solutions, where very low interfacial tensions (e.g., < 1 dyne/cm) may be encountered, further modifications will be found convenient. The first is to project an enlarged image of the drop onto a ground-glass

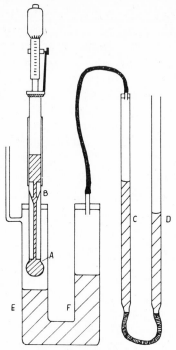

**Fig. 9.12** Drop-volume method using micrometer syringe.

screen (since the drop size is so small); the second is to use a stainless-steel tip cut as shown in Fig. 9.13. With this type of tip, particularly if made hydrophobic by rubbing with a little ferric stearate, troubles due to contact angles and wetting are largely eliminated, and with an orifice of ca. 1 mm diameter interfacial tensions less than 0.1 dyne/cm can be readily measured. The final modification is the use of a rectangular glass cell (e.g., as used in spectrophotometers) with a hole drilled in one side and a small glass or Plexiglas container cemented as shown in Fig. 9.13. This enables the glass cell to be immersed in a thermostat, keeps the liquid level constant, and allows numerous solutions to be measured very rapidly. (The aqueous solution should always be in the syringe. For oils denser than water the tip should be inverted by attaching a suitable U connection to the syringe.)

A modification of the drop-weight method which renders unnecessary the correction factors used in the Harkins-Brown method has been published by Brown and McCormick [37]. In this the flat tip of the usual apparatus is replaced by an inverted cone which is calibrated with a liquid of known surface

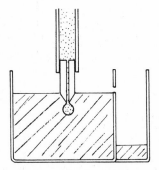

**Fig. 9.13** Tip and cell used for low interfacial tensions.

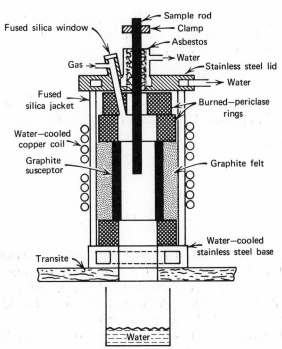

**Fig. 9.14** Induction furnace used for drop-weight surface tension measurements on refractory materials [39]. (Courtesy of McNally, Yeh, and Balasubramanian.)

tension. All unstable drops hanging from such a cone have similar shape, thus eliminating the above correction factors.

The drop-weight method has been extended by Addison et al. [38] to measurements on molten sodium up to 220°C, and by McNally, Yeh, and Balasubramanian [39] to measurements on refractory liquids melting at over 2000°C. The latter authors used a rod of the solid material as its own dropping tip (Fig. 9.14).

## Ring Method

Measurements of the force required to detach a frame, usually in the form of a ring, from the surface of a liquid form the basis of a commonly used method. Equipment using a torsion balance and a platinum ring is commercially available, being usually referred to as a "du Nouy tensiometer." The method is very rapid and simple, does not require large volumes of liquid, and for comparative purposes is probably the method of first choice. For absolute measurement of surface tension it is essential to apply correction factors to the simple theoretical treatment as detailed below.

### THEORY

The simple theory of the ring, or other detachment method, assumes that the detachment force is equal to the product of the surface tension and the periphery of the surface detached. Thus for a ring of radius $R$,

$$\text{detachment force} = 4\pi R\gamma. \tag{9.24}$$

Experiment shows that this can be very seriously in error, so that Harkins and Jordan [17] introduced a correction factor $F$, the equation for $\gamma$ then becoming

$$\gamma = \frac{Mg}{4\pi R} F, \tag{9.25}$$

or

$$\gamma = \frac{Mg}{4\pi R} f\left(\frac{R^3}{V}, \frac{R}{r}\right). \tag{9.26}$$

The correction factor $F$ is found to be a function of the variables $R^3/V$ and of $R/r$, where $V$ is the volume of liquid raised above the plane surface of the liquid by the maximum pull of the ring and $r$ is the radius of the wire of which the ring is made. $V$ is equal to the mass, $M$, of the liquid, as determined by the balance, divided by its density $\rho$.

Correction factors $(F)$ for the ring method are given in Table 9.2. An extension of the tables to cover higher densities and lower surface tensions has been provided by Fox and Chrisman [40].

EXPERIMENTAL

The maximum pull on the ring may be measured with a chainomatic balance or electro-balance if the highest accuracy is desired, but for many purposes the normal type of torsion balance is quite adequate. A simple robust device using a torsion strip with lamp and scale has been described [41]. The ring is usually of platinum and is cleaned by flaming before use. The balance is zeroed with the ring dry; the liquid is then brought into contact with the ring, and the maximum pull, developed as ring and liquid are separated, is measured. It is convenient to use a container permitting the liquid to be overflowed, so that a clean surface can be readily generated. Alternatively the surface can be cleaned by sucking off with a glass tube drawn to a fine tip and connected to a water pump; this is facilitated if a little ignited talc is first sprinkled on the surface as an indicator.

The ring should be free of kinks and as horizontal as possible, the latter being checked by observing its reflection in the surface of the liquid. (According to Harkins and Jordan [17] a departure of 1° introduces an error of 0.5%, 2° an error of 1.5%).

Since the method assumes a zero contact angle, due attention to wetting of the ring is necessary, particularly with surfactant solutions.

A number of variations of the ring method have been proposed. A differential method using two rings, one suspended from each arm of a balance, has been developed by Dole and Swartout [42]. A comparatively simple device, in which the wire is in the form of three sides of a rectangle, with the open side toward the liquid, is described by Lemonde [43]. In this the maximum net pull, after correcting for the mass of the wire, is put equal to $2\gamma l$, where $l$ is the length of the open side.

RING METHOD AT INTERFACES

The ring method is also very useful for interfacial tension studies. For oil–water systems with the oil less dense than the water it is found convenient to use a hydrophilic ring (the platinum ring is flamed and immersed in water before use), whereas with oils more dense than water a hydrophobic ring is preferable. A metal ring can be made hydrophobic by depositing a thin layer of carbon black from a smoky gas flame (not recommended for platinum), or by coating with Teflon.

Interfaces can be cleaned by the suction method described above.

## Bubble-Pressure Method

This method is one of the few which is readily amenable to remote control and is therefore of particular interest in such systems as molten metals [44, 45]. It is a quick method, requires simple apparatus, and is capable of considerable accuracy. Since the surface is renewed quite quickly (every few

**Table 9.2** Correction Factors ($F$) for the Ring Method [a]

| $R^3/V$ | $R/r=$ 30 | 32 | 34 | 36 | 38 | 40 | 42 | 44 | 46 | 48 | 50 | 52 | 54 | 56 | 58 | 60 |
|---|---|---|---|---|---|---|---|---|---|---|---|---|---|---|---|---|
| 0.30 | 1.012 | 1.018 | 1.024 | 1.029 | 1.034 | 1.038 | 1.042 | 1.046 | 1.049 | 1.052 | 1.054 | | | | | |
| 0.31 | 1.006 | 1.013 | 1.018 | 1.024 | 1.028 | 1.033 | 1.039 | 1.041 | 1.044 | 1.046 | 1.049 | | | | | |
| 0.32 | 1.001 | 1.008 | 1.012 | 1.019 | 1.023 | 1.028 | 1.033 | 1.035 | 1.039 | 1.041 | 1.045 | | | | | |
| 0.33 | 0.9959 | 1.003 | 1.008 | 1.014 | 1.018 | 1.024 | 1.028 | 1.030 | 1.035 | 1.036 | 1.040 | | | | | |
| 0.34 | 0.9918 | 0.998 | 1.003 | 1.010 | 1.014 | 1.019 | 1.023 | 1.026 | 1.031 | 1.032 | 1.036 | | | | | |
| 0.35 | 0.9865 | 0.993 | 0.999 | 1.006 | 1.008 | 1.015 | 1.019 | 1.022 | 1.026 | 1.027 | 1.031 | | | | | |
| 0.36 | 0.9824 | 0.989 | 0.995 | 1.002 | 1.005 | 1.010 | 1.015 | 1.018 | 1.022 | 1.024 | 1.027 | | | | | |
| 0.37 | 0.9781 | 0.985 | 0.991 | 0.998 | 1.001 | 1.006 | 1.011 | 1.014 | 1.018 | 1.020 | 1.024 | | | | | |
| 0.38 | 0.9743 | 0.981 | 0.987 | 0.995 | 0.998 | 1.003 | 1.007 | 1.010 | 1.015 | 1.017 | 1.020 | | | | | |
| 0.39 | 0.9707 | 0.977 | 0.983 | 0.991 | 0.994 | 0.9988 | 1.004 | 1.007 | 1.011 | 1.013 | 1.017 | | | | | |
| 0.40 | 0.9672 | 0.974 | 0.980 | 0.986 | 0.991 | 0.9959 | 1.000 | 1.004 | 1.008 | 1.010 | 1.013 | 1.016 | 1.018 | 1.020 | 1.021 | 1.022 |
| 0.41 | 0.9636 | 0.970 | 0.976 | 0.983 | 0.987 | 0.9922 | 0.997 | 1.001 | 1.005 | 1.007 | 1.010 | 1.013 | 1.015 | 1.017 | 1.019 | 1.019 |
| 0.42 | 0.9605 | 0.968 | 0.973 | 0.980 | 0.984 | 0.9892 | 0.994 | 0.998 | 1.002 | 1.004 | 1.007 | 1.010 | 1.013 | 1.014 | 1.016 | 1.017 |
| 0.43 | 0.9577 | 0.964 | 0.970 | 0.977 | 0.981 | 0.9863 | 0.991 | 0.995 | 0.999 | 1.001 | 1.005 | 1.007 | 1.010 | 1.011 | 1.014 | 1.014 |
| 0.44 | 0.9546 | 0.961 | 0.967 | 0.974 | 0.979 | 0.9833 | 0.988 | 0.992 | 0.997 | 0.998 | 1.002 | 1.005 | 1.007 | 1.009 | 1.011 | 1.011 |
| 0.45 | 0.9521 | 0.959 | 0.965 | 0.971 | 0.976 | 0.9809 | 0.986 | 0.990 | 0.993 | 0.996 | 0.9993 | 1.002 | 1.004 | 1.006 | 1.009 | 1.009 |
| 0.46 | 0.9491 | 0.956 | 0.962 | 0.969 | 0.973 | 0.9779 | 0.983 | 0.987 | 0.991 | 0.994 | 0.9968 | 1.000 | 1.002 | 1.004 | 1.006 | 1.007 |
| 0.47 | 0.9467 | 0.954 | 0.960 | 0.966 | 0.971 | 0.9757 | 0.980 | 0.985 | 0.988 | 0.992 | 0.9945 | 0.998 | 1.000 | 1.002 | 1.004 | 1.005 |
| 0.48 | 0.9443 | 0.951 | 0.957 | 0.963 | 0.968 | 0.9732 | 0.978 | 0.983 | 0.986 | 0.989 | 0.9922 | 0.995 | 0.997 | 0.999 | 1.002 | 1.003 |
| 0.49 | 0.9419 | 0.949 | 0.955 | 0.961 | 0.966 | 0.9710 | 0.976 | 0.981 | 0.984 | 0.987 | 0.9899 | 0.993 | 0.995 | 0.997 | 1.000 | 1.001 |
| 0.50 | 0.9402 | 0.946 | 0.952 | 0.959 | 0.964 | 0.9687 | 0.973 | 0.978 | 0.981 | 0.985 | 0.9876 | 0.991 | 0.993 | 0.995 | 0.997 | 0.9984 |
| 0.51 | 0.9378 | 0.944 | 0.950 | 0.956 | 0.961 | 0.9665 | 0.971 | 0.976 | 0.979 | 0.983 | 0.9856 | 0.989 | 0.991 | 0.993 | 0.995 | 0.9965 |
| 0.52 | 0.9354 | 0.942 | 0.948 | 0.954 | 0.959 | 0.9645 | 0.969 | 0.974 | 0.977 | 0.981 | 0.9836 | 0.987 | 0.989 | 0.991 | 0.994 | 0.9945 |
| 0.53 | 0.9337 | 0.940 | 0.946 | 0.952 | 0.957 | 0.9625 | 0.967 | 0.972 | 0.975 | 0.979 | 0.9815 | 0.985 | 0.987 | 0.990 | 0.992 | 0.9929 |
| 0.54 | 0.9315 | 0.938 | 0.944 | 0.950 | 0.955 | 0.9603 | 0.965 | 0.970 | 0.974 | 0.977 | 0.9797 | 0.983 | 0.986 | 0.988 | 0.990 | 0.9909 |
| 0.55 | 0.9298 | 0.936 | 0.942 | 0.948 | 0.953 | 0.9585 | 0.964 | 0.968 | 0.972 | 0.975 | 0.9779 | 0.981 | 0.984 | 0.986 | 0.988 | 0.9892 |
| 0.56 | 0.9281 | 0.934 | 0.940 | 0.946 | 0.951 | 0.9567 | 0.962 | 0.966 | 0.970 | 0.974 | 0.9763 | 0.980 | 0.982 | 0.984 | 0.986 | 0.9879 |
| 0.57 | 0.9262 | 0.932 | 0.939 | 0.944 | 0.949 | 0.9550 | 0.960 | 0.964 | 0.968 | 0.972 | 0.9745 | 0.978 | 0.980 | 0.983 | 0.984 | 0.9861 |
| 0.58 | 0.9247 | 0.930 | 0.938 | 0.942 | 0.947 | 0.9532 | 0.958 | 0.963 | 0.966 | 0.970 | 0.9730 | 0.976 | 0.979 | 0.981 | 0.982 | 0.9842 |

| | | | | | | | | | | | | | | | |
|---|---|---|---|---|---|---|---|---|---|---|---|---|---|---|---|
| 0.59 | 0.9230 | 0.929 | 0.935 | 0.940 | 0.946 | 0.9515 | 0.956 | 0.961 | 0.965 | 0.968 | 0.9714 | 0.975 | 0.977 | 0.979 | 0.981 | 0.9827 |
| 0.60 | 0.9215 | 0.927 | 0.933 | 0.939 | 0.944 | 0.9497 | 0.954 | 0.959 | 0.963 | 0.967 | 0.9701 | 0.973 | 0.976 | 0.978 | 0.979 | 0.9813 |
| 0.62 | 0.9184 | 0.924 | 0.930 | 0.936 | 0.941 | 0.9467 | 0.951 | 0.956 | 0.960 | 0.964 | 0.9669 | 0.970 | 0.973 | 0.975 | 0.976 | 0.9784 |
| 0.64 | 0.9150 | 0.921 | 0.927 | 0.932 | 0.938 | 0.9439 | 0.948 | 0.953 | 0.957 | 0.961 | 0.9643 | 0.968 | 0.970 | 0.972 | 0.973 | 0.9754 |
| 0.66 | 0.9121 | 0.918 | 0.925 | 0.930 | 0.935 | 0.9408 | 0.946 | 0.950 | 0.954 | 0.959 | 0.9614 | 0.965 | 0.967 | 0.969 | 0.971 | 0.9728 |
| 0.68 | 0.9093 | 0.915 | 0.921 | 0.927 | 0.932 | 0.9382 | 0.943 | 0.948 | 0.951 | 0.956 | 0.9590 | 0.963 | 0.965 | 0.967 | 0.968 | 0.9703 |
| 0.70 | 0.9064 | 0.912 | 0.919 | 0.924 | 0.929 | 0.9352 | 0.940 | 0.945 | 0.949 | 0.953 | 0.9563 | 0.960 | 0.962 | 0.964 | 0.966 | 0.9678 |
| 0.72 | 0.9037 | 0.910 | 0.916 | 0.921 | 0.927 | 0.9328 | 0.937 | 0.943 | 0.946 | 0.951 | 0.9542 | 0.957 | 0.960 | 0.962 | 0.964 | 0.9656 |
| 0.74 | 0.9012 | 0.907 | 0.913 | 0.919 | 0.924 | 0.9303 | 0.935 | 0.940 | 0.944 | 0.949 | 0.9519 | 0.955 | 0.958 | 0.960 | 0.962 | 0.9636 |
| 0.76 | 0.8987 | 0.905 | 0.911 | 0.916 | 0.922 | 0.9277 | 0.933 | 0.938 | 0.942 | 0.947 | 0.9459 | 0.953 | 0.956 | 0.958 | 0.960 | 0.9616 |
| 0.78 | 0.8964 | 0.902 | 0.908 | 0.914 | 0.920 | 0.9258 | 0.930 | 0.936 | 0.939 | 0.944 | 0.9457 | 0.951 | 0.954 | 0.956 | 0.958 | 0.9598 |
| 0.80 | 0.8937 | 0.900 | 0.906 | 0.912 | 0.918 | 0.9230 | 0.928 | 0.933 | 0.937 | 0.942 | 0.9454 | 0.949 | 0.952 | 0.954 | 0.956 | 0.9581 |
| 0.82 | 0.8917 | 0.898 | 0.904 | 0.909 | 0.915 | 0.9211 | 0.926 | 0.931 | 0.935 | 0.940 | 0.9436 | 0.947 | 0.950 | 0.952 | 0.954 | 0.9563 |
| 0.84 | 0.8894 | 0.895 | 0.902 | 0.907 | 0.913 | 0.9190 | 0.924 | 0.929 | 0.933 | 0.938 | 0.9419 | 0.946 | 0.949 | 0.951 | 0.953 | 0.9548 |
| 0.86 | 0.8874 | 0.893 | 0.900 | 0.905 | 0.911 | 0.9171 | 0.922 | 0.927 | 0.932 | 0.936 | 0.9402 | 0.944 | 0.947 | 0.949 | 0.951 | 0.9534 |
| 0.88 | 0.8853 | 0.891 | 0.898 | 0.903 | 0.909 | 0.9152 | 0.921 | 0.926 | 0.930 | 0.934 | 0.9384 | 0.942 | 0.945 | 0.947 | 0.950 | 0.9517 |
| 0.90 | 0.8831 | 0.889 | 0.896 | 0.902 | 0.907 | 0.9131 | 0.919 | 0.924 | 0.928 | 0.933 | 0.9367 | 0.940 | 0.943 | 0.946 | 0.948 | 0.9504 |
| 0.92 | 0.8809 | 0.887 | 0.894 | 0.900 | 0.905 | 0.9114 | 0.917 | 0.922 | 0.926 | 0.931 | 0.9350 | 0.939 | 0.942 | 0.945 | 0.947 | 0.9489 |
| 0.94 | 0.8791 | 0.885 | 0.892 | 0.898 | 0.904 | 0.9097 | 0.915 | 0.920 | 0.925 | 0.929 | 0.9333 | 0.937 | 0.940 | 0.943 | 0.945 | 0.9476 |
| 0.96 | 0.8770 | 0.883 | 0.890 | 0.896 | 0.902 | 0.9074 | 0.914 | 0.919 | 0.923 | 0.928 | 0.9320 | 0.936 | 0.939 | 0.942 | 0.944 | 0.9462 |
| 0.98 | 0.8754 | 0.882 | 0.888 | 0.894 | 0.900 | 0.9064 | 0.912 | 0.917 | 0.922 | 0.926 | 0.9305 | 0.934 | 0.937 | 0.940 | 0.943 | 0.9452 |
| 1.00 | 0.8734 | 0.880 | 0.886 | 0.892 | 0.899 | 0.9047 | 0.910 | 0.916 | 0.920 | 0.925 | 0.9290 | 0.933 | 0.936 | 0.939 | 0.941 | 0.9438 |
| 1.05 | 0.8688 | 0.875 | 0.882 | 0.888 | 0.895 | 0.9007 | 0.906 | 0.912 | 0.916 | 0.921 | 0.9253 | 0.929 | 0.932 | 0.936 | 0.938 | 0.9408 |
| 1.10 | 0.8644 | 0.871 | 0.878 | 0.885 | 0.891 | 0.8970 | 0.903 | 0.908 | 0.913 | 0.917 | 0.9217 | 0.925 | 0.929 | 0.933 | 0.935 | 0.9378 |
| 1.15 | 0.8602 | 0.867 | 0.875 | 0.881 | 0.888 | 0.8937 | 0.900 | 0.905 | 0.910 | 0.914 | 0.9183 | 0.922 | 0.926 | 0.930 | 0.933 | 0.9352 |
| 1.20 | 0.8561 | 0.864 | 0.871 | 0.878 | 0.885 | 0.8904 | 0.897 | 0.902 | 0.907 | 0.911 | 0.9154 | 0.920 | 0.923 | 0.927 | 0.930 | 0.9324 |
| 1.25 | 0.8521 | 0.860 | 0.868 | 0.875 | 0.882 | 0.8874 | 0.893 | 0.899 | 0.904 | 0.908 | 0.9125 | 0.916 | 0.920 | 0.924 | 0.927 | 0.9300 |
| 1.30 | 0.8484 | 0.856 | 0.864 | 0.871 | 0.879 | 0.8845 | 0.891 | 0.896 | 0.901 | 0.905 | 0.9097 | 0.914 | 0.917 | 0.921 | 0.925 | 0.9277 |
| 1.35 | 0.8451 | 0.853 | 0.861 | 0.868 | 0.876 | 0.8819 | 0.888 | 0.893 | 0.898 | 0.903 | 0.9068 | 0.911 | 0.915 | 0.919 | 0.922 | 0.9253 |
| 1.40 | 0.8420 | 0.850 | 0.858 | 0.866 | 0.873 | 0.8794 | 0.885 | 0.891 | 0.896 | 0.900 | 0.9043 | 0.909 | 0.913 | 0.916 | 0.920 | 0.9232 |
| 1.45 | 0.8387 | 0.847 | 0.855 | 0.863 | 0.871 | 0.8764 | 0.883 | 0.888 | 0.893 | 0.898 | 0.9014 | 0.906 | 0.910 | 0.914 | 0.918 | 0.9207 |

a W. D. Harkins and H. F. Jordan, *J. Amer. Chem. Soc.* **52**, 1751 (1930).

**Table 9.2**  (*Concluded*)

| $R^3/V$ | $R/r=$ 30 | 32 | 34 | 36 | 38 | 40 | 42 | 44 | 46 | 48 | 50 | 52 | 54 | 56 | 58 | 60 | 65 | 70 | 75 | 80 |
|---|---|---|---|---|---|---|---|---|---|---|---|---|---|---|---|---|---|---|---|---|
| 1.50 | 0.8356 | 0.844 | 0.853 | 0.861 | 0.868 | 0.8744 | 0.881 | 0.886 | 0.891 | 0.895 | 0.8995 | 0.904 | 0.908 | 0.912 | 0.916 | 0.9190 | | | | |
| 1.55 | 0.8327 | 0.841 | 0.850 | 0.858 | 0.866 | 0.8722 | 0.878 | 0.883 | 0.888 | 0.893 | 0.8970 | 0.901 | 0.906 | 0.910 | 0.914 | 0.9171 | | | | 0.9382 |
| 1.60 | 0.8297 | 0.839 | 0.848 | 0.856 | 0.863 | 0.8700 | 0.876 | 0.881 | 0.886 | 0.891 | 0.8947 | 0.899 | 0.904 | 0.908 | 0.912 | 0.9152 | 0.922 | 0.928 | 0.933 | 0.9365 |
| 1.65 | 0.8272 | 0.836 | 0.845 | 0.853 | 0.861 | 0.8678 | 0.874 | 0.879 | 0.884 | 0.889 | 0.8927 | 0.897 | 0.902 | 0.906 | 0.910 | 0.9133 | 0.921 | 0.927 | 0.931 | 0.9354 |
| 1.70 | 0.8245 | 0.834 | 0.843 | 0.851 | 0.859 | 0.8658 | 0.872 | 0.877 | 0.882 | 0.886 | 0.8906 | 0.895 | 0.900 | 0.904 | 0.909 | 0.9116 | 0.919 | 0.925 | 0.930 | 0.9341 |
| 1.75 | 0.8217 | 0.831 | 0.840 | 0.849 | 0.857 | 0.8638 | 0.870 | 0.875 | 0.880 | 0.884 | 0.8886 | 0.893 | 0.898 | 0.902 | 0.907 | 0.9097 | 0.918 | 0.924 | 0.929 | 0.9328 |
| 1.80 | 0.8194 | 0.829 | 0.838 | 0.847 | 0.855 | 0.8618 | 0.868 | 0.873 | 0.878 | 0.882 | 0.8867 | 0.891 | 0.896 | 0.900 | 0.905 | 0.9080 | 0.916 | 0.922 | 0.927 | 0.9317 |
| 1.85 | 0.8168 | 0.827 | 0.836 | 0.845 | 0.853 | 0.8596 | 0.866 | 0.871 | 0.876 | 0.881 | 0.8849 | 0.889 | 0.895 | 0.899 | 0.903 | 0.9066 | 0.915 | 0.921 | 0.926 | 0.9305 |
| 1.90 | 0.8143 | 0.824 | 0.834 | 0.843 | 0.851 | 0.8578 | 0.864 | 0.869 | 0.874 | 0.879 | 0.8831 | 0.888 | 0.893 | 0.897 | 0.902 | 0.9047 | 0.913 | 0.919 | 0.925 | 0.9291 |
| 1.95 | 0.8119 | 0.822 | 0.832 | 0.841 | 0.849 | 0.8559 | 0.862 | 0.867 | 0.872 | 0.877 | 0.8815 | 0.886 | 0.891 | 0.895 | 0.900 | 0.9034 | 0.912 | 0.918 | 0.923 | 0.9281 |
| 2.00 | 0.8098 | 0.820 | 0.830 | 0.839 | 0.847 | 0.8539 | 0.860 | 0.865 | 0.870 | 0.875 | 0.8798 | 0.884 | 0.890 | 0.893 | 0.899 | 0.9016 | 0.910 | 0.917 | 0.922 | 0.9270 |
| 2.10 | 0.8056 | 0.816 | 0.826 | 0.835 | 0.843 | 0.8502 | 0.856 | 0.862 | 0.867 | 0.872 | 0.8768 | 0.881 | 0.886 | 0.890 | 0.895 | 0.8991 | 0.908 | 0.914 | 0.920 | 0.9247 |
| 2.20 | 0.8015 | 0.812 | 0.822 | 0.831 | 0.839 | 0.8464 | 0.853 | 0.858 | 0.864 | 0.869 | 0.8738 | 0.879 | 0.883 | 0.887 | 0.892 | 0.8962 | 0.905 | 0.911 | 0.917 | 0.9226 |
| 2.30 | 0.7976 | 0.808 | 0.818 | 0.828 | 0.835 | 0.8428 | 0.849 | 0.855 | 0.861 | 0.866 | 0.8710 | 0.876 | 0.880 | 0.884 | 0.890 | 0.8935 | 0.903 | 0.909 | 0.915 | 0.9206 |
| 2.40 | 0.7936 | 0.804 | 0.814 | 0.824 | 0.832 | 0.8393 | 0.846 | 0.852 | 0.857 | 0.863 | 0.8680 | 0.873 | 0.878 | 0.882 | 0.887 | 0.8910 | 0.900 | 0.907 | 0.913 | 0.9185 |
| 2.50 | 0.7898 | 0.800 | 0.811 | 0.820 | 0.828 | 0.8360 | 0.843 | 0.849 | 0.854 | 0.860 | 0.8651 | 0.870 | 0.875 | 0.879 | 0.884 | 0.8884 | 0.897 | 0.904 | 0.910 | 0.9166 |
| 2.60 | 0.7861 | 0.797 | 0.807 | 0.817 | 0.825 | 0.8325 | 0.840 | 0.846 | 0.851 | 0.857 | 0.8624 | 0.868 | 0.872 | 0.877 | 0.882 | 0.8859 | 0.895 | 0.902 | 0.908 | 0.9145 |
| 2.70 | 0.7824 | 0.793 | 0.803 | 0.813 | 0.822 | 0.8291 | 0.836 | 0.843 | 0.848 | 0.854 | 0.8598 | 0.865 | 0.870 | 0.874 | 0.880 | 0.8837 | 0.893 | 0.900 | 0.906 | 0.9126 |
| 2.80 | 0.7788 | 0.790 | 0.800 | 0.810 | 0.818 | 0.8260 | 0.834 | 0.840 | 0.846 | 0.852 | 0.8570 | 0.862 | 0.867 | 0.872 | 0.877 | 0.8813 | 0.891 | 0.898 | 0.904 | 0.9107 |
| 2.90 | 0.7752 | 0.786 | 0.796 | 0.806 | 0.815 | 0.8230 | 0.831 | 0.837 | 0.843 | 0.849 | 0.8545 | 0.860 | 0.865 | 0.870 | 0.875 | 0.8790 | 0.889 | 0.896 | 0.902 | 0.9089 |
| 3.00 | 0.7716 | 0.783 | 0.793 | 0.803 | 0.812 | 0.8200 | 0.828 | 0.834 | 0.841 | 0.846 | 0.8521 | 0.858 | 0.863 | 0.868 | 0.873 | 0.8770 | 0.887 | 0.894 | 0.900 | 0.9068 |
| 3.10 | 0.7677 | 0.779 | 0.790 | 0.800 | 0.809 | 0.8170 | 0.825 | 0.832 | 0.838 | 0.844 | 0.8494 | 0.855 | 0.860 | 0.866 | 0.871 | 0.8750 | 0.885 | 0.892 | 0.899 | 0.9049 |
| 3.20 | 0.7644 | 0.776 | 0.787 | 0.797 | 0.806 | 0.8140 | 0.822 | 0.829 | 0.835 | 0.842 | 0.8472 | 0.853 | 0.858 | 0.864 | 0.869 | 0.8730 | 0.883 | 0.890 | 0.897 | 0.9030 |
| 3.30 | 0.7610 | 0.772 | 0.783 | 0.793 | 0.803 | 0.8113 | 0.820 | 0.827 | 0.833 | 0.840 | 0.8449 | 0.851 | 0.856 | 0.862 | 0.866 | 0.8710 | 0.881 | 0.888 | 0.895 | 0.9012 |
| 3.40 | 0.7572 | 0.769 | 0.780 | 0.790 | 0.800 | 0.8083 | 0.817 | 0.824 | 0.831 | 0.837 | 0.8424 | 0.849 | 0.854 | 0.860 | 0.864 | 0.8688 | 0.879 | 0.886 | 0.893 | 0.8993 |
| 3.50 | 0.7542 | 0.766 | 0.777 | 0.788 | 0.798 | 0.8057 | 0.814 | 0.822 | 0.829 | 0.835 | 0.8404 | 0.847 | 0.852 | 0.858 | 0.862 | 0.8668 | 0.877 | 0.884 | 0.892 | 0.8974 |

seconds) surface contamination is minimized. Contact-angle problems do not usually arise, although due care must be taken to ensure that the bubble forms on either the inside or the outside diameter of the tube.

Sugden's two-tube modification described below is particularly simple to use and is capable of an accuracy of about 1 part in 200 [46].

THEORY

If bubbles are slowly blown from the tip of a tube of radius $r$ immersed to a distance $d$ beneath the surface of a liquid then

$$\Delta P_{max} = P_{max} - P_d, \tag{9.27}$$

where $P_{max}$ is the measured maximum pressure and $P_d$ the pressure corresponding to the hydrostatic head $d$. (If the liquid wets the material of the tube, which is the usual case, then $r$ is the internal radius.)

On simple theory [(9.11)] $\Delta P_{max} = 2\gamma/r$. However, this is strictly valid only for sufficiently small tubes, and in general correction factors have to be applied.

If $\Delta P_{max}$ is expressed in terms of the corresponding height of a column of liquid, in other words, $\Delta P_{max} = \rho gh$, then the relationship becomes identical with that for the simple capillary rise. (This assumes that the density of the vapor phase is negligible.)

For other than small tubes the correction becomes significant. Sugden [46] has made use of Bashforth and Adams' tables to calculate correction factors for this method and these are given in Table 9.3. The first approximation of the capillary constant $a_1$ is calculated using the simple (9.12) (i.e., $X = r$). From the value of $r/a_1$ so obtained, the corresponding value of $X/r$ is read from Table 9.3. From the derivation of $X$ ($X = a_2/h$) the second approximation $a_2$ is obtained, and so on.

EXPERIMENTAL

The tip of the tube is prepared by grinding off the end of a piece of capillary tubing so as to give a smooth surface perpendicular to the walls. The capillary diameter, which may be as low as 0.008 cm, is measured by means of a suitable projection microscope. The tip is immersed to a measured distance below the surface of the liquid (as measured by a cathetometer), and bubbles blown at a suitable slow rate (about 35–60/min). The upper part of the capillary tube is connected to a manometer gauge, such as a U-tube containing tinted absolute alcohol, or $n$-butyl phthalate.

In Sugden's modification, two tubes with tips of widely different diameters are used, and the bubbles are formed by suction from an aspirator attached to a U-tube manometer [47]. The radius of the larger tube is usually between

**Table 9.3**   Minimum Value of $X/r$ for Values of $r/a$ from 0 to 1.50 [a]

| $r/a$ | 0.00 | 0.01 | 0.02 | 0.03 | 0.04 | 0.05 | 0.06 | 0.07 | 0.08 | 0.09 |
|---|---|---|---|---|---|---|---|---|---|---|
| 0.0 | 1.0000 | 0.9999 | 0.9997 | 0.9994 | 0.9990 | 0.9984 | 0.9977 | 0.9968 | 0.9958 | 0.9946 |
| 0.1 | 0.9934 | 0.9920 | 0.9905 | 0.9888 | 0.9870 | 0.9851 | 0.9831 | 0.9809 | 0.9786 | 0.9762 |
| 0.2 | 0.9737 | 0.9710 | 0.9682 | 0.9653 | 0.9623 | 0.9592 | 0.9560 | 0.9527 | 0.9492 | 0.9456 |
| 0.3 | 0.9419 | 0.9382 | 0.9344 | 0.9305 | 0.9265 | 0.9224 | 0.9182 | 0.9138 | 0.9093 | 0.9047 |
| 0.4 | 0.9000 | 0.8952 | 0.8903 | 0.8853 | 0.8802 | 0.8750 | 0.8698 | 0.8645 | 0.8592 | 0.8538 |
| 0.5 | 0.8484 | 0.8429 | 0.8374 | 0.8319 | 0.8263 | 0.8207 | 0.8151 | 0.8094 | 0.8037 | 0.7979 |
| 0.6 | 0.7920 | 0.7860 | 0.7800 | 0.7739 | 0.7678 | 0.7616 | 0.7554 | 0.7493 | 0.7432 | 0.7372 |
| 0.7 | 0.7312 | 0.7252 | 0.7192 | 0.7132 | 0.7072 | 0.7012 | 0.6953 | 0.6894 | 0.6835 | 0.6776 |
| 0.8 | 0.6718 | 0.6660 | 0.6603 | 0.6547 | 0.6492 | 0.6438 | 0.6385 | 0.6333 | 0.6281 | 0.6230 |
| 0.9 | 0.6179 | 0.6129 | 0.6079 | 0.6030 | 0.5981 | 0.5953 | 0.5885 | 0.5838 | 0.5792 | 0.5747 |
| 1.0 | 0.5703 | 0.5659 | 0.5616 | 0.5573 | 0.5531 | 0.5489 | 0.5448 | 0.5408 | 0.5368 | 0.5329 |
| 1.1 | 0.5290 | 0.5251 | 0.5213 | 0.5176 | 0.5139 | 0.5103 | 0.5067 | 0.5032 | 0.4997 | 0.4962 |
| 1.2 | 0.4928 | 0.4895 | 0.4862 | 0.4829 | 0.4797 | 0.4765 | 0.4733 | 0.4702 | 0.4671 | 0.4641 |
| 1.3 | 0.4611 | 0.4582 | 0.4553 | 0.4524 | 0.4496 | 0.4468 | 0.4440 | 0.4413 | 0.4386 | 0.4359 |
| 1.4 | 0.4333 | 0.4307 | 0.4281 | 0.4256 | 0.4231 | 0.4206 | 0.4181 | 0.4157 | 0.4133 | 0.4109 |
| 1.5 | 0.4085 | — | — | — | — | — | — | — | — | — |

[a] S. Sugden, *The Parachor and Valency*, Routledge, London, 1930, Table 80, p. 219.

0.1 and 0.2 cm, and that of the smaller one $<0.01$ cm (Fig. 9.15). Surface tension is calculated from the simple empirical equation

$$\gamma = k(h_N - h_W), \tag{9.28}$$

where $h_N$ and $h_W$ are the differences in the manometer levels when bubbles are blown from the narrow and wide capillaries respectively. The apparatus constant, $k$, is found by calibrating with pure water.

The method gives an accuracy of ca. 0.5% and can also be used for liquid–liquid interfaces [48].

## Methods Based on Drop or Bubble Shapes

If the interface whose tension is to be measured can be formed as a drop or bubble of one phase in the other, its geometry is determined by the interaction of gravitational forces, which decrease as the cube of the linear dimension, and surface forces, which decrease as the square of the linear dimension. For very small drops, surface forces will thus predominate, and the interface will be spherical; for larger drops, as gravitational effects become comparable with surface effects, measurement of the departure from sphericity of drop-shape permits an accurate assessment of the interfacial tension.

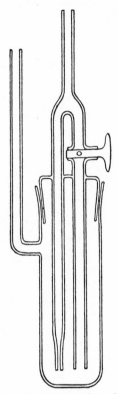

**Fig. 9.15**  Sugden apparatus as modified by Quayle and Smart.

Although this method has long been established, the advent of the digital computer has only recently permitted its high accuracy to be utilized for routine measurements. Since the drop or bubble has only to be supported in a manner which permits it to be photographed, the method is especially applicable to systems in which one phase is a molten salt or metal, and also to systems where it is desired to study the effects of changing temperature or pressure. Its advantages were early recognized by Andreas, Hauser, and Tucker [49]:

"(1) It is sufficiently simple to be subjected to a complete mathematical analysis. (2) The results are independent of the angle of contact between the fluid interface and the apparatus. (3) The method is static, and is therefore not influenced by viscosity effects. (4) Measurements are made instantaneously. (5) Successive measurements can be made on a given surface without disturbing it, thus permitting an accurate study of the aging of surfaces.

(6) Boundary tensions of any magnitude can be observed. Values as great as 370 dyne cm$^{-1}$ and as small as 0.3 dyne cm$^{-1}$ were measured with the present (pendent drop) apparatus. (7) Either surface tension or interfacial tension can be measured in any system in which at least one fluid is transparent and the fluids are of unequal density. (8) Only small samples are required. One cubic centimeter of the internal phase is usually more than enough for several measurements. (9) The apparatus is adapted to simple temperature control. (10) The photographs on which the measurements are made serve as permanent records."

The analysis by these authors of the shape of a pendent drop will serve to illustrate the general underlying theory. We take a coordinate system with origin at the drop base (Fig. 9.16).

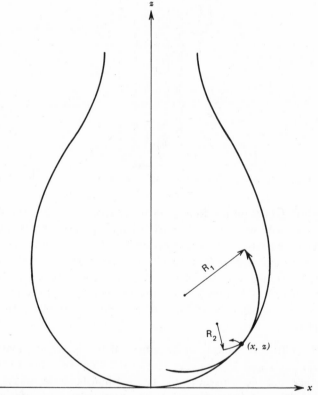

**Fig. 9.16**   Pendent drop profile. [Calculated by numerical solution of (9.31) for a small mercury drop about to detach.]

At any point $(x,z)$ with principal radii of curvature $R_1$ and $R_2$, the excess pressure, $P$, due to a surface tension $\gamma$ is given by

$$P = \gamma(R_1^{-1} + R_2^{-1}).  \qquad (9.29)$$

At the origin, $O$, from symmetry, $R_1 = R_2 = b$, say. Then the hydrostatic pressure $P'$ at $(x,z)$ is given by

$$P' = \frac{2\gamma}{b} - \rho g z,  \qquad (9.30)$$

where $\rho$ is the effective density of the drop. For hydrostatic equilibrium, $P = P'$ and equating (9.29) with (9.30) and introducing the normal formulae for the radii of curvature leads to the fundamental differential equation for the drop profile:

$$\frac{d^2z}{dx^2} + \frac{1}{x} \cdot \frac{dz}{dx}\left[1 + \left(\frac{dz}{dx}\right)^2\right] = \left(\frac{2}{b} - \frac{\rho g z}{\gamma}\right)\left[1 + \left(\frac{dz}{dx}\right)^2\right]^{3/2}.  \qquad (9.31)$$

This equation holds for pendent or sessile drops and bubbles, the appropriate solution for a particular situation depending only on the boundary conditions. Its use has also been extended to thin films in equilibrium with drops [50], but this modification will not be considered here.

The general applicability of (9.31) may be seen most easily by a different derivation [51], which considers $\gamma$ as a surface energy rather than a surface tension. For a drop of mass $m$, area $A$, and center of mass $\bar{z}$, the total energy, $E$, is given by

$$E = \gamma A + mg\bar{z} + \text{constant}.  \qquad (9.32)$$

If the drop is regarded as being formed by revolving the profile of Fig. 9.16 about the $z$-axis, then $A$, $m$, and $\bar{z}$ may be described by integral functions of the generating profile function $x(z)$.

The equilibrium configuration occurs when $E$ is minimized subject to the restriction that the volume of the drop remains constant. Application of the calculus of variations [52] leads to the Euler condition

$$\frac{-d^2x}{dz^2} + \frac{1}{x}\left[1 + \left(\frac{dx}{dz}\right)^2\right] = \left(\lambda - \frac{\rho g z}{\gamma}\right)\left[1 + \left(\frac{dx}{dz}\right)^2\right]^{3/2}.  \qquad (9.33)$$

Converting (9.33) to the form in which $z$ is a function of $x$ yields (9.31)* if the undetermined multiplier $\lambda$ is identified with $2/b$. (Alternatively, $\lambda$ may be determined from the initial conditions in the case when $g = 0$, since it is a constant independent of the other parameters.)

* This is most easily obtained by multiplying (9.33) by $(dz/dx)^3$ and noting that $-(d^2x/dz^2)(dz/dx)^3 = (d^2z/dx^2)$.

Thus, the differential equation (9.31) is nothing more nor less than the condition for a drop, constrained by surface tension in a gravitational field, to have minimum energy. Note that it is intrinsically independent of the method of support, and hence of the contact angle, although these factors can enter the actual solutions through the boundary conditions, and we shall now consider the forms of these most commonly used in surface tension measurements.

### Pendent-Drop Method

A simplified diagram of the apparatus required for this method is shown in Fig. 9.17.

The drop is suspended from an appropriate orifice (see Drop-Weight Methods) in a thermostated bath of the second phase, and illuminated by a parallel beam of monochromatic light, provided by a conventional lamp-lens-filter combination, an electronic flash, or a laser. The image is focused by a high-magnification objective onto a fast film (e.g., Polaroid ASA 3000), and the scale-factor of the photograph is determined either from the geometrical optics of the system, or by direct calibration. *Accurate alignment is essential* for measurements to have better than about 0.5% accuracy. To simplify the computation of $\gamma$, (9.31) must first be put in a dimensionless form by choosing $b$ as the unit of length. Then with $X = x/b$, and so on, and $\beta = g\rho b^2/\gamma$,

$$\frac{d^2Z}{dX^2} + \frac{1}{X} \cdot \frac{dZ}{dX}\left[1 + \left(\frac{dZ}{dX}\right)^2\right] = (2 - \beta Z)\left[1 + \left(\frac{dZ}{dX}\right)^2\right]^{3/2}. \quad (9.34)$$

Writing $d_e$ for the maximum diameter of the drop,

$$\gamma = \frac{g\rho b^2}{\beta} = \frac{g\rho d_e^2}{\beta(d_e^2/b)} = \frac{g\rho d_e^2}{H}, \quad (9.35)$$

where $H = \beta(d_e^2/b)$. Since any two such dimensions as $d_e$ and $b$ characterize the shape of a drop, a knowledge of $H$ as a function of drop shape allows determination of $\gamma$ if the shape can be measured.

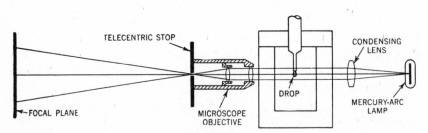

**Fig. 9.17**  Diagram of system for photographing the outline of a pendent drop.

The parameter $b$ is difficult to measure, however, and Andreas et al. [49] introduced a new shape parameter $S = d_s/d_e$ where $d_s$ is the diameter at $z = d_e$ (Fig. 9.18). These authors compiled tables of $1/H$ versus $S$ from empirical measurements on water, but it has long been recognized [53] that this is a poor standard, and much more accurate tables have since been calculated by numerical solution of (9.34) by Fordham [54], Niederhauser and Bartell [55], and Mills [56]. Stauffer [57] later extended these using digital computer techniques. For routine measurements, however, where the data will probably undergo computer analysis, tables are an inconvenient form to use. Stegemeier [58] showed that the tables could be approximately fitted to an equation of the form $1/H = aS^{-b}$, and this notion has recently been extended by Misak [59], who divided the tabulated data into six groups and fitted a modified Stegemeier equation to each group. The six equations give $1/M$ as a function of $S$, in the range $0.300 \leq S \leq 1.000$, to as great a degree of accuracy as is likely to be required in any calculation.

Several other definitions of the shape parameter $S$ have been proposed, notably those of Winkel [60], and Roe, Bacchetta, and Wong [61]. Winkel suggests the use of the maximum and minimum diameters in place of $d_s$ and $d_e$. From an experimental point of view, however, the method suffers from the disadvantage that most drops do not show a neck until the limit of stability has been passed [62], and for the usual shape (Fig. 9.19) the method is impractical. In cases where it may be applied, however, it represents a considerable improvement in the ease and accuracy of drop-profile measurements.

Roe and co-workers have made a most useful extension of the original Andreas, Hauser, and Tucker definition, and one which allows full use of modern computational power to extend the precision of the method. These authors define a *series* of shapes $S_n$ for any given drop by $S_n = d_n/d_e$, where $d_n$ is the diameter at a height $Z_n = (n/10)d_e$, $8 \leq n \leq 12$. Clearly $d_{10}$ and $S_{10}$

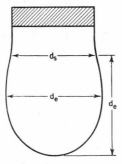

**Fig. 9.18** Shape of a particular pendent drop.

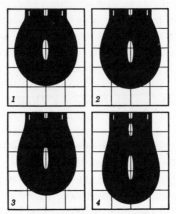

**Fig. 9.19** Shape of a drop of 0.025% aqueous solution of sodium stearate as a function of age: (*1*) aged 10 sec, $\gamma = 71.9$, $S = 0.787$; (*2*) aged 60 sec, $\gamma = 58.2$, $S = 0.818$; (*3*) aged 120 sec, $\gamma = 54.4$, $S = 0.828$; (*4*) aged 1800 sec, $\gamma = 39.2$, $S = 0.849$.

are identical with $d_s$ and $S$, respectively. Tables have been constructed of $1/H$ versus $S_n$ by numerical solution of (9.34) with appropriate boundary conditions, and by measuring several $S_n$ values from a single photograph, a check may be made on the consistency of the data. If the various $1/H$ values for a given drop differ, either the measurements themselves, or the conditions under which they were taken, are in error. There is no reason why Misak's curve-fitting procedure should not be applied to these tables to simplify computer evaluation of data.

A further major advantage of this method is that the fractional error, $P$, in $1/H$ relative to the fractional experimental error in $d_e$, has also been computed. The sample calculation below shows a typical use of the selected planes method. ($CCl_4$ in air at 25.0°; surface 15 sec old.)

$$\rho_{CCl_4} = 1.585 \text{ g/cm}^3$$

$$\rho_{air} = \underline{0.001 \text{ g/cm}^3}$$

$$\rho = 1.584 \text{ g/cm}^3$$

and

$$d_e = 0.228 \text{ cm} \pm 0.1\%$$

$$d_{10} = 0.224 \text{ cm} \pm 0.1\%$$

$$S_{10} = \frac{d_{10}}{d_e} = 0.982 \pm 0.2\%.$$

From Ref. 61, Appendix, Table 2.3b: $1/H = 0.324$, $P = 4.3\%$ relative to $1\%$ in $d_e$. Thus

$$\gamma = \frac{g\rho d_e^2}{H} = 26.14 \text{ dyne/cm} \pm 0.5\%$$

$$= 26.14 \pm 0.16 \text{ dyne/cm}.$$

The International Critical Tables (Vol. 4, p. 447), give the value as $26.15 \pm 0.1$ dyne/cm.

The error in $1/H$ may be reduced by optimizing the choice of $S_n$; in this example, using $S_{12}$ would have halved the error even though the measurements were made to the same accuracy. Greatest accuracy is obtained when $S_n$ with $n = 11$ or 12 is used for a drop having $1/H$ values between 0.3 and 0.5 (Table 9.4). Roe [63] has used the method to study surface tensions of polymeric liquids.

An interesting variation of the pendent-drop method has been suggested by Addison and Hutchinson [64], who found that for drops of identical volume the length is a monotonic function of the surface tension. By calibrating the capillary with a standard liquid, unknown surface tensions may be quickly and easily estimated to within a few percent. The method is not capable of great accuracy, however, and is limited to non-surface-active substances (or, at best, to very dilute solutions of these).

The pendent-drop method has been used to study the surface tension of surfactant solutions [65, 66] as it changes with age (Fig. 9.19), with temperature and with pressure [67–69]. It has also been applied successfully to molten metals [70]; Adamson [71] gives a table of surface-tension values obtained by this technique for these and a number of other substances.

### Sessile-Drop or Bubble Method

This is a particularly useful variation of the previous technique, being mathematically and experimentally simpler. At the expense of only slight loss of accuracy, measurements may be made directly with a cathetometer on a drop (or bubble) of the substance which is sitting on (or under) a flat plate (Fig. 9.20), and the surface tension computed with a nomogram. Simultaneously, the contact angle may be determined directly, or calculated *from the profile*. For more sophisticated measurements, the experimental cell described by Wheeler, Tartar, and Lingafelter [72] embodies all the practical requirements. The cell is made of Pyrex glass and the tubular body is 6 cm long and 4 cm in diameter (see Fig. 9.21). The windows are ground optically plane and carefully sealed to the cell, at the edges, to prevent optical distortion. The plate $A$, beneath which the bubble is formed, is 2.5 cm in diameter, 0.3 cm thick, and ground very slightly concave on the lower surface to pre-

**Table 9.4** Partial Listing of the Table Giving $1/H$ Versus $S_{12}$ [a]

| $S_{12}$ | 0 | 1 | 2 | 3 | 4 | 5 | 6 | 7 | 8 | 9 | P |
|---|---|---|---|---|---|---|---|---|---|---|---|
| 0.75 | 0.47719 | 0.47643 | 0.47568 | 0.47492 | 0.47417 | 0.47342 | 0.47267 | 0.47192 | 0.47118 | 0.47043 | 2.29 |
| 0.76 | 0.46969 | 0.46895 | 0.46821 | 0.46747 | 0.46673 | 0.46600 | 0.46526 | 0.46453 | 0.46380 | 0.46307 | 2.25 |
| 0.77 | 0.46234 | 0.46162 | 0.46089 | 0.46017 | 0.45944 | 0.45872 | 0.45800 | 0.45728 | 0.45657 | 0.45585 | 2.22 |
| 0.78 | 0.45514 | 0.45443 | 0.45372 | 0.45301 | 0.45230 | 0.45159 | 0.45089 | 0.45019 | 0.44948 | 0.44878 | 2.19 |
| 0.79 | 0.44808 | 0.44739 | 0.44669 | 0.44599 | 0.44530 | 0.44461 | 0.44392 | 0.44323 | 0.44254 | 0.44185 | 2.17 |
| 0.80 | 0.44117 | 0.44048 | 0.43980 | 0.43912 | 0.43844 | 0.43776 | 0.43708 | 0.43641 | 0.43573 | 0.43506 | 2.14 |
| 0.81 | 0.43439 | 0.43372 | 0.43305 | 0.43238 | 0.43172 | 0.43105 | 0.43039 | 0.42973 | 0.42906 | 0.42840 | 2.12 |
| 0.82 | 0.42775 | 0.42709 | 0.42643 | 0.42578 | 0.42513 | 0.42447 | 0.42382 | 0.42317 | 0.42253 | 0.42188 | 2.09 |
| 0.83 | 0.42123 | 0.42059 | 0.41995 | 0.41931 | 0.41867 | 0.41803 | 0.41729 | 0.41675 | 0.41612 | 0.41549 | 2.07 |
| 0.84 | 0.41485 | 0.41422 | 0.41359 | 0.41296 | 0.41233 | 0.41171 | 0.41108 | 0.41046 | 0.40984 | 0.40922 | 2.06 |
| 0.85 | 0.40859 | 0.40798 | 0.40736 | 0.40674 | 0.40613 | 0.40551 | 0.40490 | 0.40429 | 0.40368 | 0.40307 | 2.04 |
| 0.86 | 0.40246 | 0.40185 | 0.40125 | 0.40064 | 0.40004 | 0.39944 | 0.39884 | 0.39824 | 0.39764 | 0.39704 | 2.03 |
| 0.87 | 0.39645 | 0.39585 | 0.39526 | 0.39467 | 0.39407 | 0.39348 | 0.39289 | 0.39231 | 0.39172 | 0.39113 | 2.02 |
| 0.88 | 0.39055 | 0.38997 | 0.38938 | 0.38880 | 0.38822 | 0.38764 | 0.38706 | 0.38649 | 0.38591 | 0.38534 | 2.01 |
| 0.89 | 0.38477 | 0.38419 | 0.38362 | 0.38305 | 0.38248 | 0.38192 | 0.38135 | 0.38078 | 0.38022 | 0.37965 | 2.00 |
| 0.90 | 0.37909 | 0.37853 | 0.37797 | 0.37741 | 0.37685 | 0.37630 | 0.37574 | 0.37518 | 0.37463 | 0.37408 | 2.00 |
| 0.91 | 0.37353 | 0.37293 | 0.37243 | 0.37188 | 0.37133 | 0.37078 | 0.37024 | 0.36969 | 0.36915 | 0.36861 | 1.99 |
| 0.92 | 0.36807 | 0.36752 | 0.36698 | 0.36645 | 0.36591 | 0.36537 | 0.36484 | 0.36430 | 0.36377 | 0.36324 | 2.00 |
| 0.93 | 0.36271 | 0.36281 | 0.36165 | 0.36112 | 0.36059 | 0.36006 | 0.35954 | 0.35901 | 0.35849 | 0.35797 | 2.00 |
| 0.94 | 0.35744 | 0.35692 | 0.35640 | 0.35588 | 0.35537 | 0.35485 | 0.35433 | 0.35382 | 0.35331 | 0.35279 | 2.00 |
| 0.95 | 0.35228 | 0.35177 | 0.35126 | 0.35075 | 0.35024 | 0.34973 | 0.34922 | 0.34872 | 0.34821 | 0.34771 | 2.01 |
| 0.96 | 0.34720 | 0.34670 | 0.34620 | 0.34570 | 0.34520 | 0.34470 | 0.34420 | 0.34370 | 0.34321 | 0.34271 | 2.02 |
| 0.97 | 0.34222 | 0.34172 | 0.34123 | 0.34074 | 0.34025 | 0.33975 | 0.33926 | 0.33877 | 0.33829 | 0.33780 | 2.04 |
| 0.98 | 0.33731 | 0.33683 | 0.33634 | 0.33586 | 0.33538 | 0.33489 | 0.33441 | 0.33393 | 0.33345 | 0.33297 | 2.05 |
| 0.99 | 0.33249 | 0.33201 | 0.33153 | 0.33106 | 0.33058 | 0.33010 | 0.32963 | 0.32916 | 0.32868 | 0.32821 | 2.07 |

[a] From Ref. 61, Table 1, p. 4192.

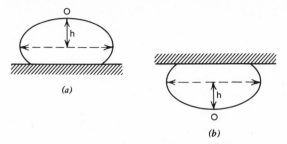

**Fig. 9.20** (*a*) Sessile drop; (*b*) bubble beneath plate.

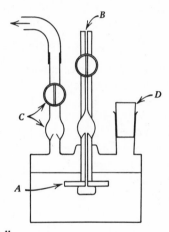

**Fig. 9.21** Sessile bubble cell.

vent the bubble from slipping. A capillary tube, *B*, is sealed perpendicular to this plate and coincident with a small hole through the center. The upper end of the tube is equipped with a stopcock. The plate is mounted in the center of the cell. An additional tube, *C*, with a stopcock, is used with suction to facilitate the formation of the bubble. The cell is filled through opening *D* which is then closed with a stopper. Bubbles are formed by lowering the pressure in the cell and permitting air to flow under the plate, through the capillary opening, until a bubble of the desired size is formed, and the system is allowed to equilibrate. The cell is easily modified for use with sessile drops.

Further experimental details are given by Kemball [73], Ziesing [74], and Smolders [75]. Either photographic or direct measurements may be used.

The only measurements required are the distance, *h*, from the equatorial plane to the apex, and the diameter of the equatorial plane. Provided the drop (or bubble; the theory is identical) is large enough to make the curvature at the apex negligible, (9.31) can be solved to give

$$\gamma = \tfrac{1}{2}h^2\rho g. \tag{9.36}$$

Since only the region between the equatorial plane and the apex is considered, this result is independent of the contact angle.

Porter [76] has tabulated corrections to be applied when the assumption of negligible curvature at the apex does not apply. These corrections are a function of $h/r$, where $r$ is the equatorial radius, but as they provide no internal check of any irregularities which might occur in the drop contour, an alternative procedure due to Smolders and Duyvis [77] is often useful. From the definition of $\beta$ [see (9.35)] it follows that

$$\gamma = \frac{\rho g b^2}{\beta} = \frac{\rho g r^2}{\beta[f(\beta)]^2}, \tag{9.37}$$

where $f(\beta) = r/b$. The factor $f(\beta)$ may be calculated as a function of $r$ from the fundamental differential equation for the drop profile [54].

Butler and Bloom [78] have described a Fortran algorithm which permits a best-fit adjustment of the parameters to the experimental measurements, by fitting the measured drop shape to the theoretical profile generated by solving (9.34). Measurements at a large number of points around the drop contour are used, and the internal consistency of the fit gives a good indication of any errors. Larken [79] has extended the profile calculations to drops on an inclined plane. Brandner and Melrose [80] have computed generalized correction factors which allow the calculation of the surface tension of different segments of the same drop, affording a good check on consistency. Taking two perpendicular photographs of the same drop would permit a further check if required.

A novel method of calculation using a nomogram (Fig. 9.22) has been proposed by Staicopolus [81]. Denoting by $x_{90°}$ and $y_{90°}$, the equatorial radius and the corresponding height above the base, respectively, the nomogram is used as follows:

1. Experimentally obtain $x_{90°}$ and $y_{90°}$ and compute their ratio $(x/y)_{90°}$. Locate this value on the $A$ scale (e.g., point 1) and draw a horizontal line* crossing curve I (e.g., at point 2).

2. From point 2 draw a vertical line† and obtain the value of $B$ on the $D$ scale (e.g., at point 3).

3. A straight line through point 2 (on curve I) and the point $P$ on the upper scale $E$ intersects curve III (at point 4) and curve IV (at point 5).

---

* A "horizontal line" is understood to be a line parallel to the abscissa.
† A "vertical line" is understood to be a line parallel to the ordinate.

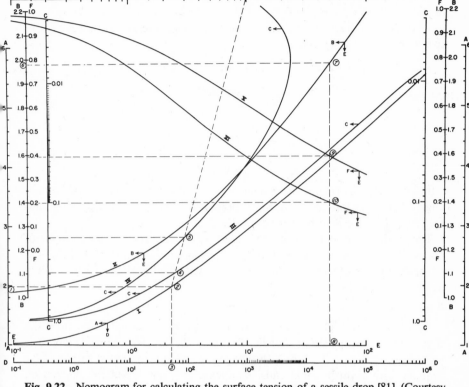

**Fig. 9.22**  Nomogram for calculating the surface tension of a sessile drop [81]. (Courtesy of Staicopolus.)

4. The intersections of horizontal lines through these points and the $C$ scale give the values of $(x/b)_{90°}$ and $(y/b)_{90°}$, respectively. From these the value of $b$ can be calculated:

$$b = \frac{x_{90°}}{(x/b)_{90°}} = \frac{y_{90°}}{(y/b)_{90°}}$$

This procedure is followed for all values of $(x/y)_{90°}$ between 1.022 and 5.5288, which correspond to $\beta$ values from 0.1 to $10^6$. For $(x/y)_{90°}$ values between 1.022 and 2.184 (corresponding to $\beta$ values from 0.1 to 100) one can obtain more accurate values for $(x/b)_{90°}$ and $(y/b)_{90°}$ if the following procedure is used:

1. Calculate $(x/y)_{90°}$ as before and locate this value on the $B$ scale (e.g., point 6).

2. A horizontal line through point 6 intersects curve II at point 7.

3. A perpendicular through this point locates the value of $\beta$ (e.g., point 8) on the $E$ scale and the points 9 and 10 on curves V and VI, respectively.

4. Horizontal lines through these points to the $F$ scale determine the respective values $(x/b)_{90°}$, $(y/b)_{90°}$.

Surface tension $\gamma$ can be calculated using either (9.38) or (9.39):

$$\gamma = \frac{\rho g(x_{90°})^2}{\beta(x/b)_{90°}{}^2} \tag{9.38}$$

or

$$\gamma = \frac{\rho g(y_{90°})^2}{\beta(y/b)_{90°}{}^2} \tag{9.39}$$

Other nomograms permit the calculation to be extended to measure contact angle. The equations used by Staicopolus [82] are of empirical origin, but they have been verified by Parvatikar [83], using solutions to (9.34).

The sessile-drop method has been used to measure the surface tensions of viscous fluids [84], of molten iron as a function of carbon content [85], and of liquid alkali metals [86]. It has also been extended to drops with an acute angle of contact [87].

### Rotating-Drop Method

A fluid drop in a glass tube containing a denser liquid will elongate if the tube is rapidly spun about its axis. Provided the drop (or bubble) length is large compared with its radius, the shape may be approximated by a circular cylinder with hemispherical ends. Vonnegut [88] developed the underlying theory to permit the use of this phenomenon in the measurement of surface tension, but apparently little notice was taken of it at the time.

Recently, however, Princen, Zia, and Mason [89] found it an extremely useful technique for studying highly viscous and viscoelastic systems. The apparatus (Fig. 9.23) is fairly simple, although accurate machining and precision-bore tubing are necessary to reduce vibration. The clamp assembly can be spun at up to 10,000 rpm by a ⅓ hp electric motor via 2 series-connected, variable-speed gear boxes.

A critical step in the measurement is the introduction of a bubble of accurately known volume. After thoroughly cleaning the cell, the stopper with the capillary is wetted with the denser (phase 2) liquid, and inserted in one end of the tube. The tube is held vertically, filled completely, and allowed to stand so that any trapped air bubbles may escape, after which it is inclined nearly horizontally with the open end up. A hypodermic needle fitted to a microburette is slowly inserted through the capillary at the lower end and a carefully measured (to $10^{-4}$ cm$^3$) volume of phase 1, which coalesces to form

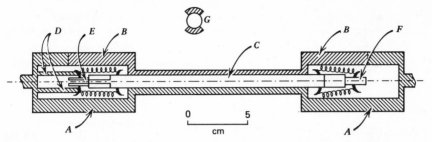

**Fig. 9.23** Rotating drop apparatus (schematic): *A*, fixed part of clamp assembly; *B*, removable part of clamp assembly; *C*, glass tubing; *D*, two pins to prevent slip between the glass tubing and clamp assembly; *E,F*, ground-glass stoppers (*E* has capillary at its center); *G*, cross-section [89].

a single bubble, is introduced. The needle is slowly withdrawn and the upper (open) end of the tube is closed with the other stopper, the excess phase 2 being expelled through the capillary. In this way the drop is introduced under very little hydrostatic pressure. The tube is then inserted in the apparatus clamp.

The bubble length is measured, using a horizontal cathetometer, over a range of speeds of rotation. After reaching equilibrium, the drop length from tip to tip is measured twice at each speed of rotation, from left to right, and then in reverse. In this way, the error caused by any possible inclination of the tube is reduced.

Readers are referred to the original paper for the calculation of interfacial tension from the measurements of drop length. Since this varies widely with angular velocity (Fig. 9.24), the method permits a check on consistency by determining γ at a number of rotation speeds, although the remarks of Borneas and co-workers [90, 91], on the effect of rotation on interfacial-tension values, should be borne in mind when interpreting the results. These authors have noted that the surface tensions of certain liquids, including water, benzene, aliphatic alcohols, and some colloidal solutions, may be altered by rotation, being either decreased or increased up to a definite limiting value.

### The Hanging-Plate (Wilhelmy) Method

In this method a thin plate, usually a microscope cover glass or a piece of platinum foil, attached to the arm of a balance (analytical or torsion type) is partially immersed in the liquid under study.

The simplest procedure is to measure the maximum pull necessary to detach the plate from the surface, in other words, to use it as a detachment method. Under these conditions, and assuming zero contact angle,

$$\text{maximum pull} = 2\gamma(l + t), \tag{9.40}$$

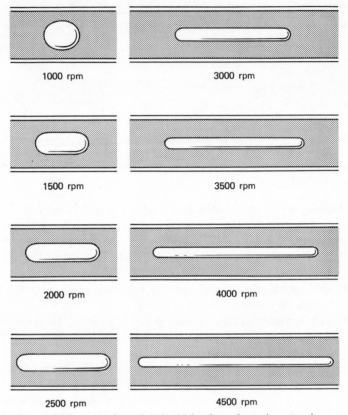

1000 rpm          3000 rpm

1500 rpm          3500 rpm

2000 rpm          4000 rpm

2500 rpm          4500 rpm

**Fig. 9.24**  Shape of a heptane drop (0.156 cm³) in glycerol rotating at various speeds [89].

where $l$ is the width of the plate and $t$ its thickness [92]. (Frequently $t$ can be neglected compared with $l$.) Equation 9.40 holds to within $0.1\%$ so no corrections are usually needed.

In the alternative procedure, which is particularly useful where a *change* in surface tension is sought (as with insoluble monolayers on water), the plate is partially immersed. From the dry ($W_{\text{plate}}$) and immersed ($W_{\text{total}}$) weights, and the depth of immersion, the surface tension can be calculated from the equation (assuming zero contact angle),

$$2\gamma(l + t) = W_{\text{total}} = (W_{\text{plate}} - b) \tag{9.41}$$

where $b$ is the bouyancy correction.

To ensure that zero contact angle is maintained it is helpful to roughen the plate slightly.

Recently Elworthy and Mysels [93] have described an ingenious modification of the plate method for measuring the surface tension of surfactant solutions.

### Surface-Potential Method

Adsorption of surface-active molecules at an interface produces a change in the phase boundary potential, termed the surface potential ($\Delta V$). The method has been used for aqueous systems and has the advantage that it causes no disturbance to the surface. The experimental technique is identical with that used in the study of insoluble monolayers, as described in Chapter 10, pp. 567–569.

### Methods Based on the Spreading of Oils

With the increasing attention now being given to surface contamination of large areas of water such as rivers, lakes, reservoirs, and so on, a simple method due to Adam [94] is particularly useful. A series of solutions of a spreading substance (e.g., dodecyl alcohol) in a nonspreading vehicle (e.g., medicinal paraffin) is prepared and the spreading pressures of the mixtures determined by placing a drop on a monolayer at different surface pressures. By noting the change from spreading to nonspreading behavior as drops of the various solutions are added to the water surface under study, the surface tension can be estimated to better than 1 dyne/cm [95]. Adam advises care when using the method on solutions of strongly adsorbed substances such as proteins.

### Method of Ripples

The velocity of propagation of waves on the surface of a liquid depends on their wavelength and on the surface tension of the liquid. From experimental studies of solutions it appears that the surface tension involved is the "static" one, showing that motion of the ripples produces no displacement of adsorbed substances [96].

Recently there has been renewed interest in the technique [97–99], and studies of surfactant solutions [100] have shown good agreement between theory and experiment.

### Polarized Mercury as One Component of the Interface

The wide range of values reported in the literature for the surface tension of polarized mercury pays tribute to the difficulties involved in surface measurements on this element. Although many of the discrepancies are traceable to small amounts of impurity, classical methods of interfacial tension measurement are unsuitable for many polarized mercury–solution systems even when the greatest care is taken in purifying the mercury [101]. Surfactant-containing solution phases, especially, seem to be inherently un-suited to measurement by the usual techniques [103, 104], although with care drop-times may be used as a measure of interfacial tension [105, 106]. The correction equations used in this method, however, are usually based on the work of Smith [107], who determined the constants empirically by numeri-cal integration of drop photographs. Since his measured drop areas are generally smaller than the area for a sphere of the same volume, the equa-tions derived are presumably in error, although this fact seems to have been overlooked by later workers.

The most accurate method for measuring mercury–solution interfacial tensions is that based on the Lippman equation for an ideally polarized dropping mercury electrode [108]:

$$\gamma = \int_{E_0}^{E} \int_{E_0}^{E} C(dE)(dE), \tag{9.42}$$

where $C$ is the capacitance of the interface, $E$ is the applied polarizing potential relative to a reference electrode, and $E_0$ the value of $E$ corresponding to zero electrode charge. $C$ may be accurately measured as a function of $E$, and the integration carried out numerically by computer.

Comparisons of surface tensions obtained using (9.42) with those obtained by other methods agree well for solutions of a number of inorganic salts, and there is growing evidence [103,109] that the values obtained by this method for surfactant solutions are more reliable than those obtained by classical means.

It is not proposed to give details of this specialized application, but it is worth noting its extension to the measurement of interfacial tensions between a number of other liquid metals and molten salts, especially by the Russian school [110].

Discussions of the theory and details of the apparatus are given in Refs. 102 to 117.

# 4   METHODS FOR DETERMINATION OF DYNAMIC SURFACE TENSION

## Oscillating-Jet Method

The changes in shape which occur along a jet of liquid issuing from a non-circular orifice are a well-known phenomenon. Surface-tension forces tend to rectify the original departure from a circular section, but the momentum of the liquid causes the jet to overshoot the circular form and so the process is repeated. From the wavelength of the oscillations the surface tension may be calculated, so that with solutions the dynamic surface tension at various points (and therefore at various surface ages) can be calculated. Extensive work on solutions has been carried out by Addison and his co-workers [118] and by Sutherland [119].

The apparatus used by Addison is shown in Fig. 9.25. The solution is contained in the reservoir, $A$, maintained at a constant pressure by the open capillary, $B$ (as shown by manometer, $M$), and flows through tube $C$ to issue from the elliptical orifice $D$. The orifice usually has a mean diameter of 0.4–1 mm, the major–minor axis ratio being as small as possible consistent with the formation of visible waves. The wave length may be measured photographically or directly by viewing through lens $L$ against the millimeter scale $E$, the jet being illuminated by the strip light $G$. The flow rate of liquid is measured directly from the time required to rise between fixed marks in the measuring cylinder $F$. The surface age is computed from the mean velocity of the jet and the distance from the orifice of the midpoint of each wave used in a calculation. (Surface ages of ca. 2 msec and greater can be studied in this way.) From the variation of wave length along the jet, the surface tension corresponding to various surface ages can be calculated, and a number of

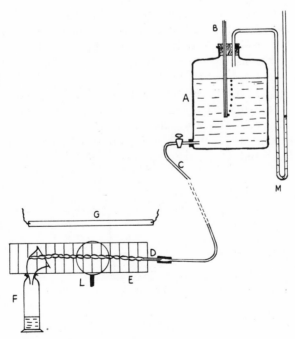

**Fig. 9.25**  Oscillating-jet technique for dynamic surface tension.

systems, chiefly alcohols in water, have been examined in this way by Addison [118], and Sutherland [119].

An analysis [120] of published data on solutions shows that the experimental results depend on the equipment used (e.g., orifice diameter and shape, speed of flow), and in view of the complicated flow and diffusion processes occurring in the jet it has not yet been possible to interpret the results at all precisely.

### Surface Potential Method

As pointed out above (p. 547) adsorption of surface-active molecules at an interface produces a change in the phase boundary or contact potential, termed the surface potential ($\Delta V$). In the case of aqueous solutions, $\Delta V = V_{\text{soln}} - V_{\text{water}}$, and it is measured by means of the techniques described in Chapter 10, pp. 567–569. The application to measurement of dynamic surface tensions is due to Posner and Alexander [121], two types of apparatus being used, depending on the time-scale involved.

For surface ages between ca 0.01 and 1 sec, the channel apparatus shown in Fig. 9.26 is used. The channel $C$ is constructed of Plexiglas or polystyrene, the groove having dimensions ca. $1.2 \times 0.15$ cm. A glass tube is cemented to the back, and this has a small side arm, $H$, to take the usual half-cell (Ag/AgCl), and also a manometer tube, $M$. Liquid from a constant-pressure reservoir flows through tube $L$, and in order to maintain a constant depth of liquid in the channel the channel is tilted slightly. The channel and air electrode, $A$, are contained in a screened box to reduce electrical interference from outside. The contact potential of the solution under test (e.g., 0.01 % sec-octyl alcohol in 0.001 $M$ HCl) is first determined as a function of flow rate with the air electrode at a fixed distance (e.g., 1 cm) from where the liquid first emerges into the air. (The rate of flow is found by timing the outflow into a graduated cylinder.) The apparatus is then thoroughly rinsed out and 0.001 $M$ HCl solution run through and the contact again determined over the same flow rates.

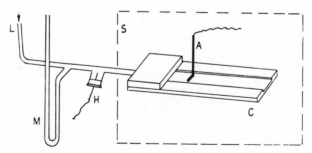

**Fig. 9.26**  Channel apparatus for dynamic surface tension.

From the differences and the flow rate the $\Delta V$–$t$ curve readily follows, $t$ being the surface age. (The velocity of the surface at the point where the contact potential is measured is taken as equal to the average velocity of the bulk liquid.) Values of $\Delta V$ are readily transformed into corresponding surface tensions by means of static measurements with appropriate solutions in a trough. This procedure is only justified when extrapolation to zero flow-rate of the dynamic $\Delta V$ versus flow-rate plot leads to the static value of $\Delta V$, and care must therefore be taken to allow for potentials developing on the Plexiglas trough itself. Usually no problems arise if both the static and dynamic measurements are taken on the same trough [121].

For studying adsorption at much shorter times, down to ca. 0.001 sec, it is necessary to reduce the volume flow; otherwise the requisite amounts become prohibitive. This is achieved by use of a fine jet issuing from a circular orifice, $J$, in a glass tube, as shown in Fig. 9.27. Liquid from the constant pressure reservoir flows through $L$ and is thermostated by water flowing through $T$. $H$ is the usual half-cell, and $S$ the screened box. In this case it is more convenient to keep the pressure head fixed and to move the air electrode, $A$, along the jet by means of a fine graduated movement. Dilute HCl is first run through, followed by the solution under test, also containing dilute HCl. The average rate of flow is measured directly, as with the channel apparatus, by timing into the measuring cylinder $B$. The jets usually employed have ranged from ca. 0.2 to 0.5 mm radius.

## Other Dynamic Methods

The dynamic method due to Maass [122] involves a vertical jet of circular section falling upon a flat surface. Standing waves are set up on the jet surface, from which the surface tension may be calculated.

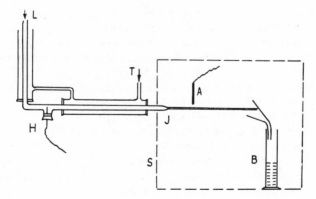

**Fig. 9.27** Jet apparatus for dynamic surface tension.

In the method due to Bond and Puls [123] two jets of circular section are caused to impinge head on, and from the dimensions of the sheet of liquid thus formed the surface tension may be calculated. This method gives an average tension over the expanding surface.

Defay and Hommelen [124] have adapted a static technique for measuring relatively slowly changing surface tensions, using the falling-meniscus method. In this method, a glass tube a few millimeters in bore, and with a capillary hole at one end, is immersed in the solution to be studied. It is then slowly withdrawn (vertically, with capillary hole uppermost) until the meniscus just breaks away from the capillary hole. If this is of radius $r$, and the height from the *highest point* of the meniscus to the bulk is $h_0$ at detachment, then surface tension is given by

$$\gamma = \frac{\rho g r}{2}\left(h_0 - \frac{2r}{3}\right). \tag{9.43}$$

To measure dynamic surface tensions, the meniscus is raised to a height greater than that corresponding to the static value of $\gamma$, and the time the meniscus takes after formation to break away is noted. A series of such measurements yields the full surface tension–time curve.

Other static methods, such as that of the pendent drop, may also be adapted to study time dependence (see Fig. 9.19).

## References

1. E. A. Guggenheim, *Thermodynamics*, North Holland, Amsterdam, 1957.
2. J. J. Kipling, *Adsorption from Solutions of Non-electrolytes*, Academic, London, 1965.
3. I. Prigogine, R. Defay, and A. Bellemans, *Surface Tension and Adsorption*, Longmans, 1966 (Translated by D. H. Everett).
4. B. E. Sundquist and R. A. Oriani, *J. Chem. Phys.* **36**, 2604 (1962).
5. J. F. Padday and N. D. Uffindell, *J. Phys. Chem.* **72** (5), 1407 (1968).
6. F. M. Fowkes, *J. Phys. Chem.* **66** (2), 382 (1962).
7. J. R. Dann, *J. Colloid Interface Sci.* **32** (2), 302 (1970); *ibid.* **32** (2) 321 (1970).
8. F. M. Fowkes, *J. Phys. Chem.* **67** (12), 2538 (1963).
9. R. J. Good, *Nature* **212**, 276 (1966).
10. J. L. Shereshefsky, *J. Colloid Interface Sci.* **24** (3), 317 (1967).
11. E. W. Hough and H. G. Warren, *Soc. Petrol. Eng. J.* **6** (4), 345 (1966).
12. I. P. Bazarov, *Russian J. Phys. Chem.* **41** (9), 1177 (1967).
13. P. Guareschi, *Atti. Accad. Ligure Sci. Lett.* (*Genova*), **19**, 109 (1962).
14. B. K. Barerji, *Indian J. Pure Appl. Phys.* **2**(3), 103 (1964).
15. P. J. Sell and A. W. Neumann, *Z. Phys. Chem.* (*Frankfurt*) **41** (3/4), 191 (1964).

16. N. K. Adam, *The Physics and Chemistry of Surfaces*, 3rd ed., Oxford University Press, Oxford, 1941.
17. W. D. Harkins and H. F. Jordan, *J. Amer. Chem. Soc.* **52**, 1751 (1930).
18. T. A. Erikson, *J. Phys. Chem.* **69**, 1809 (1965).
19. W. D. Harkins and E. C. Humphery, *J. Amer. Chem. Soc.* **38**, 228 (1916).
20. A. Ferguson and P. E. Dawson, *Trans. Faraday Soc.* **17**, 384 (1922).
21. G. Jones and W. A. Ray, *J. Amer. Chem. Soc.* **59**, 187 (1937).
22. A. Ferguson, *Proc. Phys. Soc. (London)* **36**, 37 (1923); *ibid.* **44**, 511 (1932).
23. N. O. Young, *Rev. Sci. Instr.* **26**, 561 (1955).
24. W. D. Harkins and F. E. Brown, *J. Amer. Chem. Soc.* **41**, 499 (1919).
25. W. Drost-Hansen, *Ind. Eng. Chem.* **57** (3), 38 (1965); *ibid.* **57** (4), 18 (1965).
26. J. L. Lando and H. T. Oakley, *J. Colloid Interface Sci.* **25** (4), 526 (1967).
27. S. B. Lang and C. R. Wilke, *Rev. Sci. Instr.* **36** (8), 1255 (1965).
28. J. R. Hommelen, *J. Colloid Interface Sci.* **14** (4), 385 (1959).
29. H. N. Harkins and W. D. Harkins, *J. Clin. Invest.* **7**, 263 (1929).
30. R. R. Sharma, *Indian J. Technol.* **3** (6), 180 (1965).
31. L. A. Shits, *Russian J. Phys. Chem.* **39** (7), 955 (1965).
32. S. Roffia and E. Vianello, *J. Electroanal. Chem.* **17** (1/2), 13 (1968).
33. B. Adinoff, Thermodynamics of Hydrocarbon Surfaces, Ph. D. Thesis, University of Chicago, 1943.
34. A. F. H. Ward and L. Tordai, *J. Sci. Instr.* **21**, 143 (1944).
35. N. K. Adam, *The Physics and Chemistry of Surfaces*, 3rd ed., Oxford University Press, Oxford, 1941, p. 380.
36. H. C. Parreira, *J. Colloid Interface Sci.* **20** (1), 44 (1965).
37. R. C. Brown and H. McCormick, *Phil. Mag.* **39** (6), 420 (1948).
38. C. C. Addison, W. E. Addison, D. H. Kerridge, and J. Lewis, *J. Chem. Soc.* **1955**, 2262.
39. R. N. McNally, H. C. Yeh, and N. Balasubramanian, *J. Materials Sci.* **3**, 136 (1968).
40. H. W. Fox and C. H. Chrisman, *J. Phys. Chem.* **56**, 284 (1952).
41. A. E. Alexander, *Nature* **159**, 304 (1947).
42. M. Dole and J. A. Swartout, *J. Amer. Chem. Soc.* **62**, 3039 (1940).
43. H. Lemonde, *J. Phys. Radium* **9** (7), 505 (1938).
44. C. C. Addison and J. M. Coldrey, *J. Chem. Soc.* **1961**, 468.
45. D. W. G. White,*Mines Branch Res. Rept.,R-157*, Canadian Dept. Mines Tech. Surv., 1965.
46. S. Sugden, *J. Chem. Soc.* **1922**, 858; *ibid.* **1924**, 27.
47. O. R. Quayle and K. O. Smart, *J. Amer. Chem. Soc.* **66**, 937 (1944).
48. E. Hutchinson, *Trans. Faraday Soc.* **39**, 229 (1943).
49. J. M. Andreas, E. A. Hauser, and W. B. Tucker, *J. Phys. Chem.* **42**, 1001 (1938).
50. A. Scheludko, B. Radoev, and T. Kolarov, *Trans. Faraday Soc.* **64** (8), 2213 (1968).
51. J. B. Hayter, "The Measurement of Adsorption at a Dropping Mercury Electrode," B.Sc. (Hons) Thesis, University of Sydney, 1966.
52. R. Courant, *Differential and Integral Calculus*, Vol. 2, Blackie, 1936, p. 515.

53. W. Drost-Hansen, *Ind. Eng. Chem.* **57** (3), 38 (1965); *ibid.* **57** (4), 18 (1965).
54. S. Fordham, *Proc. Roy. Soc.* (*London*) **194A**, 1 (1948).
55. D. O. Niederhauser and F. E. Bartell, *Report on Progress—Fundamental Research on the Occurrence and Recovery of Petroleum;* Publication of the American Petroleum Institute, Lord Baltimore Press, Baltimore, 1950, p. 114.
56. O. S. Mills, *Brit. J. Appl. Phys.* **4**, 247 (1953).
57. C. E. Stauffer, *J. Phys. Chem.* **69**, 1933 (1965).
58. G. L. Stegemeier, "Interfacial Tension of Synthetic Condensate Systems," Ph.D. Dissertation, The University of Texas, Austin, Tex., 1959.
59. M. D. Misak, *J. Colloid Interface Sci.* **27** (1), 141 (1968); the equation for the range $0.300 \leq S \leq 0.400$ is contained in a private communication.
60. D. Winkel, *J. Phys. Chem.* **69** (1), 348 (1965).
61. R.-J. Roe, V. L. Bacchetta, and P. M. G. Wong, *J. Phys. Chem.* **71** (13), 4190 (1967).
62. J. E. Halligan and L. E. Burkhart, *J. Colloid Interface Sci.* **27** (1), 127 (1968).
63. R.-J. Roe, *J. Phys. Chem.* **72** (6), 2013 (1968).
64. C. C. Addison and S. K. Hutchinson, *J. Chem. Soc.* **1949**, 3387.
65. E. A. Boucher, T. M. Grinchuk, and A. C. Zettlemoyer, *J. Colloid Interface Sci.* **23** (4), 600 (1967).
66. G. Petre and M. L. Schayer-Polischuk, *J. Chim. phys.* **63** (10), 1409 (1966).
67. E. A. Hauser and A. S. Michaels, *J. Phys. Colloid Chem.* **52**, 1157 (1948).
68. A. S. Michaels and E. A. Hauser, *J. Phys. Colloid Chem.* **55**, 408 (1951).
69. H. V. Jennings, Jr., *J. Colloid Interface Sci.* **24** (3), 323 (1967).
70. J. Tilly and J. C. Kelly, *Brit. J. Appl. Phys.* **14** (10), 717 (1963).
71. A. W. Adamson, *Physical Chemistry of Surfaces*, 2nd ed., Interscience, New York, 1967, p. 42.
72. D. L. Wheeler, H. V. Tartar, and E. C. Lingafelter, *J. Amer. Chem. Soc.* **67**, 2115 (1945).
73. C. Kemball, *Trans. Faraday Soc.* **42**, 526 (1946).
74. G. M. Ziesing, *Aust. J. Phys.* **6**, 86 (1953).
75. C. A. Smolders, *Rec. trav. chim.* **80**, 699 (1961).
76. A. W. Porter, *Phil. Mag.* **15**, 163 (1933).
77. C. A. Smolders and E. M. Duyvis, *Rec. trav. chim.* **80**, 635 (1961).
78. J. N. Butler and B. H. Bloom, *Surface Sci.* **4** (1), 1 (1966).
79. B. K. Larkin, *J. Colloid Interface Sci.* **23** (3), 305 (1967).
80. C. F. Brandner and J. C. Melrose, *Amer. Chem. Soc. Div. Petrol. Chem., Preprints* **10** (4), D-33 (1965).
81. D. N. Staicopolus, *J. Colloid Interface Sci.* **23** (3), 453 (1967).
82. D. N. Staicopolus, *J. Colloid Interface Sci.* **17** (5), 439 (1962); *ibid.* **18** (8), 793 (1963).
83. K. G. Parvatikar, *J. Colloid Interface Sci.* **23** (2), 174 (1967).
84. P. R. Johnson and R. C. L. Bosworth, *Proc. Roy. Soc.* (*N. S. W.*) **83**, 164 (1950).
85. P. Kosakevitch, S. Chatel, and M. Sage, *Compt. Rend.* **236**, 2064 (1953).
86. D. H. Bradhurst and A. S. Buchanan, *Aust. J. Chem.* **14** (3), 397 (1961).

87. Yu. N. Ivashchenko, V. N. Eremenko, and B. B. Bogatyrenko, *Russian J. Phys. Chem.* **39** (2), 278 (1965).

88. B. Vonnegut, *Rev. Sci. Instr.* **13**, 6 (1942).

89. H. M. Princen, I. Y. Z. Zia, and S. G. Mason, *J. Colloid Interface Sci.* **23** (1), 99 (1967).

90. M. Borneas and E. Kalman, *Compt. Rend.* **245** (19), 1710 (1957).

91. M. Borneas and I. Babutia, *Compt. Rend.* **249** (12), 1036 (1959).

92. D. O. Jordan and J. E. Lane, *Aust. J. Chem.* **17**, 7 (1964).

93. P. H. Elworthy and K. J. Mysels, *J. Colloid Interface Sci.* **21**, 331 (1966).

94. N. K. Adam, *Proc. Roy. Soc. (London)* **122B**, 134 (1937).

95. P. Pomerantz, W. C. Clinton, and W. A. Zisman, *J. Colloid Interface Sci.* **24**, 16 (1967).

96. For example, R. C. Brown, *Proc. Phys. Soc. (London)* **48**, 312 (1936).

97. F. C. Goodrich, *Proc. Roy. Soc. (London)* **260A**, 490, 503 (1961).

98. T. W. Healy and V. K. La Mer, *J. Phys. Chem.* **68**, 3535 (1964).

99. J. T. Davies, *Proc. Roy. Soc. (London)* **290A**, 515 (1966).

100. J. Lucassen and R. S. Hansen, *J. Colloid Interface Sci.* **22**, 32 (1966).

101. M. E. Nicholas, P. A. Joyner, B. M. Tissen, and M. D. Alson, *J. Phys. Chem.* **65** (8), 1373 (1961).

102. R. Parsons, in *Modern Aspects of Electrochemistry*, Vol. 1, J. O'M. Bockris and B. E. Conway, Eds., Butterworths, London, 1954, pp. 103 ff.

103. R. Parsons and P. C. Symons, *Trans. Faraday Soc.* **64** (4), 1077 (1968).

104. S. Minc and M. Brzostowska, *Roczniki Chem.* **40**, 1759 (1966).

105. R. G. Barradas and F. M. Kimmerle, *Can. J. Chem.* **45**, 109 (1967).

106. R. G. Barradas and F. M. Kimmerle, *J. Electroanal. Chem.* **9** (5/6), 483 (1965).

107. G. S. Smith, *Trans. Faraday Soc.* **47**, 63 (1951).

108. P. Delahay, *Double Layer and Electrode Kinetics*, Interscience, New York, 1965, p. 17.

109. M. Sluyters-Rehbach, W. J. A. Woittiez, and J. H. Sluyters, *J. Electroanal. Chem.* **13**, 31 (1967).

110. E. A. Ukshe, N. G. Bakun, D. I. Leikis, and A. N. Frumkin, *Electrochim. Acta* **9**, 431 (1964).

111. J. L. Duda and J. S. Vrentas, *J. Phys. Chem.* **72**, 1187–1200 (1968).

112. D. C. Grahame, *Chem. Rev.* **41**, 441 (1947).

113. M. A. V. Devanathan and B. V. K. S. R. A. Tilak, *Chem. Rev.* **65** (6), 635 (1965).

114. J. B. Hayter, *J. Electroanal. Chem.* **19**, 181 (1968).

115. P. Delahay, *J. Electroanal. Chem.* **16**, 116 (1968).

116. D. C. Grahame, *J. Amer. Chem. Soc.* **63**, 1207 (1941).

117. G. H. Nancollas and C. A. Vincent, *Electrochim. Acta* **10**, 97 (1965).

118. C. C. Addison, *J. Chem. Soc.* **1943**, 535; **1944**, 477; **1945**, 98; and later papers.

119. E. K. Rideal and K. L. Sutherland, *Trans. Faraday Soc.* **48**, 1109 (1952).

120. K. L. Sutherland, *Rev. Pure Appl. Chem. (Australia)* **1**, 35 (1951).

121. A. M. Posner and A. E. Alexander, *Trans. Faraday Soc.* **45**, 651 (1949).

122. O. Maass, *Trans. Roy. Soc. Can.* **29**, 105 (1935).

123. W. N. Bond and H. O. Puls, *Phil. Mag.* **24**, 864 (1937).

124. R. Defay and J. R. Hommelen, *J. Colloid Sci.* **14** (4), 401 (1959).

Chapter **X**

# DETERMINATION OF PROPERTIES OF INSOLUBLE MONOLAYERS AT MOBILE INTERFACES

A. E. Alexander*
and
G. E. Hibberd

* Deceased.

## 1  INTRODUCTION

Under certain conditions it is possible for a film, monomolecular in thickness, to exist between two contiguous bulk phases in which it is entirely, or largely, insoluble. Such films are termed "insoluble monolayers" and in theory could exist between interfaces of the following types: gas–liquid, liquid–liquid, gas–solid, solid–liquid, and solid–solid. However, in practice certain inherent complexities associated with solid surfaces, such as the preparation of clean reproducible surfaces and the different behavior of various crystal faces, make the study of mobile interfaces more attractive, both experimentally and theoretically. This account will therefore be largely restricted to the gas–liquid and liquid–liquid interfaces, more particularly to the air–water (A–W) and oil–water (O–W) interfaces.

The techniques of insoluble monolayers have been applied to a number of problems of a very diverse nature, as will be shown below. In addition, the results form the basis for much of our understanding of the basic physicochemical factors involved in surface phenomena, which in turn so frequently determine the macroscopic behavior of colloidal systems.

Before proceeding further, a note of caution should be sounded, particularly for the benefit of those new to the field. Many of the techniques described below appear to be simple and, in fact, are simple, as regards both apparatus and its mode of use. The simplicity can be very deceptive, however, because of certain pitfalls regarding technique and interpretation of results. Many of these points of technique are pointed out at appropriate places in the text; as regards interpretation, this frequently requires a reasonably detailed knowledge of earlier work and of the basic physicochemical factors involved.

The present discussion makes no attempt to be exhaustive experimentally, theoretically, or historically. The principal object is to enable those new to the field to realize its potentialities and limitations and to outline the more important techniques available. (For further details see, for example, Adam [1], Alexander and Johnson [2], Davies and Rideal [3], and the comprehensive monograph by Gaines [4].)

## 2 INSOLUBLE MONOLAYERS AT THE AIR–WATER INTERFACE

The requirements which enable a chemical compound to exist in the form of an insoluble monolayer at an air–water interface are, firstly, a paraffin chain or other group virtually insoluble in water, and, secondly, a polar group having considerable affinity for water (as indicated by solubility considerations for example).

The compounds first studied as insoluble monolayers were the higher fatty acids and their glycerides, these being orientated on the water surface as shown in Fig. 10.1 for the case of, say, palmitic acid. Such an orientation enables the attraction between the polar water molecules and the polar group (—COOH in this case) to be satisfied while at the same time not requiring the insoluble paraffin chain to be immersed in the water. (Fatty acids above about $C_{12}$ are virtually insoluble in water.) Subsequently a great variety of organic compounds have been investigated as indicated by the following list, which is not, however, intended to be exhaustive.

1. Paraffin chain derivatives of chain lengths greater than about $C_{12}$, such as acids, alcohols, amines, esters, ketones, ethers, glycerides, and phospholipids.

2. Ring systems containing one or more polar groups, such as certain sterols, triterpenes, porphyrins, and phthalocyanines.

3. Polymers, both natural and synthetic, for example, many proteins, certain cellulose derivatives, polyacrylates and methacrylates, nylon, synthetic polypeptides, and silicones.

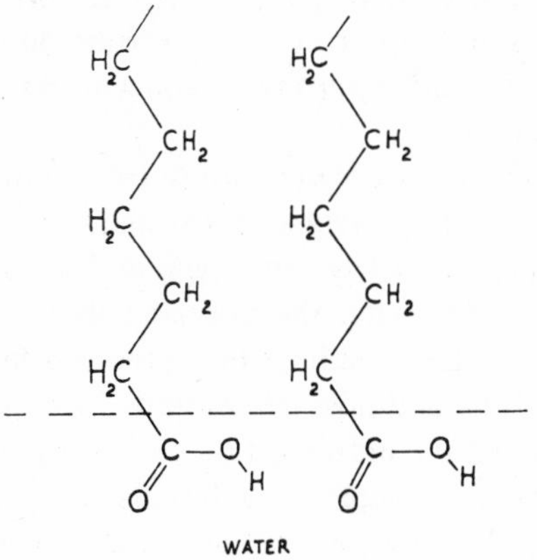

WATER

**Fig. 10.1**  Orientation of a monolayer of fatty acid on a water surface.

When the molecules contain ionized groups of high water-attracting power, such as $-SO_4^-Na^+$ in certain detergents, or $-COO^-Na^+$ in fatty acids at alkaline pH, the chain length must be at least $C_{20}$ if a stable monolayer is to be formed. Lower chain members can frequently be studied by spreading on strong salt solutions (see below). The use of strong salt solutions also makes possible the study of many other compounds where the balance between the polar and nonpolar portions is such that the monolayer is otherwise too rapidly soluble, for example $C_{10}$ alcohols, polyvinyl alcohol, polyacrylic acid, and sterols containing several polar groups.

As a key to the literature relating to the monolayer behavior of specific substances, the indexed bibliography of Stephens [5] is invaluable.

## Methods of Spreading

Certain compounds, for example, oleic acid and hexadecanol, will spread spontaneously on water to form a stable monolayer but, except for certain special purposes (e.g., the study of duplex films, Section 6), this direct spreading is seldom employed. Most compounds have to be assisted in some manner, and this is normally done by spreading from a suitable solvent. The chief requirements of a suitable solvent are that it shall spread the material rapidly and completely and then be readily lost either by evaporation (e.g., light petroleum, benzene) or by solution (e.g., aqueous alcoholic mixtures). Where organic solvents denser than water are necessary for solubility purposes (e.g., $CHCl_3$) it is advisable to reduce their density as much as possible by dilution with a medium of lower density such as light petroleum, benzene, or ether. Addition of a polar compound is also advantageous; for example, the addition of a few per cent of propyl alcohol will often give complete spreading even with very dense organic solvents. Spreading from aqueous solutions, as used with proteins and certain polymers, is also assisted by a similar addition. For most purposes a concentration of about 0.5 mg/ml will be found quite suitable.

The solution is spread on the water surface from some type of micrometer syringe, of which the "Agla" syringe (Burroughs Wellcome Co.) shown in Fig. 10.2 will be found particularly convenient and accurate. This can be read with an accuracy of $\pm.0001$ ml, the normal volumes spread being of the order of 0.05 to 0.1 ml. In spreading, the syringe is first filled, air bubbles expelled, and then it is held almost horizontally just above the center of the clean water surface while the individual drops are forced out in slow succession. When spreading from aqueous solutions or forming monolayers of polymeric materials, the precautions recommended by Crisp [6] should be observed and the spreading solution expelled from the syringe held with the needle tip in the surface. A rubber band looped from the plunger to the

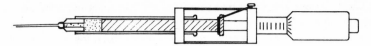

**Fig. 10.2** "Agla" micrometer syringe.

thumb-screw, as shown in Fig. 10.2, will be found helpful. The area available for spreading should be as large as convenient, preferably at least twice that of the close-packed monolayer.

Completeness of spreading can be tested for as follows.

1. By dilution of the spreading solution, preferably with a good solvent, and examination of its effect upon the area per molecule (see next section).

2. By examination of the spread monolayer in a dark field ultramicroscope, a particularly suitable type being the Leitz Ultropak. (A monolayer is invisible, but unspread material or collapsed film readily shows up.) The latter is experimentally more difficult [2, 4], and, fortunately, is seldom necessary, since with some experience the former method usually suffices.

It is essential that all solvents, substrate solutions, and the apparatus used to contain the monolayer system be free from surface active contaminants. This is particularly important when examining substances too soluble in water to give stable monolayers by spreading on strong salt solutions, for example, NaCl, $(NH_4)_2SO_4$. In such cases the salt solution should be shaken up with activated charcoal before use and it is often helpful to have a thin layer of active charcoal on the floor of the trough.

## Methods of Examination of Spread Monolayers

The extreme thinness of monolayers, often 10–30 Å, makes indirect methods of examination inevitable. Various physical properties have been used for this purpose, the principal ones being the surface pressure ($\Pi$), the surface potential ($\Delta V$), the surface viscosity ($\eta_s$), and the surface shear modulus ($G_s$). Some studies of optical properties have also been made and will be referred to again in Section 4.

In general, all the above properties are measured as a function of the area per molecule ($A$) expressed in square angstroms. (If the molecular weight is unknown, the area is usually given in square meters per milligram.)

## Measurement of Surface Pressure ($\Pi$)

The surface pressure ($\Pi$) is by definition the lowering of surface tension produced by the film, that is,

$$\Pi = \gamma_{water} - \gamma_{film}. \tag{10.1}$$

It can, therefore, in principle, be obtained from measurement of the surface tension before and after spreading the monolayer, but in practice it is usually obtained directly by some type of "film balance." Numerous film balances, some of them most elaborate, have been described in the literature [4]. The one described here was designed for ease of construction and simplicity in use, together with reasonable accuracy and sensitivity, and has been in use for over 20 years for both teaching and research purposes [7]. An accuracy of 0.1 dyne/cm can readily be obtained; this is adequate for almost all purposes except for gaseous films at low pressures. For the latter systems, which are of importance in connection with the determination of molecular weights (see Section 3), some changes in technique are necessary. (For details see, for example, Guastalla [8], Allan and Alexander [9], Harrap [10].)

Fig. 10.3 shows the essential parts of the film balance. Two modifications are possible, namely, the horizontal float (Fig. 10.3a) and the Wilhelmy or hanging plate (Fig. 10.3b). The former has a definite advantage in being comparatively insensitive to contact angles, but leakage of the film through the improper attachment of the threads, $M$, can cause considerable difficulties for the beginner. The hanging-plate method is much simpler experimentally and will be found very satisfactory for almost all purposes provided due precautions to maintain a zero contact angle are observed. Two simple directives usually ensure this: firstly, lightly rub the hanging plate (mica) in a vertical direction with fine carborundum paper (followed by a rinse in alcohol); secondly, always study the films *in compression*.

In Fig. 10.3a, $A$ is the trough, usually of dimensions about $15 \times 30 \times 1$ cm, placed on a stand with leveling screws. The float or hanging plate is attached to a torsion strip, $B$, preferably of phosphor bronze, about 10 cm long and 1–2 mm in width. Depending on the surface pressures to be measured, various strip thicknesses are used, but for most purposes about 0.15 mm will be found suitable. The float, $C$, consists of a thin strip of mica about $1 \times 14$ cm (i.e., about 1 cm less than the width of the trough) cemented by paraffin wax to a thin aluminium frame which is attached to block $D$, through which the torsion strip passes and which also carries a small mirror. Block $D$ carries an arm with a light balance pan, the dimensions of $ED$ and $CD$ being about 5 cm. By means of the lamp, $F$, and a plane mirror, $G$, which is conveniently placed about 1 m vertically above the apparatus, the small deflections of the float are greatly enlarged into deflections along scale $H$. $I, J$, and $K$ are movable slides, used to clean the surface and to limit the area of the monolayer, made conveniently of strips of window glass about $20 \times 1$ cm, coated with paraffin wax, or from strips of Plexiglas or polystyrene (unplasticized). The distance between float and movable slide $J$ is given by scale $L$. For very accurate work it is convenient for $J$ to be driven by means of a carriage which is attached to a worm drive and the whole apparatus to be contained

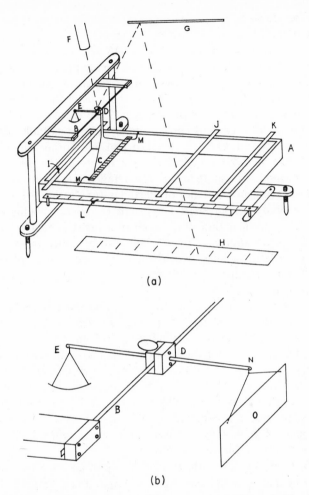

(a)

(b)

**Fig. 10.3**  (*a*) Film balance apparatus, using the horozontal-float modification; (*b*) The Wilhelmy hanging-plate modification.

in a Plexiglas box or wooden box with a glass window in the top. (Details of a suitable thermostating arrangement are given below.) With the handle of the worm drive outside, the box can be kept closed during the whole of an experiment. The film is confined between *J* and the floating threads, *M*, which consist of short lengths (1-2 cm) of thin nylon, terylene, or Vaselined cotton attached by means of small blobs of paraffin wax to the ends of the float and the adjacent sides of the trough. In attaching them care should be taken to prevent the wax from spreading onto the threads; otherwise troubles due to leaks will be encountered. Should a leak occur, this can usually be remedied by touching a heated pin at the point in question.

The essential differences between the horizontal and hanging-plate apparatus are shown in Fig. 10.3*b*. The side arm *DN* is conveniently made as long as *DE* to simplify the calibration, and the hanging plate, *O*, which is usually made of thin mica sheet, is attached to it by means of fine wire, so that it is about half immersed in the water. The dimensions of the mica plate depend on the sensitivity required, but usually 5 × 2 cm will be found convenient.

Calibration of both types is carried out by adding suitable small weights (usually up to about 1 g) to the balance pan and noting the deflections on scale *H*. Unless very large deflections are involved, the weight-deflection curve will be found to be quite linear. The detailed calculation is given below.

Troughs of glass, polystyrene, or Plexiglas (unplasticized) will be found adequate for almost all purposes. The plastic trough can readily be made up from sheet about ¼-in. in thickness, the strips being cemented to the base plate by means of plastic solution. Troughs may also be milled from a solid piece of Teflon. To avoid buckling which sometimes occurs with plastic troughs lacking structural rigidity, it is often necessary to screw the trough (from below) to a rigid base plate.

If the glass trough has a beveled edge, this must be ground flat and then made hydrophobic by careful coating of the previously heated surface with a solution of hard paraffin wax in light petroleum. (If any wax runs inside, this should be immediately removed by rubbing with filter paper.)

With polystyrene and Plexiglas the inherent hydrophobic nature of their surfaces makes this unnecessary. For most purposes, cleaning of troughs is best done by running tap water followed by careful wiping with filter paper. (With glass troughs, chromic acid is sometimes preferred.) If the movable slides are made of glass, they also have to be made hydrophobic by heating and coating with paraffin wax as above.

The surface pressure–area ($\Pi$–$A$) relationship of an insoluble monolayer is determined by means of the hanging-plate apparatus as follows. The trough is filled just to the brim with water, the mica plate hung in position (with tweezers), and the surface cleaned by sweeping several times with the movable barriers. The calibration is checked by adding a suitable small weight (e.g., 0.5 g) to the balance pan and then spreading a suitable volume of the solution of the film-forming compound on the surface as described above. The film is compressed by suitable amounts, the optical deflection being read at each point. From the known dimensions of the trough, the volume and concentration of the spread solution and from the calibration of the optical system, the surface pressure, $\Pi$, and area per molecule, $A$, at each point are readily calculated as shown below. The $\Pi$–$A$ curve is then plotted graphically.

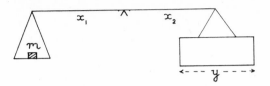

**Fig. 10.4**

The relationship between the surface pressure, $\Pi$, and the optical deflection, $d$, which can readily be deduced from Fig. 10.4, is

$$\Pi = \alpha d \qquad (10.2)$$

where $\alpha = mx_1g/2yx_2$, in which $g$ is the acceleration due to gravity; $x_1$, $x_2$, the lengths of balance arms (usually identical); $y$, the length of mica plate; $m$, the slope of weight-deflection curve (obtained by adding suitable known weights to the pan and observing the corresponding values of $d$).

For some purposes, particularly for the study of phase changes and their thermodynamics (see below), it is desirable to have a continuous record of the $\Pi$–$A$ curve, and several suitable devices have been published [4, 11–15].

The simplest systems are based upon the Wilhelmy-plate principle and employ photographic registration. Photographic registration has, however, a number of disadvantages, being rather slow and expensive if numerous curves are required. Instruments are available in which a photosensitive device is forced, under the control of a servo mechanism, to follow a light spot and a voltage proportional to the optical deflection is recorded but these are also rather expensive. Because of this disadvantage a simple direct registration apparatus was developed which has given good service in the laboratory of one of the authors [16]. A schematic diagram which indicates the principles of its action is given in Fig. 10.5.

The film is compressed by means of a small electric motor, $A$, operating through a reduction gear to drive a worm running longitudinally under the trough, the worm being attached to the carriage, $B$, carrying the movable barrier. Simultaneously, the recording drum, $C$, is caused to rotate by a thin wire rope fixed to the carriage and passing over a guide wheel to a nest of pulleys on the drum axis. These give a choice of drum speed and hence of area scale on the paper chart. The wire is kept taut by a weight attached to the free end. The surface pressure is measured as an optical deflection as described above for the manual balance, movement of the mica plate resulting in the displacement of the image of a set of cross wires along the axis of the drum. The paper chart is attached to the drum by a spring clip, and the actual recording of the force-area curve is done manually. As the cross-wire image moves over the rotating drum, a handle, $D$, is turned to maintain

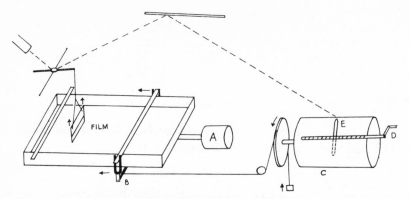

FILM

A

E

D

C

B

**Fig. 10.5**  Schematic diagram of recording film balance.

the tip of a pointer, $E$, on the cross wires. A pen fixed to the pointer and pressing against the drum then traces out the pressure–area curve. With a little practice it is possible to follow the movement of the cross wires smoothly and accurately.

Calibration is carried out as usual, the sensitivity being varied either by an alteration in the length of the mica plate or a change of the optical path length by means of the main mirror. The surface-pressure scale is linear provided the torsion system is operated with deflections within about 5° either side of the horizontal.

The whole instrument is mounted in a box with a partition separating the film balance from the recording drum and with windows in the roof for the optical system. The two free walls of the balance compartment are jacketed with copper tanks, and through these and a glass serpentine tube in the trough is pumped water from a thermostat. The thermostat tank can be heated by an immersed electrical heater or cooled by means of a refrigerating unit. For rapid control of the air temperature it is convenient to have fitted inside the box a small heater operated off a variable transformer and a small electric fan. A small sliding door on the front of the box enables the film to be spread with the minimum of temperature disturbance.

A very convenient and inexpensive system for the continuous recording of surface pressure has been obtained by one of the authors using a Sanborn direct current linear displacement transducer (DCDT) as shown in Fig. 10.6. The Sanborn DCDT is a contained unit consisting essentially of a linear variable differential transformer with a built in carrier oscillator and phase-sensitive demodulator. An excitation voltage of 20 to 28 V dc (dry cells) is required to produce a high-level dc output voltage proportional to the linear displacement of the core from a reference position (ca. 4 V/mm with an

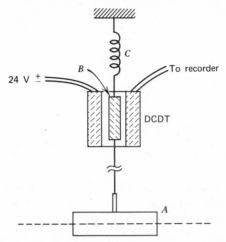

**Fig. 10.6** Schematic diagram for the continuous recording of surface pressure using a displacement transducer. $A$, Wilhelmy plate; $B$, core; $C$, coil spring.

output impedance of 2500 $\Omega$). A Wilhelmy plate, $A$, is suspended from the core, $B$, of the DCDT which is attached to one end of a coil spring, $C$. A racking device may be used to adjust the position of the DCDT relative to the core so that a null is obtained with a clean surface. Alternatively, a biasing potentiometer may be used. In so far as the coil spring is Hookean, the surface pressure, $\Pi$, is proportional to the output of the system which may be monitored on a simple strip-chart recorder. The sensitivity is determined by the dimensions of the plate, the stiffness of the spring and the characteristics of the Sanborn DCDT.

**Measurement of Surface Potential ($\Delta V$)**

The surface potential ($\Delta V$) is by definition the change in phase-boundary potential produced by the surface film; that is,

$$\Delta V = V_{\text{water}} - V_{\text{film}}. \tag{10.3}$$

Two principal methods are employed, the first using a static air electrode with an ionizing source [17], the second a vibrating plate [18]. Both methods are quite satisfactory, but the first will be found easier to construct in most laboratories and it also has certain experimental advantages.

The setup using the static air electrode is shown diagrammatically in Fig. 10.7. Any valve electrometer, $F$, of high impedance can be used, and quadrant electrometers such as the Lindemann and Compton have been used successfully. Many commercial high-impedance instruments used in the potentiometric measurement of pH with a glass electrode are entirely satisfactory and vibrating reed electrometer voltmeters are in common use.

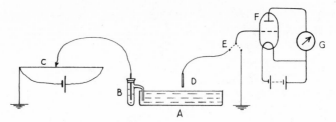

**Fig. 10.7** Diagrammatic setup for surface potentials by the radioactive air-electrode method. *A*, trough of film balance; *B*, half-cell; *C*, potentiometer; *D*, air electrode; *E*, switch; *F*, electrometer valve; *G*, galvanometer.

The half-cell, *B*, can be a calomel or Ag–AgCl electrode, although for most purposes a silver wire dipping into the solution will be found quite satisfactory. The air electrode, *D*, consists of a metal plate or wire (silver or platinum preferably) well insulated by polystyrene and located about 1–2 mm above the surface. The air gap is rendered conducting by the radiation from an $\alpha$-particle source such as polonium-210 or plutonium-234 and suitable fabricated $^{210}$Po air-ionizing sources are commercially available through both the American and British Atomic Energy authorities.

The potential with a clean water surface ($V_{water}$) is first measured by moving switch *E* to and fro until a balance is obtained on the galvanometer, *G*. The monolayer is then spread in the normal way and the process repeated, giving $V_{film}$ and hence $\Delta V$. The monolayer is compressed and the $\Delta V$–$A$ curve built up point by point.

In the vibrating-plate method a thin disk, *D*, of gold, gold-plated brass, or silver, is attached to the voice-coil of a loudspeaker, *E*, vibrating at a frequency of 200–1000 Hz and located as close as possible to the water surface. The general arrangement is shown in Fig. 10.8. The vibration of the electrode gives a condenser of variable capacity so that an ac signal is generated unless the metal and liquid surfaces happen to be at the same potential.

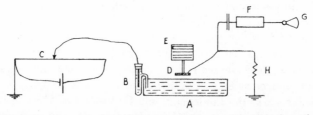

**Fig. 10.8** Diagrammatic setup for surface potentials by the vibrating plate method. *A*, trough of film balance; *B*, half-cell; *C*, potentiometer; *D*, vibrating plate; *E*, voice-coil of loudspeaker; *F*, amplifier; *G*, headphones or oscilloscope; *H*, high resistance (ca. 50 M$\Omega$).

This signal is amplified by the amplifier, $F$, and detected on either head-phones or an oscilliscope, $G$. By means of the potentiometer, $C$, the applied potential is varied until the output signal reaches a minimum, giving $V_{water}$. The monolayer is then spread and the determination repeated, giving $V_{film}$. The difference gives $\Delta V$ and hence the $\Delta V$–$A$ curve as above.

In both of the above methods adequate screening of leads is most essential, and the film balance is enclosed in a metal screened box with the minimum of openings.

The chief experimental advantage of the first method lies in ease of movement of the air electrode with respect to the surface, and hence its ability to explore the monolayer for homogeneity.

Surface potentials of monolayers are found to depend chiefly on two factors, namely, the concentration of molecules in the surface, and the nature of the polar group in the molecule. Hence $\Delta V$ results are usually written in the form:

$$\Delta V = 4\pi n\mu, \tag{10.4}$$

(formally analogous to the Helmholtz equation) where $n$ is the molecular density in molecules per square centimeter (i.e., $1/A$) and $\mu$ is a constant characterizing the polar group. The factor $\mu$ is often termed the "vertical component of the apparent surface moment" and is usually expressed in milli-Debye units (i.e., $10^{-21}$ esu).

## Measurement of Surface Viscosity $(\eta_s)$ and Surface Shear Modulus $(G_s)$

Despite their extreme thinness, monolayers exhibit appreciable viscosity and in some cases rigidity as well. A qualitative measure can be obtained in various ways [19], the simplest being by dusting a little talcum powder (this must be unscented!) on the surface and blowing gently with a "squeegee" or eye-dropper. Films exhibiting rigidity show an elastic recoil in this test, although it must be emphasized that careful blowing is necessary if the weaker rigidities are to be detected.

Turning now to quantitative techniques, the surface viscosity is usually measured by one of two methods, the choice depending on the system. For films of low viscosities the "surface canal" or "surface slit" apparatus is employed, whereas with those of high viscosities the oscillation method is to be preferred. A comprehensive review of these techniques has been given by Joly [20].

*Canal Viscometer* [20, 21]. The essential nature of the low-viscosity apparatus will be clear from Fig. 10.9. $A$ is the float of the film balance; $B$, the barrier with canal; I, initially a clean water surface; II, the film-covered surface; and III, a clean water surface. The difference of pressure between II and I, $\Delta\Pi$, is kept constant by moving barrier $D$ in Fig. 10.8$a$ or $B$ in Fig. 10.8$b$. It is

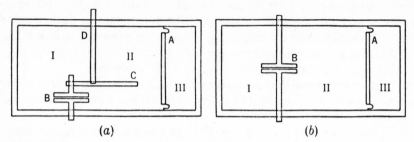

**Fig. 10.9**  Surface trough with canal (or slit) viscometer.

obvious that the horizontal or the vertical type of film balance may be used. The pressure of film I may be kept constant by the use of a piston oil, but it is much better to use a movable barrier (not shown) which automatically keeps the pressure constant. The barrier is moved by a device actuated by a film balance of the vertical type.

The surface viscosity $(\eta_s)$ is then calculated from the equation analogous to that for three-dimensional Poiseuille flow:

$$\eta_s = \frac{\Delta\Pi(d^3)}{12lQ}, \tag{10.5}$$

where $Q$ is the area of film passing through in unit time, and $d$ and $l$ are, respectively, the width and length of the channel. It has been found experimentally that this equation is obeyed only for very narrow channels, $Q$ then increasing less rapidly than expected as $d$ increases.

Several empirical modifications [4] to (10.5) have been proposed to correct for the drag of the underlying viscous liquid and the expression arising from the analysis of Harkins and Kirkwood [22] is generally applied:

$$\eta_s = \frac{\Delta\Pi(d^3)}{12lQ} - \frac{\eta^0 d}{\pi}, \tag{10.6}$$

where $\eta^0$ is the viscosity of the subphase liquid.

*Oscillation Viscometer* [20, 23].    Films in the condensed state, particularly those of proteins and polymers, tend to give high surface viscosities, and in such cases the oscillation method is employed. The oscillating system consists of a ring or disk, a plate on edge (e.g., a glass microscope slide), or a thick platinum wire, attached through a heavy bob, $A$, to a torsion wire (see Fig. 10.10). The upper end of the wire is attached to a torsion head, by means of which the system can be set into oscillation. A plastic guard tube, $B$, surrounds the torsion wire to exclude draughts. Using a lamp and scale,

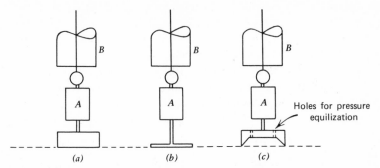

**Fig. 10.10**  Oscillating systems for surface viscosity: (*a*) glass slide just in contact with the surface; (*b*) thick platinum wire or glass disk; (*c*) a plastic ring or hollowed-out disk.

the amplitudes are measured, also the period, first with a clean water surface and then after spreading the monolayer. Some form of racking device, attached either to the top of the oscillating system or to the trough, greatly facilitates the necessary adjustments.

In the oscillating vane or needle modification, the use of a stout platinum wire as in Fig. 10.10*b* is preferred to a glass plate since it can be cleaned readily by heating in a small flame (a length of about 4 cm will be found convenient). A brass bob of moment of inertia about 80 g.cm$^2$ is used with a tungsten or steel torsion wire selected to give a period in air of about 10 sec.

The surface viscosity ($\eta_s$) is calculated from the expression [23]

$$\eta_s = 2.3(\lambda_{\text{film}} - \lambda_{\text{water}}) (4I/\tau l^2), \tag{10.7}$$

where $\lambda$ is the logarithmic decrement (the logarithm to base 10 of the ratio of successive amplitudes), $\tau$ is the period of oscillation, $l$ is the length of wire or glass plate, $I$ is the moment of inertia. ($I$ can be calculated approximately from the dimensions and weight of the bob, or more accurately by oscillation experiments in air as shown in any elementary textbook of physics.)

In addition to surface viscosity, the free-oscillation technique can give some information concerning the elasticity of a film. This can be detected by checking the period ($\tau$); with a purely viscous film $\tau$ is independent of the surface viscosity, whereas with films exhibiting viscoelastic properties $\tau$ is decreased.

With the oscillating-needle technique an elasticity index ($E_s$) has been defined [24] by

$$E_s = \frac{4\pi^2 I}{l^3}\left[\frac{1}{\tau^2} - \frac{1}{\tau_0^2}\right], \tag{10.8}$$

where the subscript refers to the clean interface. This equation will be found useful for semiquantitative and comparative purposes. If the damping is too rapid, it may be necessary to increase the moment of inertia of the bob, but with a resulting loss in sensitivity. The oscillating vane or needle modification is not amenable to hydrodynamic analysis, but may be adequate for some studies (e.g., detection of phase changes).

In the preferred modification (Fig. 10.10c) a ring or disk of radius $a$ is allowed to oscillate in a surface contained within a concentric guard ring of radius $b$. The surface viscosity $(\eta_s)$ and the surface shear modulus $(G_s)$ may be evaluated from the following equations given by Tschoegl [25]:

$$\eta_s = \frac{I}{\tau}\left(\frac{1}{a^2} - \frac{1}{b^2}\right)\left[\frac{\lambda}{4\pi^2 + \lambda^2} - \frac{\lambda_0}{4\pi^2 + \lambda_0{}^2}\right], \qquad (10.9)$$

and

$$G_s = \frac{I}{4\pi}\left(\frac{1}{a^2} - \frac{1}{b^2}\right)\left[\frac{4\pi^2 + \lambda^2}{\tau^2} - \frac{4\pi^2 + \lambda^2}{\tau_0{}^2}\right]. \qquad (10.10)$$

*Rotation Viscometers.* Fourt and Harkins [23] have determined the viscosity of films from the measurement of the torque required to maintain a uniform rate of rotation of a ring or disk. The shaft of an electric clock is used as the upper suspension; the twist between the two ends of the torsion wire is measured by two beams of light reflected from mirrors mounted on the support of the wire and on the axis of the ring, respectively. These are focused on the same scale and give a single line when the shaft is not in rotation. The torsion constant, $\kappa$ dyne/radian, may be determined by the method of oscillations. The torque may be obtained by timing the interval, $t$, between the passage of the two beams of light past a fixed mark. If the viscosity of the film is Newtonian, independent of the rate of shear, the equation for viscosity does not include a term for the rate of rotation of the ring. If $\Delta t$ is the change in the interval produced by the film, the viscosity is given by

$$\eta_s = \frac{\kappa \Delta t}{4\pi}\left[\frac{1}{a^2} - \frac{1}{b^2}\right] \qquad (10.11)$$

where $a$ is the radius of the ring or disk and $b$ is that of an outer guard ring. Anomalous viscosity or plasticity gives a dependence upon the rate of shear. In later work the interval of time, $\Delta t$, was measured by a counter started when the first beam of light passed a photocell, and stopped when the second beam fell upon the counter. The method of the rotating ring is to be preferred if the flow of the film is of the plastic type.

For the determination of shear rigidity of the more rigid films, where the rotational techniques are not satisfactory owing to severe breakdown of the film, a method based on that of Mouquin and Rideal [26] is employed [27] (see Fig. 10.11). A torque is exerted on the monolayer, $A$, by means of a disk,

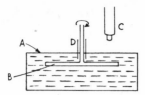

**Fig. 10.11**   Apparatus for measurement of surface shear modulus.

$B$, rotating at a suitable distance beneath the surface, and the resultant deflection produced on talcum particles in the film is measured by means of a microscope, $C$, fitted with a micrometer eyepiece. The method also indicates whether plastic flow is occurring, for in this case the particle does not return to its original position when the disk stops rotating. The disk and spindle are made of Plexiglas or polystyrene, the spindle passing through a loose glass sleeve, $D$, in the surface. By varying the speed of rotation and depth of immersion a very wide range of torques can be used. With the thicker films of proteins obtained by absorption from solutions this technique has proved very satisfactory [27].

The surface shear modulus ($G_s$) is calculated from

$$G_s = \frac{\omega \eta^\circ}{8 \delta h} (a^2 r - r^3) \qquad (10.12)$$

where $\delta$ is the displacement, $\omega$, the angular velocity of rotating disk, $h$ the depth of disk beneath the film, $r$, the distance between point of observation and center of disk, $a$, the radius of dish containing the film, and $\eta^\circ$, the viscosity of the solution.

*Surface Rheometer Using Forced Oscillations.*   A very sensitive surface rheometer [28] has been developed by one of the authors employing a dynamic technique analogous to that frequently used to study the viscoelastic properties of liquids and gels [29]. A surface film is contained in the annular spacing between two concentric rings of radius $a$ and $b$. The outer ring is clamped and the inner ring is suspended by a fine torsion wire from a coil held in the field of a permanent magnet. A sinusoidally varying current is passed through the coil and the torsion head is forced to oscillate. The resultant amplitudes of the torsion head and the oscillating ring, $\phi_0$ and $\theta_0$, respectively, are observed with the aid of two small mirrors as in the Fourt-Harkins rheometer [23]. The phase difference ($\alpha$) between the two "signals" is measured using two photo-sensitive devices and an electronic counter.

The viscoelastic properties of the film, defined in terms of the dynamic surface viscosity, $\eta_s{}'$, and the surface storage modulus, $G_s{}'$, are obtained as functions of frequency, $f$, in the range 0.008 to 1.0 Hz using the relations,

$$\frac{\kappa\phi_0}{\theta_0} \sin \alpha = 2\pi f A(\eta_s' + \eta_0') \tag{10.13}$$

$$\frac{\kappa\phi_0}{\theta_0} \cos \alpha = AG_s' + \kappa - 4\pi^2 f^2 I, \tag{10.14}$$

where $\kappa$ is the torsion constant of the suspension in dyne/radian, $I$ is the moment of inertia of the ring or disk, $A$ is a geometric factor defined as $A = 4\pi/(1/a^2 - 1/b^2)$, and $\eta_0'$ is a term to allow for the viscous drag of the underlying liquid. The apparatus constants may be determined from measurements made in the absence of a film (i.e., when $\eta_s'$ and $G_s'$ are zero).

The inertia of the system is kept to a minimum, and the torsion wire is selected to give the required sensitivity (the inertia of the driving coil is not involved). The advantages of this technique operating under steady-state conditions are several. The amplitude of the disk or ring may be kept to a minimum and nonlinearities may be detected by varying the amplitude of the oscillations. (For systems employing free oscillations it is customary to assume that the viscoelastic behavior of the film is independent of frequency and amplitude.) The ordinary steady flow viscosity is equal to the limiting value of $\eta_s'$ as the frequency approaches zero.

## 3   EXAMPLES OF INVESTIGATIONS USING AIR–WATER MONOLAYERS

For our present purposes some grouping is essential, and the following will be adopted: (a) principal types of surface films; (b) determination of structure of organic molecules; (c) determination of molecular weight of organic molecules; (d) reactions at interfaces; (e) thermodynamics of spreading; (f) kinetics of spreading; (g) diffusion through interfaces; (h) analytical applications; (i) miscellaneous applications.

### Principal Types of Surface Films

From the study of monolayers of various compounds, carried out largely by Adam and his co-workers, it has been found possible to group films into three main types: gaseous, expanded, and condensed [30]. Of these the first and last have obvious analogies in the three-dimensional world.

Gaseous (and vapor) films usually obey the equation of state, $\Pi(A - A_0) = xkT$, where $A_0$ is the "co-area" and $x$ a constant. (The Amagat equation for an imperfect gas in three-dimensions is $P(V - b) = xkT$, where $b$ is the "co-volume.") As the chain length is increased, $x$ decreases, since the van der Waals attractive forces increase. The similarity with imperfect gases is shown very clearly by plotting $\Pi A$ against $\Pi$, corresponding to the $PV$ versus $P$

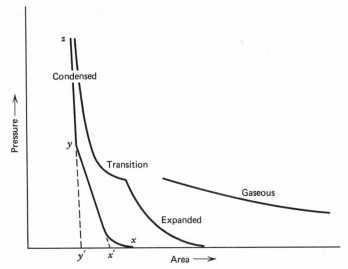

**Fig. 10.12**   Typical pressure–area curves for the condensed, expanded, and gaseous types of monolayer.

plot in three dimensions. At extremely low pressures ($<$ ca. 0.01 dyne/cm as a rule) the ideal limit, $\Pi A = kT \cong 400$ (at room temperature), is approached, and, as shown below, this enables molecular weights to be calculated.

Expanded films, occupy an intermediate position between the gaseous and condensed states (see Fig. 10.12). For example at room temperature lauric acid ($C_{12}$) and palmitic acid ($C_{16}$) give, respectively, gaseous and condensed type film, whereas, myristic acid ($C_{14}$) gives an expanded film at low pressures changing into a condensed film as the pressure is increased.

An equation of state, $(\Pi - \Pi_0)(A - A_0) = xkT$, is usually obeyed quite accurately, the value of $x$ being close to unity for un-ionized molecules and greater than unity for ionized ones. (With 1–1 electrolytes the value approaches 2.) The above equation, in which $A_0$ is the "co-area" and $\Pi_0$ is the (negative) spreading pressure arising from the attractive forces between the paraffin chains, was deduced by Langmuir by assuming that an expanded film can be regarded as a duplex film (see Section 6) of extreme thinness (15–20Å ), the upper surface consisting of hydrocarbon chains in contact with air, the lower of polar groups in two-dimensional kinetic agitation.

In the condensed type of film the molecular packing approximates that in the crystalline solid; stearic acid, for example, giving a coherent condensed monolayer from ca. 23 to ca. 19 Å$^2$, which compares with a cross-sectional area of 18.5–20 Å$^2$, (depending on the temperature) as deduced from $X$-ray diffraction measurements on the crystal. The $\Pi$–$A$ curve is frequently made up of two almost linear portions, of which the $XY$ portion (see Fig. 10.12) seems to be largely determined by the size and packing of the polar group, and the $YZ$ region by the compression of vertically orientated paraffin chains. A closer study of condensed films, using the recording film balance, has suggested that in some cases as many as five different phases may exist.

The two main condensed regions also differ in their mechanical properties. Films in the $XY$ region appear to behave as two-dimensional Newtonian liquids obeying the relation

$$\log \eta_s = \log \eta_0 + \alpha\Pi \tag{10.15}$$

where $\eta_0$ and $\alpha$ are constants and $\Pi$ is the surface pressure. In the $YZ$ region the behavior may range from that of a liquid showing anomalous viscosity (e.g., ketones and esters) to that of a solid of extreme rigidity (e.g., amides), the difference being determined largely by the interaction between the head groups, as will be shown later.

### Determination of Structure of Organic Molecules

Surface techniques have found some definite applications in this field and, as with all physical techniques, possess a number of advantages and limitations. Chief advantages are the simplicity of the apparatus and the small amounts of material (a few milligrams) required for a full monolayer and multilayer study (see also Section 4). The materials need not be crystalline nor chemically homogeneous, although naturally the latter is desirable if possible. The chief limitations arise from the requirements for stable spreading, which in general means that the substance must possess one or more polar groups and a hydrocarbon portion of sufficient magnitude to confer insolubility. By spreading on strong salt solutions (to reduce solubility limitations as mentioned earlier), by varying the temperature, or by spreading with a substance which itself forms a stable monolayer, the range of usefulness can be very considerably increased (see Section 5).

The information to be gained from surface studies can best be assessed from the actual practical examples which are listed below, but in general terms it can be said to indicate not only the general shape of a molecule but also the nature and disposition of the polar groups when more than one is present.

The following list will give some idea of the wide variety of substances which have been examined in the form of monolayers, and in may instances useful information concerning structure, stereochemistry, composition, orientation, and so on, has been forthcoming [4, 31].

1. Sterols, such as cholesterol, ergosterol, lanosterol.
2. Pentacyclic compounds, such as $\alpha$- and $\beta$-amyrin, lupeol, betuline, oleanolic acid.
3. Tricyclic compounds, such as pimaric acid, abietic acid.
4. Chlorophyll, hemin, and other porphyrins.
5. Carcinogenic compounds in mixed films with sterols.
6. Saturated fatty acids from acid-fast bacteria, and from the heartwood of *Thuja plicata* D. Don.
7. Unsaturated fatty acids from horse brain [32].
8. Polymers, such as acrylates and methacrylates, polypeptides, siloxanes.
9. Copolymers.
10. Proteins 33, 34.
11. Polymyxins.
12. Insect cuticle waxes [35].
13. Oxidized mineral oils [36].
14. Fat-soluble vitamins A, E, and K.
15. Phospholipids.

## Determination of Molecular Weight of Organic Molecules

At sufficiently low pressures all monolayers pass over into gaseous films' which eventually tend to approach the ideal gas equation, $\Pi A = kT$ (for un-ionized compounds or for ionized compounds spread on salt solutions). Experimentally this limit is usually obtained by extrapolation of the $\Pi A$ versus $\Pi$ curve, and from the value of $(\Pi A)_{\Pi=0}$ the molecular weight, if unknown, can be calculated as first shown by Guastalla [8]. (A film balance is merely a two-dimensional osmometer, and the calculation therefore follows the usual three-dimensional form except that $\Pi$ and $A$ replace, respectively, the pressure and volume terms.)

The experimental technique is not so simple as that at the higher pressures, but it has nevertheless been used in a number of cases, its main advantages being speed, the small quantity of material required, and suitability for insoluble materials. It is probably of most value with polymeric materials of intermediate molecular weight (ca. 1000–10,000), that is, in the range where most of the standard methods have their own peculiar difficulties.

Examples where this method has been used include certain proteins (e.g., egg albumin, hemoglobin, insulin), certain breakdown products from proteins (e.g., from insulin and wool keratin), the antibiotic polypeptides, polymyxins, and certain polymers [8, 9, 37]. One interesting application is in the study of the breakdown of proteins into smaller subunits, although the surface behavior does not necessarily exactly parallel the same phenomenon in bulk.

## Reactions at Interfaces

A great many reactions, in industry, in the laboratory, and in biological systems, are known to occur at an interface. Such heterogeneous reactions are frequently difficult to study quantitatively in bulk, owing to uncertainties connected with the interfacial area and its variation with time. The hydrolysis of fats by aqueous NaOH illustrates the problem very well, for as reactton proceeds the liberated soap increases the emulsification and thus increases the area available for reaction. By utilizing monolayer techniques, in those cases where applicable, this disadvantage can be overcome, and much interesting information has resulted. In addition the effect of molecular density in the surface and of molecular orientation can be followed, both of which would be difficult or impossible in three-dimensional systems.

The kinetics of the reaction can be found by following the changes in area ($A$), surface potential ($\Delta V$) or surface moment ($\mu$) with time ($t$), the variable chosen in accordance with the system. (It is preferable to study the reactions at constant $\Pi$ rather than at constant $A$, since the velocity constant frequently varies very rapidly with $\Pi$.) If the products disappear rapidly from the film, the $A$–$t$ curve is used; if the products remain ın the film, either the $A$–$t$ or the $\mu$–$t$ curve can be used. In the case of these mixed films, it is desirable to study the variation in $A$ and $\mu$ with composition, so that deviations from additivity can be allowed for if necessary.

Before outlining the types of reaction which have been studied, one general finding should be mentioned: in almost all cases the velocity constant is very dependent on the molecular orientation; that is, a true "steric factor" exists in such cases.

The types of monolayer reactions so far studied can be grouped as follows [4, 38]:

1. Oxidation of unsaturated fatty acids (*cis* and *trans* isomers; compounds containing one, two, or three unconjugated double bonds), dihydroxy fatty acids, and dialkyl sulfoxides by dilute acid permanganate.
2. Autoxidation of drying oils and conjugated unsaturated fatty acids.
3. Hydrolysis of esters and lactones by alkali.
4. Lactonization.
5. Bromination of long-chain phenols.
6. Photochemical hydrolysis of long-chain amides and of proteins.
7. Biological reactions, particularly enzymic and immunological.
8. Polymerization.
9. Exchange reactions using radio-tracer techniques.

## Thermodynamics of Spreading

Corresponding to the saturation vapor pressure in three dimensions, we have the equilibrium spreading pressure ($\Pi_s$) in two dimensions, that is, the surface pressure at which there exists equilibrium between monolayer and the bulk compound. Equilibrium spreading pressures are of considerable importance since they define the conditions under which compressed monolayers are thermodynamically stable and also enable the latent heat of spreading ($\lambda$) to be calculated from the two-dimensional Clapeyron equation:

$$\lambda = T \frac{\partial \Pi_s}{\partial T} (A_F - A_B), \qquad (10.16)$$

where $A_F$ and $A_B$ are the areas occupied by film and bulk compound, the latter being in general negligible.

Precise $\Pi-A$ and $\Pi_s-T$ measurements are essential if accurate values for $\lambda$ are to be obtained. A comprehensive study of fatty alcohols, particularly hexadecanol, has been carried out by Brooks [39].

In addition to the above, other aspects of the thermodynamics of monolayers have received some consideration, such as the free energies, entropies, and heats of spreading and expansion [39, 40]; the question of the order of phase changes and the thermodynamics of the penetration of an insoluble monolayer by soluble surface active substances have been considered [41, 42].

## Kinetics of Spreading

A quantitative measure of the *rate* at which liquid drops or crystals spread to form a monolayer is important for certain purposes. Spreading from liquids is usually much more rapid (often as high as 10–20 cm/sec) than from solids and is usually measured by cine photography, the edge of the advancing film being shown by an inert powder such as talc or lycopodium [3, 43]. Some interesting hydrodynamic studies have been made using this technique, one important result being that there is no "slip" between the spreading film and its aqueous subphase.

## Diffusion through Interfaces

The possible influence of monolayers upon diffusion across a phase boundary is of very considerable interest to the biologist and has practical application in connection with liquid–liquid extraction and with attempts to reduce evaporation from storage reservoirs in arid regions [3, 4].

The system most studied to date has been the effect of monolayers upon water evaporation, and it has been found that only condensed monolayers appear to have any marked effect. Long-chain alcohols are particularly effective, and the use of cetyl alcohol is being tried out in large-scale experiments in Australia. (The effect of thicker films upon diffusion across an interface is referred to later, in Section 6.)

Various techniques have been used to study the diffusion of foreign gases such as $O_2$, $CO_2$ and $H_2S$ through monolayers at gas–liquid interfaces [4, 44–47].

### Analytical Applications

The simplicity, rapidity, and accuracy of surface-film measurements have led to their utilization for analytical purposes in a number of cases. Examples are the measurement of very small quantities of lipoids, proteins, polypeptides, and so on, and the determination of the surface area of fine powders.

Proteins and polypeptides can be spread quantitatively on water or strong salt solutions (the latter being necessary with lower molecular weight polypeptides) as detailed in Section 2. These materials occupy about 1 m²/mg when close-packed (at, say, $\Pi = 1$ dyne/cm), so that with the simple film balance a quantity of 0.01 mg can readily be measured with an accuracy of ca. 1%. With some care this could be extended, without much loss of accuracy, down to 0.001 mg.

The determination of surface area of fine powders is of considerable industrial importance, as, for example, with paint pigments. The method may be illustrated by some recent measurements with titania, using stearic acid or octadecyl alcohol in benzene solution as the surface-active agent. To a weighed quantity of dry pigment (say 2 g) a solution of stearic acid in benzene (say 10 ml) is added and shaken until equilibrium is attained. (A few minutes is usually sufficient.) The concentration of the solution before and after adsorption is measured by spreading a known volume (diluted where necessary) on the film balance. By this means relative surface areas can readily be found. If absolute surface areas are required, it is necessary to measure the full adsorption isotherm (i.e., to repeat the above over a range of concentrations) to obtain the saturation adsorption. From this, and assuming an area of 20 Å² per molecule at saturation adsorption, the absolute surface area is readily calculated. An important point of technique is to ensure that the powders are carefully dried, since water interferes with the adsorption process.

### Miscellaneous Applications

The surface-film technique enables a study to be made of numerous problems of general interest of which two may be mentioned here, namely, the interaction of particular polar groups when immersed in an aqueous environment and the interaction between a soluble compound and an insoluble monolayer.

In view of its great relevance in protein structure, the system so far studied in most detail has been the association between the $\diagup C{=}O$ and $H{-}N\diagdown$ groups, usually ascribed to hydrogen bonding. The energetics of the process

$$\left(\diagup C{=}O\right)_{aq} + \left(H{-}N\diagdown\right)_{aq} \rightleftharpoons \left(\diagup C{=}O\cdots\cdots H{-}N\diagdown\right)_{aq}$$

as well as the influence of such factors as pH, has been studied in this way, using monolayers of long-chain ureas, amides, and so on [48, 49].

The interaction of water-soluble compounds such as soaps and synthetic detergents with monolayers of insoluble compounds such as long-chain alcohols, cholesterol, and so on, has provided much useful information on the structure of emulsions. The usual procedure is to spread the monolayer first and then to inject the water-soluble material beneath it, following the changes in area and surface potential as well as the viscosity and shear modulus where appropriate [4, 50].

## 4  MULTILAYERS ON SOLID SURFACES

Certain monolayers in the condensed state can be transferred by a simple dipping process to a slide which may be of metal, glass, mica, and so on, or even of fine wire mesh. (In the last case the film often bridges the interstices like a soap bubble.) The essentials of the process are shown in Fig. 10.13. The trough, $A$, is first filled and a Vaselined silk thread then attached to two opposite sides by means of blobs of paraffin wax. The side on which the film is to be spread, $B$, is cleaned by sweeping, and the monolayer (e.g., of methyl stearate, octadecyl acetate, stearic acid, etc.) spread in the usual way. To the other side of the thread, $C$, a drop of a "piston oil" is then added, and the monolayer compressed by the movable glass slide, $D$. (A "piston oil" is a liquid giving a constant surface pressure, common ones being oleic acid (ca. 32 dyne/cm), castor oil (ca. 16 dyne/cm), and tricresyl phosphate (ca. 10 dyne/cm).) The slide, $E$, previously made hydrophobic by polishing with a little ferric stearate or rubbing down an earlier multilayer, is then dipped slowly in and out through the monolayer. The movement of the

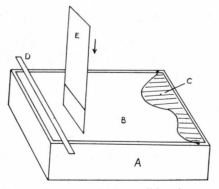

**Fig. 10.13** Transfer of a monolayer on water to a solid surface.

thread shows whether deposition is occurring and the type of deposition, $X$, $Y$, or $Z$ ($X$, deposition on immersion only; $Y$, deposition both ways, and $Z$, deposition on removal only).

Under some conditions the process can be repeated almost indefinitely, and the technique can be used to prepare thin uniform films of accurately known thickness containing an integral number of monolayer steps. (In the case of stearic acid each monolayer contributes ca. 24.5 Å) Such films have been employed in the study of boundary lubrication, as absorbing screens in nuclear physics, as electrode coatings in the study of electrical discharges and to induce crystallization in normally intractable substances such as amorphous polymers and unsaturated fatty acids. Their optical properties are of some interest, for example, in connection with their interference colors and in the preparation of "nonreflecting" glass [4, 38, 51].

One interesting application is the use of a "step film" to measure the thickness of thin transparent objects (e.g., red cell "ghosts") or of thin films on water. A "step film" is made by first depositing a certain number of layers (say 18 $Y$ layers of barium stearate) on the metal slide; this gives a yellow interference color. On further immersions the slide is dipped progressively less deeply, given the structure shown in Fig. 10.14, with interference colors which vary with the step thickness. By matching the colors of a step-film made from a known substance such as barium stearate (a $Y$ layer of which is 48.8 Å thick) against those of the unknown also in the form of a step-film, it is possible to measure the monolayer thickness of the unknown. (For accurate work a monochromatic source rather than white light is used, and the refractive index of the unknown is allowed for.) If, for various reasons a step-film cannot be built up, the layer is deposited on top of one step-film of barium stearate and compared with another step-film of barium stearate or some other compound of known thickness.

## 5   INSOLUBLE MONOLAYERS AT THE OIL–WATER INTERFACE

Interest in the behavior of films at oil–water interfaces arises chiefly in connection with emulsions and certain biological problems. The compounds which can be quantitatively examined as insoluble monolayers are compara-

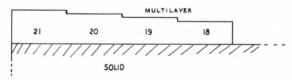

**Fig. 10.14**   Multilayer in the form of a "step-film."

tively few but include proteins, polar polymers (e.g., synthetic polypeptides, acrylates), long-chain detergents (e.g., sodium hexadecyl sulfate), and phospholipids (e.g., lecithin). In most of the work published to date comparatively nonpolar oils, such as petroleum ether, benzene, and carbon tetrachloride, have been used. These have the advantages of ready purification and comparatively high interfacial tension against water.

## Methods of Spreading

The desirable properties of the spreading solvent are that it shall be capillary-active, freely soluble in one of the two phases with a density intermediate between those of the two phases. With water-soluble substances such as proteins and detergents, addition of propyl alcohol usually suffices (e.g., 60% PrOH, 0.5 $M$ CH$_3$COONa for many proteins); with compounds soluble in polar solvents, a judicious dilution with a nonpolar liquid.

The spreading solution is injected into the interface by means of the "Agla" syringe (shown in Fig. 10.2), the needle being brought into the interface and then withdrawn so that the tip tends to extend the meniscus slightly. For this purpose it is essential that the syringe be firmly held in some form of racking device.

## Measurement of Pressure–Area (Π–A) Relationship

In the majority of studies, monolayers at an $O-W$ interface have been studied using an incremental spreading technique. In this method, the interfacial area is fixed and the surface pressure is measured for different amounts of spread material [53]. An experimental setup of adequate accuracy for most purposes is shown in Fig. 10.15. The basic torsion balance unit, $A$, is the same as that described above for air–water films, although there are some modifications. Two principal methods for measuring the surface pressure have been used. The first utilizes the ring method for interfacial tensions, a platinum ring of ca. 3-cm diameter and ca. 1-mm wire thickness being generally suitable.

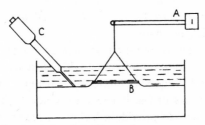

**Fig. 10.15** Apparatus for the study of $O-W$ monolayers.

The maximum pull on the ring, $B$ for the clean $O–W$ interface is first measured; then the interfacial tension ($\gamma_0$) can be calculated as shown in Chapter IX. The film is then spread by means of the Agla syringe, $C$, as described above and the operation repeated, giving $\gamma$, whence the surface pressure $\Pi = \gamma_0 - \gamma$. The volume spread being known from the Agla and the diameter of the dish being known, the area per molecule, $A$, is readily calculated. In order to reduce A, further quantities of spreading solution are injected into the interface and the measurement of $\gamma$ repeated; thus the $\Pi–A$ curve is built up point by point.

The second method of measuring the surface pressure is based on the Wilhelmy plate method as described above for $A–W$ films. An oil less dense than water is generally used, and the glass or mica plate is used in a hydrophilic state; this requires due attention to cleaning in preparation. With an oil heavier than water it is preferable to make the plate hydrophobic by coating it with a layer of carbon black. Since the normal movement of the plate (due to a reduction in $\gamma$) is upwards, there is little difficulty in maintaining a zero contact angle. In both techniques, calibration is carried out as previously described for $A–W$ monolayers.

The method involving the successive additions of the film-forming compound to an interface of constant area has two main defects: (*a*) the film is spread against an increasingly high film pressure; and (*b*) it is difficult to clean the interface satisfactorily. The alternative method in which the amount of spread material is fixed and the $\Pi–A$ relationship obtained by reducing the area of the interfacial film is preferred for more accurate studies. The interfacial film area may be varied by deforming a plastic ring within which the film has been spread [54, 55]; with this method the compression ratio is limited.

An interfacial trough, suitable for oils heavier or lighter than water has been designed by Brooks and Pethica [56] in which the interface can be swept as with a conventional $A–W$ trough (see Fig. 10.16). Two lengths of 3/16-in. Pyrex plate-glass, $A$ are held parallel to each other by fusion to glass rods, $B$. The top edges, $C$, of this frame are ground planar. Glass barriers, $D$, are used. The frame is placed in a shallow Pyrex vessel, $E$, supported on the base of a conventional surface-balance. The barriers are moved by glass arms clamped to the racking mechanism.

Before use, the cleaned trough and barriers are rendered hydrophobic by treatment with a 2% solution of dimethyldichlorosilane in carbon tetrachloride. Then, when the aqueous phase is the heavier, the inside ($F$) of the frame and the undersides of the barriers are made hydrophilic by rubbing with fine abrasive paper, to provide a leakproof seal at the barriers. Alternatively, when the oil phase is the heavier, the ground edge, $C$, and the edges

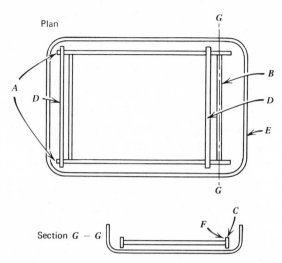

**Fig. 10.16**   The Brooks-Pethica trough for use at the $O-W$ interface. (Reproduced from *Trans. Faraday Soc.*, **60**, 209 (1964)).

of the barriers are made hydrophilic to eliminate overflow. When the apparatus is not in use, the $O-W$ interface is maintained at its normal level, so that the surfaces of the trough and barriers retain their appropriate hydrophilic or hydrophobic natures. The trough has been used successfully with several organic liquids.

Film pressure $\Pi$ is measured with a hydrophobic mica hanging-plate, held in a null position in the interface by torsion wire and a glass weight. The plate is coated with carbon-black from a butane–air flame. The plate is completely covered with a film of the organic phase, and the contact angle is zero [56].

## Measurement of Interfacial Potential ($\Delta V$)

The spreading of a film at an $O-W$ interface produces a change ($\Delta V$) in the phase boundary potential, just as for $A-W$ systems, but in this case the potential tends to decay at a rate determined by the specific resistance of the oil. With oils of high resistance, such as light petroleum and benzene, this decay process appears to be too slow to cause any serious complication.

Two methods of measurements have been employed based on those previously developed for $A-W$ monolayers (see Section 2). In the first a thin film (1–2 mm) of the oil is used with the ionizing air electrode just above the oil surface and a half-cell dipping into the aqueous phase. A large polonium electrode (ca. 1 cm² in area) and a thin oil layer are desirable to keep the resistance to a minimum. The potential ($V_0$) with the clean $O-W$ interface is measured and, after the film has been spread, the procedure is repeated giving $V_{film}$ and hence $\Delta V$.

The second method employs a vibrating plate placed in the oil phase as close to the interface as possible and attached to a suitable amplifying circuit [52].

## Measurement of Surface Viscosity and Surface Shear Modulus

The quantitative measurement of surface viscosity of $O-W$ films is best done by means of the oscillating ring or disk technique (p. 570), the needle lying in the interface with the brass bob in the air. Oscillations are first made with the clean interface; the film is then spread and the operation repeated. This method has been found particularly suitable for proteins and polymers where appreciable viscosities are encountered. For very rigid systems the Mouquin-Rideal method (described on p. 572) may be used to measure the surface shear modulus.

## Applications of Oil–Water Monolayers

Investigations of $O-W$ monolayers have so far been extremely limited; nevertheless a number of important problems have been solved, in particular the effect upon the surface properties of a monolayer when oil replaces air. (This is particularly important since previously the behavior at $O-W$ and other interfaces had to be inferred from the study of $A-W$ films.)

The $\Pi-A$ curves are usually of the gaseous type, a finding which is not surprising since the small oil molecules would tend to penetrate between the film molecules and thus reduce the van der Waals cohesive forces. Simple compounds such as detergents obey an equation of state, $(\Pi - \Pi_0)(A - A_0) = C$ (a constant). This is of the same form as that for expanded films at an $A-W$ interface, and where comparison is possible it is found that transition from an $A-W$ to an $O-W$ interface has little effect on $A_0$ (the co-area term) but reduces $\Pi_0$ (the cohesive term) to negligible values.

In some cases the $O-W$ monolayer can be compressed to about the same packing as at the A–$W$ interface, suggesting that at high values of $\Pi$ the oil molecules are largely squeezed out from the interface. This conclusion is of considerable importance in consideration of the structure of emulsions. With the uni-univalent ionized compounds (e.g., $C_{16}H_{33}SO_4^-Na^+$) the value of $C$ seems to depend on the area and salt concentration in the aqueous phase, approaching $2kT$ (i.e., ca. 800) at large areas in the absence of salt and $kT$ (i.e., ca. 400) in the presence of salt.

Values of $\Delta V$, at a given molecular area $A$, appear to be very similar to those at the $A-W$ interface, showing that the orientation of the polar groups is little influenced by the oil phase, at any rate in the systems so far studied.

As regards mechanical properties, the presence of the oil phase reduces the viscosity and shear modulus in the few systems (e.g., proteins) where comparison has so far been made.

## 6   DUPLEX FILMS

A pure paraffin oil does not spread on water but will do so if a small amount of polar organic compound, such as oleic acid, is added. Suitable addition gives an extremely thin layer, often only a few microns in thickness and showing interference colors. (The vividly colored films on puddles and wet surfaces are common examples, the polar molecules arising from oxidation or from "dopes" added to the lubrication oil.) These films are still quite thick by molecular standards and hence are termed "duplex."

Duplex films have been of considerable interest in connection with the study of adsorbed films at the $O-W$ interface and the structure of expanded films at the $A-W$ interface. They have practical importance in combating mosquito larvae in malarial regions and may find some use as a means of reducing water evaporation.

It would appear that all duplex films are thermodynamically unstable, the equilibrium state being a number of lenses surrounded by a monolayer of the polar spreader. With some polymerized spreaders, such as oxidized oleic acid, and with certain dyes, such as malachite green, the duplex film may appear to be quite stable, but this is doubtless due to the rigid interfacial films found in such systems. For thermodynamic stability the final spreading coefficient $\Pi_s'$ must be positive. $\Pi_s' = \gamma_w - (\gamma_o + \gamma_{ow})$ where $\gamma_w$ is the surface tension of the monolayer covered surface, $\gamma_0$ that of the oil, and $\gamma_{ow}$ the interfacial tension between oil and water.

## References

1. N. K. Adam, *The Physics and Chemistry of Surfaces*, 3rd ed., Oxford University Press, Oxford, 1941.
2. A. E. Alexander and P. Johnson, *Colloid Science*, Oxford University Press, Oxford, 1949.
3. J. T. Davies and E. K. Rideal, *Interfacial Phenomena*, Academic, New York, 1961.
4. G. L. Gaines, Jr., *Insoluble Monolayers at Liquid-Gas Interfaces*, Interscience, New York, 1966.
5. D. W. Stephens, *Gas/Liquid and Liquid/Liquid Interfaces: A Bibliography*, Crosfield, Warrington, England, 1962.

6. D. J. Crisp, *J. Colloid Sci.* **1**, 49, 161 (1946).
7. A. E. Alexander, *Nature* **159**, 304 (1947).
8. J. Guastalla, *Compt. Rend.* **208**, 1078 (1939).
9. A. J. G. Allan and A. E. Alexander, *Trans. Faraday Soc.* **50**, 863 (1954).
10. B. S. Harrap, *J. Colloid Sci.* **9**, 522 (1955).
11. D. G. Dervichian, *J. Phys. Radium* **6** (7), 221 (1935); Ph.D. thesis, Masson et Cie., Paris, 1936.
12. K. J. I. Andersson, S. Stallberg-Stenhagen, and E. Stenhagen, in *The Svedberg 1884–1944*, A. Tiselius et al., Eds., Almqvist and Wiksell, Uppsala, 1944, p. 11.
13. P. A. Anderson and A. A. Evett, *Rev. Sci. Instr.* **23**, 485 (1952).
14. H. J. Trurnit and W. E. Lauer, *Rev. Sci. Instr.* **30**, 975 (1959).
15. W. Rabinovitch, R. F. Robertson, and S. G. Mason, *Can. J. Chem.* **38**, 1881 (1960).
16. A. R. Gilby, Ph.D. thesis, University of New South Wales, Sydney, 1955.
17. J. H. Schulman and E. K. Rideal, *Proc. Roy. Soc.* (*London*) **130A**, 259 (1931).
18. H. G. Yamins and W. A. Zisman, *J. Chem. Phys.* **1**, 656 (1933).
19. I. Langmuir and V. J. Schaefer, *J. Amer. Chem. Soc.* **59**, 2400 (1937).
20. M. Joly, in *Recent Progress in Surface Science*, Vol. 1, J. F. Danielli, K. G. A. Pankhurst, and A. C. Riddiford, Eds., Academic, New York, 1964, p. 1.
21. D. G. Dervichian and M. Joly, *J. Phys. Radium* **8**, 471 (1937); *ibid.* **9**, 345 (1938); *J. Chem. Phys.* **6**, 226 (1938).
22. W. D. Harkins and J. G. Kirkwood, *J. Chem. Phys.* **6**, 53, 298 (1938).
23. L. Fourt and W. D. Harkins, *J. Phys. Chem.* **42**, 897 (1938).
24. L. Fourt, *J. Phys. Chem.* **43**, 887 (1939).
25. N. W. Tschoegl, *Kolloid-Z.* **181**, 19 (1962).
26. H. Mouquin and E. K. Rideal, *Proc. Roy. Soc.* (*London*) **114A**, 690 (1927).
27. C. W. N. Cumper and A. E. Alexander, *Australian J. Sci. Res.* **5**, 189 (1952).
28. G. E. Hibberd, to be published in *J. Colloid Interface Sci.*
29. J. D. Ferry, *Viscoelastic Properties of Polymers*, Wiley, New York, 1961.
30. A. E. Alexander and P. Johnson, *op. cit.*, Chap. 17.
31. E. Stenhagen, in *Determination of Organic Structures by Physical Methods*, E. A. Braude and F. C. Nachod, Eds., Academic, New York, 1955, p. 325.
32. E. D. Goddard and A. E. Alexander, *Biochem. J.* **47**, 331 (1950).
33. C. W. N. Cumper and A. E. Alexander, *Rev. Pure Appl. Chem.* (*Australia*) **1**, 121 (1951).
34. D. F. Cheesman and J. T. Davies, *Adv. Protein Chem.* **9**, 439 (1954).
35. A. R. Gilby and A. E. Alexander, *Arch. Biochem. Biophys.* **67**, 302, 307 (1957).
36. A. R. Gilby and W. Camiglieri, *J. Appl. Chem.* **6**, 219 (1956).
37. B. S. Harrap, *Australian J. Biol. Sci.* **8**, 122 (1955).
38. A. E. Alexander, *Ann. Rept. Progr. Chem.* (*Chem. Soc. London*) **41**, 5 (1944).
39. J. H. Brooks and A. E. Alexander, in *Retardation of Evaporation by Monolayers: Transport Processes*, V. K. La Mer, Ed., Academic, New York, 1962, p. 245; *J. Phys. Chem.* **66**, 1851 (1962).
40. W. D. Harkins, T. F. Young, and G. E. Boyd, *J. Chem. Phys*, **8**, 954 (1940).
41. B. A. Pethica, *Trans. Faraday Soc.* **51**, 1402 (1955).
42. P. J. Anderson and B. A. Pethica, *Trans. Faraday Soc.* **52**, 1080 (1956).

43. D. J. Crisp, *Trans. Faraday Soc.* **42,** 619 (1946).
44. M. Blank and F. J. W. Roughton, *Trans. Faraday Soc.* **56,** 1832 (1960).
45. M. Blank, in *Retardation of Evaporation by Monolayers*, p. 75.
46. J. G. Hawke and A. E. Alexander, *ibid.* p. 67.
47. J. G. Hawke and A. G. Parts, *J. Colloid Sci.* **19,** 448 (1964).
48. J. Glazer and A. E. Alexander, *Trans. Faraday Soc.* **47,** 401 (1951).
49. G. E. Hibberd and A. E. Alexander, *J. Phys. Chem.* **66,** 1854 (1962).
50. J. H. Schulman and J. A. Friend, *Kolloid-Z.* **115,** 67 (1949).
51. J. H. Schulman, *Ann. Repts. on Progr. Chem. (Chem. Soc. London)* **36,** 94 (1939).
52. J. T. Davies, *Z. Elektrochem.* **55,** 559 (1951).
53. A. E. Alexander and T. Teorell, *Trans. Faraday Soc.* **35,** 727 (1939).
54. J. H. Brooks and F. MacRitchie, *J. Colloid Sci.* **16,** 442 (1961).
55. L. Blight, C. W. N. Cumper and V. Kyte, *J. Colloid Sci.* **20,** 393 (1965).
56. J. H. Brooks and B. A. Pethica, *Trans. Faraday Soc.* **60,** 208 (1964).

## General

Adam, N. K., *The Physics and Chemistry of Surfaces*, 3rd ed., Oxford University Press, London, 1941.
Alexander, A. E., *Surface Chemistry*, Longmans, Green, London, 1951.
Alexander, A. E., *Ann. Rept. Chem. Soc.* **41,** 5 (1944).
Alexander, A. E., *Repts. Prog. Physics* **9,** 158 (1943).
Alexander, A. E., in *Advances in Colloid Science*, Vol. 3, H. Mark and E. J. W. Verwey, Interscience, New York, 1950, p. 67.
Alexander, A. E., in *Colloid Chemistry*, Vol. 7, J. Alexander, Ed., Reinhold, New York, 1950, p. 211.
Alexander, A. E., and P. Johnson, *Colloid Science*, Clarendon, Oxford, 1949.
Bikerman, J. J., *Surface Chemistry*, Academic, New York, 1948.
Cheesman, D. F., and J. T. Davies, *Adv. Protein Chem.* **9,** 439 (1954).
Cumper, C. W. N., and A. E. Alexander, *Rev. Pure Appl. Chem. (Australia)* **1,** 121 (1951).
Danielli, J. F., K. G. A. Pankhurst, and A. C. Riddiford, Eds., *Surface Phenomena in Chemistry and Biology*, Pergamon, New York, 1958.
Davies, J. T., and E. K. Rideal, *Interfacial Phenomena*, Academic, New York, 1961.
Gaines, G. L., Jr., *Insoluble Monolayers at Liquid-Gas Interfaces*, Interscience, New York, 1966.
Harkins, W. D., in *Colloid Chemistry*, Vol. 5, J. Alexander, Ed., Reinhold, New York 1944, p. 12.
Joly, M., in *Recent Progress in Surface Science*, J. F. Danielli, K. G. A. Pankhurst, and A. C. Riddiford, Eds., Vol. 1, Academic, New York, 1964, p. 1.
*Monomolecular Layers*, American Association for the Advancement of Science, Washington, D. C., 1954.
Sobotka, H., in *Medical Physics*, Vol. 2, O. Glasser, Ed., Year Book, Chicago, 1950, p. 550.
Stenhagen, E., in *Determination of Organic Structures by Physical Methods*, E. A. Braude and F. C. Nachod, Eds., Academic Press, New York, 1955, p. 325.
"Surface Chemistry" (Special supplement to *Research*), Butterworths, London, 1949.

# INDEX